LINEAR PROGRAMMING
and
THEORY OF GAMES

LINEAR PROGRAMMING *and* THEORY OF GAMES

[For three year General, Honours, Postgraduate and Engineering students. Also helpful for MBA students]

P M Karak

Ex-selection Grade Lecturer
Department of Mathematics, Sree Chaitanya College
Habra, West Bengal

New Central Book Agency (P) Ltd

LONDON
HYDERABAD ERNAKULAM BHUBANESWAR
NEW DELHI KOLKATA PUNE GUWAHATI

NCBA

REGD OFFICE
8/1 Chintamoni Das Lane, Kolkata 700 009, India
email: ncbapvtltd@ncbapvtltd.com

OVERSEAS
NCBA (UK) Ltd, 149 Park Avenue North, Northampton, NN3 2HY, UK
email: ncbauk@yahoo.com

EXPORT
NCBA Exports Pvt. Ltd, 212 Shahpur Jat, New Delhi 110 049
email: ncbaexports@ncbapvtltd.com

BRANCHES

New Central Book Agency (P) Ltd
212 Shahpur Jat, **New Delhi** 110 049
email: ncbadel@ncbapvtltd.com

House No. 3-1-315, 1st Floor
Nimboliadda, Kachiguda, **Hyderabad** 500 027
email: ncbahydb@ncbapvtltd.com

Shop Nos. 3 & 4, Vinayak Towers
681/B Budhwar Peth, Appa Balwant Chowk
Pune 411 002
email: ncbapune@ncbapvtltd.com

GSS Shopping Complex, 1st Floor
Opposite College Post Office, Convent Road
Ernakulam 682 035
0484-4021054 / 53

Shop No. 15, Bharati Towers, Ground Floor
Forest Park
Bhubaneswar 751 001
email: ncbabub@ncbapvtltd.com

Radhamayee Complex [Second Floor]
Opp. Axis Bank, Panbazar, Jaswanta Road
Guwahati 781 001
email: ncbaguwahati@ncbapvtltd.com

LINEAR PROGRAMMING AND THEORY OF GAMES ● P M Karak

First Published: 1979
Throughly Revised and Enlarged NCBA Edition: 2001
Reprinted: 2002, 2003, 2004, 2005, 2008, 2009, 2011, 2012, 2013, 2014, 2016, 2017, 2018, 2019, 2020, 2021, **2022**

PUBLISHER AND TYPESETTER
New Central Book Agency (P) Ltd
8/1 Chintamoni Das Lane, Kolkata 700 009

PRINTER
New Central Book Agency (P) Ltd
Web-Offset Division, Dhulagarh, Sankrail, Howrah

PROJECT TEAM
Abhijit Sen and Subir Paul

ISBN: 978 81 7381 405 1

Price ₹ 785.00

Preface to the Central Edition

This thoroughly revised and enlarged *Central edition* of the book which is not a new edidition but a new book, is meant for the students of three year Degree, General, Honours, Postgraduate, Engineering students of all Indian Universities. However it can be used as a basic text-book by M.B.A. students as well. The chapters of the book have been rearranged thoroughly to make the development of the subject more consistent and to solve the L.P.P. by simplex method the computations have been made in such a way as will definitely save both time and labour.

As an author of a text-book on this subject and from the teaching experience of more than two decades, I have noticed that there is some confusion and ambiguity regarding the development of the subject, particularly in the chapter *Simplex Method* or *Simplex Algorithm.* To remove such ambiguity and confusion, I have tried to explain the subject matter, explicitly, perfectly and scientifically. Whether I have succeeded or not, is a matter to be judged by the readers. The development of the Simplex Algorithm has been given in the Appendix.

At the end, I convey my sincere thanks to Mr. Amitabha Sen and the employees of New Central Book Agency (P) Ltd. for their whole hearted co-operation in publication of the book in such a nice and presentable manner.

Suggestions from any corner for further improvement of the book will be highly appreciated.

P. M. Karak

Saraswati Puja,
29th January, 2001
Samvabana
AH 59, Salt Lake City
Calcutta 700 091

Preface to the First Edition

Minimum Cost and Maximum Profit is the fundamental theme in the realm of Applied Economics, Business and Commerce. In the past, business communities tried to minimise the cost of production by using the trial and error method. In the early part of the sixth decade George Dantzig invented a scientific device based on Mathematics for minimizing the cost of production. This device is called Programming. In it, LINEAR PROGRAMMING has an immense importance.

Now the business world is coming to realise the efficacy of the application of the device in different fields of Applied Economics. As a natural consequence, its theoretical developments and practical applications have grown at a rapid rate in the last 25 years.

Keeping in mind the needs of the day, Calcutta University has rightly introduced the subject in the Pass and Honours syllabus in the newly introduced curriculum of Mathematics in the Degree Course. Burdwan University has also introduced the subject in Mathematics Honours Course. The present volume is written according to the Pass Course syllabus of Calcutta University and the Honours Course syllabus of Burdwan University. Each chapter included in the syllabus has been discussed properly and in detail. Again, as the book is written mainly for the Pass Course students, some complex theoretical discussions have been excluded. In the Appendix, proofs of some theorems including replacement theorem regarding vectors have been given in detail. Interested students may go through the proofs of these theorems. An alternative method of computing $z_j - c_j$ is also given in the appendix.

I am indebted to the authors and publishers of those books whose help I have taken at the time of the preparation of the manuscript. I appeal to my colleagues to help me in making the book a better one by constructive criticism of the same. Any detection of printing mistakes, inaccuracies or factual mistakes will be gracefully acknowledged.

July 1979
Sree Chaitanya College

P. M. Karak

Contents

Chapter 1

Mathematical Background

1.1 Introduction

The method of solving a *Linear Programming Problem* (Simply L.P.P.) mainly lies in finding out a solution set of m simultaneous linear equations with $n(n > m)$ unknowns $x_j (j = 1, 2, \cdots, n)$

$$\left.\begin{array}{l} a_{11}x_1 + a_{12}x_2 + \cdots + a_{1n}x_n = b_1 \\ a_{21}x_1 + a_{22}x_2 + \cdots + a_{2n}x_n = b_2 \\ \cdots\cdots\cdots\cdots\cdots\cdots\cdots\cdots\cdots \\ \cdots\cdots\cdots\cdots\cdots\cdots\cdots\cdots\cdots \\ a_{m1}x_1 + a_{m2}x_2 + \cdots + a_{mn}x_n = b_m \end{array}\right\} \quad (1.1.1)$$

subject to the restrictions $x_j \geq 0$, $j = 1, 2, \cdots, n$, which makes a *linear function* $z = c_1x_1 + c_2x_2 + \cdots + c_nx_n$ of all the variables (unknowns), either a maximum or minimum. The function z is known as the *objective function*. The quantities a_{ij}, $[i = 1, \cdots, m;\ j = 1, 2, \cdots, n]$; b_i's $[i = 1, 2, \cdots, m]$ and c_j's $[j = 1, 2, \cdots, n]$ are known constants. In order to solve the problem some fundamental knowledges about matrix and vector algebra are essential.

1.2 Matrices

Matrix: m, n being positive integers, $m \times n$ numbers a_{ij} $(i = 1, 2, \cdots, m;$ $j = 1, 2, \cdots, n)$ arranged in a rectangular block or array with m rows and n columns as given below

$$\begin{bmatrix} a_{11} & a_{12} & \cdots\cdots & a_{1n} \\ a_{21} & a_{22} & \cdots\cdots & a_{2n} \\ \cdots & \cdots & \cdots\cdots & \cdots \\ a_{m1} & a_{m2} & \cdots\cdots & a_{mn} \end{bmatrix}$$

is called a *matrix of order* $m \times n$ or m by n matrix. It is generally denoted by A or $[a_{ij}]_{m \times n}$. All a_{ij}'s are the elements of the matrix. A matrix is said to be real if all the elements a_{ij} are real.

Square matrix: When $m = n$, i.e., the number of rows and columns are equal, the matrix is known as a *square matrix* of order n and when $m \neq n$, the matrix is known as a *rectangular matrix*.

Diagonal matrix: In a square matrix, if all off-diagonal elements are zero, i.e., if $a_{ij} = 0$ for $i \neq j$, the matrix is known as *diagonal matrix*.

Identity matrix: In a diagonal matrix, if all non-zero elements be unity, the matrix is known as *Identity matrix* or *unit matrix*. It is usually denoted by I_n or simply by I.

Null matrix: If all elements of a matrix be zero, the matrix is known as *null matrix*.

Transpose of a matrix or transposed matrix: A matrix, formed by interchaning the elements of the rows and columns of a matrix is known as *transposed matrix* of the original one. The transpose of a $m \times n$ matrix A is denoted by A' or A^t and A^t will be a $n \times m$ matrix. The transpose of a square matrix is also a square matrix of the same order. The transpose of the transposed matrix is the original matrix, i.e., $(A')' = A$ etc.

Row matrix: Matrix having a single row is known as a *row matrix*. $1 \times n$ matrix is a row matrix.

Column matrix: A matrix having a single column is known as a *column matrix*. $m \times 1$ matrix is a column matrix.

From the definition, it is clear that the transpose of a row matrix is a column matrix and vice versa.

Equality of two matrices: Two matrices $A = [a_{ij}]$, $B = [b_{ij}]$ of same order $m \times n$ are said to be equal if $a_{ij} = b_{ij}$ for all i and j.

Addition and subtraction of matrices: The sum of two matrices $A = [a_{ij}]$, $B = [b_{ij}]$ of same order $m \times n$ is also a matrix $C = [c_{ij}]$ of the same order, where $c_{ij} = a_{ij} + b_{ij}$ for all i and j etc.

From the above definition, it is evident that

$$\begin{aligned} A + B &= B + A \quad \text{(Commutative law)} \\ (A + B) + C &= A + (B + C) \quad \text{(Associative law)} \\ (A + B)' &= A' + B'. \end{aligned}$$

The difference of two matrices $A = [a_{ij}]$, $B = [b_{ij}]$ of same order $m \times n$, is also a matrix $C = [c_{ij}]$ of the same order, where $c_{ij} = a_{ij} - b_{ij}$.

Multiplication of matrix with scalar quantity: If $A = [a_{ij}]$ be a $m \times n$ matrix and k be any scalar quantity then $kA = Ak = [ka_{ij}]$ is also a matrix of the same order.

From this we can say $k(A + B) = kA + kB$ and so on.

Multiplication of matrices: If $A = [a_{ij}]_{m \times n}$ and $B = [b_{jk}]_{n \times p}$ be two matrices of order $m \times n$ and $n \times p$ respectively, the product of the two matrices A

and B which is denoted as AB is a matrix C of order $m \times p$ such that $C = [c_{ik}]_{m \times p}$, where

$$c_{ik} = \sum_{j=1}^{n} a_{ij} b_{jk}.$$

For matrix product to be feasible, the number of columns of the first matrix must be equal to the number of rows of the second matrix. From the definition, it is evident that in general BA does not exist though AB exists, and even if both AB and BA exist, they are not equal in general. For two square matrices A and B, of same order AB and BA both exist but $AB \neq BA$ in general.

Properties of matrix multiplication: If A, B and C are all matrices then

$$\begin{aligned} (A+B)C &= AC + BC \\ C(A+B) &= CA + CB \\ (AB)C &= A(BC) \\ a(AB) &= A(aB) = (aA)B, \quad \text{where } a \text{ is a scalar.} \end{aligned}$$

$(AB)' = B'A'$ and $AI = IA = A$.

It is assumed in all the cases that the matrix product is possible.

$AB = 0$, does not necessarily follow that either A or B be zero or both of them be zero.

With the knowledge of the matrix product, the set of simultaneous linear equations given in (1.1.1) can be written with the help of matrix notation as given below in a more compact form

$$\mathbf{Ax} = \mathbf{b}, \tag{1.2.1}$$

where A is a matrix of order $m \times n$ given by

$$A = \begin{bmatrix} a_{11} & a_{12} & \cdots & a_{1n} \\ a_{21} & a_{22} & \cdots & a_{2n} \\ \cdots & \cdots & \cdots & \cdots \\ a_{m1} & a_{m2} & \cdots & a_{mn} \end{bmatrix} \tag{1.2.2}$$

$\mathbf{x}$ is column matrix of order $n \times 1$, having elements $x_1, x_2, \cdots, x_n$. $\mathbf{b}$ is column matrix of order $m \times 1$ having elements $b_1, b_2, \cdots, b_m$. A is known as the *co-efficient matrix.*

Sub-matrix: Matrix formed by omitting some rows and columns of a matrix, is known as a *sub-matrix* of the original matrix.

Determinant of a matrix: The determinant formed with the elements of a square matrix is known as *determinant of the matrix.*

If A be a square matrix of order n, $A = [a_{ij}]_{n \times n}$, then the determinant of the matrix which is denoted by $|A|$ or det A, is given by

$$|A| = \begin{bmatrix} a_{11} & a_{12} & \cdots & a_{1n} \\ a_{21} & a_{22} & \cdots & a_{2n} \\ \cdots & \cdots & \cdots & \cdots \\ a_{n1} & a_{n2} & \cdots & a_{nn} \end{bmatrix}. \tag{1.2.3}$$

Note: $|A|$ is here different from the modulus of a quantity A, where the same notation has been used.

The method of simplification of a determinant and the properties of a determinant are known to the students of the degree course. We shall not here discuss all the properties. We discuss only about the **Minor** and the **Cofactor** of an element of a determinant.

Minor: The minor of an element a_{ij} of a determinant $|A|$ is a determinant formed by omitting the ith row and the jth column of the determinant $|A|$. It is usually denoted by M_{ij}.

Cofactor: The cofactor of an element a_{ij} of a determinant $|A|$ is denoted by C_{ij} and given by $C_{ij} = (-1)^{i+j} M_{ij}$.

Singular and non-singular matrix: If the vlaue of the determinant of a square matrix be a non-zero quantity, then the matrix is said to be a *a non-singular matrix* and if the vlaue of the determinant of the matrix be zero then the matrix is said to be a *a singular matrix*. We know that det A = det A^t.

Thus if a square matrix be non-singular, then its transpose is also non-singular.

Rank of a matrix: The rank of a $m \times n$ matrix A is will be r, if all square sub-matrices of order $(r+1)$ are singular and at least one square sub-matrix of order r is non-singular. From the definition, it is clear that $r_{\max} \le \min(m, n)$. The rank of a matrix A is usually denoted by $r(A)$.

(i) $A = \begin{bmatrix} 2 & -3 \\ 3 & 4 \end{bmatrix}$, $|A| = \begin{vmatrix} 2 & -3 \\ 3 & 4 \end{vmatrix} = 2 \times 4 - (-3) \times 3 = 17 \neq 0.$

Thus $r(A) = 2$.

(ii) $A = \begin{bmatrix} 5 & 10 \\ 3 & 6 \end{bmatrix}$, $|A| = \begin{vmatrix} 5 & 10 \\ 3 & 6 \end{vmatrix} = 5 \times 6 - 10 \times 3 = 0.$

Thus $r(A) \neq 2$.

Now one square sub-matrix of order 1 is $B_1 = [5]$ and $|B_1| = 5 \neq 0$.

Thus $r(A) = 1$.

(iii) $A = \begin{bmatrix} 2 & 3 & -1 \\ 1 & 2 & 0 \end{bmatrix}$, one square sub-matrix of order 2 is $B_1 = \begin{bmatrix} 2 & 3 \\ 1 & 2 \end{bmatrix}$,

$|B_1| = \begin{vmatrix} 2 & 3 \\ 1 & 2 \end{vmatrix} = 2 \times 2 - 3 \times 1 = 1 \neq 0.$

Thus $r(A) = 2$.

(iv) $A = \begin{bmatrix} 2 & 3 & 5 \\ 4 & 6 & 10 \end{bmatrix}$, one square sub-matrix of order 2 is $B_1 = \begin{bmatrix} 2 & 3 \\ 4 & 6 \end{bmatrix}$,

$|B_1| = 0$, similarly $|B_2| = \begin{vmatrix} 2 & 5 \\ 4 & 10 \end{vmatrix} = 0$, $|B_3| = \begin{vmatrix} 3 & 5 \\ 6 & 10 \end{vmatrix} = 0.$

Thus $r(A) \neq 2$.

Now one square sub-matrix of order 1, $B_4 = [2]$ and $|B_4| = 2 \neq 0$.

Thus $r(A) = 1$.

(v) $A = \begin{bmatrix} 1 & 3 & 4 \\ 2 & -2 & 2 \\ 3 & 1 & 1 \end{bmatrix}$,

$$|A| = \begin{vmatrix} 1 & 3 & 4 \\ 2 & -2 & 2 \\ 3 & 1 & 1 \end{vmatrix} = 1\begin{vmatrix} -2 & 2 \\ 1 & 1 \end{vmatrix} - 3\begin{vmatrix} 2 & 2 \\ 3 & 1 \end{vmatrix} + 4\begin{vmatrix} 2 & -2 \\ 3 & 1 \end{vmatrix} = 40 \neq 0.$$

Thus $r(A) = 3$.

(vi) $A = \begin{bmatrix} 4 & 3 & 2 \\ 2 & 1 & 4 \\ 2 & 3 & -8 \end{bmatrix}$,

$$|A| = \begin{vmatrix} 4 & 3 & 2 \\ 2 & 1 & 4 \\ 2 & 3 & -8 \end{vmatrix} = 4\begin{vmatrix} 1 & 4 \\ 3 & -8 \end{vmatrix} - 3\begin{vmatrix} 2 & 4 \\ 2 & -8 \end{vmatrix} + 2\begin{vmatrix} 2 & 1 \\ 2 & 3 \end{vmatrix} = 0.$$

Thus $r(A) \neq 3$.

Now one square sub-matrix of order 2 is $B_1 = \begin{bmatrix} 4 & 3 \\ 2 & 1 \end{bmatrix}$, $|B_1| = -2 \neq 0$.

Thus $r(A) = 2$.

(vii) $A = \begin{bmatrix} 1 & 2 & 3 \\ 3 & 6 & 9 \\ 2 & 4 & 6 \end{bmatrix}$, $|A| = 0$.

Thus $r(A) \neq 3$.

Now one square sub-matrix of order 2 is $B_1 = \begin{bmatrix} 1 & 2 \\ 3 & 6 \end{bmatrix}$, $|B_1| = 0$.

Another square sub-matrix of order 2 is $B_2 = \begin{bmatrix} 2 & 3 \\ 4 & 6 \end{bmatrix}$, $|B_2| = 0$.

Similarly, it can be shown that all square sub-matrices of order 2 are singular. Thus $r(A) \neq 2$. Last of all, one square sub-matrix of order 1 is [1] which is non-singular. Thus $r(A) = 1$.

Note: I_n has rank n and a null matrix has rank 0.

Adjoint of a matrix: The adjoint of a square matrix A is a matrix of the same order, which is the transpose of the matrix B, where the elements of B are the cofactors of the corresponding elements of the determinant of the matrix A. It is denoted by Adj (A).

If $A = [a_{ij}]_{n\times n}$ then

$$\text{Adj}\,(A) = \begin{bmatrix} c_{11} & c_{21} & \cdots & c_{n1} \\ c_{12} & c_{22} & \cdots & c_{n2} \\ \cdots & \cdots & \cdots & \cdots \\ c_{1n} & c_{2n} & \cdots & c_{nn} \end{bmatrix}, \qquad (1.2.4)$$

where c_{ij} etc. are the cofactors of the corresponding elements a_{ij} of $|A|$ given in (1.2.3).

Inverse of a matrix: If there exist two square matrices A and B of same order such that $AB = BA = I$ (identity matrix of same order) then B is called as the *inverse matrix* of A and vice versa. The inverse matrix of A which is denoted by A^{-1} is given by

$$A^{-1} = \frac{1}{|A|} \cdot \text{Adj}\,(A). \tag{1.2.5}$$

From the definition, it is clear that $AA^{-1} = A^{-1}A = I$. It is also clear that inverse exists only in the non-singular matrices. The inverse matrix of an identity matrix is the same matrix.

Determination of the inverse of a 2 × 2 non-singular matrix:

$$A = \begin{bmatrix} a_{11} & a_{12} \\ a_{21} & a_{22} \end{bmatrix}; \quad |A| = a_{11}a_{22} - a_{12}a_{21} \neq 0.$$

Here the cofactors of $a_{11} = c_{11} = a_{22}$, $c_{12} = -a_{21}$, $c_{21} = -a_{12}$ and $c_{22} = a_{11}$; it follows that

$$\text{Adj}\,(A) = \begin{bmatrix} c_{11} & c_{21} \\ c_{12} & c_{22} \end{bmatrix} = \begin{bmatrix} a_{22} & -a_{12} \\ -a_{21} & a_{11} \end{bmatrix}$$

and thus

$$A^{-1} = \frac{1}{|A|} \text{Adj}\,(A) = \frac{1}{|A|} \begin{bmatrix} a_{22} & -a_{12} \\ -a_{21} & a_{11} \end{bmatrix}. \tag{1.2.6}$$

Note: Readers should remember the formula for A^{-1}.

Example 1.2.1 *Find the inverse of the matrix* $A = \begin{bmatrix} 2 & -1 \\ 4 & 3 \end{bmatrix}$.

Solution: $\det A = |A| = 2 \times 3 - (-1) \times 4 = 10 \neq 0$. Hence the matrix is non-singular and therefore the inverse exists. Thus using (1.2.6) we have,

$$A^{-1} = \frac{1}{10} \begin{bmatrix} 3 & 1 \\ -4 & 2 \end{bmatrix} = \begin{bmatrix} \frac{3}{10} & \frac{1}{10} \\ -\frac{2}{5} & \frac{1}{5} \end{bmatrix}$$

and finally we can verify that

$$AA^{-1} = \begin{bmatrix} 2 & -1 \\ 4 & 3 \end{bmatrix} \begin{bmatrix} \frac{3}{10} & \frac{1}{10} \\ -\frac{2}{5} & \frac{1}{5} \end{bmatrix}$$

$$= \begin{bmatrix} 2 \times \frac{3}{10} + (-1)\left(-\frac{2}{5}\right) & 2 \times \frac{1}{10} + (-1) \times \frac{1}{5} \\ 4 \times \frac{3}{10} + 3 \times \left(-\frac{2}{5}\right) & 4 \times \frac{1}{10} + 3 \times \frac{1}{5} \end{bmatrix} = \begin{bmatrix} 1 & 0 \\ 0 & 1 \end{bmatrix} = I_2.$$

Determination of the solution set of m linear equations with m variables by matrix method: Let the set of equations be

$$A\mathbf{x} = \mathbf{b},$$

where A is a square matrix of order m. We make an assumption that A is non-singular, i.e., $|A| \neq 0$. The solution $\hat{\mathbf{x}}$ is given by $\hat{\mathbf{x}} = A^{-1}\mathbf{b}$ and the solution set is unique.

► **Example 1.2.2** *Solve the linear equations by matrix method.*

$$\begin{aligned} x_1 + 2x_2 + 3x_3 &= 6 \\ 3x_1 + 2x_2 + 9x_3 &= 14 \\ 2x_1 + 4x_2 + x_3 &= 7. \end{aligned}$$

Solution: $A = \begin{bmatrix} 1 & 2 & 3 \\ 3 & 2 & 9 \\ 2 & 4 & 1 \end{bmatrix}$, $|A| = 20 \neq 0$.

$$A^{-1} = \frac{1}{|A|}\begin{bmatrix} -34 & 10 & 12 \\ 15 & -5 & 0 \\ 8 & 0 & -4 \end{bmatrix} \text{ by (1.2.5)}$$

$$\hat{\mathbf{x}} = A^{-1}\begin{bmatrix} 6 \\ 14 \\ 7 \end{bmatrix} = \frac{1}{20}\begin{bmatrix} -34 & 10 & 12 \\ 15 & -5 & 0 \\ 8 & 0 & -4 \end{bmatrix}\begin{bmatrix} 6 \\ 14 \\ 7 \end{bmatrix}, \text{ or, } \hat{\mathbf{x}} = \begin{bmatrix} 1 \\ 1 \\ 1 \end{bmatrix}$$

and the solution set is unique.

1.3 Vectors

All row and column matrices are known as *row* and *column* vectors respectively. The vectors are usually denoted by lower case bold type as $\mathbf{a}, \mathbf{b}, \mathbf{x}$ etc.

Row vector: A row matrix with n elements $a_1, a_2, \cdots, a_n$ is known as *n-component row vector.* It is usually denoted by $\mathbf{a}$ and $\mathbf{a} = (a_1, a_2, \cdots, a_n)$. To represent row vector first bracket "( )" has been used. The elements a_1, a_2 etc. are known as the *components of the vector.*

Column vector: A column matrix with n elements $b_1, b_2, \cdots, b_n$ is known as *n-component column vector* $\mathbf{b}$ and

$$\mathbf{b} = \begin{bmatrix} b_1 \\ b_2 \\ \cdots \\ b_n \end{bmatrix}.$$

In matrix notation column matrix is written in a vertical column. But a system is used to write the column vector in a horizontal row. The vector $\mathbf{b}$ is written as $\mathbf{b} = [b_1, b_2, \cdots, b_n]$ and third bracket "[ ]" is used to denote it as a column vector.

Note: Both row and column vectors are written in a horizontal row. And for row vector the first bracket "( )" is used and for column vector, the third bracket "[ ]" is used.

Analytically, both the row and column vectors may be considered as points in n-dimensional space which is known as *vector space* or *n-dimensinoal Eucledean space* which is usually denoted by $V_n(F)$ or E^n or E_n etc. Geometricaly, all vectors may be considered as the position vector of points with respect to the origin. In a vector space all vectors must be either row vectors or column vectors.

Unit vector: A vector is said to be a unit vector, if all its components are zero *except one with unit value.* $\mathbf{e}_i$ is a unit vector in n-dimensional vector space whose i-th component is 1. $\mathbf{e}_1 = (1, 0, \cdots, 0)$, $\mathbf{e}_2 = (0, 1, 0, \cdots, 0)$, $\mathbf{e}_3 = (0, 0, 1, \cdots, 0)$, $\mathbf{e}_n = (0, 0, \cdots, 1)$ are all unit vectors in n-dimensional space.

Null vector: *A* vector is said to be a null vector, if all the components of the vector be equal to zero. It is usually denoted by $\mathbf{0}$. An n-component null vector is written as $\mathbf{0} = (0, 0, \cdots, 0)$ with n zeros.

Equality of two vectors: Two n-component row vectors are said to be equal, if the corresponding components of both the vectors are equal. Similar is the case for column vectors. Two n-component row vectors $\mathbf{a}$ and $\mathbf{b}$ are equal only when $a_i = b_i (i = 1, 2, \cdots, n)$, where $\mathbf{a} = (a_1, a_2, \cdots, a_n)$ and $\mathbf{b} = (b_1, b_2, \cdots, b_n)$.

Sum of two vectors: The sum of two n-component row vectors is also a row vectors, whose i-th component is the sum of the ith components of the constituent vectors. If $\mathbf{a} = (a_1, a_2, \cdots, a_n)$ and $\mathbf{b} = (b_1, b_2, \cdots, b_n)$ be two n-component row vectors then

$$\mathbf{a} + \mathbf{b} = \mathbf{c},$$

where $\mathbf{c} = (a_1 + b_1, a_2 + b_2, \cdots, a_i + b_i, \cdots, a_n + b_n)$.

Similar is the case for two n-component column vectors.

Difference of two vectors: The difference of two n-component row vectors is also a row vector, whose ith component is the difference of the ith components of the constituent vectors. If $\mathbf{a} = (a_1, a_2, \cdots, a_n)$, $\mathbf{b} = (b_1, b_2, \cdots, b_n)$ be two n-component row vectors then $\mathbf{a} - \mathbf{b} = \mathbf{c}$, where $\mathbf{c} = (a_1 - b_1, a_2 - b_2, \cdots, a_n - b_n)$. Similar is the case for column vectors. The difference of two equal vectors is a null vector.

If $\mathbf{a} = (a_1, a_2, \cdots, a_n)$ than $\lambda\mathbf{a} = (\lambda a_1, \lambda a_2, \cdots, \lambda a_n)$, where λ is a scalar.

Inner product of two vectors: If $\mathbf{a}$ and $\mathbf{b}$ are two n-component row and column vectors respectively, then the product of the vectors $\mathbf{a}$ and $\mathbf{b}$ is known as the *inner product* (act ally matrix product) of the vectors denoted by $\mathbf{ab}$ and given by $\mathbf{ab} = a_1b_1 + a_2b_2 + \cdots + a_nb_n$, where $\mathbf{a} = (a_1, a_2, \cdots, a_n)$ and $\mathbf{b} = [b_1, b_2, \cdots, b_n]$ and $\mathbf{ab} \neq \mathbf{ba}$.

Inner product is also termed as scalar product. Inner product is a scalar quantity. Thus

$$(\mathbf{ab})' = \mathbf{b}'\mathbf{a}' = \mathbf{ab} \text{ etc.}$$

Note: In the case of inner product, the first vector must be row vector and the second vector must be column vector.

Length of a vector: If $\mathbf{a} = (a_1, a_2, \cdots, a_n)$ be an n-component row vector then the length of the vector $\mathbf{a}$ which is usually denoted by $|\mathbf{a}|$ is the square root of the inner product $\mathbf{aa}'$ or $|\mathbf{a}| = \sqrt{\mathbf{aa}'} = \sqrt{a_1^2 + a_2^2 + \cdots + a_n^2}$.

If the vector $\mathbf{b} = [b_1, b_2, \cdots, b_n]$ be an n-component column vector then

$$|\mathbf{b}| = \sqrt{\mathbf{b}' \cdot \mathbf{b}} = \sqrt{b_1^2 + b_2^2 + \cdots + b_n^2}.$$

Distance between two points: If $\mathbf{a}$ and $\mathbf{b}$ are vectors (which are considered as points) of the same vector space then the distance between the points is denoted by $|\mathbf{a} - \mathbf{b}|$ and given by

$$|\mathbf{a} - \mathbf{b}| = \sqrt{(a_1 - b_1)^2 + (a_2 - b_2)^2 + \cdots + (a_n - b_n)^2},$$

where $\mathbf{a} = (a_1, a_2, \cdots, a_n)$ and $\mathbf{b} = (b_1, b_2, \cdots, b_n)$.

Linear combination of vectors: The linear combination of set of k, n-component vectors $\mathbf{a}_1, \mathbf{a}_2, \cdots, \mathbf{a}_k$ is a vector $\mathbf{a}$ with n-components given by the relation

$$\mathbf{a} = \lambda_1 \mathbf{a}_1 + \lambda_2 \mathbf{a}_2 + \cdots + \lambda_k \mathbf{a}_k = \sum_{i=1}^{k} \lambda_i \mathbf{a}_i, \tag{1.3.1}$$

where λ_i are all scalar quantities.

Linear dependence of vectors: *A* set of k, n-component vectors $\mathbf{a}_1, \mathbf{a}_2, \cdots, \mathbf{a}_k$ are said to be *linearly dependent* if there exist scalars $\lambda_i (i = 1, 2, \cdots, k)$ with at least one $\lambda_i \neq 0$, such that

$$\lambda_1 \mathbf{a}_1 + \lambda_2 \mathbf{a}_2 + \cdots + \lambda_k \mathbf{a}_k = \mathbf{0}, \quad \text{or,} \quad \sum_{i=1}^{k} \lambda_i \mathbf{a}_i = \mathbf{0} \tag{1.3.2}$$

is satisfied, where $\mathbf{0}$ is an n-component null vector.

For example, in three dimensional vector space $\mathbf{a} = (1, 2, 3)$ and $\mathbf{b} = (4, 8, 12)$ are linearly dependent because $4\mathbf{a} - \mathbf{b} = \mathbf{0}$, where the scalar quantities are 4 and -1, at least one of which is $\neq 0$. Similarly the three vectors

$$\begin{aligned} \mathbf{a} &= (1, 3, -2), \\ \mathbf{b} &= (2, 1, -1) \\ \text{and} \quad \mathbf{c} &= (0, -5, 3) \end{aligned}$$

are linearly dependent because

$$2\mathbf{a} - \mathbf{b} + \mathbf{c} = 2(1, 3, -2) \quad (2, 1, -1) + (0, -5, 3) = (0, 0, 0) = \mathbf{0},$$

where the scalar quantities are 2, -1 and 1 and at least one of which is non-zero. It is evident that if two non-null vectors be linearly dependent, one can be expressed as a scalar multiple of the other.

Linearly independent vectors: A set of k, n-component vectors ($k \leq n$) $\mathbf{a}_1, \mathbf{a}_2, \cdots, \mathbf{a}_k$ are said to be linearly independent if the linear combination of all the vectors be equal to an n-component null vector only, when all the scalars are equal to zero. Mathematically, the set of vectors $\mathbf{a}_1, \mathbf{a}_2, \cdots, \mathbf{a}_k$ are said to be *linearly independent* if the condition

$$\sum_{i=1}^{k} \lambda_i \mathbf{a}_i = \mathbf{0} \tag{1.3.3}$$

is satisfied only when all the scalar quantities λ_i are zero.

The two vectors $(1, 3, -2)$ and $(4, 1, 3)$ are linearly independent because the condition $\lambda_1(1, 3, -2) + \lambda_2(4, 1, 3) = (0, 0, 0)$ will be satisfied if and only if $\lambda_1 = 0$ and $\lambda_2 = 0$. There are no other values of λ_1 and λ_2 which will satisfy the condition.

Similarly the three vectors $\mathbf{a} = (1, -1, 1)$, $\mathbf{b} = (1, 2, -1)$ and $\mathbf{c} = (3, 2, -5)$ are linearly independent. Because the condition

$$\lambda_1 \mathbf{a} + \lambda_2 \mathbf{b} + \lambda_3 \mathbf{c} = \mathbf{0} \quad [\mathbf{0} \text{ is a 3-component null vector}]$$

$$\text{or,} \quad \lambda_1(1, -1, 1) + \lambda_2(1, 2, -1) + \lambda_3(3, 2, -5) = (0, 0, 0)$$

will be satisfied only if $\lambda_1 = \lambda_2 = \lambda_3 = 0$. There are no other values of λ_1, λ_2 and λ_3 which will satisfy the condition. Hence the vectors $\mathbf{a}, \mathbf{b}$ and $\mathbf{c}$ are linearly independent. From the discussions it is clear that any single non-null vector is always linearly independent.

Generating set of the vectors: A set of k, n-component vectors $\mathbf{a}_1, \mathbf{a}_2, \cdots, \mathbf{a}_k$ in a vector space E^n is said to be a generating set, if all vectors in the space can be expressed as a linear combination of the vectors $\mathbf{a}_1, \mathbf{a}_2, \cdots, \mathbf{a}_k$. The set is also known as *spanning set*. All vectors of the generating set may be linearly dependent or independent.

Basis set: A generating set of the vectors, all of which are linearly independent, is known as a *basis set*. The set of all n, n-component unit vectors $[\mathbf{e}_1, \mathbf{e}_2, \cdots, \mathbf{e}_i, \cdots, \mathbf{e}_n]$ are said to form a basis set in E^n and this basis set is known as the *standard basis set*.

The vectors $\mathbf{e}_i$ $(i = 1, 2, \cdots, n)$ are *linearly independent* and all the vectors in the space can be generated by the set of vectors $\mathbf{e}_i$ $(i = 1, 2, \cdots, n)$. For example, in E^3, any vector $\mathbf{a} = (a_1, a_2, a_3) \in E^3$ can be expressed as $\mathbf{a} = a_1\mathbf{e}_1 + a_2\mathbf{e}_2 + a_3\mathbf{e}_3$. Thus in E^3, $\mathbf{e}_1 = (1, 0, 0)$, $\mathbf{e}_2 = (0, 1, 0)$ and $\mathbf{e}_3 = (0, 0, 1)$, form a basis which is the standard basis in E^3. For a particular vector space there may have more than one basis set.

Some well known properties of vectors: We state all the properties without proof.

1. A set of vectors is either linearly independent or dependent, i.e., the terms linearly dependent and linearly independent are mutually exclusive.

2. A set of vectors with a null vector are always linearly dependent.

3. A non-empty subset of linearly independent set of vectors are always linearly independent and a superset of a linearly dependent set of vectors are always linearly dependent.

4. For a vector space, there must have at least one basis set.
5. The representation of a vector by a basis set is unique.
6. In E^n, $(n+1)$ or more vectors must be linearly dependent.
7. The exact number of vectors in a basis set is n, in E^n.
8. If there are two or more basis sets in a vector space then each has the same number of vectors.
9. Any set of n linearly independent vectors in E^n, is a basis set.
10. The set of n vectors in E^n are linearly independent, if and only if the rank of the $n \times n$ matrix, formed with the components of n vectors as the elements of the matrix, be equal to n. If it is less than that, the vectors are linearly dependent, i.e., if the square matrix formed with the n vectors be non-singular, the vectors are L.I.; otherwise vectors are L.D.

The proofs of some important theorems on vectors.

Theorem 1.3.1 *The set of vectors containing a null vector are always linearly dependent.*

Proof: Let a set A of vectors be

$$A = \{\mathbf{a}_1, \mathbf{a}_2, \cdots, \mathbf{a}_k, \mathbf{0}\} \in E^n.$$

We requrie to prove that the vectors in A are linearly dependent.

The condition that

$$\lambda_1\mathbf{a}_1 + \lambda_2\mathbf{a}_2 + \cdots + \lambda_k\mathbf{a}_k + \lambda\mathbf{0} = \mathbf{0}$$

will be satisfied if $\lambda_1 = \lambda_2 = \cdots = \lambda_k = 0$ and $\lambda \neq 0$. [λ_j, $(j = 1, 2, \cdots, k)$, are all scalar quantities].

Therefore, according to the definition, the set of vectors are linearly dependent.

Theorem 1.3.2 (i) *The superset of any linear dependent set of vectors are linearly dependent and* (ii) *The non-empty subset of any linearly independent set of vectors are linearly independent.*

Proof: (i) Let A be a set of vectors given by

$$A = \{\mathbf{a}_1, \mathbf{a}_2, \cdots, \mathbf{a}_k\} \in E^n$$

and the vectors are linearly dependent. Any superset B of A is given by

$$B = \{\mathbf{a}_1, \mathbf{a}_2, \cdots, \mathbf{a}_k, \mathbf{a}_m, \cdots, \mathbf{a}_l\} \in E^n.$$

We require to prove that the vectors in the set B are linearly dependent.

As $\mathbf{a}_1, \mathbf{a}_2, \cdots, \mathbf{a}_k$ are linearly dependent, therefore

$$\sum_{i=1}^{k} \lambda_i \mathbf{a}_i = \mathbf{0}$$

will be astisfied with at least one $\lambda_i \neq 0$. [λ_i are all scalar quantities].

Hence the condition

$$\sum_{i=1}^{k} \lambda_i \mathbf{a}_i + 0 \cdot \mathbf{a}_m + \cdots + 0 \cdot \mathbf{a}_l = \mathbf{0}$$

will be satisfied with at least one $\lambda_i \neq 0$. Therefore, the vectors $\mathbf{a}_1, \mathbf{a}_2, \cdots, \mathbf{a}_k, \mathbf{a}_m, \cdots, \mathbf{a}_l$ are linearly dependent.

(ii) Let A be a set of vectors given by

$$A = \{\mathbf{a}_1, \mathbf{a}_2, \cdots, \mathbf{a}_k, \cdots, \mathbf{a}_n\} \in E^n$$

and the vectors are linearly independent.

Any non-empty subset B of A is given by

$$B = \{\mathbf{a}_1, \mathbf{a}_2, \cdots, \mathbf{a}_k\}.$$

We require to prove that the vectors in B are linearly independent.

As $\mathbf{a}_1, \mathbf{a}_2, \cdots, \mathbf{a}_k, \cdots, \mathbf{a}_n$ are linearly independent, therefore the condition that

$$\sum_{i=1}^{n} \lambda_i \mathbf{a}_i = \mathbf{0}$$

will be satisfied if and only if

$$\lambda_1 = \lambda_2 = \cdots = \lambda_k = \cdots = \lambda_n = 0.$$

Hence the condition that

$$\sum_{i=1}^{k} \lambda_i \mathbf{a}_i = \mathbf{0}$$

will be satisfied if and only if

$$\lambda_1 = \lambda_2 = \cdots = \lambda_k = 0.$$

Thus the vectors in the set B are linearly independent.

Theorem 1.3.3 *The representation of a vector by a basis set is unique.*

Proof: Let the set A of vectors be given by

$$A = \{\mathbf{a}_1, \mathbf{a}_2, \cdots, \mathbf{a}_n\}$$

is a basis set in E_n and let $\mathbf{a}$ be any vector in the same vector space E_n.

Let us make an assumption that the representation is not unique. Therefore, $\mathbf{a}$ can be expressed as the linear combination of n independent vectors $\mathbf{a}_1, \mathbf{a}_2, \cdots, \mathbf{a}_n$ given below:

$$\mathbf{a} = \lambda_1\mathbf{a}_1 + \lambda_2\mathbf{a}_2 + \cdots + \lambda_n\mathbf{a}_n \tag{1.3.4}$$

$$\text{and} \quad \mathbf{a} = \mu_1\mathbf{a}_1 + \mu_2\mathbf{a}_2 + \cdots + \mu_n\mathbf{a}_n. \tag{1.3.5}$$

(1.3.4)–(1.3.5) gives

$$\mathbf{0} = (\lambda_1 - \mu_1)\mathbf{a}_1 + (\lambda_2 - \mu_2)\mathbf{a}_2 + \cdots + (\lambda_n - \mu_n)\mathbf{a}_n. \tag{1.3.6}$$

As the vectors $\mathbf{a}_1, \mathbf{a}_2, \cdots, \mathbf{a}_n$ are linearly independent therefore the condition (1.3.6) will be satisfied if and only if

$$\lambda_1 - \mu_1 = \lambda_2 - \mu_2 = \cdots = \lambda_n - \mu_n = 0$$

which indicates that $\lambda_1 = \mu_1$, $\lambda_2 = \mu_2, \cdots$ and $\lambda_n = \mu_n$.

Hence the representation is unique.

Theorem 1.3.4 Replacement theorem: *In E^n, if the set of vectors $\mathbf{a}_1, \mathbf{a}_2, \cdots, \mathbf{a}_r$ be a basis set and $\mathbf{b}$ is a vector in the same space, $\mathbf{b} \neq 0$, $\mathbf{b} \neq \mathbf{a}_i$ $(i = 1, 2, \cdots, r)$ and if $\mathbf{b} = \sum_{i=1}^{r} \lambda_i\mathbf{a}_i$, where $\lambda_i \neq 0$ for some i, then $\mathbf{b}$ can replace $\mathbf{a}_l$ to form a new basis for $\lambda_l \neq 0$, i.e., if $\lambda_l \neq 0$, then $\mathbf{a}_1, \mathbf{a}_2, \cdots, \mathbf{a}_{l-1}, \mathbf{b}, \mathbf{a}_{l+1}, \cdots, \mathbf{a}_r$ form a new basis set.*

Proof: The set of vectors $\mathbf{a}_1, \mathbf{a}_2, \cdots, \mathbf{a}_r$ is a basis set in E^n. $\mathbf{b}$ is a vector given by

$$\mathbf{b} = \sum_{i=1}^{r} \lambda_i\mathbf{a}_i \text{ with } \lambda_l \neq 0. \tag{1.3.7}$$

We are to prove that the set of vectors $\mathbf{a}_1, \mathbf{a}_2, \mathbf{a}_3, \cdots, \mathbf{a}_{l-1}, \mathbf{b}, \mathbf{a}_{l+1}, \cdots, \mathbf{a}_r$ is also a basis set in E^n, i.e., we are to prove that the vectors $\mathbf{a}_1, \mathbf{a}_2, \cdots, \mathbf{a}_{l-1}, \mathbf{b}, \mathbf{a}_{l+1}, \cdots, \mathbf{a}_r$ are linearly independent and each vector in E^n can be expressed as a linear combination of the set of vectors $\mathbf{a}_1, \mathbf{a}_2, \cdots, \mathbf{a}_{l-1}, \mathbf{b}, \mathbf{a}_{l+1}, \cdots, \mathbf{a}_r$.

Let us make an assumption that the vectors are not linearly independent. Therefore, the condition that

$$\sum_{\substack{i=1\\i\neq l}}^{r} \delta_i\mathbf{a}_i + \delta\mathbf{b} = \mathbf{0} \tag{1.3.8}$$

will be satisfied with at least one of the scalars δ_i $[i = 1, 2, \cdots, r, i \neq l]$ and δ, non-zero.

Now $\delta \neq 0$; because if $\delta = 0$, then from condition (1.3.8) we get

$$\sum_{\substack{i=1\\i\neq l}}^{r} \delta_i\mathbf{a}_i = \mathbf{0}$$

with at least one scalar δ_i not equal to zero which is impossible. [as the vectors $\mathbf{a}_1, \mathbf{a}_2, \cdots, \mathbf{a}_{l-1}, \mathbf{a}_{l+1}, \cdots, \mathbf{a}_r$ are linearly independent]

Now putting the value of $\mathbf{b}$ from (1.3.7) in (1.3.8) we get

$$\sum_{\substack{i=1,\\ i\neq l}}^{r} \delta_i \mathbf{a}_i + \delta \sum_{i=1}^{r} \lambda_i \mathbf{a}_i = \mathbf{0}, \quad \text{or,} \quad \sum_{\substack{i=1\\ i\neq l}}^{r} (\delta_i + \delta\lambda_i)\mathbf{a}_i + \delta\lambda_l \mathbf{a}_l = \mathbf{0} \tag{1.3.9}$$

Since $\delta \neq 0, \lambda_l \neq 0$, therefore $\delta\lambda_l \neq 0$, and hence the vectors $\mathbf{a}_1, \mathbf{a}_2, \cdots, \mathbf{a}_l, \cdots, \mathbf{a}_r$ are linearly dependent. This contradicts the fact. Therefore, the vectors $\mathbf{a}_1, \mathbf{a}_2, \cdots, \mathbf{a}_{l-1}, \mathbf{b}, \mathbf{a}_{l+1}, \cdots, \mathbf{a}_r$ are linearly independent.

Next, let $\mathbf{x}$ be any vector in E^n. Therefore,

$$\mathbf{x} = \sum_{i=1}^{r} \mu_i \mathbf{a}_i \quad [\mu_i \text{ are all scalars}] \tag{1.3.10}$$

From (1.3.7) we get

$$\mathbf{a}_l = \frac{\mathbf{b}}{\lambda_l} - \frac{1}{\lambda_l} \sum_{\substack{i=1\\ i\neq l}}^{r} \lambda_i \mathbf{a}_i. \tag{1.3.11}$$

Putting the value of $\mathbf{a}_l$ in (1.3.10) we find that $\mathbf{x}$ can be expressed as a linear combination of the vectors $\mathbf{a}_1, \mathbf{a}_2, \cdots, \mathbf{a}_{l-1}, \mathbf{b}, \mathbf{a}_{l+1}, \cdots, \mathbf{a}_r$. Hence the theorem is proved.

For example, the three vectors $\mathbf{a} = (1, 2, 4)$, $\mathbf{b} = (1, -1, 1)$ and $\mathbf{c} = (-1, 3 - 1)$ from a basis set E^3.

$\mathbf{d} = (3, 0, 6)$ is another vector in the same space; $\mathbf{d} \neq 0$, $\mathbf{d} \neq \mathbf{a}$ or $\mathbf{b}$ or $\mathbf{c}$ and $\mathbf{d}$ can be expressed as $\mathbf{d} = \lambda_1\ (1, 2, 4) + \lambda_2(1, -1, 1) + \lambda_3(-1, 3, -1)$, where $\lambda_1 = 1$, $\lambda_2 = 2$ and $\lambda_3 = 0$ [at least one of which is not zero].

Hence

$$\mathbf{d} = \lambda_1 \mathbf{a} + \lambda_2 \mathbf{b} + \lambda_3 \mathbf{c}$$

when $\lambda_1 \neq 0$, $\lambda_2 \neq 0$ and $\lambda_3 = 0$.

Therefore, according to the above theorem $\mathbf{d}$ can replace $\mathbf{a}$ and $\mathbf{b}$ to form a new basis, i.e., $(\mathbf{d}, \mathbf{b}, \mathbf{c})$ and $(\mathbf{a}, \mathbf{d}, \mathbf{c})$ form new basis but as $\lambda_3 = 0$, $\mathbf{d}$ cannot replace $\mathbf{c}$ to form a new basis, i.e., $(\mathbf{a}, \mathbf{b}, \mathbf{d})$ is not a basis in E^3.

Theorem 1.3.5 *In E^n, the exact number of vectors in a basis is n.*

Proof: This can be proved by applying the above theorem.

We know that $(\mathbf{e}_1, \mathbf{e}_2, \cdots, \mathbf{e}_n)$ form a basis (standard) in E^n, the number of vectors of which is n. Now by using replacement theorem, by suitable replacement of one of the vectors by a vector, a new basis can be formed, the number of vectors of which is exactly n. In the same way it can be shown that in all cases the number of vectors in a basis is n.

Worked out examples

1. Find out the rank of matrices:

$$\text{(a) } A = \begin{bmatrix} 1 & 2 & 1 \\ 3 & 1 & 5 \\ 3 & 4 & 7 \end{bmatrix} \quad \text{(b) } A = \begin{bmatrix} 2 & -3 & -1 \\ 4 & 1 & 5 \\ 6 & -2 & 4 \end{bmatrix}.$$

Solution: (a) The determinant of the matrix A, given by

$$|A| = \begin{vmatrix} 1 & 2 & 1 \\ 3 & 1 & 5 \\ 3 & 4 & 7 \end{vmatrix} = -16 \neq 0.$$

Hence A, the square matrix of order three, is non-singular. Therefore, the rank of the matrix is 3.

(b) $|B| = 0$. Hence the rank of the matrix B is less than 3. Now we take any square sub-matrix B_1 of order two and $B_1 = \begin{bmatrix} 2 & -3 \\ 4 & 1 \end{bmatrix}$; $|B_1| = 14 \neq 0$. Hence the rank of the matrix B is 2.

2. Prove that the vectors (a) $\mathbf{a}_1 = (2,3)$, $\mathbf{a}_2 = (-1,2)$, $\mathbf{a}_3 = (3,5)$ are linearly dependent and (b) $\mathbf{b}_1 = (4,5)$, $\mathbf{b}_2 = (6,2)$, $\mathbf{b}_3 = (8,10)$ are linearly dependent.

Solution: In each case there are three vectors in E^3. Hence the vectors are linearly dependent.

Now in (i) let there be three constants λ_1, λ_2 and λ_3 [with at least one of them non-zero] such that the condition

$$\lambda_1\mathbf{a}_1 + \lambda_2\mathbf{a}_2 + \lambda_3\mathbf{a}_3 = 0 \text{ be satisfied}$$
$$\text{or, } \lambda_1(2,3) + \lambda_2(-1,2) + \lambda_3(3,5) = (0,0).$$

Equating we get

$$2\lambda_1 - \lambda_2 + 3\lambda_3 = 0 \qquad \text{(i)}$$
$$3\lambda_1 + 2\lambda_2 + 5\lambda_3 = 0. \qquad \text{(ii)}$$

From (i) and (ii)

$$\frac{\lambda_1}{-5-6} = \frac{\lambda_2}{9-10} = \frac{\lambda_3}{4+3} = k = 1 \text{ (say)}.$$

Therefore, $\lambda_1 = -11$, $\lambda_2 = -1$ and $\lambda_3 = 7$. Hence the vectors are linearly dependent. [Putting $k = 0$ of course $\lambda_1 = \lambda_2 = \lambda_3 = 0$ is also a solution which is called a *trivial solution.*]

(b) As in (a) we have,

$$\lambda_1\mathbf{b}_1 + \lambda_2\mathbf{b}_2 + \lambda_3\mathbf{b}_3 = \mathbf{0} \text{ [with at least one of } \lambda_1, \lambda_2, \lambda_3 \text{ non-zero]}$$
$$\text{or, } \lambda_1(4,5) + \lambda_2(6,2) + \lambda_3(8,10) = (0,0).$$

Equating we get

$$4\lambda_1 + 6\lambda_2 + 8\lambda_3 = 0 \quad \text{(i)}$$
$$5\lambda_1 + 2\lambda_2 + 10\lambda_3 = 0. \quad \text{(ii)}$$

From (i) and (ii)

$$\frac{\lambda_1}{60-16} = \frac{\lambda_2}{40-40} = \frac{\lambda_3}{8-30} = k = 1 \text{ (say)}.$$

Therefore, $\lambda_1 = 44$, $\lambda_2 = 0$, $\lambda_3 = -22$ [Hence the vectors are L.D.]

3. Prove that the vectors (4, 3, 2), (2, 1, 4) and (2, 3, −8) are linearly dependent.

Solution: Square matrix A, from with the components of the three vectors is

$$A = \begin{bmatrix} 4 & 3 & 2 \\ 2 & 1 & 4 \\ 2 & 3 & -8 \end{bmatrix} \quad \text{and} \quad |A| = \begin{vmatrix} 4 & 3 & 2 \\ 2 & 1 & 4 \\ 2 & 3 & -8 \end{vmatrix} = 0.$$

Hence the rank of the matrix A is less than 3.

Therefore, the vectors are linearly dependent. [by prop. (10)]

2nd method: Let the vectors be linearly dependent and it is possible to find three scalar quantities $\lambda_1, \lambda_2, \lambda_3$ [with at least one of them non-zero] such that the relation $\lambda_1(4,3,2) + \lambda_2(2,1,4) + \lambda_3(2,3,-8) = (0,0,0)$ holds.

Equating we get,

$$4\lambda_1 + 2\lambda_2 + 2\lambda_3 = 0 \quad \text{(i)}$$
$$3\lambda_1 + \lambda_2 + 3\lambda_3 = 0 \quad \text{(ii)}$$
$$2\lambda_1 + 4\lambda_2 - 8\lambda_3 = 0. \quad \text{(iii)}$$

From (i) and (ii)

$$\frac{\lambda_1}{6-2} = \frac{\lambda_2}{6-12} = \frac{\lambda_3}{4-6} = k = 1 \text{ (say)}.$$
$$\text{or, } \lambda_1 = 4, \lambda_2 = -6, \lambda_3 = -2.$$

With these values of λ_1, λ_2 and λ_3 equation (iii) is also satisfied.

Hence the three vectors are L.D. and we may take $\lambda_1 = 2$, $\lambda_2 = -3$, $\lambda_3 = -1$ [dividing by 2].

4. Prove that the vectors (1, 2, 0), (0, 3, 1), (−1, 0, 0) arc linearly independent.

Solution: Square matrix A formed with the components of the three vectors is

$$A = \begin{bmatrix} 1 & 2 & 0 \\ 0 & 3 & 1 \\ -1 & 0 & 0 \end{bmatrix} \quad \text{and} \quad \det A = |A| = \begin{vmatrix} 1 & 2 & 0 \\ 0 & 3 & 1 \\ -1 & 0 & 0 \end{vmatrix} = -2 \neq 0.$$

Hence the rank of the square matrix A is 3. Therefore, the vectors are linearly independent.

2nd method: Let the vectors be linearly dependent and it is possible to find three scalar quantities λ_1, λ_2, λ_3 [at least one of them non-zero] such that $\lambda_1(1,2,0)+\lambda_2(0,3,1)+\lambda_3(-1,0,0)=(0,0,0)$ be satisfied.

Equating we get

$$\lambda_1 + 0\cdot\lambda_2 - \lambda_3 = 0 \quad \text{(i)}$$

$$2\lambda_1 + 3\lambda_2 + 0\lambda_3 = 0 \quad \text{(ii)}$$

$$0\cdot\lambda_1 + \lambda_2 - 0\cdot\lambda_3 = 0. \quad \text{(iii)}$$

From (i) and (ii)

$$\frac{\lambda_1}{3} = \frac{\lambda_2}{-2} = \frac{\lambda_3}{3} = k = 1 \text{ (say)}.$$

Therefore, $\lambda_1 = 3$, $\lambda_2 = -2$, $\lambda_3 = 3$. With these set of values, equation (iii) will not be satisfied. Therefore, it is not possible to find three scalar quantities λ_1, λ_2, λ_3 [with at least one of them non-zero] such that the condition $\lambda_1(1,2,0)+\lambda_2(0,3,1)+\lambda_3(-1,0,0)=(0,0,0)$ is satisfied. The conditions will be satisfied only for $\lambda_1=\lambda_2=\lambda_3=0$. Hence the vectors are linearly independent.

3rd method: Let us assume that the vectors be linearly dependent and it is possible to find three scalar quantities λ_1, λ_2, λ_3 [at least one of them non-zero] such that the relation $\lambda_1(1,2,0)+\lambda_2(0,3,1)+\lambda_3(-1,0,0)=(0,0,0)$ be satisfied.

Equating we get

$$\lambda_1 + 0\lambda_2 - \lambda_3 = 0$$

$$2\lambda_1 + 3\lambda_2 + 0\lambda_3 = 0$$

$$0\lambda_1 + \lambda_2 + 0\lambda_2 = 0.$$

Now the coefficient matrix $= \begin{bmatrix} 1 & 0 & -1 \\ 2 & 3 & 0 \\ 0 & 1 & 0 \end{bmatrix} = A^t$ [as shown in the first method]

and $\det A^t = \det A = -2 \neq 0$. Then by Cramer's rule or by matrix inverse method it can be said that there exists unique solution for λ_1, λ_2, λ_3 and only solution is $\lambda_1=\lambda_2=\lambda_3=0$. Hence our assumption is wrong and thus the vector and not L.D. and therefore they are L.I.

5. (a) Prove that $\mathbf{a}=(2,3)$ and $\mathbf{b}=(4,-1)$ from a basis in E^2 and express $(10,8)$ as the linear combination of the basis vectors.

(b) Prove that the vectors (1, 2, 3), (2, 4, 1), (3, 2, 9) form a basis in three dimensional vector space E^3 and also prove that $(2,16/3,1)$ is a vector which may replace any one the three vectors to form a new basis set.

(c) Prove that the vectors (1, 1, 0) and (3, 0, 1) are L.I. and express (5, 2, 1) as the L.C. of the vectors (1, 1, 0) and (3, 0, 1). **[C.U. 1983]**

(d) Prove that in E^3, the vectors (1, 0, −1) and (0, 2, 3) are L.I. but the vector (1, 2, 3) cannot be expressed as the L.C. of the above vectors.

Solution: (a) There are two vectors in E^2 and $B = \begin{bmatrix} 2 & 3 \\ 4 & -1 \end{bmatrix}$ and $|B| = -14 \neq 0$. Hence the vectors are linearly independent and vector $(a_1, a_2) \in E^2$ can be expressed as the L.C. of the vector (2, 3) and (4, −1) as $(a_1, a_2) = c_1(2,3) + c_2(4,-1)$, c_1, c_2 being scalars $\Rightarrow a_1 = 2c_1 + 4c_2$, $a_2 = 3c_1 - c_2$ has unique solution for c_1 and c_2 whatever may be the value of a_1, a_2 $[\because \begin{vmatrix} 2 & 4 \\ 3 & -1 \end{vmatrix} = -14 \neq 0]$ and evidently they form a basis in E^2.

Now (10, 8) $= \lambda_1 \mathbf{a} + \lambda_2 \mathbf{b}$ (say)

$= \lambda_1(2,3) + \lambda_2(4,-1)$.

Equating we get

$$10 = 2\lambda_1 + 4\lambda_2$$

$$\text{and} \quad 8 = 3\lambda_1 - \lambda_2.$$

Solving we get

$$\lambda_1 = 3, \lambda_2 = 1.$$

Thus (10, 8) $= 3\mathbf{a} + \mathbf{b}$.

(b) $A = \begin{bmatrix} 1 & 2 & 3 \\ 2 & 4 & 1 \\ 3 & 2 & 9 \end{bmatrix}$ and $|A| = \begin{vmatrix} 1 & 2 & 3 \\ 2 & 4 & 1 \\ 3 & 2 & 9 \end{vmatrix} = -20 \neq 0$, thus $r(A) = 3$.

Hence the three vectors are linearly independent.

Now let any vector $\mathbf{a} = (a_1, a_2, a_3) \in E^3$. We shall prove that $\mathbf{a}$ can be expressed as the L.C. of the vectors (1, 2, 3), (2, 4, 1), (3, 2, 9).

Let $(a_1, a_2, a_3) = c_1(1,2,3) + c_2(2,4,1) + c_3(3,2,9)$, c_1, c_2, c_3 being three scalars.

Equating we get

$$\left.\begin{aligned} a_1 &= c_1 + 2c_2 + 3c_3 \\ a_2 &= 2c_1 + 4c_2 + 2c_3 \\ a_3 &= 3c_1 + c_2 + 9c_3 \end{aligned}\right\}. \qquad \text{(i)}$$

Now the coefficient matrix is

$$\begin{bmatrix} 1 & 2 & 3 \\ 2 & 4 & 2 \\ 3 & 1 & 9 \end{bmatrix} = A^t \text{ and } \det A^t = \det A = -20 \neq 0.$$

Therefore, there exists unique solution for c_1, c_2 and c_3 whatever may be the value of a_1, a_2 and a_3 and the vector $\mathbf{a} = (a_1, a_2, a_3)$ can be expressed as the L.C. of the vectors (1, 2, 3), (2, 4, 1) and (3, 2, 9). Hence the vectors from a basis for E^3.

Important remark: In E^3, any three vectors which are linearly independent from a basis for E^3. This property may be used in answering short answer type questions.

Let

$$(2, 16/3, 1) = \lambda_1(1,2,3) + \lambda_2(2,4,1) + \lambda_3(3,2,9).$$

Then

$$\left.\begin{aligned} \lambda_1 + 2\lambda_2 + 3\lambda_3 &= 2 \\ 2\lambda_1 + 4\lambda_2 + 2\lambda_3 &= 16/3 \\ 3\lambda_1 + \lambda_2 + 9\lambda_3 &= 1 \end{aligned}\right\}.$$

Solving these three equations we get

$$\lambda_1 = 1, \lambda_2 = 1, \lambda_3 = -1/3.$$

Hence

$$(2, 16/3, 1) = 1(1,2,3) + 1(2,4,1) - 1/3(3,2,9).$$

As (2, 16/3, 1) is not a null vector, $\lambda_1 \neq 0$, $\lambda_2 \neq 0$ and $\lambda_3 \neq 0$ therefore the vector (2, 16/3, 1) and any other two vectors will form a new basis set for E^3.

(c) The two vectors are L.I. as the vector (1, 1, 0) cannot be expressed in the manner (1, 1, 0) $= c(3,0,1)$ for any scalar value of c and evidently the vectors are not L.D. and thus they are L.I.

Let (5, 2, 1) $= c_1(1,1,0) + c_2(3,0,1)$, c_1, c_2 being scalars.

Equating we get

$$5 = c_1 + 3c_2, 2 = c_1 + 0 \cdot c_2, 1 = 0 \cdot c_1 + c_2.$$

For $c_1 = 2$ and $c_2 = 1$ all three conditions are satisfied and then (5, 2, 1) = 2 (1, 1, 0) + 1 (3, 0, 1).

(d) The vectors $(1,0,-1)$ and (0, 2, 3) are L.I. as $(1,0,-1)$ cannot be expressed by the relation $(1,0,-1) = c(0,2,3)$, c being any scalar. Thus the vectors are not L.D. and hence they are L.I.

Let (1, 2, 3)$= c_1(1,0,-1) + c_2(0,2,3)$, c_1, c_2 being scalars.

Equating we get $1 = c_1 + 0 \cdot c_2$, $2 = 0 \cdot c_1 + 2c_2$ and $3 = -c_1 + 3c_2$.

From the first two conditions we get $c_1 = 1, c_2 = 1$. But with these values the third condition will not be satisfied. Hence we cannot find two scalar quantities c_1 and c_2 such that the relation (1, 2, 3) $= c_1(1,0,-1) + c_2(0,2,3)$ holds. Hence $(1,2,3)$ cannot be expressed as the L.C. of the vectors.

Remark: Any vector in E^3 can be expressed as the L.C. of the set of three vectors in E^3 provided they form a basis for E^3. Here the two vectors do not form a basis for E^3.

Objective and Short Answer type questions with Answers

Vector

1. Define: (a) vector, (b) linear combination of a set of vectors, (c) Linearly dependent set of vectors, (d) Linearly independent set of vectors, (e) Basis set, (f) Standard basis or unit basis. [*Ans:* See sec. 1.3.]

2. Which of the statements given below are true?

   (a) A set of vectors are either linearly dependent or linearly independent.

   (b) A set of vectors containing a null vector are always linearly dependent.

   (c) The number of vectors in a basis set is $2n$ in E^n.

   (d) The representation of a vector by a basis is unique.

   (e) Any superset of a linearly independent of set of vector is linearly independent.

   (f) Any superset of a linearly dependent set of vectors is linearly dependent.

   (g) All basis sets in E^n are generating sets.

   (h) The transpose of a row vector is a column vector.

   (i) All generating sets are basis sets.

   [*Ans:* (a) true, (b) true, (c) false, (d) true, (e) false, (f) true, (g) true, (h) true, (i) false.]

3. Define the inner product or scalar product of two vectors. State the nature of the vectors associated with the inner product.

   [*Ans:* In the case of inner product of two vectors **a** and **b** (having same number of components) the first vector must be a row vector and the second one is a column vector and $\mathbf{a}\cdot\mathbf{b} \equiv \mathbf{b}'\mathbf{a}'$, where $\mathbf{a}'$ is is the transpose of **a** etc.]

4. A vector **a** in E^3 is given by $\mathbf{a} = a_1\mathbf{e}_1 + a_2\mathbf{e}_2 + a_3\mathbf{e}_3$, where $\mathbf{e}_1, \mathbf{e}_2$ etc. have their usual meanings. What are the components of **a**? [*Ans:* $\mathbf{a} = (a_1, a_2, a_3)$.]

5. Define a unit vector in E^n.

6. Vectors $\mathbf{a}_1, \mathbf{a}_2, \mathbf{a}_3$ form a basis in E^3. **a** is also a non-null vector in E^3. **a** is expressed as the linear combination given by $\mathbf{a} = 2\ \mathbf{a}_1 + 0\cdot \mathbf{a}_2 - 5\mathbf{a}_3$.

   (a) Do the vectors $\mathbf{a}, \mathbf{a}_2, \mathbf{a}_3$ form a new basis in E^3?

   (b) Do the vectors $\mathbf{a}, \mathbf{a}_1, \mathbf{a}_2$ form a new basis in E^3?

   (c) Do the vectors $\mathbf{a}, \mathbf{a}_1, \mathbf{a}_3$ form a new basis in E^3?

   Give reasons for your answer in each case.

   [*Ans:* As the coefficient of $\mathbf{a}_2$ in the above linear combination is zero, **a** cannot replace $\mathbf{a}_2$ to form a new basis.

   Hence the vectors $\mathbf{a}_1, \mathbf{a}, \mathbf{a}_3$ cannot form a new basis. In other two cases they form a new basis.]

7. Does a basis contain a null vector? Give reasons for your answer.

[*Ans:* No; because a set of vectors containing a null vector are always linearly dependent. Hence the vectors do not form a basis.]

8. (a) Give a suitable example of generating set as well as basis in E^3

(b) Prove that vectors $(1, 4, -1), (2, 5, -2)$ are linearly independent in E^3 but the vector $(1, 7, -1)$ cannot be expressed as the linear combination of the above two vectors.

(c) Prove that the vector $(5, 1, -3)$ and $(2, 1, 5)$ are linearly independent in E^3 but the vector (7, 2, 4) cannot be expressed as the linear combination of the above vectors. Give reasons for your answer.

(d) Prove that the following vectors in E^4 are L.D. (i) $(2, 4, -1, 2)$ and (ii) $(8, 16, -4, 8)$.

(e) Prove that vectors $(1, 1, 0, -2)$ and $(0, 2, 4, -1)$ in E^4 are L.I.

(f) Prove that the vectors (1, 2, 3) and (3, 2, 1) are linearly independent in E^3 but they do not form a basis set of vectors. Give reasons for your answer.

[*Ans:* (a) The standard basis set (1, 0, 0), (0, 1, 0) and (0, 0, 1) is a suitable example of basis set in E^3. As the set of vector form a basis set, evidently the set is a generating set.

(b) $(1, 4, -1)$ cannot be expressed as $(1, 4, -1) = c(2, 5, -2)$ [c being any scalar.] Hence they are L.I. $(1, 7, -1) = 3(1, 4, -1) - (2, 5, -2)$.

(c) The vectors are L.I. but they are two in numbers. Hence they cannot form a basis in E^3. Thus any vector in E^3 cannot be expressed as the linear combination of the above vectors.

(d) $(2, 4, -1, 2) = 1/2(8, 16, -4, 8)$. Hence the vectors are L.D.

(e) (1, 1, 0, 2) cannot be expressed as $(1, 1, 0, -2) = c(0, 2, 4, -1)$ [c being any scalar.] Thus the vectors are L.I.

(f) The reason is same as given in (c).]

9. (a) Do the vectors (4, 3, 2), (2, 1, 4) and $(2, 3, -8)$ form a basis for E^3? Give reasons for your answer.

(b) Do the vectors (1, 2, 3), (2, 4, 1) and (3, 2, 9) constitute a basis for E^3? Give reasons.

(c) Prove that the vectors (1, 1, 0), $(1, -1, 0)$, (0, 0, 1) form a basis for E^3. **[C.U.(P) 1983]**

(d) Do the vectors (1, 2, 3), (4, 5, 6) and (3, 6, 9) form a basis for E^3? **[C.U.(P) 1981, 1995]**

(e) Prove that the vectors $\mathbf{a} = (1, 2, 1), \mathbf{b} = (2, 3, 0)$ and $\mathbf{c} = (1, 2, 2)$ form a basis for E^3. **[C.U.(P) 1987]**

(f) Prove that the vectors (1, 1, 0), (0, 1, 1) and (1, 2, 1) form a basis of E^3. **[C.U.(P) 1990]**

[*Ans:* (a) Vectors are linearly dependent; thus they do not form a basis. (b) There are three vectors in E^3 and they are linearly independent. Hence they constitute a basis in E^3. (c) Reason is same as (b).]

10. (a) Prove that a set of vector containing only one non-null vector is always linearly independent.

(b) If a set of vectors are linearly dependent at least one of them can be expressed as a linear combination of the others.

(c) If two non-null vectors are linearly dependent then one of them is a scalar multiple of the other.

[*Ans:* (a) Let **a** be a non-null vector; therefore $\lambda\mathbf{a} = \mathbf{0}$ will be satisfied if and only if $\lambda = 0$ which indicates that **a** is linearly independent.

(b) Let $\mathbf{a}_1, \mathbf{a}_2, \cdots, \mathbf{a}_n$ be linearly dependent. They $\lambda_1\mathbf{a}_1 + \lambda_2\mathbf{a}_2 + \cdots + \lambda_n\mathbf{a}_n = 0$ will be satisfied, with at least one of $\lambda_1, \cdots \lambda_n$ not equal to zero. Let $\lambda_1 \neq 0$ then

$$\mathbf{a}_1 = -\frac{\lambda_2}{\lambda_1}\mathbf{a}_2 - \frac{\lambda_3}{\lambda_1}\mathbf{a}_3 - \cdots - \frac{\lambda_n}{\lambda_1}\mathbf{a}_n.$$

Hence $\mathbf{a}_1$ can be expressed as a linear combination of the others.

(c) If two non-null vectors $\mathbf{a}_1$ and $\mathbf{a}_2$ are linearly dependent then $\lambda_1\mathbf{a}_1 + \lambda_2\mathbf{a}_2 = \mathbf{0}$ [with at least one of λ_1 and λ_2 not zero] will be satisfied if and only if both $\lambda_1 \neq 0$, $\lambda_2 \neq 0$.

Therefore,

$$\mathbf{a}_1 = -\frac{\lambda_2}{\lambda_1}\mathbf{a}_2 \quad \text{and} \quad \mathbf{a}_2 = -\frac{\lambda_1}{\lambda_2}\mathbf{a}_1 \text{ etc.}$$

11. (a) Express $\mathbf{x} = [4, 5]$ as a linear combination of $\mathbf{a} = [1, 3], \mathbf{b} = [2, 2]$. [C.U.(P) 1983]

(b) Prove that the vectors (4, 5), (6, 2), (8, 10) are linearly dependent and express null vector as a linear combination of these vectors.

[*Ans:* (a) $\mathbf{x} = \frac{1}{2}\mathbf{a} + \frac{7}{4}\mathbf{b}$. (b) The vectors are linearly dependent as there are (2+ 1) vectors in E^2.]

Exercise 1

1. Prove that if two non-null vectors are linearly dependent then one of them is a scalar multiple of the other.

2. Prove that the following sets of vectors are linearly independent:

   (a) $\mathbf{a} = (2, 1, 4), \mathbf{b} = (1, -1, -2), \mathbf{c} = (3, 1, -2)$.

   (b) $\mathbf{a} = (1, 2, 1), \mathbf{b} = (3, 1, 5), \mathbf{c} = (3, 4, 7)$.

   (c) $\mathbf{a} = (1, 2, -1), \mathbf{b} = (3, -1, 2), \mathbf{c} = (2, -2, 3)$.

   (d) $\mathbf{a} = (1, 2, 3), \mathbf{b} = (3, -2, 1), \mathbf{c} = (4, 2, 1)$.

   (e) $(1, -1, 1, 2), (1, 2, -1, 4)$ and $(1, 1, -1, -5)$.

3. Prove that the following sets of vectors are linearly dependent:

   (a) (2, 3), (3, 4), (10, 15).

   (b) $(1, 2, -4), (3, 2-1), (6, 8, -13)$.

   (c) $\mathbf{a} = (-1, 3, -5), \mathbf{b} = (2, 0, 4), \mathbf{c} = (1, 3, -1)$.

   (d) $(2, 2, 8), (1, 0, 4), (1, 2, 4)$. [C.U.(P) 1989]

   (e) $\mathbf{a} = (1, 1, 2, 4), \mathbf{b} = (2, -1, -5, 2), \mathbf{c} = (2, 1, 1, 6)$.

4. (a) Prove that the vectors $\mathbf{a} = (2, 3)$ and $\mathbf{b} = (1, 4)$ form a basis for E^2 and express the vector $\mathbf{c} = (5, 15)$ as the linear combination of $\mathbf{a}$ and $\mathbf{b}$.
   (b) Prove that the vectors $\mathbf{a} = (1, 0, 1)$, $\mathbf{b} = (1, 2, 3)$ and $\mathbf{c} = (2, 3, 4)$ form a basis in E^3 and express $\mathbf{d} = (5, 5, 9)$ as the linear combination of the vectors.
   (c) Express (5, 0) as the linear combination of the vectors (4, 1) and (3, 2).
   (d) Prove that the vectors $\mathbf{a} = (-1, 2, 2)$, $\mathbf{b} = (2, -3, 4)$ and $\mathbf{c} = (4, 6, 3)$ form a basis in E^3 and express $\mathbf{d} = (4, 7, 11)$ as the linear combination of the basis vectors.
   (e) Do the vectors $\mathbf{a} = (4, -1)$ and $\mathbf{b} = (-2, 1)$ form a basis in E^2? If so, express $\mathbf{c} = (10, -5)$ as the linear combination of the basis vectors.

5. (a) Prove the vectors: (i) $(1, 0, 0)$, $(0, 1, 0)$, $(0, 0, 1)$ and (ii) $(1, 0, 0)$, $(1, 1, 0)$, $(1, 1, 1)$ are two basis sets in the vector space E^3.
   (b) Prove that the vectors (1, 1, 0), (0, 1, 1) and (1, 2, 1) form a basis in E^3. [C.U.(P) 1990]
   (c) Prove that the vectors (1, 1, 0), $(1, -1, 0)$ and (0, 0, 1) forms a basis for E^3. [C.U.(P) 1983]

6. Prove that vectors (1, 2, 3), (2, 4, 1) and (3, 2, 9) is a basis set in $V_3(F)$ and prove that (2, 6, 1) is a vector which may replace any one of the three vectors to form a new basis.

7. Prove that the vectors (2, 4, 7), (4, 8, 9) are linarly independent.

   [*Hints:* One is not a scalar multiple of the other.]

8. If a set of vectors are linearly dependent then prove that at least one of them can be expressed as the linear combination of the remaining others.

9. (a) Prove that the vectors $\mathbf{a} = (2, -3)$ and $\mathbf{b} = (1, 4)$ form a basis in E^2 and express $\mathbf{c} = (3, 1)$ as the linear combination of the basis vectors. Hence prove that $(\mathbf{a}, \mathbf{c})$ and $(\mathbf{b}, \mathbf{c})$ form two basis in E^2.
   (b) Express $\mathbf{d} = (2, 0, -3)$ as the linear combination of the basis vectors $\mathbf{a} = (1, 0, 0)$, $\mathbf{b} = (0, 1, 0)$ and $\mathbf{c} = (0, 0, 1)$ and hence prove that the vectors $\mathbf{a}, \mathbf{c}, \mathbf{d}$ will not be a basis, but the set of vectors $\mathbf{a}, \mathbf{b}, \mathbf{d}$ and $\mathbf{c}, \mathbf{b}, \mathbf{d}$ will form basis sets.

   [*Hints:* $\mathbf{a}, \mathbf{b}, \mathbf{c}$ form a basis in E^3 and $\mathbf{d} = 2\mathbf{a} + 0\mathbf{b} - 3\mathbf{c}$, $\mathbf{d}$ is non-null and thus $\mathbf{d}, \mathbf{a}, \mathbf{c}$ are not L.I. but $\mathbf{d}, \mathbf{a}, \mathbf{b}$ and $\mathbf{d}, \mathbf{c}, \mathbf{b}$ are L.I. (by replacement theorem).]

10. (a) Prove that in E^3, the set of vectors (2, 1, 4), (1, 0, 2) are linearly independent are express (4, 1, 8) as the linear combination of the above vectors.
   (b) Prove that in E^3, though the vectors $(0, -1, 2)$ and (1, 2, 0) are linearly independent, but the vectors $(0, -2, 1)$ cannot be expressed as the linear combination of the above vectors. Give reasons.
   (c) Given the basis vectors $\mathbf{a}_1 = (1, 0, 0)$, $\mathbf{a}_2 = (0, 1, 1)$ and $\mathbf{a}_3 = (0, 0, 1)$ for E^3. Indicate one vector which can be removed from the basis and be replaced by $\mathbf{b} = (4, 3, 6)$ while still maintaining basis. [C.U.(H) 1990]

Answers

4. (a) $\mathbf{c} = \mathbf{a} + 3\mathbf{b}$, (b) $\mathbf{d} = 2\mathbf{a} + \mathbf{b} + \mathbf{c}$, (c) $2\mathbf{a} - \mathbf{b}$, (d) $\mathbf{d} = \mathbf{a} + \mathbf{b} + \mathbf{c}$. 9. (a) $\mathbf{c} = \mathbf{a} + \mathbf{b}$.
10. (a) (2,1,4)+2(1,0,2), (c) Any one of the three vectors.

Chapter 2

Motivation and Mathematical Formulation of Linear Programming Problems

2.1 Introduction

With the limited available resources (such as raw materials, manpower, capital, power and technical appliance etc.) the main object of an industry is to produce different products in such a way that maximum profit may be earned by selling them at market prices. Similarly, the main aim of a housewife is to buy the goodgrains, vegetables, fruits and other food materials at a minimum cost which will satisfy the minimum need (regarding food values, calories, proteins, vitamins etc.) of the members of her family. All these (the nature of production of different commodities etc.) can be done mathematically by formulating a problem which is known as a *programming problem.* Some *restrictions* or *constraints* are to be adopted to formulate the problem. The function which is to be optimized (such as profit, cost etc. either maximized or minimized) is known as the *objective function.* Almost in all types of problems, the objective function and the constraints are of linear type, and these problems are known as the *Linear programming problems.*

There are different types of problems which can be put in the form of linear programming problems. The problems discussed above are known as *production allocation problem* and *diet problem* respectively.

Consider the problem of transporting various amounts of a single product from different warehouses to different destinations with the help of different transport systems such that the cost of transport be minimum. This type of problem is known as *transportation problem.*

2.2 Motivation of the subject

Motivation of the subject is to make overall maximum profit or minimum loss satisfying the constraints. Initially Mathematicians try to solve by using Calculus, but they failed to do so. In the last part of the fifth decade Mr. Danzig and other Mathematicians try to solve the problem and they together discovered the simplex method to solve the problems. Very recently, Dr. Nagendra Karmakar discovered a method to solve the problems more easily and that method is known as *Karmakar Algorithm.* But this method has not been discussed in the book.

2.3 Mathematical formulation of the problems

Almost in each and every sphere of life, all practical problems can be put in the form of programming problem and if the objective function be a linear function and if the constraints be all linear constraints, then the programming problems are called *linear programming problems.* Collecting all information or data about the problem initially these are expressed in ordinary language in a descriptive manner and from this, the problem are expressed strictly in mathematical form before solving them by any method. Below given are some illustrative examples from which we can get some idea, how the problems can be put in strictly mathematical form.

2.3.1 Production allocation problem

▶ **Example 2.3.1** *Four different metals namely, iron, copper, zinc and manganese are required to produce three commodities* A, B *and* C. *To produce one unit of* A, 40 kg *iron,* 30 kg *copper,* 7 kg *zinc and* 4 kg *manganese are needed. Similarly to produce one unit of* B, 70 kg *iron,* 14 kg *copper and* 9 kg *manganese are needed and for producing one unit of* C, 50 kg *iron,* 18 kg *copper and* 8 kg *zinc are required. The total available quantities of metals are* 1 metric ton *iron,* 5 quintals *of copper,* 2 quintals *of zinc and manganese each. The profits are* Rs. 300, Rs. 200 *and* Rs. 100 *in selling per one unit of A, B and C respectively. Formulate the problem mathematically.*

Solution: Let z be the total profit and the problem is to maximize z.

z is known as the *objective function.*

All the available quantities and the quantities required to produce different commodities are given below in the tabular form:

	Iron	Copper	Zinc	Manganese
Total	1000 kg	500 kg	200 kg	200 kg
A	40 kg	30 kg	7 kg	4 kg
B	70 kg	14 kg	0 kg	9 kg
C	50 kg	18 kg	8 kg	0 kg

To get the maximum profit let x_1 units of A, x_2 units of B and x_3 units of C are to be produced.

Then total quantity of iron needed is

$$(40x_1 + 70x_2 + 50x_3) \text{ kg.}$$

Similarly total quantity of copper needed is

$$(30x_1 + 14x_2 + 18x_3) \text{ kg.}$$

Zinc

$$(7x_1 + 0x_2 + 8x_3) \text{ kg.}$$

Manganese

$$(4x_1 + 9x_2 + 0x_3) \text{ kg.}$$

and from the condition of the problem

$$\begin{aligned} 40x_1 + 70x_2 + 50x_3 &\leq 1000 \\ 30x_1 + 14x_2 + 18x_3 &\leq 500 \\ 7x_1 + 0x_2 + 8x_3 &\leq 200 \\ 4x_1 + 9x_2 + 0x_3 &\leq 200 \end{aligned}$$

The objective function

$$z = 300x_1 + 200x_2 + 100x_3$$

which is to be maximized. Hence the problem can be formulated as,

$$\text{Maximize,} \quad z = 300x_1 + 200x_2 + 100x_3$$

subject to

$$\begin{aligned} 40x_1 + 70x_2 + 50x_3 &\leq 1000 \\ 30x_1 + 14x_2 + 18x_3 &\leq 500 \\ 7x_1 + 0x_2 + 8x_3 &\leq 200 \\ 4x_1 + 9x_2 + 0x_3 &\leq 200 \end{aligned}$$

As none of the commodity produced be negative in number then $x_1 \geq 0$, $x_2 \geq 0$, $x_3 \geq 0$.

All these inequations are known as *constraints* or *restrictions*.

2.3.2 Diet problem

► **Example 2.3.2** *A patient needs daily* 5 mg, 20 mg *and* 15 mg *of vitamins A, B and C respectively. The vitamins available from a mango, an orange and an apple, are* 0.5 mg *of A,* 1 mg *of B,* 1 mg *of C;* 2 mg *of B,* 3 mg *of C;* 0.5 mg *of A,* 3 mg *of B and* 1 mg *of C respectively. If the cost of a mango, an orange and an apple be* Rs. 0.50, Rs. 0.25 *and* Rs. 0.40 *respectively, find the minimum cost of collecting the fruits so that daily requirement of the patient be met. Formulate the problem mathematically.*

Solution: The problem is to find the minimum cost of buying the food materials. Let z be the objective function.

Let x_1 mangoes, x_2 oranges and x_3 apples be bought to get the minimum daily requirement of vitamins so that the cost be minimum. Objective function z is given by

$$z = 0.50x_1 + 0.25x_2 + 0.40x_3.$$

The quantity of vitamin A, available from the fruits, is

$$(0.5x_1 + 0x_2 + 0.5x_3) \text{ mg.}$$

Similarly the quantity of vitamin B available, is

$$(1x_1 + 2x_2 + 3x_3) \text{ mg.}$$

and vitamin C is

$$(1x_1 + 3x_2 + 1x_3) \text{ mg.}$$

From the condition of the problem

$$\begin{aligned} 0.5x_1 + 0x_2 + 0.5x_3 &\geq 5 \\ x_1 + 2x_2 + 3x_3 &\geq 20 \\ x_1 + 3x_2 + x_3 &\geq 15 \end{aligned}$$

and $x_1 \geq 0$, $x_2 \geq 0$, $x_3 \geq 0$.

Hence the problem is, Minimize,

$$z = 0.50x_1 + 0.25x_2 + 0.40x_3$$

subject to

$$\begin{aligned} 0.5x_1 + 0x_2 + 0.5x_3 &\geq 5 \\ x_1 + 2x_2 + 3x_3 &\geq 20 \\ x_1 + 3x_2 + x_3 &\geq 15 \end{aligned}$$

and $x_1, x_2, x_3 \geq 0$.

2.3.3 Transportation problem

► **Example 2.3.3** *Three different types of lorries A, B and C have been used to transport* 60 tons *solid and* 35 tons *liquid substance. A type lorry can carry* 7 tons *solid and* 3 tons *liquid. B type lorry can carry* 6 tons *solid and* 2 tons *liquid and C type lorry can carry* 3 tons *solid and* 4 tons *liquid. The cost of transport are* Rs. 500, Rs. 400 *and* Rs. 450 *per lorry of A, B and C type respectively. Find the minimum transport cost. Formulate the problem mathematically.*

Solution: The problem is to minimize the transport cost. Let z be the objective function.

Let x_1 lorries of A type, x_2 lorries of B type and x_3 lorries of C types be used to transport the materials so that the cost of transport be minimum. The objective function z is given by

$$z = 500x_1 + 400x_2 + 450x_3$$

The quantity of solid, transported by the lorries is

$$(7x_1 + 6x_2 + 3x_3) \text{ tons.}$$

The quantity of liquid, transported by the lorries is

$$(3x_1 + 2x_2 + 4x_3) \text{ tons.}$$

From the condition of the problem

$$7x_1 + 6x_2 + 3x_3 \geq 60$$
$$3x_1 + 2x_2 + 4x_3 \geq 35.$$

Hence the problem is, Minimize,

$$z = 500x_1 + 400x_2 + 450x_3$$

subject to

$$7x_1 + 6x_2 + 3x_3 \geq 60$$
$$3x_1 + 2x_2 + 4x_3 \geq 35.$$

and $x_1, x_2, x_3 \geq 0$.

2.3.4 Miscellaneous Problems

▶ **Example 2.3.4** *An electronic company manufactures two radio models each on a separate production line. The daily capacity of the first line is* 60 *radios and that of the second is* 75 *radios. Each unit of the first model uses* 10 *pieces of a certain electronic component, where as each unit of the second model requires* 8 *pieces of the same component. The maximum daily availability of the special components is* 800 *pieces. The profit per unit of models* 1 *and* 2 *are* Rs. 500 *and* Rs. 400 *respectively. Determine graphically or otherwise the optimal daily production of each model.* **[C.U.(H) '91,'94,'96]**

Solution: The problem is a problem of maximization.

Let x_1 and x_2 be the number of two radio models each on a separate production line. Therefore the objective function is

$$z = 500x_1 + 400x_2$$

which is to be maximized. [Since profit per of unit of models 1 and 2 are Rs. 500 and Rs. 400 respectively.]

Since the daily capacity of the first line and the second line are 60 and 75 radios respectively, then we have

$$x_1 \le 60$$
$$x_2 \le 75.$$

The total electronic components required $= 10x_1 + 8x_2$.

From the condition of the problem $10x_1 + 8x_2 \le 800$.

Hence the problem is to maximize,

$$z = 500x_1 + 400x_2$$

subject to

$$x_1 \le 60$$
$$x_2 \le 75$$
$$10x_1 + 8x_2 \le 800$$

and $x_1, x_2 \ge 0$.

By solving it using the graphical method (Chapter 7) it can be shown that the maximum profit will Rs. 40,000 but there will be different types of production for which the profit will be maximum such as for $(x_1 = 60, x_2 = 25)$; $(x_1 = 20, x_2 = 75)$, etc. profit will be maximum.

▶ **Example 2.3.5** *An agricultural firm has* 180 tons *of Nitrogen fertilisers,* 50 tons *of Phosphate and* 220 tons *of Potash. It will be able to sell* 3 : 3 : 4 *mixtures of these substances at a profit of* Rs. 15 *per ton and* 2 : 4 : 2 *mixtures at a profit of* Rs. 12 per *ton respectively. Pose a linear programming problem to how many tons of these two mixtures should be prepared to obtain the maximum profit?*

[C.U.(P)'93, 97]

Solution: Let the 3 : 3 : 4 mixture be called A and that of 2 : 4 : 2 be called B.

Let x_1 and x_2 tons of these two mixtures be produced to get maximum overall profit. Thus the objective function is

$$z = 15x_1 + 12x_2$$

which is to be maximized.

Let us denote nitrogen, phosphate and potash as N, Ph, P respectively, in the mixture A and,

$$\frac{N}{3} = \frac{Ph}{3} = \frac{P}{4} = k_1 \text{ (say)}.$$

Hence

$$N = 3k_1,\ Ph = 3k_1,\ P = 4k_1, \text{ so that } x_1 = 10k_1.$$

Similarly for the mixture B,

$$N = 2k_2,\ Ph = 4k_2,\ P = 2k_2 \text{ so that } x_2 = 8k_2.$$

Thus the constraints are

$$\frac{3}{10}x_1 + \frac{1}{4}x_2 \leq 180$$

[$\because$ In the mixture A, amount of nitrogen $= \frac{3k_1}{10k_1}x_1 = (3/10)x_1$]

Similarly,

$$\frac{3}{10}x_1 + \frac{1}{2}x_2 \leq 250$$

$$\text{and} \quad \frac{2}{5}x_1 + \frac{1}{4}x_2 \leq 220, \quad x_1, x_2 \geq 0.$$

Thus the constraints are

$$\begin{aligned} 6x_1 + 5x_2 &\leq 3600 \\ 3x_1 + 5x_2 &\leq 2500 \\ 8x_1 + 5x_2 &\leq 4400, \quad x_1, x_2 \geq 0 \end{aligned}$$

and the problem is, maximize $z = 15x_1 + 12x_2$.

► **Example 2.3.6** *A coin to be minted contains* 40% *silver,* 50% *copper,* 10% *nickel. The mint has available alloys A, B, C and D having the following composition and costs:*

	% Silver	% Copper	% Nickel	Costs
A	30	60	10	Rs. 11.00
B	35	35	30	Rs. 12.00
C	50	50	0	Rs. 16.00
D	40	45	15	Rs. 14.00

Present the problem of getting the alloys with specific composition at minimum cost in the form of a L.P.P. (Solution is not necessary.) **[C.U.(P)'94, '99]**

Solution: The question is incomplete since the amount of the metals are not given. We assume that the total quantity of metals is 1000 kg. and further assume that the prices of the alloys given in the table are per kg.

Let x_1, x_2, x_3, x_4 kgs. be the quantities of alloys A, B, C, D used for the purpose in kg. so that

$$x_1 + x_2 + x_3 + x_4 \leq 1000.$$

The objective function is

$$z = 11x_1 + 12x_2 + 16x_3 + 14x_4$$

and the constraints are

$$\begin{aligned} &0.3x_1 + 0.35x_2 + 0.5x_3 + 0.4x_4 \leq 400 \text{ for silver} \\ &0.6x_1 + 0.35x_2 + 0.5x_3 + 0.45x_4 \leq 500 \text{ for copper} \\ &0.1x_1 + 0.3x_2 + 0.15x_4 \leq 100 \text{ for nickel}, \quad x_1, x_2, x_3, x_4 \geq 0. \end{aligned}$$

Thus the L.P.P. is, minimize

$$z = 11x_1 + 12x_2 + 16x_3 + 14x_4$$

subject to

$$\begin{aligned}
0.3x_1 + 0.35x_2 + 0.5x_3 + 0.4x_4 &\le 400\\
0.6x_1 + 0.35x_2 + 0.5x_3 + 0.45x_4 &\le 500\\
0.1x_1 + 0.3x_2 \qquad\qquad + 0.15x_4 &\le 100\\
x_1 + x_2 + x_3 + x_4 &\le 1000 \quad x_1, x_2, x_3, x_4 \ge 0.
\end{aligned}$$

▶ **Example 2.3.7** *A soft drink plant has two bottling machines A and B. It produces and sells* 500 ml. *and* 800 ml. *bottles. The following data are available.*

Machine	500 ml	800 ml
A	100/minute	40/minute
B	60/minute	75/minute

The machines can be run 8 *hours per day and* 5 *days per week. Weekly productions of the drink cannot exceed* 30,00,000 ml. *and the market can absorb* 25,000 *bottles of* 500 ml *and* 7,000 *bottles of* 800 ml. *per week. Profit on two types of bottles is* 15 *paise and* 25 *paise respectively. The planner wishes to maximize his profit to all the productions and marketing restrictions. Formulate it as a linear programming problem.* **[C.U.(P)'98]**

Solution: Let x_1 and x_2 be number of 500 ml and 800 ml bottles produced to get an over all maximum profit. Then the profit is

$$\left(x_1 \times \frac{15}{100} + x_2 \times \frac{25}{100}\right) \text{Rs.} = (0.15x_1 + 0.25x_2) \text{ Rs.} \quad \text{(say)}.$$

From the market condition, we get

$$x_1 \le 25{,}000$$
$$x_2 \le 7{,}000$$

Total production of 500 ml bottles $= (100 + 60) \times 60 \times 8 \times 5$ per week.

Total production of 800 ml bottles $= (40 + 75) \times 60 \times 8 \times 5$ per week.

Thus the production of 500 ml and 800 ml bottles per week are 3,84,000 and 2,76,000 respectively. As the number of bottles that can be produced are greater than 25,000 and 7,000 respectively then these conditions are redundant.

The amount of soft drinks is

$$500x_1 + 800x_2 \text{ ml.}$$

Then

$$500x_1 + 800x_2 \le 30{,}00{,}000.$$

Thus the problem is,

$$\text{Maximize,} \quad z = 0.15x_1 + 0.25x_2$$

subject to

$$\begin{aligned} 500x_1 + 800x_2 &\leq 30{,}00{,}000 \\ x_1 &\leq 25{,}000 \\ x_2 &\leq 7{,}000, \quad x_1, x_2 \geq 0. \end{aligned}$$

▶ **Example 2.3.8** *A hospital has the following minimum requirement for nurses.*

Period	Clock time (24 hours day)	Minimum number of nurses required
1	6 A.M.–10 A.M.	60
2	10 A.M.– 2 P.M.	70
3	2 P.M.– 6 P.M.	60
4	6 P.M.–10 P.M.	50
5	10 P.M.– 2 A.M.	20
6	2 A.M.– 6 A.M.	30

Nurses report to the hospital wards at the beginning of each period and work for eight consecutive hours. The hospital wants to determine the minimum number of nurses so that there may be sufficient number of nurses available for each period. Formulate this as a L.P.P. **[C.U.(H)'83, '97]**

Solution: This is a minimization problem.

Let $x_1, x_2, \cdots, x_6$ be the number of nurses required for the period $1, 2, \cdots, 6$, etc. Then the objective function is

$$\text{Minimize,} \quad z = x_1 + x_2 + \cdots + x_6.$$

Now the constraints can be written in the following manner.

x_1 nurses work for the period 1 and 2 and x_2 nurses work for the period 2 and 3 etc.

Thus for the period 2

$$x_1 + x_2 \geq 70.$$

Similarly,

$$\begin{aligned} x_2 + x_3 &\geq 60 \\ x_3 + x_4 &\geq 50 \\ x_4 + x_5 &\geq 20 \\ x_5 + x_6 &\geq 30 \\ x_6 + x_1 &\geq 60 \end{aligned}$$

and $x_j \geq 0$, $j = 1, 2, \cdots, 6$.

► **Example 2.3.9** *A businessman has the option of investing his money in two plans. Plan A guarantees that each rupee invested will earn seventy paise a year hence while the plan B guarantees that each rupee invested will earn two rupees two year hence. In plan B, only investments for periods that are multiple of two years are allowed. The problem is how he invest ten thousand rupees in order to maximize the earnings at the end of the three years. Formulate the problem as a linear programming model.* **[C.U.(H) '81]**

Solution: Let x_{1A} and x_{1B} be the amounts invested for plan A and plan B respectively at the beginning of the first year.

From the condition of the problem

$$x_{1A} + x_{1B} \leq 10000. \tag{i}$$

At the end of the first year the businessman received $1.7x_{1A}$. Then if x_{2A} and x_{2B} be the amount of investment at the beginning of the second year then $x_{2A} + x_{2B} \leq 1.7x_{1A}$

$$\Rightarrow x_{2A} + x_{2B} - 1.7x_{1A} \leq 0 \tag{ii}$$

At the end of the second year he will receive back $1.7x_{2A}$ and $x_{1B}+2x_{1B} = 3x_{1B}$.

Then for the beginning of the third year

$$x_{3A} + x_{3B} \leq 1.7x_{2A} + 3x_{1B} \tag{iii}$$

In (iii) we may take $x_{3B} = 0$, since x_{3B} will fetch no money at the end of the third year.

Then

$$x_{3A} \leq 1.7x_{2A} + 3x_{1B}. \tag{iv}$$

At the end of third year his total return is $1.7x_{3A} + 3x_{2B}$. Hence the problem is to maximize

$$z = 1.7x_{3A} + 3x_{2B}.$$

subject to the condition (i), (ii), (iii) and (iv) and $x_{iA}, x_{iB} \geq 0$, $i = 1, 2, 3$.

2.4 Mathematical formulation of a L.P.P.

From the above discussions of the mathematical formulation of Linear Programming Problems (L.P.P.), general linear programming problem can be stated mathematically as follows.

Find out a set of values $x_1, x_2, \cdots, x_n$ which will optimize (either maximize or minimize) the linear function

$$z = c_1x_1 + c_2x_2 + c_3x_3 + \cdots + c_nx_n$$

subject to the restrictions

$$\left.\begin{array}{l} a_{11}x_1 + a_{12}x_2 + \cdots + a_{1n}x_n \ (\leq = \geq) \ b_1 \\ a_{21}x_1 + a_{22}x_2 + \cdots + a_{2n}x_n \ (\leq = \geq) \ b_2 \\ \cdots\cdots\cdots\cdots\cdots\cdots\cdots\cdots\cdots\cdots \\ a_{m1}x_1 + a_{m2}x_2 + \cdots + a_{mn}x_n \ (\leq = \geq) \ b_m \end{array}\right\} \quad (2.4.1)$$

and the non-negative restrictions $x_j \geq 0$, $j = 1, 2, \cdots, n$,

where a_{ij}, c_j and b_i's $(i = 1, \cdots, m)$ are all constants and x_j's are variables. Each of the linear expressions on the left side connected to the corresponding constants on the right side, by only one of the signs "≤", "=" and "≥", is known as a *constraint.* A constraint is either an equation or an inequation.

The linear function

$$z = c_1x_1 + c_2x_2 + \cdots + c_nx_n$$

is known as the *objective function.*

By using the matrix and vector notations the problem can be expressed in a compact form given below.

Optimize $z = \mathbf{cx}$ subject to the restrictions

$$A\mathbf{x}(\leq=\geq)\mathbf{b}, \quad \mathbf{x} \geq \mathbf{0}, \qquad (2.4.2)$$

where $A = [a_{ij}]$ is a $m \times n$ co-efficient matrix.

$\mathbf{c} = (c_1, c_2, \cdots, c_n)$ is a n-component row vector, which is known as a *cost* or *price* vector as it determines the cost of production or the price of a commodity.

$\mathbf{x} = [x_1, x_2, \cdots, x_n]$ is a n-component column vector which is known as *decision variable or legitimate variable vector.*

$\mathbf{b} = [b_1, b_2, \cdots, b_m]$ is a m-component column vector which is known as *requirement vector* and

$\mathbf{0} = [0, \cdots, 0]$ is a n-component *null column vector.*

If all constraints are equations then the L.P.P. is reduced to, Optimize $z = \mathbf{cx}$ or simply optimize, $\mathbf{cx}$ subject to

$$A\mathbf{x} = \mathbf{b}, \quad \mathbf{x} \geq \mathbf{0}. \qquad (2.4.3)$$

This form is called the *standard form.*

[See (3–15) and (3–16) of page 76 [Linear Programming by Hadly] and page 48 of [Linear Programming by Gass]].

In all practical discussions $b_i \geq 0$, $i = 1, 2, \cdots, m$. If some of them be negative, then make them positive, multiplying both sides of the linear constraints, having negative values of b_i, by (-1).

Feasible solution to a L.P.P. : A set of values of the variables, which satisfy all the constraints and all the non-negative restrictions of variables, is known as the *feasible solution* (F.S.) to the L.P.P.

Optimal solution to a L.P.P. : A feasible solution to a L.P.P. which makes the objective function an optimum is known as the *optimal solution* to the L.P.P.

There are various methods of finding the optimal solution of a L.P.P. : (i) *Geometrical method or graphical method* and (ii) *Algebraical method (simplex method).*

Before solving a L.P.P. by simplex method, we shall discuss the method of finding out the solution set of a system of m linear equations with n variables $(n > m)$ in the Chapter 4.

2.5 Problem of maximization and minimization

A particular L.P.P. is either a problem of maximization or a problem of minimization. The problem of minimization of the objective function z is nothing but the problem of maximization of the function $(-z)$ and vice versa and

$$\min z = -\max(-z)$$

with same set of constraints and the same solution set etc.

We shall prove this theorem in some later chapter.

Short Answer type questions

1. What is a linear programming problem?
2. Write down the mathematical form of a general L.P.P.

   [*Ans.* See Art. 2.4.1.]
3. What is the relation between the optimum value of maximization and minimization problem ?
4. Write down the standard form of a general L.P.P.

   [*Ans.* See Art. 2.4.3]

2.6 Formulation of the problems mathematically

1. A particular company manufactures two products A and B. These products are processed on the same machine. It takes 25 minutes to process one unit of product A and 15 minutes for one unit of product B and the machine operates for a maximum of 35 hours in a week. Product A requires 1 kg. and product B, 2.5 kg. of raw-material per unit, the supply of which is 170 kgs per week. If the net profit from the product are Rs. 100 and Rs. 450 per unit respectively, find how much of each product should be produced per week, in order to get maximum profit. [Formulate it mathematically].

2. A manufacturer of lether belts makes three types of belts A, B and C which are processed on three machines M_1, M_2 and M_3. Belt A requires 2 hours on machine M_1 and 3 hours on machine M_3. Belt B requires 3 hours on machine M_1, 2 hours on machine M_2 and 2 hours on machine M_3 and Belt C requires 5 hours on machine M_2 and 4 hours on machine M_3. There are 8 hours of time per day available on machine M_1, 10 hours of time per day available on machine M_2 and 15 hours of time per day available on machine M_3. The profit per unit of A, B and C are Rs. 3.00, Rs. 5.00 and Rs. 4.00 respectively. Find out the daily production of each type of belts such that the profit be maximum.

3. A firm can produce three types of cloth say A, B and C. Three kinds of wool are required for it, say red wool, green wool and blue wool. One unit length of type A cloth needs 2 yards of red wool and 3 yards of blue wool; one unit length of type B cloth needs 3 yards of red wool, 2 yards of green wool and 2 yards of blue wool; and one unit length of type C cloth needs 5 yards of green wool and 4 yards of blue wool. The firm has only a stock of 8 yards of red wool, 10 yards of green wool and 15 yards of blue wool. It is assumed that the income obtained from one unit length of type A cloth is Rs. 3.00, of type B cloth is Rs. 5.00 and that of type C cloth is Rs. 4.00. Formulate the above problem as a linear programming problem.

4. A person requires 10, 12 and 12 units of chemical A, B and C respectively for his garden. A liquid product contains 3, 2 and 1 unit of A, B and C respectively per jar. A dry product contains 1, 2 and 4 units of A, B and C per packet. If the liquid product sells for Rs. 3.00 per jar and the dry product sells for Rs. 2 per packet then formulate the problem as a linear programming problem. [C.U.(P)'87]

5. Food X contains 6 units of vitamin A per gram and 7 units of vitamin B per gram and cost 12 paise per gram. Food Y contains 8 units of vitamin A per gram and 12 units of vitamin B and cost 20 paise per gram. The daily minimum requirement of vitamin A and B are 100 units and 120 units respectively. Find the minimum cost of product mix. (Formulate the problem). [C.U.(H)'93, (P)'95]

6. A manufacturer of furniture makes two products, chairs and tables. Processing of these products is done on two machines A and B. A chairs requires 2 hours on machine A and 6 hours on machine B. A table requires 5 hours on machine A and no time on machine B. There are 16 hours of time per day available on machine A and 20 hours on machine B. Profit gained by the manufacturer from a chair and table is Rs. 10 and Rs. 50 respectively. What should be the daily production of each of the two products to obtain maximum profit? (Formulate the problem).

7. A television company has three major departments for the manufacture of two models, A and B. Monthly capacities are given as follows.

	Per unit time requirement (hours)		Hours available
	Model A	Model B	this month
Department I	4.0	2.0	1600
Department II	2.5	1.0	1200
Department III	4.5	1.5	1600

The marginal profit of model A is Rs. 400 each and that of model B is Rs. 100 each. Assuming that the company can sell any quantity of either product due to favourable market conditions, determine the highest possible profit for this month. (Formulate it).

8. A company sells two different products A and B. The company makes a profit of Rs. 40 and Rs. 30 per unit of products A and B respectively. The two products are produced in a common production process and are sold in two different markets. The production process has a capacity of 30,000 man hours. It takes 3 hours to produce one unit of A and one hour to produce one unit of B. The market has been surveyed and the company officials feel that the maximum number of unit of A that can be sold is 8,000 and the maximum of B is 12,000 units. Subject to these limitations, formulate the problem as a L.P.P.

9. A furniture manufacturer wishes to determine the number of tables and chairs to be made by him in order to optimise the use of his available resources. These products utilize two different types of timber and he has in hand 1,500 board feet of the first type and 1,000 board feet of the second type. He has 800 man hours available for the total job. Each table and chair requires 5 and 1 board feet respectively of the first type timber and 2 and 3 board feet of the second type. 3 man hours are required to make a table and 2 man hours are needed to make a chair. The manufacturer makes a profit of Rs. 12 on a table and Rs. 5 on a chair. Write down the complete linear programming formulation of the problem in terms of maximising the profit. **[C.U.(P)'86]**

10. A tailor has 80 sq m of cotton material and 120 sq m of woollen material. A suit requires 1 sq m of cotton and 3 sq m of woollen material and a dress requires 2 sq m of each. A suit sells for Rs. 500 and a dress sells for Rs. 400. Pose a L.P.P. in terms of maximizing the income. **[B.U.(P)'96]**

11. A farmer has a 100 acre farm. He can sell all tomatoes, lettuce or radishes he can raise. The price he can obtain is Rs. 1.00 per kg. for tomatoes Rs. 0.75 a head for lettuce and Rs. 2.00 per kg. for radishes. The average yield per acre is 2,000 kgs of tomatoes, 3,000 heads of lettuce and 1,000 kgs of radishes. Fertilizer is available at Rs. 0.50 per kg. and the amount per acre is 100 kgs each for tomatoes and lettuce and 50 kgs for radishes. Labour required for sowing, cultivating and harvesting per acre is 5 man days for tomatoes and radishes and 6 man-days for lettuce. A total of 400 man-days of labour are available at Rs. 20.00 per man-day. Formulate this problem as a linear programming model to maximize the total profit.

12. A factory is engaged in manufacturing three products, A, B and C which involve lathe work, grinding and assembling. The cutting, grinding and assembling times required for one unit of A are 2, 1 and 1 hours respectively. Similarly they are 3, 1, 3 hours for one unit of B and 1, 3, 1 hours for one unit of C. The profits on A, B and C are Rs. 2, Rs. 2 and Rs. 4 per unit respectively. Assuming that there are available 300 hours of lathe time, 300 hours of grinder time and 240 hours of assembly time, how many units of each product should be produced to maximize profit? (Formulate mathematically).
[C.U.(H)'87,2000,(P)'80]

13. A manufacturer makes red and blue pen. A red pen takes twice as much as time to make a blue pen one. If the manufacturer makes only blue pens, 500 can be made in a day. A red pen sells for Rs. 8 and at most 150 can be sold in a day. A blue pen sells for Rs. 5 and at most 250 can be sold in a day. The manufacturer desires to maximize his revenue. Formulate the manufacturer's problem as a linear programme. **[C.U. M.Com.'89]**

[*Hints:* Let there be 500 units of time available and 2 and 1 unit of time required to produce red and blue pen. Thus, if x_1 and x_2 be the number of red and blue pen then the problem is to maximize $8x_1 + 5x_2$ s.t. $2x_1 + x_2 \leq 500$, $x_1 \leq 150$, $x_2 \leq 250$, $x_1, x_2 \geq 0$.]

14. At a cattle breeding firm, it is prescribed that the food ration for one animal must contain at least 14, 22 and 11 units of nutrients A, B and C respectively. Two different kinds of fodder are available. Each unit weight of these two contains the following amounts of the three nutrients. **[C.U.(H)'88, (P)'92]**

	Fodder 1	Fodder 2
Nutrient A	2	1
Nutrient B	2	3
Nutrient C	1	1

It is given that the cost of fodder 1 and 2 are 3 and 2 monetary units respectively. Formulate the problem of finding the minimum cost of purchasing the fodders as a L.P.P.

15. ABC company wishes to plan its advertising strategy. There are two media under consideration, call them magazine I and magazine II. Magazine I has a reach of 2,000 potential customers and magazine II has a reach of 3,000 potential customers. The cost per page of advertising is Rs. 400 and Rs. 600 in magazines I and II respectively. The firm has a monthly budget of Rs. 6,000. There is an important requirement that the total reach for the income group under Rs. 20,000 per annum, should not exceed 4,000 potential customers. (The reach in magazine I and II for this income group is 400 and 200 potential customers). The firm wants to buy as many pages for different advertisements in the two magazines so as to maximise the total reach. Formulate the above as L.P.P.
[C.U.(H)'86]

16. A medicine firm manufacturing two types of medicines A and B, can make a profit of Rs. 20 per bottle of A and Rs. 30 per bottle of B. Both A and B need

for their production two essential chemicals C and D. Each bottle of A requires 3 litres of C and 2 litres of D and each bottle of B requires 2 litres of C and 4 litres of D. The total supply of these chemicals are restricted to 210 litres of C and 300 litres of D. Type B medicine contains alcohol and so its manufacture is restricted to 65 bottles per day. Construct a linear programming problem in terms of maximizing the daily profit of the product. [C.U.(P)'91]

17. A manufacturer of furniture makes two products; chairs and tables. Processing of the products is done on two machines A and B. A chair requires 2 hours an machine A and 6 hours on machine 13. A table requires 5 hours on machine A and 2 hours on machine B. There are 16 hours of time per day available on machine A and 22 hours on machine B. Profit gained by the manufacturer from a chair and a table is Re. 1 and Rs. 5 respectively. Formulate a linear programming problem to maximize profit per day. [C.U.(P)'90]

Answers

1. Maximize $z = 100x_1 + 450x_2$ subject to $5x_1 + 3x_2 \leq 420$, $2x_1 + 5x_2 \leq 340$, $x_1 \geq 0$, $x_2 \geq 0$.

2. Maximize $z = 3x_1+5x_2+4x_3$ subject to $2x_1+3x_2 \leq 8$, $2x_2+5x_3 \leq 10$, $3x_1+2x_2+4x_3 \leq 15$, $x_1, x_2, x_3 \geq 0$.

3. Same as Example 2.

4. Minimize $z = 3x_1 + 2x_2$ subject to $3x_1 + x_2 \geq 10$, $2x_1 + 2x_2 \geq 12$, $x_1 + 4x_2 \geq 12$, $x_1 \geq 0$, $x_2 \geq 0$.

5. Minimize $z = 12x_1 + 20x_2$ subject to $6x_1 + 8x_2 \geq 100$, $7x_1 + 12x_2 \geq 120$, $x_1 \geq 0$, $x_2 \geq 0$.

6. Maximize $z = 10x_1 + 50x_2$ subject to $2x_1 + 5x_2 \leq 16$, $6x_1 \leq 20$, $x_1 \geq 0$, $x_2 \geq 0$.

7. Maximize $z = 400x_1 + 100x_2$ subject to $4x_1 + 2x_2 \leq 1,600$, $2.5x_1 + x_2 \leq 1,200$, $4.5x_1 + 1.5x_2 \leq 1,600$, $x_1 \geq 0$, $x_2 \geq 0$.

8. Maximize $z = 40x_1 + 30x_2$ subject to $3x_1 + x_2 \leq 3,0000$, $x_1 \leq 8,000$, $x_2 \leq 12,000$, $x_1 \geq 0$, $x_2 \geq 0$.

9. Maximize $z = 12x_1 + 5x_2$ subject to $5x_1 + x_2 \leq 1,500$, $2x_1 + 3x_2 \leq 1,000$, $3x_1 + 2x_2 \leq 800$, $x_1, x_2 \geq 0$.

10. Maximize $z = 500x_1 + 400x_2$ subject to $x_1 + 2x_2 \leq 80$, $3x_1 + 2x_2 \leq 120$, $x_1, x_2 \geq 0$.

11. Maximize $z = 1,850x_1 + 2,080x_2 + 1,875x_3$ subject to $x_1 + x_2 + x_3 \leq 100$, $5x_1 + 6x_2 + 5x_3 \leq 400$, $x_1 \geq 0$, $x_2 \geq 0$, $x_3 \geq 0$.

12. Maximize $z = 2x_1 + 2x_2 + 4x_3$ subject to $2x_1 + 3x_2 + x_3 \leq 300$, $x_1 + x_2 + 3x_3 \leq 300$, $x_1 + 3x_2 + x_3 \leq 240$, $x_1 \geq 0$, $x_2 \geq 0$, $x_3 \geq 0$.

14. Minimize $z = 3x_1 + 2x_2$ subject to $2x_1 + x_2 \geq 14$, $2x_1 + 3x_2 \geq 22$, $x_1 + x_2 \geq 11$, $x_1, x_2 \geq 0$.

15. Maximize $z = 2,000x_1 + 3,000x_2$ subject to $400x_1 + 600x_2 \leq 6,000$, $400x_1 + 200x_2 \leq 4,000$, $x_1, x_2 \geq 0$.

16. Maximize $z = 20x_1 + 30x_2$ subject to $3x_1 + 2x_2 \leq 210$, $2x_1 + 4x_2 \leq 300$, $x_2 \leq 65$, $x_1, x_2 \geq 0$.

17. Maximize $z = x_1 + 5x_2$ subject to $2x_1 + 5x_2 \leq 16$, $6x_1 + 2x_2 \leq 22$, $x_1, x_2 \geq 0$.

Chapter 3

Slack and Surplus Variables

3.1 Introduction

There are two methods of solving a L.P. Problem. (1) *Geometrical method* and (2) *Algebraical method.* In algebraical method, the problem can be solved only when all constraints are equations. We now show how the constraints can be converted into equations. The algebraical method is called the *simplex method* or *simplex algorithm.*

3.2 Slack variables

When the constraints are inequations connected by the sign "$\leq$", in each inequation, a variable is added to the left hand side of it to convert it into an equation. These variables are known as *slack variables.* For example,

$$x_1 - 3x_2 + 4x_3 \leq 4$$

is an inequation connected with the sign "$\leq$". Then a variable x_4 is added to the left hand side of the constraint inequation and then it is converted into an equation

$$x_1 - 3x_2 + 4x_3 + x_4 = 4.$$

The variable x_4 is known as a *slack variable.* Similarly, a slack variable x_5 is added to the left hand side of the constraint inequation

$$3x_1 + 2x_2 - x_3 \leq 10$$

to convert it into an equation

$$3x_1 + 2x_2 - x_3 + x_5 = 10.$$

From the discussion it is clear that the *slack variables are non-negative quantities*

3.3 Surplus variables

When the constraints are inequations connected by the sign "≥" then in each inequation a variable is subtracted from the left hand side to convert it into an equation. These variables are known as the *surplus variables*. For example,

$$x_1 + 3x_2 + x_3 \geq 12$$

is an inequation connected with the sign "≥" and subtracting a variable x_6 from the left hand side of the inequation, it is being converted into an equation

$$x_1 + 3x_2 + x_3 - x_6 = 12.$$

The variable x_6 is the surplus variable.

$$x_1 - 3x_2 + 7x_3 \geq -20$$

is being converted into an equation

$$x_1 - 3x_2 + 7x_3 - x_7 = -20$$

subtracting the surplus variable x_7 from the left hand side of the inequation. The surplus variables are also *non-negative quantities*.

But if we multiply both sides of the equation

$$x_1 - 3x_2 + 7x_3 - x_7 = -20$$

by (−1) to make the constant at the right hand side positive, the equation will be

$$-x_1 + 3x_2 - 7x_3 + x_7 = 20$$

and in that case the surplus variable will change into a slack variable.

Let a general L.P.P., containing r variables and m constraints be

$$\text{Optimize,} \quad z = c_1x_1 + c_2x_2 + \cdots + c_rx_r$$

subject to

$$a_{i1}x_1 + a_{i2}x_2 + a_{ij}x_j + \cdots + a_{ir}x_r \leq = \geq b_i, \ i = 1, 2, \cdots, m;$$

$$x_j \geq 0, \ j = 1, 2, \cdots, r, \tag{3.3.1}$$

where one and only one of the signs $\leq = \geq$ holds for each constraint; but the sign may vary from one constraint to another. Let out of m constraints $k (0 \leq k \leq m)$ constraints be inequations. Then introducing k variables (slack or surplus) $x_{r+1}, x_{r+2}, \cdots, x_{r+k}$ one to each of the inequations by using the technique discussed above, all constraints can be converted into equations containing $r + k = n$ (say) variables. We further assume that $n \geq m$. Now the objective function

$$z = c_1x_1 + c_2x_2 + \cdots + c_rx_r$$

is similarly accommodated with k slack and surplus variables $x_{r+1}, x_{r+2}, \cdots, x_n$ assuming the cost components of all slack and surplus variables, zero. Then the adjusted z is equivalent to

$$z_{ad} = c_1x_1 + c_2x_2 + \cdots + c_rx_r + 0x_{r+1} + 0x_{r+2} + \cdots + 0x_n$$

and the problem (3.3.1) can be written in matrix notation as,

Optimize (maximize or minimize) $z_{ad} = \mathbf{cx}$ subject to

$$A\mathbf{x} = \mathbf{b}, \ \ \mathbf{x} \geq \mathbf{0}, \tag{3.3.2}$$

where A is a $m \times n$ matrix, known as *coefficient matrix* given by

$$A = (\mathbf{a}_1, \mathbf{a}_2, \cdots, \mathbf{a}_n),$$

where $\mathbf{a}_j = [a_{1j}, a_{2j}, \cdots, a_{mj}]$ is a column vector associated with the variables $x_j [j = 1, 2, \cdots, n]$

$\mathbf{c} = (c_1, c_2, \cdots, c_r, \overbrace{0, 0, \cdots, 0}^{k-\text{components}})$ an n-component row vector.

$\mathbf{x} = [x_1, x_2, \cdots, x_r, x_{r+1}, \cdots, x_n]$ an n-component column vector.

$\mathbf{b} = [b_1, b_2, \cdots, b_m]$ an m-component column vector.

The components of $\mathbf{b}$ can be made positive by proper adjustment.

It is interesting to note that the column vectors associated with all slack variables are unit vectors.

As the cost components associated with the slack and surplus variables are zero, then it can be easily verified that the solution set which will optimize z_{ad} with constraints given in (3.3.2) will also optimize z with constraints given in (3.3.1) and the optimum value of z and z_{ad} will be same. The proof is left to the readers. Hence in order to solve the L.P.P. given in (3.3.1) we shall solve the L.P.P. given in (3.3.2). For further discussion hence we shall use the same notation for z and z_{ad}.

► **Example 3.3.1** *Transform the following L.P.P. accordingly such that the simplex method may be applicable.*

$$\textit{Maximize } z = 2x_1 + 3x_2 - 4x_3$$

subject to

$$\begin{aligned} 4x_1 + 2x_2 - \ \ x_3 &\leq 4 \\ -3x_1 + 2x_2 + 3x_3 &\geq 6 \\ x_1 + \ \ x_2 - 3x_3 &= 8, \quad x_1 \geq 0, x_2 \geq 0, x_3 \geq 0. \end{aligned}$$

Solution: First and second constraints are inequations. Hence we shall have to add two variables. First one can be converted into an equation by adding slack variable x_4 and second one can be converted into an equation by subtracting surplus variable x_5 from the left hand side of the constraints. Hence the transformed problem can be written as

$$\text{Maximize } z = 2x_1 + 3x_2 - 4x_3 + 0x_4 + 0x_5$$

subject to

$$\begin{aligned} 4x_1 + 2x_2 - x_3 + x_4 \quad &= 4 \\ -3x_1 + 2x_2 + 3x_3 \quad - x_5 &= 6 \\ x_1 + x_2 - 3x_3 \quad &= 8, \quad x_j \geq 0, \ [j = 1, 2, \cdots, 5] \end{aligned}$$

It can be easily verified that the column vector associated with the slack variable x_4 is a unit vector $\mathbf{e}_1 = [1, 0, 0]$

► **Example 3.3.2**

$$\textit{Maximize } z = x_1 - x_2 + x_3$$

subject to

$$\begin{aligned} x_1 + x_2 - 3x_3 &\geq 4 \\ 2x_1 - 4x_2 + x_3 &\geq -5 \\ x_1 + 2x_2 - 2x_3 &\leq 3, \quad x_1 \geq 0, \ x_2 \geq 0, \ x_3 \geq 0. \end{aligned}$$

Solution: The problem can be transformed as given below

$$\text{Maximize } z = x_1 - x_2 + x_3 + 0x_4 + 0x_5 + 0x_6$$

subject to

$$\begin{aligned} x_1 + x_2 - 3x_3 - x_4 \quad &= 4 \\ 2x_1 - 4x_2 + x_3 \quad - x_5 &= -5 \\ x_1 + 2x_2 - 2x_3 \quad + x_6 &= 3, \quad x_j \geq 0, \ [j = 1, 2, \cdots, 6] \end{aligned}$$

x_4 and x_5 are surplus variables and x_6 is a slack variable. Changing the second component of **b** vector, -5 to 5, the second equation can be written as

$$-2x_1 + 4x_2 - x_3 + x_5 = 5$$

and in that case the surplus variable x_5 is converted into a slack variable.

► **Example 3.3.3**

$$\textit{Maximize } z = 2x_1 + 3x_2$$

subject to

$$\begin{aligned} x_1 + x_2 \quad &\leq 10 \\ 2x_1 + 3x_2 + x_3 &\geq 28, \quad x_1 \geq 0, \ x_2 \geq 0, \ x_3 \geq 0. \end{aligned}$$

Solution:

$$\text{Maximize } z = 2x_1 + 3x_2 + 0x_3 + 0x_4 + 0x_5$$

subject to

$$\begin{aligned} x_1 + x_2 + 0x_3 + x_4 \quad &= 10 \\ 2x_1 + 3x_2 + x_3 \quad - x_5 &= 28, \quad x_j \geq 0 \ [j = 1, 2, \cdots, 5] \end{aligned}$$

Note. Here the coefficient of x_3 in the objective function is zero. But x_3 is not a slack or surplus variable.

► **Example 3.3.4** *Express the following minimization problem as standard maximization problem by introducing slack and surplus variables.*

$$\text{Minimize } z = 4x_1 - x_2 + 2x_3$$

subject to

$$\begin{aligned} 4x_1 + x_2 - x_3 &\leq 7 \\ 2x_1 - 3x_2 + x_3 &\leq 12 \\ x_1 + x_2 + x_3 &= 8 \\ 4x_1 + 7x_2 - x_3 &\geq 16, \quad x_1, x_2, x_3 \geq 0. \end{aligned}$$

Solution: After the introduction of the slack variables in the first two constraints and surplus variable in the fourth constraint, the converted problem is

$$\text{Minimize } z = 4x_1 - x_2 + 2x_3 + 0x_4 + 0x_5 + 0x_6$$

subject to

$$\begin{aligned} 4x_1 + x_2 - x_3 + x_4 \qquad\qquad &= 7 \\ 2x_1 - 3x_2 + x_3 \qquad + x_5 \qquad &= 12 \\ x_1 + x_2 + x_3 \qquad\qquad\qquad &= 8 \\ 4x_1 + 7x_2 - x_3 \qquad\qquad - x_6 &= 16, \quad x_1, x_2, \cdots, x_6 \geq 0; \end{aligned}$$

x_1, x_2, x_3 legitimate variables, x_4, x_5 slack variables and x_6 is a surplus variable. Now this problem is to be written in standard maximization problem

$$\text{Maximize } (-z) = z^* = -4x_1 + x_2 - 2x_3 + 0x_4 + 0x_5 + 0x_6$$

subject to

$$\begin{aligned} 4x_1 + x_2 - x_3 + x_4 \qquad\qquad &= 7 \\ 2x_1 - 3x_2 + x_3 \qquad + x_5 \qquad &= 12 \\ x_1 + x_2 + x_3 \qquad\qquad\qquad &= 8 \\ 4x_1 + 7x_2 - x_3 \qquad\qquad - x_6 &= 16, \quad x_1, x_2, \cdots, x_6 \geq 0 \end{aligned}$$

and Minimum value of z = negative value of the maximum of $(-z)$ with the same solution set [Proof is given at some later chapter].

Remark: Note that the two problems will have the same optimal solution set (if it exists at all) but the optimal values of the objective functions will be quite different which has been stated in the last two lines of the answer. Thus without mentioning it in the second problem cannot be considered as the standard maximising form of the first problem.

► **Example 3.3.5** *Rewrite the L.P.P. in standard maximization form by supplying slack and surplus variables.* **[C.U.(P)'86]**

$$\text{Minimize } z = 3x_1 - 2x_2 + 4x_3$$

subject to

$$\begin{aligned} x_1 - x_2 + 3x_3 &\geq 1 \\ 2x_1 + 3x_2 - 5x_3 &\geq -3 \\ 4x_1 + 2x_2 \qquad &\geq 2, \quad x_1, x_2, x_3 \geq 0. \end{aligned}$$

State which are the slack and which are surplus variables.

Solution: For the general convention, we shall make $\mathbf{b} \geq 0$, thus the second constraint is

$$-2x_1 - 3x_2 + 5x_3 \leq 3.$$

Now the standard maximization form is

$$\text{Maximize } (-z) = -3x_1 + 2x_2 - 4x_3 + 0x_4 + 0x_5 + 0x_6$$

subject to

$$\begin{aligned} x_1 - x_2 + 3x_3 - x_4 &= 1 \\ -2x_1 - 3x_2 + 5x_3 + x_5 &= 3 \\ 4x_1 + 2x_2 + x_6 &= 2, \quad x_j \geq 0,\ j = 1, 2, \cdots, 6 \end{aligned}$$

where the minimum value of z = negative of the maximum value of $(-z)$ with the same solution set or simply

$$\min z = -\max(-z).$$

x_4 is a surplus and x_5, x_6 are two slack variables.

3.4 Variable unrestricted in sign

The difference of two non-negative variables is a variable unrestricted in sign. Let x_1 and x_2 be two non-negative variables. The difference of these two variables is a variable x_3 which is unrestricted in sign. If $x_1 > x_2$, then $x_3 > 0$; if $x_1 < x_2$, then $x_3 < 0$ and if $x_1 = x_2$, then $x_3 = 0$.

► **Example 3.4.1** *Write down the following L.P.P., where the variables are non-negative and in standard form.*

$$\textit{Maximize} \quad z = 2x_1 + 3x_2 - x_3$$

subject to

$$\begin{aligned} 4x_1 + x_2 + x_3 &\geq 4 \\ 7x_1 + 4x_2 - x_3 &\leq 25, \quad x_1, x_3 \geq 0, \end{aligned}$$

x_2 is unrestricted in sign.

Solution: Introducing slack and surplus variables etc. and writing $x_2 = x_2' - x_2''$, where $x_2' \geq 0$, $x_2'' \geq 0$, the problem in the standard form

$$\text{maximize} \quad z = 2x_1 + 3x_2' - 3x_2'' - x_3 + 0x_4 + 0x_5$$

subject to

$$\begin{aligned} 4x_1 + x_2' - x_2'' + x_3 - x_4 &= 4 \\ 7x_1 + 4x_2' - 4x_2'' - x_3 + x_5 &= 25, \quad x_1, x_1', x_2'', x_3 \geq 0, \end{aligned}$$

x_4, surplus and x_5, slack variable and all variables are non-negative.

Exercise 3

1. Transform the following constraints into equations by using slack or surplus variables.

$$\begin{aligned} x_1 + 2x_2 - 3x_3 &\geq 2 \\ x_1 + x_2 + 4x_3 &\leq 16 \\ x_1 + x_2 - x_3 &= 4, \quad x_1 \geq 0, x_2 \geq 0, x_3 \geq 0. \end{aligned}$$

2. Transform the following L.P.P. into the form where all constraints are of equality type. [C.U.(H)'80]

(a) Maximize $z = 2x_1 + x_2 - 6x_3 - x_4$

subject to
$$\begin{aligned} 3x_1 + x_4 &\leq 25 \\ x_1 + x_2 + x_3 + x_4 &= 20 \\ 4x_1 + 6x_2 &\geq 5 \\ 2 \leq 2x_1 + 3x_3 + 2x_4 &\leq 30, \quad x_j \geq 0, j = 1,2,3,4. \end{aligned}$$

(b) Minimize $z = x_1 + x_2 - x_3$

subject to
$$\begin{aligned} x_1 + x_2 - x_3 &\geq 4 \\ 2x_1 + x_2 - 3x_3 &\leq 24 \\ 2x_1 - x_2 + x_3 &\geq 10, \quad x_1 \geq 0, x_2 \geq 0, x_3 \geq 0. \end{aligned}$$

(c) Maximize $z = 4x_1 + x_2 - x_3$

subject to
$$\begin{aligned} x_1 - x_2 + x_3 &\geq 16 \\ 2x_1 + x_2 + x_3 &\geq 18, \quad x_1 \geq 0, x_2 \geq 0, x_3 \geq 0. \end{aligned}$$

3. Express the following minimization problems into standard maximization problems.

(a) Minimize $z = 2x_1 - x_2 + x_3$

subject to
$$\begin{aligned} 4x_1 + x_2 + x_3 &= 6 \\ 7x_1 + 3x_2 + 2x_3 &\geq 20 \\ 4x_1 + 7x_2 - 3x_3 &\leq 10, \quad x_1, x_2, x_3 \geq 0. \end{aligned}$$

(b) Minimize $z = 3x_1 + 2x_2 - x_3$

subject to
$$\begin{aligned} 2x_1 + x_2 + x_3 &\geq 6 \\ 7x_1 - 13x_2 - 2x_3 &\leq 2 \\ 9x_1 + 2x_2 + x_3 &= 4, \quad x_1, x_2, x_3 \geq 0. \end{aligned}$$

Answers

$$\begin{aligned} x_1 + 2x_2 - 3x_3 - x_4 &= 2 \\ x_1 + x_2 + 4x_3 + x_5 &= 16 \\ x_1 + x_2 - x_3 &= 4, \quad x_j \geq 0, j = 1, 2, \cdots, 5 \end{aligned}$$

2. (a) Maximize $z = 2x_1 + x_2 - 6x_3 - x_4 + 0x_5 + 0x_6 + 0x_7 + 0x_8$

$$\text{subject to} \quad \begin{array}{l} 3x_1 + x_4 + x_5 = 25 \\ x_1 + x_2 + x_3 + x_4 = 20 \\ 4x_1 + 6x_2 - x_6 = 5 \\ 2x_1 + 3x_3 + 2x_4 - x_7 = 2 \\ 2x_1 + 3x_3 + 2x_4 + x_8 = 30, \end{array}$$

$$x_j \geq 0,\ j = (1, 2, \cdots, 8)$$

(b) Minimize $z = x_1 + x_2 - x_3 + 0x_4 + 0x_5 + 0x_6$

$$\text{subject to} \quad \begin{array}{l} x_1 + x_2 - x_3 - x_4 = 4 \\ 2x_1 + x_2 - 3x_3 + x_5 = 4 \\ 2x_1 - x_2 + x_3 - x_6 = 10, \quad x_1, x_2, x_3 \geq 0, \end{array}$$

x_4 and x_6 surplus variables and x_5 slack variable, all of them ≥ 0.

(c) Maximize $z = 4x_1 + x_2 - x_3 + 0x_4 + 0x_5$

$$\text{subject to} \quad \begin{array}{l} x_1 - x_2 + x_3 - x_4 = 16 \\ 2x_1 + x_2 + x_3 - x_5 = 18, \quad x_1, x_2, x_3 \geq 0, \end{array}$$

x_4 and x_5 are surplus variables ≥ 0.

3. (a) Maximize $(-z) = -2x_1 + x_2 - x_3 + 0x_4 + 0x_5$

$$\text{subject to} \quad \begin{array}{l} 4x_1 + x_2 + x_3 = 6 \\ 7x_1 + 3x_2 + 2x_3 - x_4 = 20 \\ 4x_1 + 7x_2 - 3x_3 + x_5 = 10, \quad x_1, x_2, x_3 \geq 0, \end{array}$$

x_4 surplus and x_5 slack variable all ≥ 0, where

minimum value of $z = -$ maximum value of $(-z)$

with the same solution set or simply min $z = -\max(-z)$.

(b) Maximize $(-z) = -3x_1 - 2x_2 + x_3 + 0x_4 + 0x_5$

$$\text{subject to} \quad \begin{array}{l} 2x_1 + x_2 + x_3 - x_4 = 6 \\ 7x_1 - 13x_2 - 2x_3 + x_5 = 2 \\ 9x_1 + 2x_2 + x_3 = 4, \quad x_1, x_2, x_3 \geq 0 \end{array}$$

x_4 surplus and x_5 slack variable all ≥ 0, where

minimum value of $z = -$ maximum value of $(-z)$

with the same solution set or simply min $z = -\max(-z)$.

Chapter 4

Basic Solutions of a Set of Simultaneous Linear Equations

4.1 Introduction

Let us consider m linear equations with n variables ($n > m$) and let the set of equations be

$$\begin{aligned}
a_{11}x_1 + a_{12}x_2 + \cdots + a_{1j}x_j + \cdots + a_{1n}x_n &= b_1 \\
a_{21}x_1 + a_{22}x_2 + \cdots + a_{2j}x_j + \cdots + a_{2n}x_n &= b_2 \\
\cdots\cdots\cdots\cdots\cdots\cdots\cdots\cdots\cdots\cdots\cdots\cdots & \\
\cdots\cdots\cdots\cdots\cdots\cdots\cdots\cdots\cdots\cdots\cdots\cdots & \\
a_{i1}x_1 + a_{i2}x_2 + \cdots + a_{ij}x_j + \cdots + a_{in}x_n &= b_i \\
\cdots\cdots\cdots\cdots\cdots\cdots\cdots\cdots\cdots\cdots\cdots\cdots & \\
\cdots\cdots\cdots\cdots\cdots\cdots\cdots\cdots\cdots\cdots\cdots\cdots & \\
a_{m1}x_1 + a_{m2}x_2 + \cdots + a_{mj}x_j + \cdots + a_{mn}x_n &= b_m
\end{aligned}$$

These set of equations can be written in compact form

$$A\mathbf{x} = \mathbf{b}, \quad \text{where} \tag{4.1.1}$$

$A =$ coefficient matrix $= [a_{ij}]_{\substack{i=1,2,\cdots,m \\ j=1,2,\cdots,n}}$

$\mathbf{x} = [x_1, x_2, \cdots, x_j, \cdots, x_n]$

$\mathbf{b} = [b_1, b_2, \cdots, b_i, \cdots, b_m]$

We further assume that $R(A) = m$, which indicates that all equations are linearly independent and none of them are redundant.

These set of equations can also be written in the form

$$\mathbf{a}_1x_1 + \mathbf{a}_2x_2 + \cdots + \mathbf{a}_jx_j + \cdots + \mathbf{a}_nx_n = \mathbf{b}, \tag{4.1.2}$$

where $\mathbf{a}_j = [a_{1j}, a_{2j}, \cdots, a_{ij}, \cdots, a_{mj}]$ an m-component column vector and all $\mathbf{a}_j$ vectors are non-null vectors. These vectors are called the *activity vectors.* From the theory of linear algebra involving matrix, we can say that here infinitely many solutions exist. Now we find a particular type solutions of the set of equations, which are finite in number, the discussions of which are given below.

4.2 Basic solutions

From the set of n column vectors $\mathbf{a}_j (j = 1, 2, \cdots, n)$ we select arbitrarily a set of m vectors $(m < n)$ which are linearly independent [there exists at least one such set of vectors since $R(A) = m$]. These set of vectors constitute a basis B. The vectors which are not included in the basis are called *non-basis vectors.* Assuming that all variables attached with non-basis vectors, zero, we get a set of m equations with m variables and since the matrix is a basis matrix, then B is non-singular, i.e., $\det B \neq 0$ and there exists a unique solution for the set of m equations containing m variables. This solution is called a *basic solution.* The variables attached with the linearly independent set of vectors $\mathbf{a}_j$ are called *basic variables* and the number of basic variables will be always m and $(n - m)$ variables whose values are assumed to be zero, attached with non-basis vectors are called *non-basic variables.* With the assumption that all non-basic variables be zero, the set of equations reduces to

$$B\mathbf{x}_B = \mathbf{b}, \tag{4.2.1}$$

where B is the basic matrix, $\mathbf{x}_B$, the column vectors having m components (which are called the *basic variables*). Now using the matrix inverse method of finding the solution of the set of equations

$$(B^{-1}B)\mathbf{x}_B = B^{-1}\mathbf{b}$$
$$\text{or, } \quad I_m\mathbf{x}_B = \mathbf{x}_B = B^{-1}\mathbf{b} \tag{4.2.2}$$

$\mathbf{x}_B$, the m component column vector is written $\mathbf{x}_B = [x_{B1}, x_{B2}, \cdots, x_{Bi}, \cdots, x_{Bm}]$. Practically the general solution $\mathbf{x}_B$ is written in the manner.

$\mathbf{x}_B = [B^{-1}\mathbf{b}, \mathbf{0}]$, where $\mathbf{0}$ vector is a $(n - m)$ component null column vector which contains the values of all non-basic variables and the basic and non-basic variables must be posted in their respective positions.

Since out of n vectors, m vectors constitute a basis, then theoretically maximum number of basis matrices be ${}^nC_m$ and hence the maximum number of basic solutions will be ${}^nC_m$. It may be less than ${}^nC_m$ in a particular problem and for finite n, ${}^nC_m$ is finite and thus basic solutions are finite in number. Now we formally define the basic solution in the manner.

Basic solution: Given a system of m simultaneous linear equations containing n variables $(n > m)$ and the set of equations be $A\mathbf{x} = \mathbf{b}$, $R(A) = m$. If any $m \times m$, non-singular matrix be arbitrarily selected from A and if we assume all $(n - m)$ variables zero, which are not associated with the column vectors of the matrix, the solution so obtained is called a *basic solution*. The variables which are attached with m-vectors of the non-singular matrix are called *basic variables* and remaining $(n-m)$ variables whose values are assumed to be zero, are called *non-basic variables*. The values of m-components of the basic solution may be positive, negative or zero. From this we can conclude that a solution is said to a basic solution if the vectors $\mathbf{a}_j$ associated with non-zero variables be linearly independent. This condition is necessary and sufficient. The following solved problems will give us a clear idea about basic solutions. We write down the inverse of the (2×2) basis matrix which will be immensely helpful in solving the problems. if the matrix be $A = \begin{bmatrix} a_{11} & a_{12} \\ a_{21} & a_{22} \end{bmatrix}$ and the matrix be non-singular, i.e., $\det A = a_{11}a_{22} - a_{12}a_{21} \neq 0$, then

$$A^{-1} = \frac{1}{\det A}\begin{bmatrix} a_{22} & -a_{12} \\ -a_{21} & a_{11} \end{bmatrix}.$$

This is only true for 2×2 non-singular matrix. For example,

$$A = \begin{bmatrix} 5 & 1 \\ 4 & 2 \end{bmatrix}, \quad \det A = 6 \neq 0.$$

$$A^{-1} = \frac{1}{6}\begin{bmatrix} 2 & -1 \\ -4 & 5 \end{bmatrix} = \begin{bmatrix} 1/3 & -1/6 \\ -2/3 & 5/6 \end{bmatrix} \text{ etc.}$$

4.3 Definitions

4.3.1 Non-degenerate Basic Solution

If all components of a solution set corresponding to the basic variables are non-zero quantities then the basic solution is known as the *non-degenerate basic solution*. In the worked out Example 4.3.1 all basic solutions are non-degenerate as in each basic solution the components of solution set corresponding to the basic variables are non-zero quantities.

4.3.2 Degenerate Basic Solution

If some components of the solution set corresponding to the basic variables are zero, the basic solution is known as a *degenerate basic solution*.

In a linear programming problem, the variables are essentially non-negative quantities, i.e., variable vector $\mathbf{x} \geq \mathbf{0}$.

4.3.3 Feasible solution to a Linear Programming Problem

If all components of a solution set are non-negative quantities then the solution is known as a *feasible solution* (F.S.) *of the* L.P.P.

4.3.4 All Feasible Solution to a L.P.P.

If all components of a solution set of L.P.P. are positive integers including zero then the solution is known as *all feasible solution* (A.F.S.) of the L.P.P. In the worked out Example 4.3.1, $\mathbf{x}_1$ and $\mathbf{x}_3$ are the feasible solutions but $\mathbf{x}_2$ is not a feasible solution. The solution set (2, 3, 4) is not only a feasbile solution but also an *all feasible solution.*

4.3.5 Basic Feasible Solution (B.F.S.) of a L.P.P.

The solution set of a L.P.P. which is feasbile as well as basic is known as the *basic feasible solution* of the problem. A solution will be B.F.S. if all components of the solution set corresponding to the basic variables are non-negative quantities. $\mathbf{x}_1$ and $\mathbf{x}_3$ are B.F.S. of the worked out Example 4.3.1.

4.3.6 Non-degenerate B.F.S.

The solution of a L.P.P. of which all components corresponding to the basic variables are positive quantities, is known as *non-degenerate* B.F.S. In the worked out example 4.3.1 $\mathbf{x}_1$ and $\mathbf{x}_3$ are non-degenerate B.F.S. also.

4.3.7 Degenerate B.F.S.

The feasible solution set of L.P.P. of which some components corresponding to the basic variables are zero, is known as the *degenerate* B.F.S.

From the defintion, it is clear that all components corresponding to the basic variables are non-negative quantities in case of a B.F.S. If all of them are positive, then it is a non-degenerate B.F.S. and if at least one of them is zero then it is a degenerate B.F.S.

▶ **Example 4.3.1** *Find the basic solution or solutions, if there be any, of the set of equations*

$$2x_1 + 4x_2 - 2x_3 = 10$$
$$10x_1 + 3x_2 + 7x_3 = 33$$

Solution: The set of equations can be written in the manner

$$\mathbf{a}_1 x_1 + \mathbf{a}_2 x_2 + \mathbf{a}_3 x_3 = \mathbf{b}, \qquad \text{(i)}$$

where $\mathbf{a}_1 = \begin{bmatrix} 2 \\ 10 \end{bmatrix}$, $\mathbf{a}_2 = \begin{bmatrix} 4 \\ 3 \end{bmatrix}$, $\mathbf{a}_3 = \begin{bmatrix} -2 \\ 7 \end{bmatrix}$, and $\mathbf{b} = \begin{bmatrix} 10 \\ 33 \end{bmatrix}$;

$A = \begin{bmatrix} 2 & 4 & -2 \\ 10 & 3 & 7 \end{bmatrix}$; $R(A) = 2$.

Thus the equations are linearly independent.

The three square sub-matrices from A are

$$B_1 = (\mathbf{a}_1, \mathbf{a}_2) = \begin{bmatrix} 2 & 4 \\ 10 & 3 \end{bmatrix},$$

$$B_2 = (\mathbf{a}_1, \mathbf{a}_3) = \begin{bmatrix} 2 & -2 \\ 10 & 7 \end{bmatrix}$$

$$\text{and} \quad B_3 = (\mathbf{a}_2, \mathbf{a}_3) = \begin{bmatrix} 4 & -2 \\ 3 & 7 \end{bmatrix}$$

$$\det B_1 = -34 \neq 0, \quad \det B_2 = 34 \neq 0, \quad \det B_3 = 34 \neq 0.$$

All three square sub-matrices are non-singular and therefore all of them are basis matrices and thus there exist three basic solutions. For basis B_1,

$$B_1^{-1} = \frac{1}{-34}\begin{bmatrix} 3 & -4 \\ -10 & 2 \end{bmatrix}$$

and the basic solution

$$\mathbf{x}_{B_1} = B_1^{-1}\mathbf{b} = \frac{1}{-34}\begin{bmatrix} 3 & -4 \\ -10 & 2 \end{bmatrix}\begin{bmatrix} 10 \\ 33 \end{bmatrix} = \begin{bmatrix} 3 \\ 1 \end{bmatrix}.$$

Hence $\mathbf{x}_{B_1} = [x_{B_1}, x_{B_2}, 0] = [x_1 = 3, x_2 = 1, x_3 = 0]$, x_1 and x_2 are basic variables and $x_3 = 0$ is non-basic. For basis, B_2, $B_2^{-1} = \dfrac{1}{34}\begin{bmatrix} 7 & 2 \\ -10 & 2 \end{bmatrix}$.

Basic solution $\mathbf{x}_{B_2} = B_2^{-1}\mathbf{b} = \dfrac{1}{34}\begin{bmatrix} 7 & 2 \\ -10 & 2 \end{bmatrix}\begin{bmatrix} 10 \\ 33 \end{bmatrix} = \begin{bmatrix} 4 \\ -1 \end{bmatrix}$.

Hence $\mathbf{x}_{B_2} = [x_{B_1}, 0, x_{B_2}] = [x_1 = 4, x_2 = 0, x_3 = -1]$, x_1 and x_3 are basic variables and $x_2 = 0$, non-basic. For basis B_3, $B_3^{-1} = \dfrac{1}{34}\begin{bmatrix} 7 & 2 \\ -3 & 4 \end{bmatrix}$.

Basic solution $\mathbf{x}_{B_3} = B_3^{-1}\mathbf{b} = \dfrac{1}{34}\begin{bmatrix} 7 & 2 \\ -3 & 4 \end{bmatrix}\begin{bmatrix} 10 \\ 33 \end{bmatrix} = \begin{bmatrix} 4 \\ 3 \end{bmatrix}$.

Thus $\mathbf{x}_{B_3} = [0, x_{B_1}, x_{B_2}] = [x_1 = 0, x_2 = 4, x_3 = 3]$. Here x_2 and x_3 are basic variables and $x_1 = 0$ is non-basic.

Note 1. $\mathbf{x}_{B_1}$ and $\mathbf{x}_{B_3}$ are basic feasible solutions and $\mathbf{x}_{B_2}$ is basic but non-feasible.

2. For $\mathbf{x}_{B_1}$, if we take $x_3 = 0$, the equations reduce to

$$\begin{aligned} 2x_1 + 4x_2 &= 10 \\ \text{and} \quad 10x_1 + 3x_2 &= 33 \end{aligned}$$

and using the method of solving the two linear simultaneous equations with two unknowns, which we have done in school algebra, we get $x_1 = 3$, $x_2 = 1$ and the basic solution is $\mathbf{x}_{B_1} = [x_1 = 3, x_2 = 1, x_3 = 0]$, x_3 is non-basic. But before testing, that x_1, x_2 be basic variables, i.e., x_3 to be non-basic variable, we cannot at random assume $x_3 = 0$ and in that case we may put into a confusion which will be discussed in problems given afterwards. Using this procedure, we can find all basic solutions, but the students are advised not to use this method and insisted them to use the matrix inverse method to find the solutions, since only this method will truely help the students to understand the procedure of solving L.P.P. by simplex method.

▶ **Example 4.3.2** *Two linear simultaneous equations (linearly independent) with four unknowns (variables) are given below.*

$$\begin{aligned} 4x_1 + 2x_2 + 3x_3 - 8x_4 &= 6 \\ 3x_1 + 5x_2 + 4x_3 - 6x_4 &= 8 \end{aligned}$$

(a) *How many basic solutions are there?*

(b) *Find all of them.*

(c) *Discuss the nature of each and every basic solution.*

Solution: Here the equations can be written in the manner

$$\mathbf{a}_1 x_1 + \mathbf{a}_2 x_2 + \mathbf{a}_3 x_3 + \mathbf{a}_4 x_4 = \mathbf{b},$$

where the column vectors associated with the variables are

$$\mathbf{a}_1 = \begin{bmatrix} 4 \\ 3 \end{bmatrix}, \ \mathbf{a}_2 = \begin{bmatrix} 2 \\ 5 \end{bmatrix}, \ \mathbf{a}_3 = \begin{bmatrix} 3 \\ 4 \end{bmatrix}, \ \mathbf{a}_4 = \begin{bmatrix} -8 \\ -6 \end{bmatrix} \text{ and } \mathbf{b} = \begin{bmatrix} 6 \\ 8 \end{bmatrix}$$

(a) From the four vectors $\mathbf{a}_1, \cdots, \mathbf{a}_4$, taking two at a time, ${}^4C_2 = 6$ square sub-matrices are

$$B_1 = (\mathbf{a}_1, \mathbf{a}_2) = \begin{bmatrix} 4 & 2 \\ 3 & 5 \end{bmatrix}, \ B_2 = (\mathbf{a}_1, \mathbf{a}_3) = \begin{bmatrix} 4 & 3 \\ 3 & 4 \end{bmatrix}, \ B_3 = (\mathbf{a}_1, \mathbf{a}_4) = \begin{bmatrix} 4 & -8 \\ 3 & -6 \end{bmatrix}$$

$$B_4 = (\mathbf{a}_2, \mathbf{a}_3) = \begin{bmatrix} 2 & 3 \\ 5 & 4 \end{bmatrix}, \ B_5 = (\mathbf{a}_2, \mathbf{a}_4) = \begin{bmatrix} 2 & -8 \\ 5 & -6 \end{bmatrix}, \ B_6 = (\mathbf{a}_3, \mathbf{a}_4) = \begin{bmatrix} 3 & -8 \\ 4 & -6 \end{bmatrix}$$

$$\begin{aligned} &\det B_1 = 14 \neq 0, \quad \det B_2 = 7 \neq 0, \quad \det B_3 = 0, \\ &\det B_4 = -7 \neq 0, \quad \det B_5 = 28 \neq 0, \quad \det B_6 = 14 \neq 0. \end{aligned}$$

Here five square sub-matrices are non-singular and hence there exist five basis matrices and corresponding to which there exist exactly five basic solutions.

(b) Now, for basis matrix B_1, $B_1^{-1} = \dfrac{1}{14}\begin{bmatrix} 5 & -2 \\ -3 & 4 \end{bmatrix}$ and the basic solution

$$\mathbf{x}_1 = B_1^{-1}\mathbf{b} = \frac{1}{14}\begin{bmatrix} 5 & -2 \\ -3 & 4 \end{bmatrix}\begin{bmatrix} 6 \\ 8 \end{bmatrix} = \begin{bmatrix} 1 \\ 1 \end{bmatrix}.$$

Hence the basic solution $\mathbf{x}_1 = [x_1 = 1,\ x_2 = 1,\ x_3 = x_4 = 0]$, x_1, x_2 basic and x_3 and x_4 are non-basic variables.

For basis matrix B_2, $B_2^{-1} = \frac{1}{7}\begin{bmatrix} 4 & -3 \\ -3 & 4 \end{bmatrix}$,

$$\mathbf{x}_2 = \frac{1}{7}\begin{bmatrix} 4 & -3 \\ -3 & 4 \end{bmatrix}\begin{bmatrix} 6 \\ 8 \end{bmatrix} = \begin{bmatrix} 0 \\ 2 \end{bmatrix}.$$

Hence the basic solution

$\mathbf{x}_2 = [x_1 \neq 0$ (basic), $x_2 = 0$ (non-basic) $x_3 = 2$ (basic), $\mathbf{x}_4 = 0$ (non-basic)].

$\mathbf{x}_2$ is a degenerate basic solution and for a degenerate basic solution, basic and non-basic variables must be stated clearly.

Similarly, proceeding the same way, we can say for basis B_4,

$\mathbf{x}_4 = [x_1 = 0$ (non-basic), $x_2 = 0$ (basic), $x_3 = 2$ (basic), $x_4 = 0$ (non-basic)].

For basis B_5, $\mathbf{x}_5 = [x_1 = 0,\ x_2 = 1,\ x_3 = 0,\ x_4 = -1/2]$.

For basis B_6, $\mathbf{x}_6 = [x_1 = x_2 = 0$ (non-basic), $x_3 = 2$ (basic), $x_4 = 0$ (basic)].

Since B_3 is not a basis which contains the vectors $\mathbf{a}_1$ and $\mathbf{a}_4$, then there exists no basic solution corresponding to the matrix B_3 which indicates that the two variables x_1 and x_4 together cannot be considered as basic variables; on the other hand the set of two variables x_2 and x_3 simultaneously cannot be treated as non-basic.

(c) $\mathbf{x}_1$ is a non-degenerate B.F.S.

$\mathbf{x}_2$ is a degenerate B.F.S.

$\mathbf{x}_4$ is a degenerate B.F.S.

$\mathbf{x}_5$ is a non-degenerate B.S. but not feasible.

$\mathbf{x}_6$ is a degenerate B.F.S.

► **Example 4.3.3** *Show that* $x_1 = 5,\ x_2 = 0,\ x_3 = -1$ *is a basic solution of the system of equations*

$$\begin{aligned} x_1 + 2x_2 + x_3 &= 4 \\ 2x_1 + x_2 + 5x_3 &= 5 \end{aligned}$$

Find the other basic solutions, if there be any. **[C.U.(H) '85; (P) '91, '99]**

Solution: The set of values $x_1 = 5$, $x_2 = 0$, $x_3 = -1$ satisfy the two equations and the vectors $\mathbf{a}_1 = \begin{bmatrix} 1 \\ 2 \end{bmatrix}$, $\mathbf{a}_3 = \begin{bmatrix} 1 \\ 5 \end{bmatrix}$ associated with non-zero variables x_1 and x_3 are linearly independent since $B = (\mathbf{a}_1, \mathbf{a}_3) = \begin{bmatrix} 1 & 1 \\ 2 & 5 \end{bmatrix}$ and $\det B = 3 \neq 0$. Hence the solution is basic but not feasible. Now, two other square matrices are (taking two at a time from $\mathbf{a}_1, \mathbf{a}_2$ and $\mathbf{a}_3$). $B_1 = (\mathbf{a}_1, \mathbf{a}_2) = \begin{bmatrix} 1 & 2 \\ 2 & 1 \end{bmatrix}$ $\det B_1 = -3 \neq 0$.

$$B_2 = (\mathbf{a}_2, \mathbf{a}_3) = \begin{bmatrix} 2 & 1 \\ 1 & 5 \end{bmatrix}$$ and $\det B_2 = 9 \neq 0$. Thus B_1 and B_2 are both basis and there are other two B.S. corresponding to B_1 and B_2. Now, B.S. $\mathbf{x}_1$ corresponding to B_1 is

$$B_1^{-1}\mathbf{b} = -\frac{1}{3}\begin{bmatrix} 1 & -2 \\ -2 & 1 \end{bmatrix}\begin{bmatrix} 4 \\ 5 \end{bmatrix} = \begin{bmatrix} 2 \\ 1 \end{bmatrix}, \quad \text{i.e.,} \quad \mathbf{x}_1 = [2,\ 1,\ 0]$$

and B.S. corresponding to B_2 is

$$B_2^{-1}\mathbf{b} = \frac{1}{9}\begin{bmatrix} 5 & -1 \\ -1 & 2 \end{bmatrix}\begin{bmatrix} 4 \\ 5 \end{bmatrix} = \begin{bmatrix} 5/3 \\ 2/3 \end{bmatrix}, \quad \text{i.e.,} \quad \mathbf{x}_2 = [0,\ 5/3,\ 2/3].$$

[**Note:** The last two B.S. are feasible and non-degenerate.]

► **Example 4.3.4** *The two linearly independent equations with three variables are given below.*

$$\begin{aligned} 2x_1 - 3x_2 + \ \ 5x_3 &= 10 \\ 4x_1 + \ \ x_2 + 10x_3 &= 20 \end{aligned}$$

Find, if possible, a basic solution with x_2, a non-basic variable.

Solution: The equations are linearly independent. But the vectors

$$\mathbf{a}_1 = \begin{bmatrix} 2 \\ 4 \end{bmatrix}, \quad \mathbf{a}_3 = \begin{bmatrix} 5 \\ 10 \end{bmatrix}$$

are not linearly independent since

$$B = (\mathbf{a}_1, \mathbf{a}_3) = \begin{bmatrix} 2 & 5 \\ 4 & 10 \end{bmatrix}, \quad \det B = 0.$$

Thus, if we initially assume x_2 as non-basic whose value is zero the two equations reduce to

$$\begin{aligned} 2x_1 + \ \ 5x_3 &= 10 \\ 4x_1 + 10x_3 &= 20. \end{aligned}$$

Here the two equations are consistent and they are the same equation. Infinite number of solutions satisfy the equations. For example, $x_1 = 5/2$, $x_3 = 1$ is a solution but $[x_1 = 5/2,\ x_2 = 0,\ x_3 = 1]$ is not a basic solution. Similarly, $x_1 = -5$, $x_3 = 4$ is a solution. But $[x_1 = -5,\ x_2 = 0,\ x_3 = 4]$ is not a basic solution.

► **Example 4.3.5** *Find all basic solutions of the set of equations given below. (Problem is taken from Linear Programming, G. Hadley. page* 70, 2–14)

$$\begin{aligned} x_1 + 2x_2 + 3x_3 + 4x_4 &= 7 \\ 2x_1 + \ \ x_2 + \ \ x_3 + 2x_4 &= 3 \end{aligned}$$

Solution: Here $R(A) = 2$, the equations are linearly independent. The set of equations can be written in the manner

$$\mathbf{a}_1 x_1 + \mathbf{a}_2 x_2 + \mathbf{a}_3 x_3 + \mathbf{a}_4 x_4 = \mathbf{b},$$

where $\mathbf{a}_1 = \begin{bmatrix} 1 \\ 2 \end{bmatrix}$, $\mathbf{a}_2 = \begin{bmatrix} 2 \\ 1 \end{bmatrix}$, $\mathbf{a}_3 = \begin{bmatrix} 3 \\ 1 \end{bmatrix}$, $\mathbf{a}_4 = \begin{bmatrix} 4 \\ 2 \end{bmatrix}$ and $\mathbf{b} = \begin{bmatrix} 7 \\ 3 \end{bmatrix}$.

$$\begin{aligned}
B_1 &= (\mathbf{a}_1, \mathbf{a}_2) = \begin{bmatrix} 1 & 2 \\ 2 & 1 \end{bmatrix}, \quad \det B_1 = -3 \neq 0, \\
B_2 &= (\mathbf{a}_1, \mathbf{a}_3) = \begin{bmatrix} 1 & 3 \\ 2 & 1 \end{bmatrix}, \quad \det B_2 = -5 \neq 0, \\
B_3 &= (\mathbf{a}_1, \mathbf{a}_4) = \begin{bmatrix} 1 & 4 \\ 2 & 2 \end{bmatrix}, \quad \det B_3 = -6 \neq 0, \\
B_4 &= (\mathbf{a}_2, \mathbf{a}_3) = \begin{bmatrix} 2 & 3 \\ 1 & 1 \end{bmatrix}, \quad \det B_4 = -1 \neq 0, \\
B_5 &= (\mathbf{a}_2, \mathbf{a}_4) = \begin{bmatrix} 2 & 4 \\ 1 & 2 \end{bmatrix}, \quad \det B_5 = 0, \\
B_6 &= (\mathbf{a}_3, \mathbf{a}_4) = \begin{bmatrix} 3 & 4 \\ 1 & 2 \end{bmatrix}, \quad \det B_6 = 2 \neq 0.
\end{aligned}$$

Five matrices are non-singular except the matrix B_5. Thus there exist five basic solutions of the above problem.

For B_1, the basic solution is

$$\mathbf{x}_1 = \frac{1}{-3}\begin{bmatrix} 1 & -2 \\ -2 & 1 \end{bmatrix}\begin{bmatrix} 7 \\ 3 \end{bmatrix} = \begin{bmatrix} -1/3 \\ 11/3 \end{bmatrix}.$$

$$\text{Then } \mathbf{x}_1 = \left[-\frac{1}{3}, \frac{11}{3}, 0, 0 \right].$$

For B_2, the basic solution is

$$\mathbf{x}_2 = -\frac{1}{5}\begin{bmatrix} 1 & -3 \\ -2 & 1 \end{bmatrix}\begin{bmatrix} 7 \\ 3 \end{bmatrix} = \begin{bmatrix} 2/5 \\ 11/5 \end{bmatrix}.$$

$$\text{Then } \mathbf{x}_2 = \left[\frac{2}{5}, 0, \frac{11}{5}, 0 \right].$$

For B_3, the basic solution is

$$\mathbf{x}_3 = -\frac{1}{6}\begin{bmatrix} 2 & -4 \\ -2 & 1 \end{bmatrix}\begin{bmatrix} 7 \\ 3 \end{bmatrix} = \begin{bmatrix} -1/3 \\ 11/6 \end{bmatrix}.$$

$$\text{Then } \mathbf{x}_3 = \left[-\frac{1}{3}, 0, 0, \frac{11}{6} \right]$$

For B_4, the basic solution is

$$\mathbf{x}_4 = \frac{1}{-1}\begin{bmatrix} 1 & -3 \\ -1 & 2 \end{bmatrix}\begin{bmatrix} 7 \\ 3 \end{bmatrix} = \begin{bmatrix} 2 \\ 1 \end{bmatrix}.$$

$$\text{Then } \mathbf{x}_4 = \begin{bmatrix} 0, & 2, & 1, & 0 \end{bmatrix}$$

For B_6, the basic solution is

$$\mathbf{x}_6 = \frac{1}{2}\begin{bmatrix} 2 & -4 \\ -1 & 3 \end{bmatrix}\begin{bmatrix} 7 \\ 3 \end{bmatrix} = \begin{bmatrix} 1 \\ 1 \end{bmatrix}$$

$$\text{Then } \mathbf{x}_6 = \begin{bmatrix} 0, & 0, & 1, & 1 \end{bmatrix}$$

► **Example 4.3.6** *How many basic solutions are there in the following linearly independent set of equations? Find all of them.*

$$\begin{aligned} 2x_1 - x_2 + 3x_3 + x_4 &= 6 \\ 4x_1 - 2x_2 - x_3 + 2x_4 &= 10 \end{aligned}$$

Solution: The equations are linearly independent. There are two equations with four variables. Then there are at most ${}^4C_2 = 6$ basic solutions, with two basic variables and $4 - 2 = 2$ non-basic variables.

The column vectors

$$\mathbf{a}_1 = \begin{bmatrix} 2 \\ 4 \end{bmatrix}, \mathbf{a}_2 = \begin{bmatrix} -1 \\ -2 \end{bmatrix}, \mathbf{a}_3 = \begin{bmatrix} 3 \\ -1 \end{bmatrix}, \mathbf{a}_4 = \begin{bmatrix} 1 \\ 2 \end{bmatrix} \text{ and } \mathbf{b} = \begin{bmatrix} 6 \\ 10 \end{bmatrix}$$

The six squares matrices, taking two at a time from $\mathbf{a}_1, \mathbf{a}_2, \cdots, \mathbf{a}_4$ are

$$B_1 = (\mathbf{a}_1, \mathbf{a}_2) = \begin{bmatrix} 2 & -1 \\ 4 & -2 \end{bmatrix}, \quad B_2 = (\mathbf{a}_1, \mathbf{a}_3) = \begin{bmatrix} 2 & 3 \\ 4 & -1 \end{bmatrix},$$

$$B_3 = (\mathbf{a}_1, \mathbf{a}_4) = \begin{bmatrix} 2 & 1 \\ 4 & 2 \end{bmatrix}, \quad B_4 = (\mathbf{a}_2, \mathbf{a}_3) = \begin{bmatrix} -1 & 3 \\ -2 & -1 \end{bmatrix},$$

$$B_5 = (\mathbf{a}_2, \mathbf{a}_4) = \begin{bmatrix} -1 & 1 \\ -2 & 2 \end{bmatrix}, \quad B_6 = (\mathbf{a}_3, \mathbf{a}_4) = \begin{bmatrix} 3 & 1 \\ -1 & 2 \end{bmatrix},$$

$$\det B_1 = 0, \quad \det B_2 = -14 \neq 0, \quad \det B_3 = 0,$$
$$\det B_4 = 7 \neq 0, \quad \det B_5 = 0, \quad \det B_6 = 7 \neq 0.$$

There are only three non-singular matrices and corresponding to these matrices (which are considered to be basis matrices) there exist only three basic solutions.

For basis matrix B_2, $B_2^{-1} = -\frac{1}{14}\begin{bmatrix} -1 & -3 \\ -4 & 2 \end{bmatrix}$ and the basic solution

$$\mathbf{x}_{B_2} = B_2^{-1}\mathbf{b} = -\frac{1}{14}\begin{bmatrix} -1 & -3 \\ -4 & 2 \end{bmatrix}\begin{bmatrix} 6 \\ 10 \end{bmatrix} = \begin{bmatrix} 18/7 \\ 2/7 \end{bmatrix}$$

Hence B.S. = [18/7, 0, 2/7, 0].

For basis matrix B_4,

$$B_4^{-1} = \frac{1}{7}\begin{bmatrix} -1 & -3 \\ 2 & -1 \end{bmatrix} \text{ and } \mathbf{x}_{B_4} = \frac{1}{7}\begin{bmatrix} -1 & -3 \\ 2 & -1 \end{bmatrix}\begin{bmatrix} 6 \\ 10 \end{bmatrix} = \begin{bmatrix} -36/7 \\ 2/7 \end{bmatrix}$$

Hence the B.S. corresponding to the basis B_4 is $[0, -36/7, 2/7, 0]$.

For basis matrix B_6,

$$B_6^{-1} = \frac{1}{7}\begin{bmatrix} 2 & -1 \\ 1 & 3 \end{bmatrix} \text{ and } \mathbf{x}_{B_6} = \frac{1}{7}\begin{bmatrix} 2 & -1 \\ 1 & 3 \end{bmatrix}\begin{bmatrix} 6 \\ 10 \end{bmatrix} = \begin{bmatrix} 2/7 \\ 36/7 \end{bmatrix}$$

Hence the B.S. corresponding to the basis B_6 is [0, 0, 2/7, 36/7].

Note: Of the three B.S., two are feasible and one is non-feasible and all the three B.S. are non-degenerate.

► **Example 4.3.7** *Two linearly independent equations with three variables are given below*

$$\begin{aligned} 3x_1 + x_2 - 6x_3 &= 1 \\ 2x_1 - 3x_2 - 4x_3 &= 5 \end{aligned}$$

Find, if possible, the basic solution with x_2 as a non-basic variable.

Solution: There are two equations (linearly independent) with three variables. Then in each basic solution, there will have two basic variables and one non-basic variable. If x_2 is taken as non-basic, then the remaining x_1 and x_3 will be considered basic provided the vectors $\mathbf{a}_1 = \begin{bmatrix} 3 \\ 2 \end{bmatrix}$ and $\mathbf{a}_3 = \begin{bmatrix} -6 \\ -4 \end{bmatrix}$ associated with the variables x_1 and x_3 respectively be linearly independent. Now $B = (\mathbf{a}_1, \mathbf{a}_3) = \begin{bmatrix} 3 & -6 \\ 2 & -2 \end{bmatrix}$ and $\det B = 0$. Thus the two vectors are not L.I. and hence x_1 and x_3 simultaneously cannot be considered as basic variables and hence there exists no B.S. with x_2 as non-basic variable.

► **Example 4.3.8** (a) *Prove that $x_1 = 2$, $x_2 = 0$ and $x_3 = 1$ is a basic solution to set of equations,*

$$\begin{aligned} 2x_1 + x_2 - x_3 &= 3 \\ x_1 + 2x_2 + 3x_3 &= 5 \end{aligned}$$

(b) *Prove that $x_1 = 2$, $x_2 = -1$ and $x_3 = 0$ is a solution set but not a basic solution to the set of equations,*

$$\begin{aligned} 3x_1 - 2x_2 + x_3 &= 8 \\ 9x_1 - 6x_2 + 4x_3 &= 24 \end{aligned}$$

Solution: (a) $A = \begin{bmatrix} 2 & 1 & -1 \\ 1 & 2 & 3 \end{bmatrix}$, $\mathbf{b} = \begin{bmatrix} 3 \\ 5 \end{bmatrix}$, $R(A) = 2$.

Hence the two equations are linearly independent. The set of values satisfy the equations. Hence the set is a solution set.

Now the vectors $\mathbf{a}_1 = \begin{bmatrix} 2 \\ 1 \end{bmatrix}$ and $\mathbf{a}_3 = \begin{bmatrix} -1 \\ 3 \end{bmatrix}$ associated with non-zero variables x_1 and x_3 are L.I. Hence the solution is a B.S.

To test that $\mathbf{a}_1$ and $\mathbf{a}_3$ are L.I., we have $B = (\mathbf{a}_1, \mathbf{a}_3) = \begin{bmatrix} 2 & -1 \\ 1 & 3 \end{bmatrix}$ and $|B| = 7 \neq 0$. Hence the vectors are L.I.

(b) $A = \begin{bmatrix} 3 & -2 & 1 \\ 9 & -6 & 4 \end{bmatrix}$ and $\mathbf{b} = \begin{bmatrix} 8 \\ 24 \end{bmatrix}$, $R(A) = 2$.

Hence the equations are linearly independent. The set of values satisfy the equations. Hence the set is a solution set. But the vectors $\mathbf{a}_1$ and $\mathbf{a}_2$ associated with non-zero variables are not L.I. Hence the solution set is not basic.

► **Example 4.3.9** *Prove that x_3 will be a non-basic variable in a basic solution of the set of equations,*

$$\begin{aligned} x_1 + 4x_2 - x_3 &= 3 \\ 5x_1 + 2x_2 + 3x_3 &= 4 \end{aligned}$$

and find that B.S. **[C.U.(P) '93,'86]**

Solution: Here, there are two linearly independent equations with three variables. Then for x_3 non-basic, x_1 and x_2 must be basic, the condition for which is $\mathbf{a}_1 = \begin{bmatrix} 1 \\ 5 \end{bmatrix}$, $\mathbf{a}_2 \begin{bmatrix} 4 \\ 2 \end{bmatrix}$ must be L.I. Here the vectors $\mathbf{a}_1$ and $\mathbf{a}_2$ are L.I. As the matrix $B = (\mathbf{a}_1, \mathbf{a}_2) = \begin{bmatrix} 1 & 4 \\ 5 & 2 \end{bmatrix}$ and $\det B = -18 \neq 0$. Hence x_1 and x_2 together can be considered as basic variables and the corresponding B.S.

$$\mathbf{x} = B^{-1}\mathbf{b} = \frac{1}{-18}\begin{bmatrix} 2 & -4 \\ -5 & 1 \end{bmatrix}\begin{bmatrix} 3 \\ 4 \end{bmatrix} = \begin{bmatrix} 5/9 \\ 11/18 \end{bmatrix}.$$

Hence the B.S. is $\mathbf{x} = [5/8, 11/18, 0]$ with x_3 non-basic.

Exercise 4

1. (a) Define a basic solution of a system of m linearly independent equations with n unknowns $(n > m)$.

   (b) How many basic solutions are there in the given set of linearly independent equations

$$\begin{aligned} 2x_1 - 5x_2 + 6x_3 &= 9 \\ 6x_1 + x_2 + 18x_3 &= 11 \end{aligned}$$

   Find all of them.

2. Find out all basic solutions of the set of equations

(a) $x_1 + 2x_2 - x_3 = 4$
$2x_1 + x_2 + x_3 = 10$

(b) $3x_1 + x_2 - x_3 = 5$
$6x_1 + 4x_2 - 2x_3 = 2$

(c) How many basic solutions are there in the following set of equations. Find all of them.

$$x_1 + 3x_2 + 2x_3 + 3x_4 = 10$$
$$2x_1 - x_2 + 4x_3 + 6x_4 = 16$$

(d) $2x_1 + x_2 + 4x_3 = 11$
$3x_1 + x_2 + 5x_3 = 14$? **[C.U.(P)'82]**

3. Find out all basic solutions to the following equations identifying in each case the basis vectors

$$x_1 + x_2 + x_3 = 4$$
$$2x_1 + 5x_2 - 2x_3 = 3$$

[C.U.(H)'81]

4. Find all the basic solutions of the system of equations,

$$x_1 + x_2 + 2x_3 = 9$$
$$3x_1 + 2x_2 + 5x_3 = 22$$

[C.U.(P)'80]

5. Determine all basic feasible solutions of the set of equations

$$2x_1 + 6x_2 + 2x_3 + x_4 = 3$$
$$6x_1 + 4x_2 + 4x_3 + 6x_4 = 2$$

Identify the basic and non-basic variables in each case. **[C.U.(P)'88]**

6. Find out all basic feasible solutions of the set of equations

$$2x_1 + 3x_2 - x_3 + 4x_4 = 8$$
$$-x_1 + 2x_2 - 6x_3 + 7x_4 = 3$$

[J.U.M.Sc.'81]

7. Prove that x_1 and x_3 cannot be simultaneously considered as the basic variables of the set of equations

$$2x_1 + x_2 - x_3 + 4x_4 = 6$$
$$6x_1 + 3x_2 - 3x_3 + x_4 = 10$$

Find all the basic solutions with x_1 as one basic variable.

8. How many basic solutions are there in the following set of equations

$$2x_1 - 5x_2 + x_3 + 3x_4 = 4$$
$$3x_1 - 10x_2 + 2x_3 + 6x_4 = 12 \text{ ?}$$

Find all basic solution or solutions with x_1 as one of the basic variables.

9. Prove that $x_1 = 2$, $x_2 = 1$, $x_3 = 0$, $x_4 = 2$ is a solution of the set of equations but not basic.

$$\begin{aligned} 2x_1 - 3x_2 + x_3 \quad\quad &= 1 \\ 4x_1 - 6x_2 - x_3 \quad\quad &= 2 \\ 3x_1 + 4x_2 + 2x_3 + x_4 &= 12 \end{aligned}$$

Find one basic solution, where x_3 and x_4 are two basic variables.

10. For the set of equations

$$\begin{aligned} x_1 + x_2 + 2x_3 &= 9 \\ 3x_1 + 2x_2 + 5x_3 &= 22 \end{aligned}$$

Show that the feasible solution (2, 3, 2) is not basic. Find all the basic feasible solutions. **[C.U.(P)'87]**

11. Find the basic feasible solutions of the following equation

$$2x_1 + 3x_3 - x_3 = 6.$$

Answers

1. 2, [2, −1, 0], [0, −1, 2/3]. 2. (a) [16/3, −2/3, 0], [14/3, 0, 2/3], [0, 14/3, 16/3]; (b) [3, −4, 0], [0, −4, −9]; (c) 3, [58/7, 4/7, 0, 0], [0, 4/7, 23/7, 0]; (d) [3, 5, 0], [1/2, 0, 5/2], [0, −1, 3]. 3. [17/3, −5/3, 0], [0, 11/7, 17/7]. 4. [4, 5, 0], [−1, 0, 5], [0, 1; 4]. 5. [$x_1 = 0, x_2 = 1/2, x_3 = x_4 = 0$ (non-basic)]. [$x_2 = 1/2$, $x_3 = 0$, $x_1 = x_4 = 0$ (non-basic)]; [$x_2 = 1/2$, $x_4 = 0$, $x_1 = x_3 = 0$ (non-basic)]. All the three solutions are degenerate. 6. [1, 2, 0, 0], [22/9, 0, 0, 7/9], [0, 45/16, 7/16, 0], [0, 0, 44/17, 45/17]. 7. Det $\begin{bmatrix} 2 & -1 \\ 6 & -3 \end{bmatrix} = 0$. So x_1 and x_3 cannot be considered simultaneously as the basic variables. 8. 3, [−4, −12/5, 0, 0], [−4, 0, 12, 0], [−4, 0, 0, 4]. 9. [$x_1 = 0$, $x_2 = -1/3$, $x_3 = 0$ (basic), $x_4 = 40/3$]. 10. [4, 5, 0], [0, 1, 4]. 11. [3, 0, 0], [0, 2, 0].

Chapter 5

N-Dimensional Euclidean Space and Convex Set

5.1 Introduction

There are two types of vectors, row and column vectors. All n-component row vectors may be considered as points in an n-dimensional space which is known as *n-dimensional vector space* or *n-dimensional Euclidean space.* Similarly all n-component column vectors may be considered as points in n-dimensional vector space. The vector space is generally denoted by V_n or E^n or R^n etc. The points in a particular vector space are either all row or column vectors.

Point set : Point sets are sets whose elements are all points in E^n. Point set is usually denoted by capital letter X or S. If the set X contains m points $\mathbf{x}_1, \mathbf{x}_2, \cdots, \mathbf{x}_m$, then X is given by

$$X = \{\mathbf{x} \quad \mathbf{x} = \mathbf{x}_j, \quad [j = 1, 2, \cdots, m]\}.$$

Line : If $\mathbf{x}_1 = (x_{11}, x_{12}, \cdots, x_{1n})$ and $\mathbf{x}_2 = (x_{21}, x_{22}, \cdots, x_{2n})$ be two points in E^n, then the line joining the points $\mathbf{x}_1$ and $\mathbf{x}_2$, $[\mathbf{x}_2 \neq \mathbf{x}_1]$ is a set X of points given by

$$X = \{\mathbf{x} : \mathbf{x} = \lambda \mathbf{x}_1 + (1 - \lambda)\mathbf{x}_2, \quad \text{for all real } \lambda\}.$$

For different values of λ, different points will be obtained which all lie in a line which is a line in E^n.

Line segment : If $\mathbf{x}_1$ and $\mathbf{x}_2$ be two distinct points in E^n, then the line segment joining these two points is a set X of points given by

$$X = \{\mathbf{x} : \mathbf{x} = \lambda \mathbf{x}_1 + (1 - \lambda)\mathbf{x}_2, 0 \leq \lambda \leq 1\}.$$

Hyperplane : A hyperplane in E^n is a set X of points given by

$$X = \{\mathbf{x} : \mathbf{cx} = k\},$$

where $\mathbf{c}$ is a row vector, given by $\mathbf{c} = (c_1, c_2, \cdots, c_n)$ and not all $c_j = 0$ and $\mathbf{x} = [x_1, x_2, \cdots, x_n]$ is an n-component column vector.

A hyperplane can be defined as a set of points which will satisfy

$$c_1x_1 + c_2x_2 + \cdots + c_nx_n = k.$$

A hyperplane divides the space E^n into three mutually exclusive disjoint sets given by

$$\begin{aligned} X_1 &= \{\mathbf{x} : \mathbf{cx} > k\} \\ X_2 &= \{\mathbf{x} : \mathbf{cx} = k\} \\ X_3 &= \{\mathbf{x} : \mathbf{cx} < k\}. \end{aligned}$$

The sets X_1 and X_2 are called as *open half spaces.*

The objective function $z = \mathbf{cx}$ and the constraints with equality sign $A\mathbf{x} = \mathbf{b}$ are all hyperplanes in a L.P.P.

Hypersphere : A hypersphere in E^n with centre at $\mathbf{a} = (a_1, a_2, \cdots, a_n)$ and radius $\varepsilon > 0$ is defined to be a set X of points given by

$$X = \{\mathbf{x} : |\mathbf{x} - \mathbf{a}| = \varepsilon\}, \text{ where } \mathbf{x} = (x_1, x_2, \cdots, x_n).$$

The equation can be written as

$$(x_1 - a_1)^2 + (x_2 - a_2)^2 + \cdots + (x_n - a_n)^2 = \varepsilon^2.$$

The hypersphere in two dimension is a circle and in three dimension a sphere.

An ε-neighbourhood : An ε-neighbourhood about a point $\mathbf{a}$ is defined as the set X of points lying inside the hypersphere with centre at $\mathbf{a}$ and radius $\varepsilon > 0$, i.e., the ε-neighbourhood about the point $\mathbf{a}$ is a set of points given by

$$X = \{\mathbf{x} : |\mathbf{x} - \mathbf{a}| < \varepsilon\}.$$

An interior point of a set : A point $\mathbf{a}$ is said to be an interior point of the set S if there exists an ε-neighbourhood about the point $\mathbf{a}$ which contains only points of the same set. From the definition it is clear that an interior point of a set S must be an element of the set S.

Boundary point of a set : A point $\mathbf{a}$ is said to be a boundary point of a set S if every ε-neighbourhood about $\mathbf{a}$ ($\varepsilon > 0$, however, small it may be) containts points which are in the set S and the points which are not in the set S. A boundary point may or may not be an element of the set.

Open set : A set S is said to be open if it contains only interior points.

Closed set : A set S is said to be closed set if it contains all its boundary points. [For more precise definitions, consult any Text book of Real Analysis.]

Bounded set : A set S is said to be strictly bounded set if there exists a positive number r such that for any point $\mathbf{x}$ belonging to S, $|\mathbf{x}| \leq r$. If for every $\mathbf{x}$, all its components have lower limits only then the set is bounded from below.

5.2 Convex Combination and Convex Sets

Convex combination of a set of points : The convex combination of a set of **k** points $\mathbf{x}_1, \mathbf{x}_2, \cdots, \mathbf{x}_k$ in a space E^n is also a point $\mathbf{x}$ in the same space, given by

$$\mathbf{x} = \lambda_1 \mathbf{x}_1 + \lambda_2 \mathbf{x}_2 + \cdots + \lambda_k \mathbf{x}_k,$$

where λ_i is real scalar and ≥ 0 for all i and $\sum_{i=1}^{k} \lambda_i = 1$.

For different values of the scalar quantities λ_i $[i = 1, 2, \cdots, k]$ satisfying the conditions $\sum_{i=1}^{k} \lambda_i = 1$ and $\lambda_i \geq 0$, $\forall i$, a set of points will be obtained from the convex combinations of the set of k finite points which is a point set in E^n. This point set is known as the *convex polyhedron.*

The point set, convex polyhedron X is given by

$$X = \left\{ \mathbf{x} : \mathbf{x} = \sum_{i=1}^{k} \lambda_i \mathbf{x}_i, \sum_{i=1}^{k} \lambda_i = 1 \text{ and } \lambda_i \geq 0, \ \forall i \right\}.$$

From the definition it is also clear that line segment is a convex combination of the two distinct points in the same vector space.

Convex set : A point set is said to be a convex set if the convex combination of any two points of the set is in the set. In other words if the line segment joining any two distinct points of the set is in the set then the set is known as a *convex set;* otherwise the set is a non-convex set. Convex set is usually denoted by C or X.

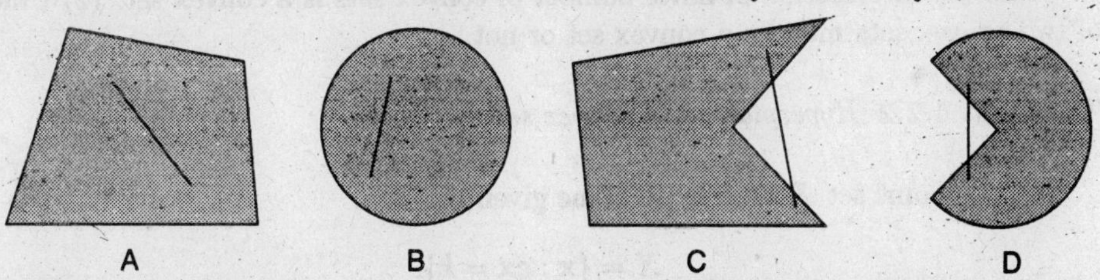

Fig. 5.1: (i) A and B : Convex Set (ii) C and D : Non-convex Set

Extreme points of the convex set : A points $\mathbf{x}$ is said to be an extreme point of the convex set C if it cannot be expressed as the convex combination of any other two distinct points $\mathbf{x}_1$ and $\mathbf{x}_2$ of the set C, i.e., $\mathbf{x}$ cannot be expressed in the form $\mathbf{x} = \lambda \mathbf{x}_1 + (1 - \lambda)\mathbf{x}_2$, $0 < \lambda < 1$. From the definition, it is clear that all extreme points of the convex set C are the boundary points but all boundary points may not be necessarily extreme points. Every point of the boundary of a circle is an extreme point of the convex set which includes the boundary and the interior of the circle. The extreme points of a rectangle are its four vertices.

Convex hull : If X be a point set, then convex hull of X which is denoted by $C(X)$, is the set of all convex combinations of sets of points from X. If the set X consists of a finite number of points then the convex hull $C(X)$ is called as a *convex polyhedron.* We shall prove that convex polyhedron is a convex set. If X is a point set, consists of only eight vertices of a cube then $C(X)$ is the whole cube which is also a convex polyhedron. A circle with all its interior points is a convex hull because all boundary points are extreme points which are not finite. Thus a circle with all its interior points is a convex hull but not a convex polyhedron. Hence the conclusion is that all convex polyhedrons are convex hull but all convex hulls may not be convex polyhedron.

For a convex polyhedron, any point in the set can be expressed as a convex combination of its extreme points.

Simplex : A : simplex is an n-dimensional convex polyhedron having exactly $(n+1)$ vertices. A simplex in zero dimension is a point; in one dimension, a line segment; in two dimension, a triangle and in three dimension, a tetrahedron.

Theorem 5.2.1 *Intersection of two convex sets is also a convex set :*

Let X_1 and X_2 be two convex sets and $X = X_1 \cap X_2$. It is required to prove that X is also a convex set.

Let $\mathbf{x}_1$ and $\mathbf{x}_2$ be any two distinct points of X. Then $\mathbf{x}_1$ and $\mathbf{x}_2$ are also points of the sets X_1 and X_2. Let $\mathbf{x}_3$ be a point given by

$$\mathbf{x}_3 = \lambda\mathbf{x}_1 + (1-\lambda)\mathbf{x}_2, \quad 0 \le \lambda \le 1.$$

As $\mathbf{x}_3$ is a convex combination of the two points $\mathbf{x}_1$ and $\mathbf{x}_2$ then $\mathbf{x}_3$ is a point of X_1 as well as a point of X_2 which indicates that $\mathbf{x}_3$ is a point on $X_1 \cap X_2 = X$. But $\mathbf{x}_3$ is the convex combination of two distinct points of X. Hence X is a convex set.

N.B. (1) Intersection of finite number of convex sets is a convex set. (2) Union of two convex sets may be a convex set or not.

Theorem 5.2.2 *Hyperplane is a convex set.*

Let the point set X be a hyperplane given by

$$X = \{\mathbf{x} : \mathbf{cx} = k\}.$$

Let $\mathbf{x}_1$ and $\mathbf{x}_2$ be any two distinct points of $\mathbf{cx} = k$.

Then $\mathbf{cx}_1 = k$ and $\mathbf{cx}_2 = k$.

Let $\mathbf{x}_3$ be a point given by

$$\mathbf{x}_3 = \lambda\mathbf{x}_1 + (1-\lambda)\mathbf{x}_2, \quad 0 \le \lambda \le 1.$$

Therefore,

$$\mathbf{cx}_3 = \lambda\mathbf{cx}_1 + (1-\lambda)\mathbf{cx}_2 = \lambda k + (1-\lambda)k = k$$

which indicates that $\mathbf{x}_3$ is also a point of $\mathbf{cx} = k$.

But $\mathbf{x}_3$ is the convex combination of two distinct points $\mathbf{x}_1$ and $\mathbf{x}_2$ of X. Hence X is a convex set. Set X is closed also.

Note : Similarly it can be proved that $X = \{\mathbf{x} : \mathbf{cx} >< k\}$ are also convex sets.

Theorem 5.2.3 *Convex polyhedron is a convex set.*

Let S be a point set consisting of finite number of points $\mathbf{x}_1, \mathbf{x}_2, \cdots, \mathbf{x}_k$ in R^n.

Convex polyhedron $\mathbf{C}(S) = X = \left\{ \mathbf{x} : \mathbf{x} = \sum_{i=1}^{k} \lambda_i \mathbf{x}_i, \lambda_i \geq 0, \sum_{i=1}^{k} \lambda_i = 1 \right\}$.

It is required to prove that X is a convex set.

Let $\mathbf{u}$ and $\mathbf{v}$ be any two distinct points of X given by

$$\mathbf{u} = \sum_{i=1}^{k} a_i \mathbf{x}_i, \quad a_i \geq 0, \quad \sum_{i=1}^{k} a_i = 1.$$

$$\mathbf{v} = \sum_{i=1}^{k} b_i \mathbf{x}_i, \quad b_i \geq 0, \quad \sum_{i=1}^{k} b_i = 1.$$

Consider $\mathbf{w}$, where

$$\mathbf{w} = \lambda \mathbf{u} + (1 - \lambda)\mathbf{v}, \quad 0 \leq \lambda \leq 1$$

then

$$\mathbf{w} = \lambda \sum_{i=1}^{k} a_i \mathbf{x}_i + (1 - \lambda) \sum_{i=1}^{k} b_i \mathbf{x}_i = \sum_{i=1}^{k} \{\lambda a_i + (1 - \lambda) b_i\} \mathbf{x}_i = \sum_{i=1}^{k} c_i \mathbf{x}_i,$$

where $c_i = \lambda a_i + (1 - \lambda) b_i$.

Now

$$\sum_{i=1}^{k} c_i = \lambda \sum_{i=1}^{k} a_i + (1 - \lambda) \sum_{i=1}^{k} b_i = 1$$

and $c_i \geq 0$ as $a_i \geq 0$, $b_i \geq 0$ and $0 \leq \lambda \leq 1$.

Hence $\mathbf{w}$ is also a point of X, which is a convex combination of two distinct points $\mathbf{u}$ and $\mathbf{v}$ of X. Hence X is a convex set.

Theorem 5.2.4 *The set of all feasible solutions to a L.P.P.* $A\mathbf{x} = \mathbf{b}$, $\mathbf{x} \geq \mathbf{0}$ *is a closed convex set.*

Let X be the point set of all the feasible solutions of the problem $A\mathbf{x} = \mathbf{b}$, $\mathbf{x} \geq \mathbf{0}$.

If the set X has only one point then there is nothing to prove.

If X has at least two distinct points $\mathbf{x}_1$ and $\mathbf{x}_2$ then $A\mathbf{x}_1 = \mathbf{b}$, $\mathbf{x}_1 \geq \mathbf{0}$ and $A\mathbf{x}_2 = \mathbf{b}$, $\mathbf{x}_2 \geq \mathbf{0}$.

Consider a point $\mathbf{x}_3$ such that

$$\mathbf{x}_3 = \lambda \mathbf{x}_1 + (1 - \lambda)\mathbf{x}_2, \quad 0 \leq \lambda \leq 1.$$

Thus,

$$A\mathbf{x}_3 = \lambda A\mathbf{x}_1 + (1 - \lambda) A\mathbf{x}_2 = \lambda \mathbf{b} + (1 - \lambda)\mathbf{b} = \mathbf{b}.$$

Again $\mathbf{x}_3 \geq \mathbf{0}$ as $\mathbf{x}_1 \geq \mathbf{0}$, $\mathbf{x}_2 \geq \mathbf{0}$ and $0 \leq \lambda \leq 1$.

Then $\mathbf{x}_3$ is also a feasible solution to the problem $A\mathbf{x} = \mathbf{b}$, $\mathbf{x} \geq \mathbf{0}$.

But $\mathbf{x}_3$ is the convex combination of two distinct points $\mathbf{x}_1$ and $\mathbf{x}_2$ of the set X. Thus X is a convex set.

Now the finite number of constraints represented by $A\mathbf{x} = \mathbf{b}$ are closed sets and also the set of inequations (finite) represented by $\mathbf{x} \geq \mathbf{0}$ are closed sets and therefore the intersection of finite number of closed sets which is the set of all feasible solutions is a closed set.

Similarly we can establish that the set of feasible solutions to a L.P.P. $A\mathbf{x}(\leq = \geq)\mathbf{b}$, $\mathbf{x} \geq \mathbf{0}$ is a closed convex set.

Note : (1) For different values of λ, satisfying the condition $0 \leq \lambda \leq 1$, an infinite set of points will be obtained. Therefore, if a L.P.P. has at least two feasible solutions then it has infinite number of feasible solutions.

(2) There are two types of convex sets of F.S. of a L.P.P. (i) strictly bounded or convex polyhedron and (ii) bounded from below only or convex polytope.

Theorem 5.2.5 *All B.F.S. of the set of equations* $A\mathbf{x} = \mathbf{b}$, $\mathbf{x} \geq \mathbf{0}$ *are extreme points of the convex set of feasible solutions of the equations and conversely.*

Proof. Let $A = (\mathbf{a}_1, \mathbf{a}_2, \cdots, \mathbf{a}_n)$ be the coefficient matrix of order $m \times n$ $(n > m)$ and let us assume that B be a basis matrix $B = (\mathbf{a}_1, \mathbf{a}_2, \cdots, \mathbf{a}_m)$, where $\mathbf{a}_1, \mathbf{a}_2, \cdots, \mathbf{a}_m$ are the column vectors corresponding to first m variables $x_1, x_2, \cdots, x_m$.

Let $\mathbf{x}$ be a B.F.S. and $\mathbf{x}$ is given by

$$\mathbf{x} = [\mathbf{x}_B, \mathbf{0}], \tag{5.2.1}$$

where $\mathbf{x}_B = B^{-1}\mathbf{b}$ and $\mathbf{0}$ is a $(n-m)$ component null column vector.

It is required to prove that $\mathbf{x}$ is an extreme point of the convex set X of the feasible solution of the equations $A\mathbf{x} = \mathbf{b}$, $\mathbf{x} \geq \mathbf{0}$.

Let $\mathbf{x}$ be not an extreme point of the convex set X.

Thus there exist two points $\mathbf{x}_1$ and $\mathbf{x}_2 [\mathbf{x}_1 \neq \mathbf{x}_2]$ in X such that it is possible to express $\mathbf{x}$, as given by

$$\mathbf{x} = \lambda\mathbf{x}_1 + (1-\lambda)\mathbf{x}_2, \quad 0 < \lambda < 1 \quad \text{[note the limit of } \lambda\text{]},$$

where $\mathbf{x}_1$ and $\mathbf{x}_2$ are given by

$$\mathbf{x}_1 = [\mathbf{u}_1, \mathbf{v}_1] \text{ and } \mathbf{x}_2 = [\mathbf{u}_2, \mathbf{v}_2], \tag{5.2.2}$$

where $\mathbf{u}_1$ contains m components of $\mathbf{x}_1$, corresponding to the variables $x_1, x_2, \cdots, x_m$ and $\mathbf{v}_1$ contains the remaining $(n-m)$ components of $\mathbf{x}_1$. Similarly $\mathbf{u}_2$ and $\mathbf{v}_2$ contain the first m and the remaining $(n-m)$ components of $\mathbf{x}_2$ respectively.

Thus

$$\begin{aligned}\mathbf{x} &= \lambda[\mathbf{u}_1, \mathbf{v}_1] + (1-\lambda)[\mathbf{u}_2, \mathbf{v}_2] \\ &= [\lambda\mathbf{u}_1 + (1-\lambda)\mathbf{u}_2, \lambda\mathbf{v}_1 + (1-\lambda)\mathbf{v}_2].\end{aligned}$$

As $\mathbf{x} = [\mathbf{x}_B, \mathbf{0}]$, then equating the components corresponding to the last $(n-m)$ variables, we get

$$\lambda \mathbf{v}_1 + (1-\lambda)\mathbf{v}_2 = \mathbf{0}$$

which is possible only when $\mathbf{v}_1 = \mathbf{0}$ and $\mathbf{v}_2 = \mathbf{0}$ [as $\mathbf{v}_1 \geq \mathbf{0}, \mathbf{v}_2 \geq \mathbf{0}$ and $0 < \lambda < 1$]. Thus

$$\mathbf{x}_1 = [\mathbf{u}_1, \mathbf{0}] \text{ and } \mathbf{x}_2 = [\mathbf{u}_2, \mathbf{0}].$$

Hence $\mathbf{u}_1$ and $\mathbf{u}_2$ are the m components of the solution set corresponding to the basic variables $x_1, x_2, \cdots, x_m$ for which the basis matrix is B. Then $\mathbf{u}_1 = B^{-1}\mathbf{b}$ and $\mathbf{u}_2 = B^{-1}\mathbf{b}$. Hence $\mathbf{x}_B = \mathbf{u}_1 = \mathbf{u}_2$. Therefore, three points $\mathbf{x}, \mathbf{x}_1$ and $\mathbf{x}_2$ are not different and thus $\mathbf{x}$ cannot be expressed as the convex combination of two distinct points. So a B.F.S. $\mathbf{x}$ is also an extreme point. Similar is the case for all B.F.S.

Part II : Let $\mathbf{x}$ be an extreme point of the convex set X of all feasible solutions of the equations

$$A\mathbf{x} = \mathbf{b}, \quad \mathbf{x} \geq \mathbf{0}.$$

It is required to prove that $\mathbf{x}$ is a B.F.S.

Let

$$\mathbf{x} = [x_1, x_2, \cdots, x_k, \overbrace{0, \cdots, 0}^{n-k}]$$

where $x_j \geq 0$ for $j = 1, 2, \cdots, k$.

If the column vectors $\mathbf{a}_1, \mathbf{a}_2, \cdots, \mathbf{a}_k$ associated with the variables $x_1, x_2, \cdots, x_k$ respectively are L.I. (which is possible only for $k \leq m$) then $\mathbf{x}$, the extreme point of the convex set, is a B.F.S. and we have nothing to prove.

If $\mathbf{a}_1, \mathbf{a}_2, \cdots, \mathbf{a}_k$ are not L.I. then

$$\sum_{j=1}^{k} \mathbf{a}_j x_j = \mathbf{b} \tag{5.2.3}$$

$$\text{and} \sum_{j=1}^{k} \mathbf{a}_j \lambda_j = \mathbf{0} \tag{5.2.4}$$

with at least one $\lambda_j \neq 0$.

Let $\delta > 0$, then from (5.2.3) and (5.2.4) we get

$$\sum_{j=1}^{k} (x_j \pm \delta\lambda_j)\mathbf{a}_j = \mathbf{b}. \tag{5.2.5}$$

Consider δ in the interval $0 < \delta < l$, where

$$l = \min_j \left(\frac{x_j}{|\lambda_j|} \right), \quad \lambda_j \neq 0.$$

Then $x_j \pm \delta\lambda_j \geq 0$ for $j = 1, 2, \cdots, k$.

Hence the two points

$$\mathbf{x}_1 = [x_1 + \delta\lambda_1, x_2 + \delta\lambda_2, \cdots, x_k + \delta\lambda_k, \overbrace{0, \cdots, 0}^{n-k}]$$

$$\text{and} \quad \mathbf{x}_2 = [x_1 - \delta\lambda_1, x_2 - \delta\lambda_2, \cdots, x_k - \delta\lambda_k, \overbrace{0, \cdots, 0}^{n-k}]$$

are the points of the convex set X.

Now

$$\frac{1}{2}\mathbf{x}_1 + \frac{1}{2}\mathbf{x}_2 = \mathbf{x}.$$

Then $\mathbf{x}$ can be expressed as

$$\mathbf{x} = \lambda\mathbf{x}_1 + (1-\lambda)\mathbf{x}_2, \quad \text{where } \lambda = 1/2.$$

Thus $\mathbf{x}$ is being expressed as the convex combination of two distinct points $\mathbf{x}_1$ and $\mathbf{x}_2$ in X which is against the assumption that $\mathbf{x}$ is an extreme point. Thus the column vectors $\mathbf{a}_1, \mathbf{a}_2, \cdots, \mathbf{a}_k$ are L.I. and hence $\mathbf{x}$ is a B.F.S.

Note : There is one to one correspondence between the extreme points and B.F.S. in the case of non-degenerate B.F.S.

► **Example 5.2.1** *Extreme points are finite in number.*

There are at most ${}^nC_m$ B.S. to a set of m equations with n unknowns. Hence the maximum number of B.F.S. to a L.P.P. is ${}^nC_m$ which is finite provided n is finite. This proves the theorem.

Theorem 5.2.6 *Any point of a convex polyhedorn can be expressed as a convex combination of its extreme points.*

Proof. Let the finite number of extreme points of the convex polyhedron be $\mathbf{x}_1, \mathbf{x}_2, \cdots, \mathbf{x}_n$. Let $\mathbf{x}$ be any point of the convex polyhedorn. We take any extreme point (say) $\mathbf{x}_1$ and join it with $\mathbf{x}$ by a line segment and extend it to meet either of a boundary line or a boundary surface (section of a hyperplane). Let the point of intersection be $\mathbf{x}^*$. Now $\mathbf{x}^*$ can always be expressed as the convex combination of some of the extreme points of the bounding hyperplane or line. Therefore, $\mathbf{x}^*$ can be expressed as

$$\mathbf{x}^* = \sum_{i=1}^{k} \mu_i \mathbf{x}_i \geq 0; \qquad \sum_{i=1}^{k} \mu_i = 1, \ k \leq n.$$

Now $\mathbf{x}$ can be expressed as

$$\mathbf{x} = \lambda_1\mathbf{x}_1 + \lambda_2\mathbf{x}^*, \quad \lambda_1 \geq 0, \lambda_2 \geq 0, \lambda_1 + \lambda_2 = 1,$$

$$\text{or,} \quad \mathbf{x} = \lambda_1\mathbf{x}_1 + \lambda_2 \sum_{i=1}^{k} \mu_i \mathbf{x}_i.$$

Here $\lambda_1 \geq 0$, $\lambda_2\mu_i \geq 0$ for $i = 1, 2, \cdots, k$ and

$$\begin{aligned}\lambda + \lambda_2\mu_1 + \lambda_2\mu_2 + \cdots + \lambda_2\mu_k &= \lambda_1 + \lambda_2(\mu_1 + \mu_2 + \cdots + \mu_k) \\ &= \lambda_1 + \lambda_2 = 1.\end{aligned}$$

Therefore, $\mathbf{x}$ can be expressed as the convex combination of some or all the extreme points of the convex polyhedron.

Theorem 5.2.7 *If the convex set of F.S. of a L.P.P. be a convex polyhedron, then any objective function has both finite maximum and minimum corresponding to some extreme points of the convex polyhedron.*

Proof. Let X, convex set of F.S. of a L.P.P. be a convex polyhedron with finite number of extreme points $\mathbf{x}_1, \mathbf{x}_2, \cdots, \mathbf{x}_k$ and the objective function be $z = \mathbf{cx}$.

Therefore, z has definitely finite optimal value [maximum or minimum].

It is required to prove that z attains its optimal value corresponding to one of the values $\mathbf{x}_1, \mathbf{x}_2, \cdots, \mathbf{x}_k$. Let the objective function z is to be maximised and for a point $\hat{\mathbf{x}}$ of $X, \hat{z} = \mathbf{c}\hat{\mathbf{x}}$ If $\hat{\mathbf{x}}$ is an extreme point then there is nothing to prove. If $\hat{\mathbf{x}}$ be not an extreme point then

$$\hat{\mathbf{x}} = \lambda_1\mathbf{x}_1 + \cdots + \lambda_k\mathbf{x}_k = \sum_{i=1}^{k} \lambda_i\mathbf{x}_i, \lambda_i \geq 0, \sum_{i=1}^{k} \lambda_i = 1.$$

Then

$$\hat{z} = \mathbf{c}\hat{\mathbf{x}} = \mathbf{c}(\lambda_1\mathbf{x}_1 + \lambda_2\mathbf{x}_2 + \cdots + \lambda_k\mathbf{x}_k) = \lambda_1\mathbf{cx}_1 + \lambda_2\mathbf{cx}_2 + \cdots + \lambda_k\mathbf{cx}_k.$$

If the maximum of $\mathbf{cx}_i$ be $\mathbf{cx}_1$ (say) $[i = 1, 2, \cdots, k]$.

Then

$$\hat{z} \leq \mathbf{cx}_1(\lambda_1 + \lambda_2 + \cdots + \lambda_k), \quad \text{or,} \quad \hat{z} \leq \mathbf{cx}_1.$$

But the maximum value of z is $\hat{z} = \mathbf{c}\hat{\mathbf{x}}$.

Then $\mathbf{c}\hat{\mathbf{x}} = \mathbf{cx}_1$ and therefore $\hat{\mathbf{x}} = \mathbf{x}_1$ (an extreme point).

Hence the maximum of the objective function z is attained at one of the extreme points which proves the theorem.

Similar is the proof when the objective function z is to be minimized.

Note (1) : When the convex set of F.S. of a L.P.P. is strictly bounded, i.e., a convex polyhedron, any objective function has both finite maximum and minimum values which will be obtained from the extreme points of the convex set of F.S. But if the convex set of F.S. is bounded from below only, then an objective function may have either maximum or minimum but not the both, which will also be obtained from the extreme points of the set. Further a particular objective function may not have any optimum value, maximum or minimum, when the convex set is a convex polytope.

Note (2) : In the syllabus of C.U. statement of the theorem is wrong. The convex set will be definitely a convex polyhedron. Otherwise, the theorem is meaningless (See Theorem 2, page 51 Linear Programming : Saul I. Gass).

Definition 5.2.1 *Multiple optimal solutions : If a L.P.P. has more than one optimal solution, then the problem is said to have multiple optimal solutions.*

Theorem 5.2.8 *If a L.P.P. has at least two optimal feasible solutions, then there are infinite number of optimal solutions, which are the convex combination of the initial optimal solutions.*

Let $\mathbf{x}_1$ and $\mathbf{x}_2$ be $(\mathbf{x}_1 \neq \mathbf{x}_2)$ be two optimal feasible solutions of a L.P.P. which will maximize the objective function $z = \mathbf{cx}$ subject to $A\mathbf{x} = \mathbf{b}$, $\mathbf{x} \geq \mathbf{0}$.

Then $\hat{z} = \mathbf{cx}_1$ and $\hat{z} = \mathbf{cx}_2$ and $A\mathbf{x}_1 = \mathbf{b}$, $A\mathbf{x}_2 = \mathbf{b}$.

Let

$$\mathbf{x}_3 = \lambda\mathbf{x}_1 + (1-\lambda)\mathbf{x}_2, \quad 0 \leq \lambda \leq 1.$$
$$A\mathbf{x}_3 = \lambda A\mathbf{x}_1 + (1-\lambda)A\mathbf{x}_2 = \mathbf{b}$$

which indicates that $\mathbf{x}_3$ is also a solution set of $A\mathbf{x} = \mathbf{b}$.

Again

$$\mathbf{cx}_3 = \lambda\mathbf{cx}_1 + (1-\lambda)\mathbf{cx}_2 = \lambda\hat{z} + (1-\lambda)\hat{z} = \hat{z}$$

and $\mathbf{x}_3 \geq \mathbf{0}$ as $\mathbf{x}_1, \mathbf{x}_2 \geq \mathbf{0}$ and $0 \leq \lambda \leq 1$.

Therefore, $\mathbf{x}_3$ is also an optimal feasible solution of the L.P.P. which is the convex combination of two distinct points $\mathbf{x}_1$ and $\mathbf{x}_2$ which indicates that there are infinite optimal solutions.

Note : From this we can say that if a L.P.P. has $k (k \geq 2)$ optimal feasible solutions then the problem has an infinite optimal solutions which are the convex combination of k initial optimal solutions. This statement is valid both for basic and non-basic optimal feasible solutions.

5.3 Fundamental theorem of L.P.P. [Geometrical approach]

Theorem 5.3.1 *If a L.P.P. admits an optimal solution, then the objective function assumes the optimum value at an extreme point of the convex set generated by the set of all feasible solutions.*

Let X be the convex set of all F.S. of a L.P.P. and $\mathbf{cx}$, an objective function of a maximization problem and we have assumed that $\max(\mathbf{cx}) = z^*$ corresponding to a point or points in X. We are to prove that there exists at least one extreme point of X which makes the objective function a maximum. The hyperplane $\mathbf{cx} = z^*$ is called the optimal hyperplane. Let T be the intersection of X and $\mathbf{cx} = z^*$. T is not empty [as by assumption, there is at least one point which is both in X and $\mathbf{cx} = z^*$]. Let $\mathbf{x}^*$ be an extreme point of T. Thus $\mathbf{cx}^* = z^*$ and there do not exist two different points in T such that $\mathbf{x}^*$ can be expressed as the convex combination of these two points. Again as $\mathbf{x}^* \in T$, $\mathbf{x}^* \in X$. We assume that $\mathbf{x}^*$ is not an extreme point of X. Thus there exist two points $\mathbf{x}_1$ and $\mathbf{x}_2$ in X such that

$$\mathbf{x}^* = \lambda\mathbf{x}_1 + (1-\lambda)\mathbf{x}_2, \quad 0 < \lambda < 1 \tag{5.3.1}$$
$$\text{or,} \quad \mathbf{cx}^* = \lambda\mathbf{cx}_1 + (1-\lambda)\mathbf{cx}_2 = z^*. \tag{5.3.2}$$

But $\mathbf{cx}_1 \leq z^*$ and $\mathbf{cx}_2 \leq z^*$ [as z^* is the maximum value of $\mathbf{cx}$]. Thus (2) will be satisfied if and only if $\mathbf{cx}_1 = z^*$ and $\mathbf{cx}_2 = z^*$ which indicate that $\mathbf{x}_1$ and $\mathbf{x}_2$ are on the hyperplane $\mathbf{cx} = z^*$, i.e., $\mathbf{x}_1$ and $\mathbf{x}_2$ are on T. Hence x^* can be expressed as the convex combination of two distinct points of T is also an extreme point of X and hence there exists at least one extreme point of X which makes the objective function $\mathbf{cx}$, maximum. Finally we are to show that T has at least one extreme point. T is the intersection of two closed convex sets, one ($\mathbf{cx} = z^*$) is unbounded and the other X [convex set of F.S.] is at least bounded from below. Hence T is a closed convex set containing only F.S. which is at least bounded from below and hence T has at least one extreme point. Similar is the proof for a minimization problem.

Note (1). The very existence of the optimal solution is assumed in the statement of the theorem. But the mere existence of extreme points of a convex set of F.S. do not give definite assurance of having an optimal solution of a given objective function.

5.4 Some important difficult theorems on convex set

Supporting hyperplane. A hyperplane $\mathbf{cx} = z$ is said to be an optimal hyperplane of a convex set X of F.S. of a L.P.P. if for an optimal solution $\mathbf{x}^*$, $\mathbf{cx}^* = z^*$ and for all other points $\mathbf{u}$ belonging to X, $\mathbf{cu} \leq z^*$ for a maximization problem and $\mathbf{cu} \geq z^*$ for a minimization problem.

Theorem 5.4.1 *The optimal hyperplane is a supporting hyperplane in the convex set of F.S. of a L.P.P.*

We consider a maximization problem. Let X be the convex set of F.S. of a L.P.P. and $\mathbf{cx} = z^*$ be an optimal hyperplane and $\mathbf{x}^*$ be an optimal solution such that $\mathbf{cx}^* = z^*$. We shall have to prove that no point of the hyuperplane $\mathbf{cx} = z^*$ is an interior point of X.

Let $\mathbf{x}^*$ be an interior point of X. Then there exists $\varepsilon > 0$ such that every ε neighbourhood of $\mathbf{x}^*$ is in X. Thus the point $\mathbf{x}_1 = \mathbf{x}^* + \frac{\varepsilon}{2}\frac{\mathbf{c}'}{|\mathbf{c}|}$ is in X [As $\mathbf{x}_1$ is a column vector then $\mathbf{c}'$ is taken as $\mathbf{c}$ is a row vector.].

Thus

$$z_1 = \mathbf{cx}_1 = \mathbf{cx}^* + \frac{\varepsilon}{2}\frac{\mathbf{cc}'}{|\mathbf{c}|}$$

$$= z^* + \frac{\epsilon}{2}|\mathbf{c}| \quad [\because \mathbf{cx}^* = z^* \text{ and } \mathbf{cc}' = |\mathbf{c}|^2]$$

Now $\varepsilon > 0$ and $|\mathbf{c}| > 0$ then $z_1 > z^*$ which is against the assumption that z^* is the maximum value of the objective function. Thus $\mathbf{x}^*$ cannot be an interior point. It will be at best a boundary point and for all other points $\mathbf{u}$ in X, $\mathbf{cu} \leq z^*$. Hence the optimal hyperplane is a supporting hyperplane also.

Theorem 5.4.2 *A closed convex set which is bounded from below has extreme points in every supporting hyperplane.*

Let X be a closed convex set bounded from below and $\mathbf{cx} = z^*$ be a supporting hyperplane at a point $\mathbf{x}^*$ of X such that $\mathbf{cx}^* = z^*$. Let T be the intersection of the sets X and $\mathbf{cx} = z^*$. T is not empty as at least $\mathbf{x}^*$ is in T and T is also bounded from below and closed. We are to prove that any extreme point of T is also an extreme point of X.

Let $\mathbf{t}$ be any extreme point of T and we assume that $\mathbf{t}$ will not be an extreme point of X.

Thus $\mathbf{ct} = z^*$ and there do not exist two different points in T such that $\mathbf{t}$ can be expressed as the convex combination of these two points. Again as $\mathbf{t} \in T$, $\mathbf{t} \in X$. As $\mathbf{t}$ is not an extreme point of X, then there exist two different points $\mathbf{x}_1$ and $\mathbf{x}_2$ in X such that $\mathbf{t}$ can be expressed as given by

$$\mathbf{t} = \lambda\mathbf{x}_1 + (1-\lambda)\mathbf{x}_2, 0 < \lambda < 1 \tag{5.4.1}$$

$$\text{or,}\quad \mathbf{ct} = \lambda\mathbf{cx}_1 + (1-\lambda)\mathbf{cx}_2 = z^* \tag{5.4.2}$$

But as $\mathbf{cx} = z^*$ is supporting hyperplane, then either both $\mathbf{cx}_1 \geq z^*$ and $\mathbf{cx}_2 \geq z^*$ or both $\mathbf{cx}_1 \leq z^*$ and $\mathbf{cx}_2 \leq z^*$. But the relation (2) will be satisfied if and only if $\mathbf{cx}_1 = z^*$ and $\mathbf{cx}_2 = z^*$ which indicates that $\mathbf{x}_1$ and $\mathbf{x}_2$ are on the hyperplane $\mathbf{cx} = z^*$, i.e., $\mathbf{x}_1$ and $\mathbf{x}_2 \in T$. Hence $\mathbf{t}$ can be expressed as the convex combination of two distinct points of T which is against our assumption. Thus any extreme point of T is also an extreme point of X. Now we are to show that T has at least one extreme point. T is the intersection of two closed convex set, one ($\mathbf{cx} = z^*$) is unbounded and other X is bounded from below. Then T is also a closed convex set which is bounded from below.

Now select a point in T which has the first component minimum (algebraic) and such point or points exist in T as T is bounded from below. If the point be not unique, then select the point which has the first two components minimum and if not unique try to go further in search of a point which has first j components minimum (algebraic) values and for a particular value of j we shall get a unique point as one such point always exists. Let the unique point be $\mathbf{t}^*$ and it will be an extreme point of T. If not, then there exist two distinct points $\mathbf{t}_1$ and $\mathbf{t}_2$ in T such that

$$\mathbf{t}^* = \lambda\mathbf{t}_1 + (1-\lambda)\mathbf{t}_2, 0 < \lambda < 1. \tag{5.4.3}$$

Let $\mathbf{t}_i^*$ be any component of $\mathbf{t}^*$ previous to $\mathbf{t}_j^*$, $i = 1, 2, \cdots, (j-1)$. Thus equating for the ith component we get

$$\mathbf{t}_i^* = \lambda t_{i1} + (1-\lambda)t_{i2}, 0 < \lambda < 1 \tag{5.4.4}$$

where t_{i1} etc. is the ith component of $\mathbf{t}_1$ etc.

Let us assume that $t_{i1} \neq t_{i2}$ and $t_{i1} > t_{i2}$. Then from (4) we get $\mathbf{t}_i^* > t_{i2} + (1-\lambda)$ $t_{i2} = t_{i2}$. Again if $t_{i1} < t_{i2}$ then $\mathbf{t}_i^* > \lambda t_{i1} + (1-\lambda)\ t_{i1} = t_{i1}$. Both the conditions are impossible because $\mathbf{t}_i$ is the minimum algebraic value. The only possible case is then $t_i = t_{i1} = t_{i2}$, $i = 1, 2, \cdots, (j-1)$.

Similarly from (3) we observe that jth component requires that $t_j^* > \min[t_{j1}, t_{j2}]$ or $t_j^* = t_{j1} = t_{j2}$. But either of these alternatives contradicts the fact that t_j^* is the unique minimum for the jth component when the previous $(j-1)$ components are at their minimum values. Hence t^* cannot be expressed as the convex combination of two distinct points in T and as such t^* is an extreme point and the theorem is proved.

▶ **Example 5.4.1** *Prove that in E^2, the set $S = \{(x_1, x_2)/x_1 - 2x_2 = 2\}$ is a convex set.*

Solution: The set is not a null set. Let (x_{11}, x_{21}) and (x_{12}, x_{22}) be any two points of the set. Then

$$\left.\begin{aligned} x_{11} - 2x_{21} &= 2 \\ \text{and} \quad x_{12} - 2x_{22} &= 2 \end{aligned}\right\}. \tag{i}$$

Now the convex combination of the two points is a point which is

$$[\lambda x_{11} + (1-\lambda)x_{12}, \lambda x_{21} + (1-\lambda)x_{22}], \quad 0 \le \lambda \le 1.$$

Now

$$\begin{aligned} &\lambda x_{11} + (1-\lambda)x_{12} - 2\{\lambda x_{21} + (1-\lambda)x_{22}\} \\ &= \lambda(x_{11} - 2x_{21}) + (x_{12} - 2x_{22}) - \lambda(x_{12} - 2x_{22}) \\ &= 2\lambda + 2 - 2\lambda = 2 \quad \text{[From (i)]} \end{aligned}$$

Hence the point, which is the convex combination of any two points is a point of the set. Thus the set is a convex set.

▶ **Example 5.4.2** *Prove that in E^2, the set $X = \{(x, y) | x + 2y \le 5\}$ is a convex set.*

Solution: The set is not a null set. Let (x_1, y_1) and (x_2, y_2) be any two points of the set X.

Then

$$\left.\begin{aligned} x_1 + 2y_1 &\le 5 \\ \text{and} \quad x_2 + 2y_2 &\le 5 \end{aligned}\right\}. \tag{i}$$

Now the convex combination of the two points is a point given by

$$[\lambda x_1 + (1-\lambda)x_2, \lambda y_1 + (1-\lambda)y_2], \quad 0 \le \lambda \le 1.$$

Now

$$\begin{aligned} \lambda x_1 + (1-\lambda)x_2 + 2\{\lambda y_1 + (1-\lambda)y_2\} &= \lambda(x_1 + 2y_1) + x_2 + 2y_2 - \lambda(x_2 + 2y_2) \\ &\le 5\lambda + 5 - 5\lambda = 5 \quad [\because 0 \le \lambda \le 1]. \end{aligned}$$

that the point is a point of the set. Thus the set is a convex set.

▶ **Example 5.4.3** *Prove that the set defined by* $X = \{\mathbf{x} : |\mathbf{x}| \leq 2\}$ *is a convex set.*

Solution: The set X is not a null set. Let x_1 and x_2 be any two points belonging to the set X. Then

$$|x_1| \leq 2, \quad |x_2| \leq 2. \qquad \text{(i)}$$

The convex combination of the points x_1 and x_2 is a point

$$\mathbf{x}^* = \lambda x_1 + (1-\lambda)x_2, \quad 0 \leq \lambda \leq 1.$$

Now

$$\begin{aligned} |\lambda x_1 + (1-\lambda)x_2| &\leq |\lambda x_1| + |(1-\lambda)x_2| \\ &= \lambda|x_1| + (1-\lambda)|x_2| \quad [\because \lambda, 1-\lambda \geq 0] \\ &\leq 2\lambda + 2(1-\lambda) = 2 \quad \text{[From (i)]} \end{aligned}$$

$\Rightarrow |\lambda x_1 + (1-\lambda)x_2| \leq 2$ which indicates that $\mathbf{x}^* \in X$ and the set is a convex set.

▶ **Example 5.4.4** *Prove that the point* $(1, 0, 2, -2)$ *is a point of the open half space* $2x_1 + 3x_2 + x_3 - 3x_4 < 13$ *but the point* $(2, 2, -1, -4)$ *is not a point of the open half space* $2x_1 + 3x_2 + x_3 - 3x_4 < 13$.

Solution: For the point $(1, 0, 2, -2)$, the value of the expression

$$\begin{aligned} 2x_1 + 3x_2 + x_3 - 3x_4 &= 2 \times 1 + 3 \times 0 + 1 \times 2 - 3(-2) \\ &= 10 < 13. \end{aligned}$$

Hence the point lies on the open half space

$$2x_1 + 3x_2 + x_3 - 3x_4 < 13.$$

For the point $(2, 2, -1-4)$, the value of the expression

$$\begin{aligned} 2x_1 + 3x_2 + x_3 - 3x_4 &= 2 \times 2 + 3 \times 2 - 1 - 3 \times (-4) \\ &= 21 > 13. \end{aligned}$$

Hence the point $(2, 2, -1, -4)$ does not belong to the open half space

$$2x_1 + 3x_2 + x_3 - 3x_4 < 13.$$

▶ **Example 5.4.5** *Prove that in* E^2, *the set* $X = \{(x, y)/x^2 + y^2 \leq 4\}$ *is a convex set.*

(i) **Geometrical approach :** The set of points represents a circle of radius 2 with the boundary and all its interior points. Hence the set is a convex set.

(ii) **Algebraical approach :** Let (x_1, y_1) and (x_2, y_2) be any two points of the set X.

Then

$$\left.\begin{aligned} x_1^2 + y_1^2 \le 4 \\ x_2^2 + y_2^2 \le 4 \end{aligned}\right\}. \tag{i}$$

Now any convex combination of the points is a point which is

$$\{\lambda x_1 + (1-\lambda)x_2, \lambda y_1 + (1-\lambda)y_2\}, 0 \le \lambda \le 1.$$

Again,

$$\begin{aligned} &\{\lambda x_1 + (1-\lambda)x_2\}^2 + \{\lambda y_1 + (1-\lambda)y_2\}^2 \\ &= \lambda^2(x_1^2+y_1^2) + (1-\lambda)^2(x_2^2+y_2^2) + 2\lambda(1-\lambda)(x_1x_2+y_1y_2) \\ &\le \lambda^2(x_1^2+y_1^2) + (1-\lambda)^2(x_2^2+y_2^2) + 2\lambda(1-\lambda)\left[\frac{x_1^2+x_2^2+y_1^2+y_2^2}{2}\right] \\ &\left[\because \quad x_1x_2 \le \frac{x_1^2+x_2^2}{2}, \quad y_1y_2 \le \frac{y_1^2+y_2^2}{2}\right] \\ &\le 4\lambda^2 + 4(1-\lambda)^2 + 8\lambda(1-\lambda) \quad [\because \quad x_1^2+x_2^2+y_1^2+y_2^2 \le 8] \\ &= 4\lambda^2 + 4 + 4\lambda^2 - 8\lambda + 8\lambda - 8\lambda^2 \\ &= 4 \Rightarrow \text{ the point } \in X. \text{ Thus the set is a convex set.} \end{aligned}$$

► **Example 5.4.6** *Prove that in E^2, set $X = \{(x,y)/|x| \le 2, |y| \le 1\}$ is a convex set.*

(1) **Geometrical approach :** The set represents a rectangle with all its interior points having the four extreme points $(2,1)(2,-1)$, $(-2,-1)$ and $(-2,1)$. Hence the set is a convex set.

(2) **Algebraical approach :** The set is not empty. Let (x_1, y_1) and (x_2, y_2) be any two points belonging to the set X. Then

$$|x_1| \le 2, |x_2| \le 2 \tag{i}$$

$$\text{and} \quad |y_1| \le 1, |y_2| \le 1. \tag{ii}$$

The convex combination of the points which is a point given by

$$\{\lambda x_1 + (1-\lambda)x_2, \ \lambda y_1 + (1-\lambda)y_2\}, \quad 0 \le \lambda \le 1.$$

Now,

$$\begin{aligned} |\lambda x_1 + (1-\lambda)x_2| &\le |\lambda x_1| + |(1-\lambda)x_2| \qquad\qquad \text{(iii)} \\ &= \lambda|x_1| + (1-\lambda)|x_2| \quad [\because \quad \lambda \text{ and } 1-\lambda \ge 0] \\ &\le 2\lambda + 2(1-\lambda) \\ &= 2 \\ &\Rightarrow |\lambda x_1 + (1-\lambda)x_2| \le 2 \end{aligned}$$

In the same way we can prove that

$$|\lambda y_1 + (1-\lambda)y_2| \le 1 \tag{iv}$$

Thus the convex combination of any two points $\in X$. Thus the set X is a convex set.

▶ **Example 5.4.7** *Prove that in E^2, the set $X = \{(x, y)/y^2 \leq x\}$ is a convex set.*

Solution: The set is not empty. Let (x_1, y_1) and (x_2, y_2) be any two points of the set. Then

$$y_1^2 \leq x_1 \text{ and } y_2^2 \leq x_2. \quad \text{(v)}$$

The convex combination of the points which is also a point is

$$\{\lambda x_1 + (1-\lambda)x_2, \lambda y_1 + (1-\lambda)y_2\}, 0 \leq \lambda \leq 1.$$

Now $\{\lambda y_1 + (1-\lambda)y_2\}^2 = \lambda^2 y_1^2 + (1-\lambda)^2 y_2^2 + 2\lambda(1-\lambda)y_1 y_2$

$$\leq \lambda^2 y_1^2 + (1-\lambda)^2 y_2^2 + \lambda(1-\lambda)(y_1^2 + y_2^2) \quad [\because y_1^2 + y_2^2 \geq 2y_1 y_2]$$
$$= y_2^2 - 2\lambda y_2^2 + \lambda y_1^2 + \lambda y_2^2$$
$$= \lambda y_1^2 + y_2^2(1-\lambda)$$
$$\leq \lambda x_1 + (1-\lambda)x_2 \quad [\because \lambda \geq 0]$$

$\Rightarrow$ the point $\{\lambda x_1 + (1-\lambda)x_2, \lambda y_1 + (1-\lambda)y_2\} \in X$. Hence the set X is a convex set.

▶ **Example 5.4.8** *Prove that in E^2, the set $X = \{(x, y)|y^2 \geq 4x\}$ is not a convex set.*

Solution: $(0,0)$ and $(1,2)$ are two points $\in X$. A particular convex combination of the two points is

$$\left\{\frac{1}{2}\cdot 0 + \frac{1}{2}\cdot 1, \frac{1}{2}\cdot 0 + \frac{1}{2}\cdot 2\right\} = \left(\frac{1}{2}, 1\right) \text{ for } \left[\lambda = \frac{1}{2}\right]$$

Now the point $(\frac{1}{2}, 1) \notin X$ as $1^2 \ngeq 4 \cdot \frac{1}{2} = 2$.

Thus the set is not a convex set.

Short Answer Type Questions with Answers

1. Define : (a) point set, (b) line, (c) line segment, (d) hyper-plane in E^n, (e) Euclidean space, (f) Hyper-sphere, (g) Convex-polyhedron, (h) Convex hull. [*Ans.* See 5.1 and 5.2]

2. What is a convex set? What are its extreme points? [*Ans.* See 5.2]

3. (a) Give two examples of convex sets in R^2. **[C.U.'82]**

   (b) Give two examples of convex set in R^3.

   [*Ans.* (a) Triangle, Rectangle, square, circle with their interior points are the examples of convex set in E^2.

   (b) Cube and sphere with all its interior points.]

4. Give an example of a convex set which has no extreme point.

[*Ans.* Hyper-plane, line etc.]

5. (a) Give an example of a convex set whose all boundary points are extreme points.

[*Ans.* Circle with all its interior points and boundary points. Here all boundary points are extreme points.]

(b) Explain the terms : (i) convex hull, (ii) convex polyhedron and (iii) Simplex. State giving reasons which of the following are the examples of one or more of these : A line segment, a circle, a triangle, a quadrilateral.

[C.U.(H)'83]

[*Ans.* Line segment, triangle; convex polyhedron [as well as simplex]. circle [convex hull], convex quadrilateral [Convex polyhedron].

6. What is a simplex? Name one simplex each in two and three dimension respectively.

7. Find out the extreme points (if any) of the following convex sets :

(a) $S = \{(x, y)|x^2 + y^2 \leq 25\}$ **[C.U.'82]**

(b) $S = \{(x, y)||x|| \leq a, |y| \leq b\}$

(c) $S = \{(x, y)||x| \leq 1, y \leq 1\}$

(d) Set $\mathbf{cx} = k$. **[C.U.'82]**

[*Ans.* (a) A point (x_1, y_1) which satisfy the condition $x_1^2 + y_1^2 = 25$ is an extreme point. From that it is evident that this set has infinite number extreme points

(b) $[x_1 = a, y_1 = b]$, $[x_2 = a, y_2 = -b]$, $[x_3 = -a, y_3 = b]$, $[x_4 = -a, y_4 = -b]$.

(c) $[x_1 = 1, y_1 = 1]$, $[x_2 = -1, y_2 = 1]$.

(d) This set has no extreme point at all.]

8. (a) Prove that hyperplane is a convex set.

(b) Prove that set of F.S. of $A\mathbf{x} = \mathbf{b}$, $x \geq \mathbf{0}$ is a closed convex set.

[*Ans.* See theorem 5.2.4]

9. What is the relation between B.F.S. and extreme point solution of a convex set of the feasible solutions of a L.P.P.?

[*Ans.* Each B.F.S. is a extreme point solution and vice-versa and there is one to one correspondance between a B.F.S. and extreme point solution in the absence of degeneracy.]

10. Prove that extreme points are finite in number. [*Ans.* See theorem 5.2.8]

11. A hyperplane is given by the equation $3x_1 + 2x_2 + 4x_3 + 6x_4 = 7$. Find in which half space are the points $(-6, 1, 7, 2)$ and $(1, 2, -4, 1)$? **[C.U.(P)'83,'99]**

[*Ans.* First point is in the space $3x_1 + 2x_2 + 4x_3 + 6x_4 > 7$ and the second point is in the space $3x_1 + 2x_2 + 4x_3 + 6x_4 < 7$.]

12. Are the points $(4, 1, 0, 2)$ and $(-3, 2, 4, 4)$ of the same side of the hyperplane $2x_1 + x_1 + 3x_3 + x_4 = 10$? [*Ans.* Yes]

13. (a) Is the point $(1, 10)$ in the convex set of F.S. determined by the constraints $2x_1 + 5x_2 \leq 40, x_1 + x_2 \leq 11, x_2 \geq 4, x_1 \geq 0, x_2 \geq 0$?

    Give reasons for your answer.

    [*Ans.* No. Because the point satisfies the second and third constraints but does not satisfy the first constraint.]

    (b) Is the point $(3, 1)$ in the convex set of F.S. determined by the constraints $x_1 + x_2 \leq 4$, $2x_1 + x_2 \leq 6$, $4x_1 + 3x_2 \leq 28, x_1 \geq 0, x_2 \geq 0$? Give reasons for your answer.

    [*Ans.* No. The point does not satisfy the second constraint.]

    (c) Is the point $(4, 3)$ in the convex set of feasible solution determined by the set of constraints $2x_1 + 7x_2 \geq 22$, $x_1 + x_2 \geq 6$, $3x_1 + x_2 \geq 10$, $x_1 \geq 0$, $x_2 \geq 0$?

    [*Ans.* Yes. Because the point satisfies all the five constraints.]

14. Write a short note about the nature of the convex set of the feasible solutions of a L.P.P.

    [*Ans.* The convex set of F.S. of a L.P.P. is either strictly bounded convex polyhedron or a convex set bounded from below only. Of course, the convex set may be a null set.

15. A particular objective function has both finite maximum and finite minimum subject to certain constraints. What is the nature of the convex set of feasible solutions of the L.P.P.?

    [*Ans.* Strictly bounded convex polyhedron.]

16. The convex set of F.S. of a particular L.P.P. is a strictly bounded convex polyhedron with the extrene points $(0, 4)$, $(7, 4)$, $(5, 6)$, $(1, 8)$. The objective function is $2x_1 - 3x_2$. Find the maximum and minimum value of the objective function.

    [*Ans.* $\max z = 2$ at (7, 4) and $\min z = -22$ at (1, 8).]

17. The convex set of F.S. of a particular L.P.P. is bounded from below only. It has three extreme points (5, 1), (1, 6) and (4, 8). The objective function $x_1 - 3x_2$ is known to have finite maximum. Find the maximum value of the objective function.

    [*Ans.* $\max z = 2$ at $(5, 1)$.]

18. The convex set of feasible solutions of a L.P.P. is bounded from below only and it has three extreme points $(4, 2)$, $(1, 5)$ and $(4, 8)$. The objective function $x_1 - x_2$ is known to have finite minimum. Find the minimum value of the objective function.

    [*Ans.* $\min z = -4$ at $(1, 5)$ and $(4, 8)$ which indicates that infinite optimal solutions exist.

19. Find graphically the feasible space, if any for the following case,

$$2x_1 + 3x_2 \leq 6,\ 2x_1 + 3x_2 \geq 6,\ x_1,\ x_2 \geq 0.$$ **[C.U.(P)'83]**

[*Ans.* Segment of the line $2x_1 + 3x_2 = 6$ with extreme points $(3,0)$ and $(0,2)$.]

Exercise 5

1. Define the boundary points and extreme points of a convex set. Prove that all the extreme points are boundary points but the converse is not necessarily true. Can there be any convex set without any extreme point?

2. Give an example of convex set, where all the boundary points are extreme points.

3. Prove that the set $X = \{(x_1, x_2) : x_1 x_2 \leq 1, x, \geq 0, x_2 \geq 0\}$ is not a convex set.

4. (a) Prove that the set $z = \{(x_1, x_2) : x_1^2 + x_2^2 \leq 25\}$ is a convex set.
   (b) Prove that the set $S = \{(x_1, x_2) | x_1^2 + x_2^2 = 4\}$ is not a convex set.
   (c) Prove that $[2, 0, 1, 0]$ is an extreme point of the convex set of F.S. of the set of equations

$$\begin{aligned} 3x_1 - 5x_2 + x_3 - 2x_4 &= 7 \\ 6x_1 - 10x_2 - x_3 + 5x_4 &= 11, \quad x_j \geq 0, \quad j = 1, \cdots, 4. \end{aligned}$$

5. Prove that extreme points of the convex set of F.S. of a L.P.P. are finite in number [number of variables are finite].

6. A hyperplane is given by the equation

$$2x_1 + 3x_2 + 4x_3 - x_4 = 6.$$

Find in which half space are the points $(4, -3, 2, 1)$, $(1, 2 - 3, 1)$?

7. (a) Prove that hyperplane is a convex set.
   (b) Prove that the open half spaces $A\mathbf{x} < \mathbf{b}$ and $A\mathbf{x} > \mathbf{b}$ are convex sets.
   (c) Prove that the convex polyhedron is a convex set.

8. The vertices of a triangle are $(0,0)$, $(0,4)$, $(3,0)$. Express the points $(1,1)$, $(2,1)$, $(1,1.5)$ as the convex combination of the extreme points.

9. Prove that every extreme point of the convex set of F.S. to $A\mathbf{x} = \mathbf{b}$ is a B.F.S. and conversely. **[C.U.M.Sc.(Appl.Math.)'77]**

10. Prove that if $\mathbf{x}_1, \mathbf{x}_2, \cdots, \mathbf{x}_k$ be k different optimal feasible solutions to a L.P.P. then any convex combination of $\mathbf{x}_1, \mathbf{x}_2, \cdots, \mathbf{x}_k$ is also an optimal solution. **[C.U.M.Sc.(Appl.Math.)'76]**

11. Find the extreme points of the convex set of F.S. given by

$$2x_1 + x_2 + 3x_3 + 2x_4 = 11$$
$$3x_1 - 3x_2 + 5x_3 + x_4 = 17, \quad x_j \geq 0,\ j = 1, \cdots, 4.$$

[*Ans.* [4, 0, 1, 0], [0, 2/7, 25/7, 0], [0, 0, 23/7, 4/7]]

12. Define :

    (a) A supporting hyperplane of a convex et X.

    (b) Optimal hyperplane of a convex set of the feasible solutions of a L.P.P.

13. Prove that an optimal hyperplane is a supporting hyperplane of the convex set of F.S. of L.P.P.

14. Prove that a closed convex set hich is bounded from below has extreme ponts in every supporting hyperplane.

Chapter 6

Fundamental Properties of Simplex Method

6.1 Reduction of a F.S. to a B.F.S. (Reduction Theorem)

Important theorem 6.2.1 *If a linear programming problem,* $A\mathbf{x} = \mathbf{b}$, $\mathbf{x} \geq \mathbf{0}$, *where* A *is the* $m \times n$ *coefficient matrix* $(n > m)$, $r(A) = m$ *has one feasible solution then it has at least one basic feasible solution.*

Proof: Let

$$\mathbf{x} = [x_1, x_2, \cdots, x_n]$$

be a feasible solution to the set of equations $A\mathbf{x} = \mathbf{b}$, $\mathbf{x} \geq \mathbf{0}$. Out of n components of the feasible solution, let k components be positive and the remaining $(n-k)$ components be zero $(1 \leq k \leq n)$ and let us also make an assumption that the first k components are positive and the last $(n-k)$ components are zero.

Then

$$\mathbf{x} = [x_1, x_2, \cdots, x_k, \overbrace{0, \cdots, 0}^{n-k}]$$

and if $\mathbf{a}_1$, $\mathbf{a}_2$, $\cdots$, $\mathbf{a}_k$ be the column vectors corresponding to the variables x_1, x_2, $\cdots$, x_k then

$$x_1\mathbf{a}_1 + x_2\mathbf{a}_2 + \cdots + x_k\mathbf{a}_k = \mathbf{b}, \quad \text{or,} \quad \sum_{j=1}^{k} x_j\mathbf{a}_j = \mathbf{b}. \tag{6.1.1}$$

Now there are three cases to discuss:

(i) $k \leq m$ and the column vectors $\mathbf{a}_1$, $\mathbf{a}_2$, $\cdots$, $\mathbf{a}_k$ are L.I.

(ii) $k > m$.

(iii) $k \leq m$ and the column vectors $\mathbf{a}_1$, $\mathbf{a}_2$, $\cdots$, $\mathbf{a}_k$ are not L.I.

(i) If $k \leq m$ and the column vectors $\mathbf{a}_1, \mathbf{a}_2, \cdots, \mathbf{a}_k$ are L.I. then by definition, the feasible solution is the basic feasible solution. If $k = m$ then the solution is a non-degenerate B.F.S. and if $k < m$ then the solution is a degenerate B.F.S.

(ii) When $k > m$ and the column vectors $\mathbf{a}_1, \mathbf{a}_2, \cdots, \mathbf{a}_k$ are not L.I. and hence the solution is not basic. But by applying a technique which is given below, the number of positive variables can be reduced and ultimately k will be m, keeping all the m non-zero variables positive and the solution will be non-degenerate B.F.S. provided the column vectors corresponding to the positive variables are L.I.

Procedure: As the column vectors $\mathbf{a}_1, \mathbf{a}_2, \cdots, \mathbf{a}_k$ are not linearly independent then there exist scalar quantities λ_j $[j = 1, 2, \cdots, k]$ not all zero such that

$$\lambda_1 \mathbf{a}_1 + \lambda_2 \mathbf{a}_2 + \cdots + \lambda_k \mathbf{a}_k = \mathbf{0}, \quad \text{or,} \quad \sum_{j=1}^{k} \lambda_j \mathbf{a}_j = \mathbf{0}. \tag{6.1.2}$$

Now at least one λ_j is positive [if it is negative, multiply both sides by -1 to make it positive].

Let

$$\nu = \max_j \left(\frac{\lambda_j}{x_j} \right) \quad [j = 1, 2, \cdots, k].$$

As all $x_j > 0$ and $\max \lambda_j > 0$ then ν is essentially a positive quantity.

From (6.2.1) and (6.2.2) we get

$$\sum_{j=1}^{k} \left(x_j - \frac{\lambda_j}{\nu} \right) \mathbf{a}_j = \mathbf{b}. \tag{6.1.3}$$

which indicates that

$$\mathbf{x}' = \left[x_1 - \frac{\lambda_1}{\nu}, \; x_2 - \frac{\lambda_2}{\nu}, \cdots, x_k - \frac{\lambda_k}{\nu}, 0, \cdots, 0 \right]$$

is a solution set of the equations $A\mathbf{x} = \mathbf{b}$.

Now

$$\nu \geq \frac{\lambda_j}{x_j} \quad [\text{the sign of equality holds for maximum } \lambda_j / x_j]$$

then

$$x_j \geq \frac{\lambda_j}{\nu}, \quad \text{or,} \quad x_j - \frac{\lambda_j}{\nu} \geq 0 \quad \text{and} \quad x_j - \frac{\lambda_j}{\nu} = 0 \text{ for at least one } j.$$

Then

$$x'_j = x_j - \frac{\lambda_j}{\nu} \geq 0 \quad [j = 1, 2, \cdots, k]$$

and at least one of them is equal to zero.

Therefore

$$\mathbf{x}' = [x'_1, x'_2, \cdots, x'_k, 0, \cdots, 0,]$$

is also a feasible solution of $A\mathbf{x} = \mathbf{b}$ with maximum number of positive variables $k - 1$. By applying the method repeatedly, ultimately a feasible solution with only

m positive values of the variables will be obtained. Now if the column vectors corresponding to the positive variables [we assume it for the time being] are L.I. then the solution will be a non-degenerate B.F.S.

(iii) If $k = m$ and the column vectors $\mathbf{a}_1, \mathbf{a}_2, \cdots, \mathbf{a}_m$ are not L.I. then the feasible solution is not a basic solution. Applying the procedure given in case (6.1.2the number of non-zero values of variables can be reduced to $p(p < m)$ of which all the non-zero values are positive and there exists a value $p(1 \leq p < m)$, such that $\mathbf{a}_1$, $\mathbf{a}_2, \cdots, \mathbf{a}_p$ are linearly independent and the solution is a degenerate B.F.S. This procedure is applicable also to the case (6.1.2) when $\mathbf{a}_1, \mathbf{a}_2, \cdots, \mathbf{a}_k (k = m)$ are not L.I.

Alternative statement: If a set of m independent linear equations with n variables $(n > m)$ has a F.S. then it has at least one B.F.S.

Note (a): $x_1 = 4, x_2 = 0, x_3 = 1$ is a feasible solution of the set of equations

$$\begin{aligned} 2x_1 - x_2 - x_3 &= 7 \\ \text{and} \quad 4x_1 + 3x_2 - 2x_3 &= 14 \end{aligned}$$

Here the number of positive variables in F.S. is 2 and the number of equations is 2, i.e., $k = 2 = m \cdot r(A) = 2$ but the column vectors of the positive variables x_1 and x_3, i.e., $\mathbf{a}_1$ and $\mathbf{a}_3$ are not linearly independent.

Hence the solution $x_1 = 4, x_2 = 0$ and $x_3 = 1$ is not a basic solution.

$$\mathbf{a}_1 = \begin{bmatrix} 2 \\ 4 \end{bmatrix} \quad \text{and} \quad \mathbf{a}_3 = \begin{bmatrix} -1 \\ -2 \end{bmatrix}.$$

Then

$$\mathbf{a}_1 = -2\mathbf{a}_3, \quad \text{or,} \quad \mathbf{a}_1 + 2\mathbf{a}_3 = \mathbf{0}.$$

Then $\lambda_1 = 1$ and $\lambda_3 = 2$,

$$\nu = \max_j \left(\frac{\lambda_j}{x_j}, \lambda_j > 0 \right) = \max \left[\frac{1}{4}, \frac{2}{1} \right] = 2.$$

Hence

$$x_1' = 4 - \frac{1}{2} = \frac{7}{2}, \quad x_3' = 1 - \frac{2}{2} = 0$$

and therefore $\mathbf{x}' = [\frac{7}{2}, 0, 0]$ is a degenerate basic feasible solution of the set of equations

$$\begin{aligned} 2x_1 - x_2 - x_3 &= 7 \\ \text{and} \quad 4x_1 + 3x_2 - 2x_3 &= 14 \end{aligned}$$

with x_1 and x_2 as basic and x_3 non-basic variables. [$\mathbf{a}_1$ and $\mathbf{a}_2$ are linearly independent.]

Note (b): If all λ_j are non-negative or non-positive only one B.F.S. can be obtained by using the theory.

An important theorem (without proof):

Theorem 6.1.2 *The necessary and the sufficient condition, that all basic solutions will exist and non-degenerate, is that every set of m column vectors of the augmented matrix [Ab] is L.I.*

▶ **Example 6.1.1** $x_1 = 1$, $x_2 = 3$, $x_3 = 2$ *is a feasible solution of the equations*

$$\begin{aligned} 2x_1 + 4x_2 - 2x_3 &= 10 \\ 10x_1 + 3x_2 + 7x_3 &= 33 \end{aligned}$$

Reduce the above F.S. to a B.F.S. by reduction theorem.

Solution: Three vectors

$$\mathbf{a}_1 = \begin{bmatrix} 2 \\ 10 \end{bmatrix}, \quad \mathbf{a}_2 = \begin{bmatrix} 4 \\ 3 \end{bmatrix} \quad \text{and} \quad \mathbf{a}_3 = \begin{bmatrix} -2 \\ 7 \end{bmatrix},$$

$$A \begin{bmatrix} 2 & 4 & -2 \\ 10 & 3 & 7 \end{bmatrix} \quad \text{and} \quad r(A) = 2.$$

Hence the two equations are linearly independent. But the three vectors $\mathbf{a}_1$, $\mathbf{a}_2$ and $\mathbf{a}_3$ are not linearly independent. Hence there exist three constants λ_1, λ_2, λ_3 (at least one of them non-zero) such that

$$\lambda_1\mathbf{a}_1 + \lambda_2\mathbf{a}_2 + \lambda_3\mathbf{a}_3 = \mathbf{0} \text{ will be satisfied,}$$

$$\text{or,} \qquad \lambda_1 \begin{bmatrix} 2 \\ 10 \end{bmatrix} + \lambda_2 \begin{bmatrix} 4 \\ 3 \end{bmatrix} + \lambda_3 \begin{bmatrix} -2 \\ 7 \end{bmatrix} = \mathbf{0}. \tag{i}$$

From (i) we get

$$\begin{aligned} 2\lambda_1 + 4\lambda_2 - 2\lambda_3 &= 0 \\ 10\lambda_1 + 3\lambda_2 + 7\lambda_3 &= 0. \end{aligned}$$

By cross-multiplication we get

$$\frac{\lambda_1}{34} = \frac{\lambda_2}{-34} = \frac{\lambda_3}{-34} = k = -\frac{1}{34} \text{ (say)}.$$

Then we get $\lambda_1 = -1$, $\lambda_2 = 1$, $\lambda_3 = 1$.

Hence

$$-\mathbf{a}_1 + \mathbf{a}_2 + \mathbf{a}_3 = \mathbf{0}. \tag{ii}$$

Taking $\lambda_1 = -1$, $\lambda_2 = 1$ and $\lambda_3 = 1$ we have

$$\nu = \max_j \left(\frac{\lambda_j}{x_j},\ \lambda_j > 0\right) = \max\left(\frac{1}{3}, \frac{1}{2}\right) = \frac{1}{2}.$$

Hence a feasible solution given by

$$\mathbf{x}_1' = \left[x_1 - \frac{\lambda_1}{\nu},\ x_2 - \frac{\lambda_2}{\nu},\ x_3 - \frac{\lambda_3}{\nu}\right] \quad \left[\textbf{Note: } |B| = \begin{vmatrix} 2 & 4 \\ 10 & 3 \end{vmatrix} = -34 \neq 0\right]$$

$$= [1+2,\ 3-2,\ 2-2]$$

$$= [3,\ 1,\ 0]$$

is a basic feasible solution, because the positive variables are two and the column vectors $\mathbf{a}_1$ and $\mathbf{a}_2$ corresponding to the positive variables x_1 and x_2 are linearly independent.

Again (ii) can be written as

$$\mathbf{a}_1 - \mathbf{a}_2 - \mathbf{a}_3 = \mathbf{0};$$

then $\lambda_1 = 1$, $\lambda_2 = -1$, $\lambda_3 = -1$.

$$\nu = \max_j \left(\frac{\lambda_j}{x_j},\ \lambda_j > 0\right) = \max\left(\frac{1}{1}\right) = 1.$$

Hence the feasible solution given by

$$\mathbf{x}_2' = \left[x_1 - \frac{\lambda_1}{v},\ x_2 - \frac{\lambda_2}{v},\ x_3 - \frac{\lambda_3}{v}\right]$$

$$= [1-1,\ 3+1,\ 2+1]$$

$$= [0,\ 4,\ 3]$$

is a B.F.S. as the column vectors $\mathbf{a}_2$, $\mathbf{a}_3$ corresponding to the positive variables x_2 and x_3 are linearly independent.

Note: There are two basic feasible solutions from the above procedure, both of which are non -degenerate. Compare the answers with the worked out Example (4.3.1) where B.S. are obtained from the stand point of definition.

The next problem (6.4.2) will be solved by using a new technique not shown in the above problem.

► **Example 6.1.2** *Given* $(1, 1, 2)$ *is a basic feasible solution to the set of equations*

$$\begin{aligned} x_1 + 2x_2 + 3x_3 &= 9 \\ 2x_1 - x_2 + x_3 &= 3 \end{aligned}$$

Reduce the above feasible solution to one or more basic feasible solutions.

[C.U.(P)'86]

Solution: $\mathbf{a}_1 = \begin{bmatrix} 1 \\ 2 \end{bmatrix}$, $\mathbf{a}_2 = \begin{bmatrix} 2 \\ -1 \end{bmatrix}$, $\mathbf{a}_3 = \begin{bmatrix} 3 \\ 1 \end{bmatrix}$ and $\mathbf{b} = \begin{bmatrix} 9 \\ 3 \end{bmatrix}$

The vectors $\mathbf{a}_1, \mathbf{a}_2, \mathbf{a}_3$ associated with the three non-zero (here positive) variables x_1, x_2, x_3 are L.D. [$\because$ There are three, two component vectors]. Hence the given feasible solution is not basic.

The set of equations are

$$\mathbf{a}_1 x_1 + \mathbf{a}_2 x_2 + \mathbf{a}_3 x_3 = \mathbf{b}. \tag{i}$$

Now $(1, 1, 2)$ is a solution set. Therefore

$$\mathbf{a}_1 + \mathbf{a}_2 + 2\mathbf{a}_3 = \mathbf{b}. \tag{ii}$$

The three vectors $\mathbf{a}_1$, $\mathbf{a}_2$, $\mathbf{a}_3$ are L.D. Therefore, there exist three scalar quantities λ_1, λ_2 and λ_3 [at least one of them be non-zero] such that

$$\lambda_1 \mathbf{a}_1 + \lambda_2 \mathbf{a}_2 + \lambda_3 \mathbf{a}_3 = \mathbf{0} \quad \text{relation holds}$$

$$\text{or,} \quad \lambda_1 \begin{bmatrix} 1 \\ 2 \end{bmatrix} + \lambda_2 \begin{bmatrix} 2 \\ -1 \end{bmatrix} + \lambda_3 \begin{bmatrix} 3 \\ 1 \end{bmatrix} = \begin{bmatrix} 0 \\ 0 \end{bmatrix} \text{ holds.} \tag{iii}$$

Equating we get

$$\left.\begin{aligned} \lambda_1 + 2\lambda_2 + 3\lambda_3 &= 0 \\ 2\lambda_1 - \lambda_2 + \lambda_3 &= 0 \end{aligned}\right\}. \tag{iv}$$

By cross-multiplication we get

$$\frac{\lambda_1}{5} = \frac{\lambda_2}{5} = \frac{\lambda_3}{-5} = k \neq 0 = \frac{1}{5} \quad \text{(say)}.$$

Then $\lambda_1 = 1$, $\lambda_2 = 1$, $\lambda_3 = -1$.

Putting the values of λ_1, λ_2, λ_3 in (iii) we get

$$\mathbf{a}_1 + \mathbf{a}_2 - \mathbf{a}_3 = \mathbf{0}. \tag{v}$$

Now

$$\max_j \left(\frac{\lambda_j}{x_j}, \lambda_j > 0 \right) = \max \left(\frac{\lambda_1}{x_1}, \frac{\lambda_2}{x_2} \right) = \max \left(\frac{1}{1}, \frac{1}{1} \right) = 1$$

which occurs at $j = 1, 2$.

Thus we shall have to eliminate either $\mathbf{a}_1$ or $\mathbf{a}_2$ from (ii) and (v) to get a feasible solution with number of positive variables less than three.

From (ii) and (v) we get

$$\begin{aligned} \mathbf{a}_1 + \mathbf{a}_2 + 2\mathbf{a}_3 &= \mathbf{b} \\ \mathbf{a}_1 + \mathbf{a}_2 - \mathbf{a}_3 &= \mathbf{0}. \end{aligned}$$

Subtracting

$$0\mathbf{a}_1 + 0\mathbf{a}_2 + 3\mathbf{a}_3 = \mathbf{b}. \tag{vi}$$

From (vi) we get $x_1 = 0$, $x_2 = 0$, $x_3 = 3$ is a feasible solution of the set of equations.

Now the vector $\mathbf{a}_3$ associated with non-zero variable x_3 is L.I.

Hence the solution is a B.F.S. Here [$x_1 = 0$ (basic), $x_2 = 0$ (non-basic), $x_3 = 3$ (basic)] and [$x_1 = 0$ (non-basic), $x_2 = 0$ (basic) $x_3 = 3$ (basic)] are two basic feasible solutions as both $(\mathbf{a}_3, \mathbf{a}_1)$ and $(\mathbf{a}_3, \mathbf{a}_2)$ are L.I.

Again (v) can be written as

$$-\mathbf{a}_1 - \mathbf{a}_2 + \mathbf{a}_3 = \mathbf{0} \tag{vii}$$

and

$$\max_j \left(\frac{\lambda_j}{x_j}, \lambda_j > 0\right) = \max\left(\frac{1}{2}\right) = \frac{1}{2}$$

which occurs at $j = 3$.

Then the third vector is to be eliminated from (ii) and (vii) and to get a F.S.

From (ii) and (vii) we get

$$\begin{aligned} \mathbf{a}_1 + \mathbf{a}_2 + 2\mathbf{a}_3 &= \mathbf{b} \\ -2\mathbf{a}_1 - 2\mathbf{a}_2 + 2\mathbf{a}_3 &= \mathbf{0} \end{aligned}$$

Subtracting,

$$3\mathbf{a}_1 + 3\mathbf{a}_2 + 0\mathbf{a}_3 = \mathbf{b}. \tag{viii}$$

From (viii) we get $x_1 = 3, x_2 = 3, x_3 = 0$ is a feasible solution of the problem and the solution is basic as the vectors $\mathbf{a}_1$ and $\mathbf{a}_2$ associated with the positive variables are L.I.

▶ **Example 6.1.3** $x_1 = 1, x_2 = 2, x_3 = 1$ *and* $x_4 = 0$ *is a F.S. to the set of equations*

$$\begin{aligned} 11x_1 + 2x_2 - 9x_3 + 4x_4 &= 6 \\ 15x_1 + 3x_2 - 12x_3 + 6x_4 &= 9 \end{aligned}$$

Reduce the F.S. to more than one B.F.S. and prove that one of them is non-degenerate and the other solutions are degenerate.

$$A = \begin{bmatrix} 11 & 2 & -9 & 4 \\ 15 & 3 & -12 & 6 \end{bmatrix} \quad \textit{and} \quad r(A) = 2.$$

Solution: Hence the equations are independent. Again the set of equations can be written as

$$\mathbf{a}_1 x_1 + \mathbf{a}_2 x_2 + \mathbf{a}_3 x_3 + \mathbf{a}_4 x_4 = \mathbf{b}, \tag{i}$$

where

$$\mathbf{a}_1 = \begin{bmatrix} 11 \\ 15 \end{bmatrix}, \quad \mathbf{a}_2 = \begin{bmatrix} 2 \\ 3 \end{bmatrix}, \quad \mathbf{a}_3 = \begin{bmatrix} -9 \\ -12 \end{bmatrix},$$

$$\mathbf{a}_4 = \begin{bmatrix} 4 \\ 6 \end{bmatrix} \quad \text{and} \quad \mathbf{b} = \begin{bmatrix} 6 \\ 9 \end{bmatrix}.$$

$x_1 = 1$, $x_2 = 2$, $x_3 = 1$ and $x_4 = 0$ is a solution set of (i) and hence (i) can be written as

$$\mathbf{a}_1 + 2\mathbf{a}_2 + \mathbf{a}_3 = \mathbf{b}. \tag{ii}$$

The solution set given abobe is not basic because the number 3 of positive variables are greater than 2, the number of independent equations and evidently $\mathbf{a}_1$, $\mathbf{a}_2$ and $\mathbf{a}_3$ are linearly dependent. The vectors $\mathbf{a}_1$, $\mathbf{a}_2$ and $\mathbf{a}_3$ (associated with the positive variables) can be expressed as

$$\mathbf{a}_1 - \mathbf{a}_2 + \mathbf{a}_3 = \mathbf{0}. \tag{iii}$$

Hence $\lambda = 1$, $\lambda_2 = -1$ and $\lambda_3 = 1$.

$$\max_{j=1,2,3}\left(\frac{\lambda_j}{x_j},\ \lambda_j > 0\right) = \max\left(\frac{\lambda_1}{x_1},\ \frac{\lambda_3}{x_3}\right) = \max\left(\frac{1}{1},\ \frac{1}{1}\right) = 1$$

which happen for $j = 1$ and 3. Hence eliminating either $\mathbf{a}_1$ or $\mathbf{a}_3$ from (ii) and (iii) we get a new F.S. Now eliminating $\mathbf{a}_1$ we get

$$3\mathbf{a}_2 = \mathbf{b}. \tag{iv}$$

Hence $x_1 = 0$, $x_2 = 3$, $x_3 = 0$, $x_4 = 0$ is a new F.S. which we obtain from (iv) [The coefficient of $\mathbf{a}_2$ is 3 which is x_2; and all other variables are zero]. The solutions is basic as the vector $\mathbf{a}_2$ associated with the positive variable x_2 is linearly independent.

Vectors $\mathbf{a}_2$, $\mathbf{a}_1$ are L.I.; Vectors $\mathbf{a}_2, \mathbf{a}_3$ are L.I.; Vectors $\mathbf{a}_2$, $\mathbf{a}_4$ are not L.I. Hence the basic feasible solutions are [$x_1 = 0$, $x_2 = 3$; $x_3 = x_4 = 0$ (non-basic)]; [$x_1 = 0$ (non-basic)], $x_2 = 3$, $x_3 = 0$, $x_4 = 0$ (non-basic). The variables x_2 and x_4 together cannot be considered as basic variables. All B.F.S. are degenerate.

Again (iii) can be written as

$$-\mathbf{a}_1 + \mathbf{a}_2 - \mathbf{a}_3 = \mathbf{0}. \tag{v}$$

Thus $\lambda_1 = -1$, $\lambda_2 = 1$, $\lambda_3 = -1$.

$$\max_{j}\left(\frac{\lambda_j}{x_j},\ \lambda_j > 0\right) = \max\left(\frac{1}{2}\right) = \frac{1}{2}$$

which happens for $j = 2$.

Thus eliminating vectors $\mathbf{a}_2$ from (ii) and (v) we get

$$3\mathbf{a}_1 + 3\mathbf{a}_3 = \mathbf{b} \tag{vi}$$

from which we get a new F.S.

$x_1 = 3$, $x_2 = 0$, $x_3 = 3$, $x_4 = 0$. The solution is basic also as the vectors $\mathbf{a}_1$ and $\mathbf{a}_3$ associated with positive variables are L.I. and the solution is non-degenerate.

► **Example 6.1.4** $x_1 = 1$, $x_2 = 2$, $x_3 = 1$ *is a feasible solution of the following set of linearly independent equations.*

$$\begin{aligned} 2x_1 + 3x_2 + 5x_3 &= 13 \\ 3x_1 - x_2 + 3x_3 &= 4 \end{aligned}$$

Reduce the F.S. to a B.F.S.

Solution: Here

$$\mathbf{a}_1 = \begin{bmatrix} 2 \\ 3 \end{bmatrix}, \quad \mathbf{a}_2 = \begin{bmatrix} 3 \\ -1 \end{bmatrix}, \quad \mathbf{a}_3 = \begin{bmatrix} 5 \\ 3 \end{bmatrix} \quad \text{and} \quad \mathbf{b} = \begin{bmatrix} 13 \\ 4 \end{bmatrix}.$$

The set of equations can be written as

$$\mathbf{a}_1 x_1 + \mathbf{a}_2 x_2 + \mathbf{a}_3 x_3 = \mathbf{b}. \tag{i}$$

Since $x_1 = 1$, $x_2 = 2$, $x_3 = 1$ is a solution of (i) then

$$\mathbf{a}_1 + 2\mathbf{a}_2 + \mathbf{a}_3 = \mathbf{b}. \tag{ii}$$

The vectors $\mathbf{a}_1$, $\mathbf{a}_2$, $\mathbf{a}_3$ are linearly dependent.

[Since there are three vectors each having two components]. Thus there exist three scalars, λ_1, λ_2, λ_3 at least one of which is non-zero such that

$$\lambda_1 \begin{bmatrix} 2 \\ 3 \end{bmatrix} + \lambda_2 \begin{bmatrix} 3 \\ -1 \end{bmatrix} + \lambda_3 \begin{bmatrix} 5 \\ 3 \end{bmatrix} = \begin{bmatrix} 0 \\ 0 \end{bmatrix} \text{ be satisfied.}$$

Now, equating we get

$$\begin{aligned} 2\lambda_1 + 3\lambda_2 + 5\lambda_3 &= 0 \\ 3\lambda_1 - \lambda_2 + 3\lambda_3 &= 0 \end{aligned}$$

By cross multiplication, we get

$$\frac{\lambda_1}{9+5} = \frac{\lambda_2}{15-6} = \frac{\lambda_3}{-2-9} = k = 1 \text{ (say)}.$$

Then $\lambda_1 = 14$, $\lambda_2 = 9$, $\lambda_3 = -11$ and hence we can write

$$14\mathbf{a}_1 + 9\mathbf{a}_2 - 11\mathbf{a}_3 = \mathbf{0}. \tag{iii}$$

Now

$$\max\left(\frac{\lambda_j}{x_j},\ \lambda_j > 0\right) = \left(\frac{14}{1},\ \frac{9}{2}\right) = 14$$

which occurs for $j = 1$. Then the first vector is to be removed from (i) and (iii). We have from (ii)

$$\left.\begin{aligned} 14\mathbf{a}_1 + 28\mathbf{a}_2 + 14\mathbf{a}_3 &= 14\mathbf{b} \\ \text{and} \quad 14\mathbf{a}_1 + 9\mathbf{a}_2 - 11\mathbf{a}_3 &= \mathbf{0} \end{aligned}\right\}. \tag{iv}$$

Subtracting we get

$$\begin{aligned} 19\mathbf{a}_2 + 25\mathbf{a}_3 &= 14\mathbf{b} \\ \text{or,} \quad 0\mathbf{a}_1 + 19\mathbf{a}_2 + 25\mathbf{a}_3 &= 14\mathbf{b} \\ \text{or,} \quad \frac{0}{14}\mathbf{a}_1 + \frac{19}{14}\mathbf{a}_2 + \frac{25}{14}\mathbf{a}_3 &= \mathbf{b} \end{aligned} \tag{v}$$

From (v) we can say that $(0, 19/14, 25/14)$ is a feasible solution set of the above equations. Now the equation is basic also, since the vectors, $\mathbf{a}_2$ and $\mathbf{a}_3$ associated with the positive variables are

$$\mathbf{a}_2 = \begin{bmatrix} 3 \\ -1 \end{bmatrix}, \quad \mathbf{a}_3 = \begin{bmatrix} 5 \\ 3 \end{bmatrix} \quad \text{and} \quad \det \begin{vmatrix} 3 & 5 \\ -1 & 3 \end{vmatrix} = 14 \neq 0.$$

Thus the solution $(0, 19/14, 25/14)$ is a basic feasible solution.

Again (iii) can be written in the manner

$$-14\mathbf{a}_1 - 9\mathbf{a}_2 + 11\mathbf{a}_3 = \mathbf{0}. \tag{vi}$$

$$\max_j \left(\frac{\lambda_j}{x_j}, \lambda_j > 0 \right) = \max \left(\frac{11}{2} \right) = \frac{11}{2}$$

which occurs for $j = 3$.

Therefore, the third vector is to be eliminated from (ii) and (iii)

$$\begin{aligned} 11\mathbf{a}_1 + 22\mathbf{a}_2 + 11\mathbf{a}_3 &= 11\mathbf{b} \\ 14\mathbf{a}_1 + 9\mathbf{a}_2 - 11\mathbf{a}_3 &= \mathbf{0} \end{aligned}$$

Adding we get

$$25\mathbf{a}_1 + 31\mathbf{a}_2 + 0\mathbf{a}_3 = 11\mathbf{b}$$

$$\text{or,} \quad \frac{25}{11}\mathbf{a}_1 + \frac{31}{11}\mathbf{a}_2 + \frac{0}{11}\mathbf{a}_3 = \mathbf{b}$$

from which we conclude that $(25/11, 31/11, 0)$ is a feasible solution of the set of equations, and the solution is also basic, since the vectors $\mathbf{a}_1$ and $\mathbf{a}_2$ associated with the positive variables are L.I.

► **Example 6.1.5** *Prove that $x_1 = 3$, $x_2 = 2$, $x_3 = 4$, $x_4 = 0$ is a feasible solution of the set of equations but not basic.*

$$\begin{aligned} 2x_1 + 5x_2 - 3x_3 + x_4 &= 4 \\ 6x_1 + 16x_2 - 9x_3 + 5x_4 &= 14 \end{aligned}$$

Reduce the F.S. to a B.F.S. by using reduction theory and prove that by using the theory, it is not possible to find more than one B.F.S. though the problem may have more than one B.F.S.

Solution: The set of equations are

$$\mathbf{a}_1 x_1 + \mathbf{a}_2 x_2 + \mathbf{a}_3 x_3 + \mathbf{a}_4 x_4 = \mathbf{b}, \tag{i}$$

where

$$\mathbf{a}_1 = \begin{bmatrix} 2 \\ 6 \end{bmatrix}, \quad \mathbf{a}_2 = \begin{bmatrix} 5 \\ 16 \end{bmatrix}, \quad \mathbf{a}_3 = \begin{bmatrix} -3 \\ -9 \end{bmatrix},$$

$$\mathbf{a}_4 = \begin{bmatrix} 1 \\ 5 \end{bmatrix}, \quad \text{and} \quad \mathbf{b} = \begin{bmatrix} 4 \\ 14 \end{bmatrix}.$$

The three vectors associated with positive variables are linearly dependent. Hence the solution is not a basic solution though the solution is feasible since all components of the solution set are non-negative and satisfy the two equations.

Putting the values of $x_1, x_2, \cdots, x_4$ in (i) we get

$$3\mathbf{a}_1 + 2\mathbf{a}_2 + 4\mathbf{a}_3 = \mathbf{b}. \tag{ii}$$

Again only by mere inspection, we can say that

$$3\mathbf{a}_1 + 0\mathbf{a}_2 + 2\mathbf{a}_3 = \mathbf{0}. \tag{iii}$$

Thus $\lambda_1 = 3,\ \lambda_2 = 0,\ \lambda_3 = 2.$

$$\max\left(\frac{\lambda_j}{x_j},\ \lambda_j > 0\right) = \max\left(\frac{\lambda_1}{x_1},\ \frac{\lambda_3}{x_3}\right) = \max\left(\frac{3}{3},\ \frac{2}{4}\right) = 1$$

which occurs for $j = 1$.

Then the first vector is to be eliminated from (ii) and (iii).

We have

$$\begin{aligned} 3\mathbf{a}_1 + 2\mathbf{a}_2 + 4\mathbf{a}_3 &= \mathbf{b} \\ 3\mathbf{a}_1 + 0\mathbf{a}_2 + 2\mathbf{a}_3 &= \mathbf{0} \end{aligned}$$

Subtracting we get

$$0 \cdot \mathbf{a}_1 + 2\mathbf{a}_2 + 2\mathbf{a}_3 = \mathbf{b} \tag{iv}$$

which implies that $(0, 2, 2, 0)$ is a feasible solution of the set of equations. The vectors associated with the positive variables are $\mathbf{a}_2$ and $\mathbf{a}_3$ which are linearly independent since

$$\begin{vmatrix} 5 & -3 \\ 16 & -9 \end{vmatrix} = 3 \neq 0.$$

Hence the solution $(0, 2, 2, 0)$ obtained by reduction method is basic feasible.

Note (1) Since all λ_j are non-negative then by reduction method, it is not possible to obtain more than one B.F.S. though [0, 2/3, 0, 2/3] is another B.F.S.

► **Example 6.1.6** *Consider the system of equations,*

$$\begin{aligned} x_1 + 2x_2 + 4x_3 + x_4 &= 7 \\ 2x_1 - x_2 + 3x_3 - 2x_4 &= 4 \end{aligned}$$

Here $x_1 = 1$, $x_2 = 1$, $x_3 = 1$ and $x_4 = 0$ is a feasible solution.

Reduce the feasible solution to two different basic feasible solutions.

[D.U.B.Sc.(Math.)'77]

Solution: $A = \begin{bmatrix} 1 & 2 & 4 & 1 \\ 2 & -1 & 3 & -2 \end{bmatrix}$ and $R(A) = 2$.

Hence the equations are linearly independent. Again the set of equations can be written as

$$\mathbf{a}_1 x_1 + \mathbf{a}_2 x_2 + \mathbf{a}_3 x_3 + \mathbf{a}_4 x_4 = \mathbf{b}, \tag{i}$$

where

$$\mathbf{a}_1 = \begin{bmatrix} 1 \\ 2 \end{bmatrix}, \quad \mathbf{a}_2 = \begin{bmatrix} 2 \\ -1 \end{bmatrix}, \quad \mathbf{a}_3 = \begin{bmatrix} 4 \\ 3 \end{bmatrix},$$

$$\mathbf{a}_4 = \begin{bmatrix} 1 \\ -2 \end{bmatrix} \quad \text{and} \quad \mathbf{b} = \begin{bmatrix} 7 \\ 4 \end{bmatrix}.$$

As $x_1 = 1, x_2 = 1, x_3 = 1$ and $x_4 = 0$ is a solution set then (i) can be written as

$$\mathbf{a}_1 + \mathbf{a}_2 + \mathbf{a}_3 = \mathbf{b}. \tag{ii}$$

Here the vectors $\mathbf{a}_1$, $\mathbf{a}_2$, $\mathbf{a}_3$ associated with the non-zero variables are L.D

Hence the solution given in the problem is not basic.

Now we can find that

$$2\mathbf{a}_1 + \mathbf{a}_2 - \mathbf{a}_3 = \mathbf{0}. \tag{iii}$$

[Find out the values of λ_1, λ_2 etc. by previous knowledge]

Hence $\lambda_1 = 2, \lambda_2 = 1, \lambda_3 = -1$.

$$\max_j \left(\frac{\lambda_j}{x_j}, \lambda_j > 0 \right) = \max \left(\frac{2}{1}, \frac{1}{1} \right) = 2$$

which occurs at $j = 1$.

Thus eliminating $\mathbf{a}_1$ from (ii) and (iii) we shall get a new feasible solution. From (ii) and (iii) we get

$$\frac{1}{2}\mathbf{a}_2 + \frac{3}{2}\mathbf{a}_3 = \mathbf{b}$$

from which we conclude that

$x_1 = 0$, $x_2 = 1/2$, $x_3 = 3/2$ and $x_4 = 0$ is a new feasible solution to the problem. The solution is basic as the vectors $\mathbf{a}_2$ and $\mathbf{a}_3$ associated with the positive variables are L.I. Again (iii) can be written as

$$-2\mathbf{a}_1 - \mathbf{a}_2 + \mathbf{a}_3 = \mathbf{0}. \tag{iv}$$

Thus $\lambda_1 = -2, \lambda_2 = -1, \lambda_3 = 1$.

$$\max_j \left(\frac{\lambda_j}{x_j}, \lambda_j > 0 \right) = \max \left(\frac{1}{1} \right) = 1$$

which occurs for $j = 3$.

Thus eliminating $\mathbf{a}_3$ from (ii) and (iv) we get

$$3\mathbf{a}_1 + 2\mathbf{a}_2 = \mathbf{b}$$

from which we conclude that $x_1 = 3$, $x_2 = 2$, $x_3 = 0$, $x_4 = 0$ is a new feasible solution and the solution is basic also as the vectors associated with the positive variables are L.I.

Thus two B.F.S. are

$$\mathbf{x}_1 = \left[0,\ \frac{1}{2},\ \frac{3}{2},\ 0\right] \quad \text{and} \quad \mathbf{x}_2 = [3,\ 2,\ 0,\ 0].$$

► **Example 6.1.7** *Prove that $x_1 = 3$, $x_2 = 0$, $x_3 = 0$ is a F.S. to the following set of equations*

$$\begin{aligned} 4x_1 + \ x_2 - 3x_3 &= 12 \\ 6x_1 + \tfrac{3}{2}x_2 + \ x_3 &= 18 \end{aligned}$$

Is the solution basic? If so, which are the basic variables?

Solution: $\mathbf{a}_1 = \begin{bmatrix} 4 \\ 6 \end{bmatrix}$, $\mathbf{a}_2 = \begin{bmatrix} 1 \\ 3/2 \end{bmatrix}$, $\mathbf{a}_3 = \begin{bmatrix} -3 \\ 1 \end{bmatrix}$, and $R(A) = 2$.

Since the solution set $x_1 = 3$, $x_2 = 0$, $x_3 = 0$ satisfy both the equations then the solution is a F.S. [As none of them negative].

The column vector $\mathbf{a}_1$ associated with non-zero variable x_1 is linearly independent.

Hence the solution is basic. Now let us assume that x_1 and x_2 are basic feasible variables. The matrix of the corresponding vectors is

$$B_1 = (\mathbf{a}_1, \mathbf{a}_2) = \begin{bmatrix} 4 & 1 \\ 6 & 3/2 \end{bmatrix} \quad \text{and} \quad |B_1| = 0.$$

Hence the matrix is not a basis matrix and therefore x_1 and x_2 together cannot be considered as basic variables.

Now assume x_1 and x_3 are basic variables. The square matrix of corresponding vectors is

$$B_2 = (\mathbf{a}_1, \mathbf{a}_3) = \begin{bmatrix} 4 & -3 \\ 6 & 1 \end{bmatrix} \quad \text{and} \quad |B_2| = 22 \neq 0.$$

Hence the matrix is a basis matrix. Thus x_1 and x_3 are the basic variables. Therefore, the solution is a basic feasible solution with x_1 and x_3 basic variables and x_2 non-basic variable. Of course, the solution is degenerate.

► **Example 6.1.8** $x_1 = 1, x_2 = 2, x_3 = 1, x_4 = 1$ *is a F.S. of the following set of equations*

$$\begin{aligned} 2x_1 + 3x_2 - 5x_3 + 2x_4 &= 5 \\ x_1 + 2x_2 - 3x_3 + 3x_4 &= 5. \end{aligned}$$

Reduce the F.S. to a B.F.S.

Solution. The set of equation can be written as

$$\mathbf{a}_1x_1 + \mathbf{a}_2x_2 + \mathbf{a}_3x_3 + \mathbf{a}_4x_4 = b \tag{i}$$

where $\mathbf{a}_1 = \begin{bmatrix} 2 \\ 1 \end{bmatrix}, \mathbf{a}_2 = \begin{bmatrix} 3 \\ 2 \end{bmatrix}, \mathbf{a}_3 = \begin{bmatrix} -5 \\ -3 \end{bmatrix}, \mathbf{a}_4 = \begin{bmatrix} 2 \\ 3 \end{bmatrix}; \mathbf{b} = \begin{bmatrix} 5 \\ 5 \end{bmatrix}$ only by observations we can write

$$\mathbf{a}_1 + \mathbf{a}_2 + \mathbf{a}_3 = 0 \tag{ii}$$

where $\mathbf{a}_1, \mathbf{a}_2, \mathbf{a}_3$ and $\mathbf{a}_4$ forms a linearly dependable set.

From (1) we get

$$\mathbf{a}_1 + 2\mathbf{a}_2 + \mathbf{a}_3 + \mathbf{a}_4 = b \tag{iii}$$

[$\because$ $x_1 = 1$, $x_2 = 2$, $x_3 = 1$, $x_4 = 1$ is a solution of (i)]

$\max\limits_j \left(\dfrac{\lambda_j}{x_j}, x_j > 0\right) = \left(\dfrac{1}{1}, \dfrac{1}{2}, \dfrac{1}{1}\right) = 1$ which occurs for $j = 1$ and $j = 3$. Thus we shall have to eliminate either $\mathbf{a}_1$ or $\mathbf{a}_3$

$$\begin{array}{lrcll} & \mathbf{a}_1 + 2\mathbf{a}_2 + \mathbf{a}_3 + \mathbf{a}_4 &=& \mathbf{b} & \text{[from (i)]} \\ & \mathbf{a}_1 + \mathbf{a}_2 + \mathbf{a}_3 \phantom{+ \mathbf{a}_4} &=& 0 & \text{[from (ii)]} \\ \hline \text{Substracting we get} & 0\mathbf{a}_1 + \mathbf{a}_2 + 0\mathbf{a}_3 + \mathbf{a}_4 &=& \mathbf{b} & \text{(iv)} \end{array}$$

From (iv) we can say that $x_1 = 0, x_2 = 1, x_3 = 0, x_4 = 1$ is F.S. of the given problem. The vectors $\mathbf{a}_2 = \begin{bmatrix} 3 \\ 2 \end{bmatrix}$ and $\mathbf{a}_4 = \begin{bmatrix} 2 \\ 3 \end{bmatrix}$ associated with positive variables are L.I.; since

$$\det \begin{bmatrix} 2 & 2 \\ 2 & 3 \end{bmatrix} = 5 \neq 0.$$

Hence the solution $x_1 = 0, x_2 = 1, x_3 = 0, x_4 = 1$ is a B.F.S.

Remarks. This type of problem is very difficult one. But I adjust the problem in such a manner such that two vectors $\mathbf{a}_1$ and $\mathbf{a}_3$ eliminate in a single operation. There may have more than one B.F.S. I do not try to find other basic feasible solutions.

6.2 Fundamental theorem of Linear programming (Algebraical approach)

Theorem 6.2.1 *If a L.P.P. Optimize, $z = \mathbf{cx}$ subject to $A\mathbf{x} = \mathbf{b}$, $\mathbf{x} \geq \mathbf{0}$, where A is the $(m \times n)$ coefficient matrix $(m < n)$ and $r(A) = m$, has an optimal solution then there exists at least one B.F.S., which will be optimal.*

Proof. Case I. Problem of maximization

Let $\mathbf{x} = [x_1, x_2, \cdots, x_n]$ be an optimal feasible solution of the problem which makes the objective function a maximum. Out of $x_1, x_2, \cdots, x_n$ let k components $(1 \leq k \leq n)$ be positive and the remaining $(n - k)$ components are zero. We further make an assumption that the first k components are positive and the last $(n-k)$ components are zero. Thus the optimal solution is

$$\hat{\mathbf{x}} = \left[x_1, x_2, \cdots, x_k, \overbrace{0, \cdots, 0}^{n-k} \right] \quad \text{and} \quad \hat{z} = \widehat{\mathbf{cx}} = \sum_{\mathbf{j}=1}^{k} c_j x_j.$$

If $\mathbf{a}_1, \mathbf{a}_2, \cdots, \mathbf{a}_k$ be the column vectors associated with the variables $x_1, x_2, \cdots, x_k$ then the optimal solution will be a B.F.S. provided the vectors $\mathbf{a}_1, \mathbf{a}_2, \cdots, \mathbf{a}_k$ are L.I. which is possible only for $k \leq m$. If $\mathbf{a}_1, \mathbf{a}_2, \cdots, \mathbf{a}_k$ are not L.I., then the solution is not a B.F.S.

Now

$$\sum_{j=1}^{k} x_j \mathbf{a}_j = \mathbf{b}, \quad x_j \geq 0,\ j = 1, 2, \cdots, k \tag{6.2.1}$$

and as $\mathbf{a}_1, \mathbf{a}_2, \cdots, \mathbf{a}_k$ are not L.I. then

$$\sum_{j=1}^{k} \lambda_j \mathbf{a}_j = 0 \tag{6.2.2}$$

with at least one $\lambda_j > 0$.

Taking

$$\nu = \max_j \left(\frac{\lambda_j}{x_j} \right)$$

which is a positive quantity, the solution set

$$\hat{x}' = [x_1', x_2', \cdots, x_k', 0, \cdots, 0]$$

where $x_j' = x_j - \dfrac{\lambda_j}{\nu} \geq 0$ $[j = 1, 2, \cdots, k]$ is also a feasible solution which contains the maximum $(k-1)$ positive variables. Let $x_k' = 0$, then

$$\hat{x}' = \left[x_1', x_2', \cdots, x_{k-1}', \overbrace{0, \cdots, 0}^{x-k+1} \right].$$

The value of the objective function for the solution set $\hat{x}'$ is

$$\begin{aligned} \hat{z}_1 &= \sum_{j=1}^{k-1} c_j x_j' = \sum_{j=1}^{k} c_j x_j' \quad [\text{as } x_k' = 0] \\ &= \sum_{j=1}^{k} c_j \left(x_j - \frac{\lambda_j}{\nu} \right) = \sum_{j=1}^{k} c_j x_j - \frac{1}{\nu} \sum_{j=1}^{k} c_j \lambda_j \\ &= \hat{z} - \frac{1}{\nu} \sum_{j=1}^{k} c_j \lambda_j \end{aligned} \tag{6.2.3}$$

If $\sum_{j=1}^{k} c_j \lambda_j = 0$ [which is always the fact] then $\hat{z}_1 = \hat{z}$ and the solution set $\hat{x}'$ is also an optimal solution. If the column vectors corresponding to the variables x_1',

$x'_2, \cdots, x'_{k-1}$ are L.I. then the solution is a B.F.S. which is also an optimal solution. If the column vectors are not L.I. then repeating the process of reducing the number of non-zero variables for a finite number of times, ultimately an optimal solution is obtained which is a B.F.S. and then the theorem is proved.

Now we are to prove that $\sum_{j=1} c_j\lambda_j = 0$.

If $\sum_{j=1}^{k} c_j\lambda_j \neq 0$ then $\sum_{j=1}^{k} c_j\lambda_j > 0$ or, < 0.

Multiply $\sum_{j=1}^{k} c_j\lambda_j$ by a quantity $\delta (\neq 0)$, such that

$$\delta \sum_{j=1}^{k} c_j\lambda_j > 0. \tag{6.2.4}$$

Hence

$$\sum_{j=1}^{k} c_j x_j + \delta \sum_{j=1}^{k} c_j\lambda_j > \sum_{j=1}^{k} c_j x_j \quad \text{or,} \quad \sum_{j=1}^{k} c_j(x_j + \delta\lambda_j) > \hat{z}. \tag{6.2.5}$$

Now multiplying (6.2.2) by δ, adding it with (6.2.1) we get

$$\sum_{j=1}^{k} (x_j + \delta\lambda_j)\mathbf{a}_j = \mathbf{b} \tag{6.2.6}$$

which indicates that $x_j + \delta\lambda_j (j = 1, 2, \cdots, k)$ is a solution set of the system $A\mathbf{x} = \mathbf{b}$. If $x_j + \delta\lambda_j \geq 0 (j = 1, 2, \cdots, k)$ for particular values of δ[1] (which is always possible) then the solution set $x_j + \delta\lambda_j$ $(j = 1, 2, \cdots, k)$ is a feasible solution of $A\mathbf{x} = \mathbf{b}$. Now from (6.1.6) it is clear that the set of values $x_j + \delta\lambda_j (j = 1, 2, \cdots, k)$ makes the objective function which is greater than $\hat{z}$. Hence the assumption that for $\hat{\mathbf{x}} = [x_1, x_2, \cdots, x_k, 0, \cdots, 0]$, the objective function z attains its maximum, is not true. Hence $\Sigma c_j\lambda_j = 0$.

Case II. Problem of minimization

If for $\hat{\mathbf{x}}$, the objective function attains its minimum $\hat{z}$ then also $\hat{z}_1 = \hat{z}$ where $\hat{z}_1$ is the value of the objective function for the solution set $\hat{\mathbf{x}}'$. Proceeding similarly as in above case the theorem can be established in the problem of minimization also.

Note. From the above theorem we can conclude that if the L.P.P. has an optimal solution, then the optimum value of the objective function will be obtained,

---

[1]Value of δ is given by the relationship

$$\max_{\substack{j \\ (\lambda_j > 0)}} \left(-\frac{x_j}{\lambda_j}\right) \leq \delta \leq \min_{\substack{j \\ (\lambda_j < 0)}} \left(-\frac{x_j}{\lambda_j}\right)$$

as for $\lambda_j > 0$, and $x_j + \delta\lambda_j \geq 0$ gives $\delta \geq -\frac{x_j}{\lambda_j}$ and for $\lambda_j < 0$ and $x_j + \delta\lambda_j \geq 0$ gives $\delta \leq -\frac{x_j}{\lambda_j}$ and for $\lambda_j = 0$, δ is unrestricted.

corresponding to a B.F.S. But the above discussion fails to give correct information about the existence of the finite optimal solution.

▶ **Example 6.2.1** *Maximize,* $z = x_1 - x_2 + x_3$ *subject to*

$$\begin{aligned} x_1 - 2x_2 + x_3 \quad\quad &= 3 \\ 2x_1 + x_2 \quad\quad + x_4 &= 2, \quad x_j \geq 0, \ j = 1, 2, \cdots, 4 \end{aligned}$$

with the assumption that the optimal solution exists.

Solution: As the problem has an optimal solution then at least one B.F.S. is optimal. The B.S. are

$$\mathbf{x}_1 = \left[\frac{7}{5}, \ -\frac{4}{5}, \ 0, \ 0\right], \quad \mathbf{x}_2 = [1, \ 0, \ 2, \ 0], \quad \mathbf{x}_3 = [3, \ 0, \ 0, \ -4],$$

$$\mathbf{x}_4 = [0, \ 2, \ 7, \ 0], \quad \mathbf{x}_5 = \left[0, \ -\frac{3}{2}, \ 0, \ \frac{7}{2}\right] \quad \mathbf{x}_6 = [0, \ 0, \ 3, \ 2]$$

(B.S. are to be determined by the method given previously)

Of these, only $\mathbf{x}_2$, $\mathbf{x}_4$, and $\mathbf{x}_6$ are B.F.S.

$$\begin{aligned} z &= 1 - 0 + 2 = 3 \text{ for } \mathbf{x} = \mathbf{x}_2 \\ z &= 0 - 2 + 7 = 5 \text{ for } \mathbf{x} = \mathbf{x}_4 \\ z &= 0 - 0 + 3 = 3 \text{ for } \mathbf{x} = \mathbf{x}_6. \end{aligned}$$

Hence the optimal value of $z = 5$ for $\mathbf{x} = \mathbf{x}_4 = [0, \ 2, \ 7, \ 0]$.

▶ **Example 6.2.2** *Find the optimal value and the optimal solution or solutions of the problem,*

$$\textit{Maximize} \quad z = x_1 - x_2 + 2x_3 + 3x_4$$

subject to

$$\begin{aligned} 2x_1 + x_2 + 3x_3 + 2x_4 &= 11 \\ 3x_1 - 3x_2 + 5x_3 + x_4 &= 17, \quad x_j \geq 0, \ j = 1, 2, \cdots, 4 \end{aligned}$$

with the assumption that finite optimal solution exists.

Solution: The set of constraints are

$$\mathbf{a}_1 x_1 + \mathbf{a}_2 x_2 + \mathbf{a}_3 x_3 + \mathbf{a}_4 x_4 = \mathbf{b}, \tag{i}$$

where $\mathbf{a}_1 = \begin{bmatrix} 2 \\ 3 \end{bmatrix}$, $\mathbf{a}_2 = \begin{bmatrix} 1 \\ -3 \end{bmatrix}$, $\mathbf{a}_3 = \begin{bmatrix} 3 \\ 5 \end{bmatrix}$, $\mathbf{a}_4 = \begin{bmatrix} 2 \\ 1 \end{bmatrix}$ and $\mathbf{b} = \begin{bmatrix} 11 \\ 17 \end{bmatrix}$.

Six square matrices taking two at a time from the vectors $\mathbf{a}_1, \cdots, \mathbf{a}_4$, etc, are

$$B_1 = (\mathbf{a}_1, \mathbf{a}_2) = \begin{bmatrix} 2 & 1 \\ 3 & -3 \end{bmatrix}, \quad B_2 = (\mathbf{a}_1, \mathbf{a}_3) = \begin{bmatrix} 2 & 3 \\ 3 & 5 \end{bmatrix},$$

$$B_3 = (\mathbf{a}_1, \mathbf{a}_4) = \begin{bmatrix} 2 & 2 \\ 3 & 1 \end{bmatrix}, \quad B_4 = (\mathbf{a}_2, \mathbf{a}_3) = \begin{bmatrix} 1 & 3 \\ -3 & 5 \end{bmatrix},$$

$$B_5 = (\mathbf{a}_2, \mathbf{a}_4) = \begin{bmatrix} 1 & 2 \\ -3 & 1 \end{bmatrix}, \quad B_6 = (\mathbf{a}_3, \mathbf{a}_4) = \begin{bmatrix} 3 & 2 \\ 5 & 1 \end{bmatrix},$$

$$\det B_1 = -9 \neq 0, \quad \det B_2 = 1 \neq 0, \quad \det B_3 = -4 \neq 0,$$

$$\det B_4 = 14 \neq 0, \quad \det B_5 = 7 \neq 0, \quad \det B_6 = -7 \neq 0.$$

The six square sub-matrices are non-singular and therefore all the matrices are basis matrices and there exist six basic solutions.

For B_1, B.S. is

$$\mathbf{x}_1 = B_1^{-1}\mathbf{b} = -\frac{1}{9}\begin{bmatrix} -3 & -1 \\ -3 & 2 \end{bmatrix}\begin{bmatrix} 11 \\ 17 \end{bmatrix}\begin{bmatrix} 50/9 \\ -1/9 \end{bmatrix}.$$

Thus

$$\mathbf{x}_1 = \left[\frac{50}{9}, \ -\frac{1}{9}, \ 0, \ 0\right]$$

Other basic solutions which are to be calculated in the same manner given below are

$$\mathbf{x}_2 = [4, \ 0, \ 1, \ 0], \quad \mathbf{x}_3 = \left[\frac{23}{4}, \ 0, \ 0, \ -\frac{1}{4}\right], \quad \mathbf{x}_4 = \left[0, \ \frac{2}{7}, \ \frac{25}{7}, \ 0\right],$$

$$\mathbf{x}_5 = \left[0, \ -\frac{23}{7}, \ 0, \ \frac{50}{7}\right] \quad \text{and} \quad \mathbf{x}_6 = \left[0, \ 0, \ \frac{23}{7}, \ \frac{4}{7}\right].$$

Of all B.S. only $\mathbf{x}_2$, $\mathbf{x}_4$ and $\mathbf{x}_6$ are feasible.

The value of z for $\mathbf{x}_2$ is $1 \times 4 - 1 \times 0 + 2 \times 1 + 3 \times 0 = 6$;

for $\mathbf{x}_4$ is $1 \times 0 - 1 \times \frac{2}{7} + 2 \times \frac{25}{7} + 3 \times 0 = \frac{48}{7}$;

for $\mathbf{x}_6$ is $1 \times 0 - 1 \times 0 + 2 \times \frac{23}{7} + 3 \times \frac{4}{7} = \frac{58}{7}$.

Thus $\max z = \frac{58}{7}$ for $x_1 = 0$, $x_2 = 0$, $x_3 = \frac{23}{7}$, $x_4 = \frac{4}{7}$, i.e., the optimal value is $\frac{58}{7}$ and the optimal solution is $[0, \ 0, \ \frac{23}{7}, \ \frac{4}{7}]$.

N.B. The problem has been solved in Example (8.5.1) in the simplex theory, where it has been shown that the problem has finite optimal value and no assumption regarding the existence of finite optimal solution is needed there.

▶ **Example 6.2.3** *Can you solve the following L.P.P.,*

Maximise, $z = 5x_1 + x_2 - 2x_3 + x_4$ *subject to*

$$\begin{aligned} x_1 + 5x_2 - 8x_3 + 3x_4 &= 6 \\ 3x_1 - x_2 + x_3 + x_4 &= 2, \quad x_j \geq 0, j = 1, \cdots, 4 \end{aligned}$$

by merely finding the basic feasible solutions of the problem? If so, then solve the problem; if not, give the proper reasons.

Solution. We cannot find the optimal value of z until and unless we know that the problem has the finite value of z. Here the problem is said to have an unbounded solution. On the other hand this problem has four B.F.S.

$$\mathbf{x}_1 = [1,1,0,0],\ \mathbf{x}_3 = [x_1 = 0 \text{ (basic)},\ x_2 = x_3 = 0 \text{ (non basic)},\ x_4 = 2]$$
$$\mathbf{x}_5 = [x_1 = 0 = x_3 \text{ (non basic)},\ x_2 = 0 \text{ (basic)}, x_4 = 2]$$
$$\mathbf{x}_6 = [x_1 = 0 = x_2 \text{ (non basic)}, x_3 = 0 \text{ (basic)}, x_4 = 2]$$

If we try to find the optimal value from the B.F.S. we get $\max z = 6$ at $x_1 = x_2 = 1$, $x_3 = 0$, $x_4 = 0$ which leads to us to a wrong conclusion and thus we cannot solve such type of problems until the boundedness of the optimal z be assumed clearly.

Exercise 6

1. The L.P.P., Minimize $z = 2x_1 - 3x_2 + x_4$

$$\text{subject to}\quad \begin{aligned} 3x_1 + 2x_2 + x_3 \qquad &= 15 \\ 2x_1 + 4x_2 \qquad + x_4 &= 8, \quad x_j \ge 0,\ j = 1,2,3,4. \end{aligned}$$

has an optimal solution. Using the property of the fundamental theroem of L.P. find the minimum value of z and the corresponding solution set.

2. (a) The L.P.P. Maximize $z = -x_1 + 4x_2 + x_3$

$$\text{subject to}\quad \begin{aligned} x_1 + 3x_2 + x_3 \qquad &= 6 \\ -3x_1 + 4x_2 \qquad + x_4 &= 4, \quad x_j \ge 0,\ j = 1,2,3,4 \end{aligned}$$

has an optimal solution. Find the maximim value of the objective function after finding all B.F.S.

(b) With the assumption that the finite value of the objective function exists, find out the optimal value of the problem, maximize, $z = x_1 + 5x_2 + x_3$

$$\text{subject to}\quad \begin{aligned} x_1 + x_2 + x_3 &= 9 \\ 2x_1 - 4x_2 + 3x_3 &= 4, \quad x_1, x_2, x_3 \ge 0 \end{aligned}$$

3. (a) The L.P.P. Maximize $z = 2x_1 - x_2 + 5x_3$

$$\text{subject to}\quad \begin{aligned} x_1 + 2x_2 + 2x_3 \qquad &= 16 \\ 4x_1 + 7x_2 \qquad + x_4 &= 6, \quad x_j \ge 0,\ j = 1,2,3,4 \end{aligned}$$

has an optimal solution. Find the maximum value of the objective function from the Basic feasible solutions.

(b) Maximize $6x_1 - 4x_2 + 4x_3 + x_4$

$$\text{subject to}\quad \begin{aligned} 2x_1 - x_2 + 3x_3 + x_4 &= 6 \\ 4x_1 - 2x_2 - x_3 + 2x_4 &= 10, \quad x_j \ge 0,\ j = 1,2,3,4 \end{aligned}$$

[Assume the existence of an optimal solution.]

4. $x_1 = 2$, $x_2 = 3$, $x_3 = 1$ is a feasible solution of the equations

$$\begin{aligned} 2x_1 + x_2 + 4x_3 &= 11 \\ 3x_1 + x_2 + 5x_3 &= 14 \end{aligned}$$

Reduce the F.S. to a B.F.S. **[C.U.(H)'90,'92,'95, '99] [V.U.(P)'91]**

5. $x_1 = 1$, $x_2 = 3$, $x_3 = 2$ is a feasible solution of the equations

$$\begin{aligned} 5x_1 + 4x_2 - x_3 &= 15 \\ 2x_1 + 7x_2 - 4x_3 &= 15 \end{aligned}$$

Reduce the F.S. to B.F.S. **[V.B.U.(H)'85; C.U.(H)'87]**

6. (a) Prove that $x_1 = 2$, $x_2 = 3$, $x_3 = 0$ is a feasible solution but not a basic feasible of the set of equations

$$\begin{aligned} 3x_1 + 5x_2 - 7x_3 &= 21 \\ 6x_1 + 10x_2 + 3x_3 &= 42 \end{aligned}$$

Reduce the feasible solution to a basic feasible solution and show that the solution is degenerate. **[C.U.(P)'89]**

(b) $x_1 = 2$, $x_2 = 4$, $x_3 = 1$ is a feasible solution to the set of equations

$$\begin{aligned} 2x_1 - x_2 + 2x_3 &= 2 \\ x_1 + 4x_2 \qquad &= 18 \end{aligned}$$

Reduce the F.S. to a B.F.S. one. **[C.U.(P)'88; V.U.(P)'91]**

7. $x_1 = 1$, $x_2 = 0$, $x_3 = 2$, $x_4 = 1$ is a feasible solution to the set of equations

$$\begin{aligned} 2x_1 + 3x_2 + 3x_3 - x_4 &= 7 \\ x_1 + 5x_2 + 2x_3 + x_4 &= 6 \end{aligned}$$

Reduce the feasible solution to one or more basic feasible solutions.

8. $x_1 = 3$, $x_2 = 1$, $x_3 = 1$ and $x_4 = 2$ is a feasible solution of the set of equations

$$\begin{aligned} x_1 + x_2 + x_3 + x_4 &= 7 \\ 2x_1 + x_2 + 2x_3 + x_4 &= 11 \\ x_2 + x_3 + 2x_4 &= 6 \end{aligned}$$

Reduce the feasible solution to one or more basic feasible solutions.

9. Prove that $x_1 = 4$, $x_2 = 1$, $x_3 = 3$ is a feasible solution to the set of equations (linearly independent).

$$\begin{aligned} 2x_1 - 3x_2 + x_3 &= 8 \\ x_1 + 2x_2 + 3x_3 &= 15 \end{aligned}$$

Is the solution set, a basic feasible solution? If not, reduce the feasible solution to one or more basic feasible solutions.

10. $x_1 = 3$, $x_2 = 1$, $x_3 = 0$, $x_4 = 3$ is a feasible solution to the set of equation (linearly independent)

$$\begin{aligned} 4x_1 + x_2 - x_3 + x_4 &= 16 \\ x_1 + 2x_2 + 3x_3 + 2x_4 &= 11 \end{aligned}$$

Find all possible B.F.S. by using reduction theorem.

11. Prove that $x_1 = 1, x_2 = 1, x_3 = 3$ is a basic feasible solution of the following linearly independent set of equations

$$\begin{aligned} 2x_1 + 3x_2 + x_3 &= 8 \\ 5x_1 - 2x_2 &= 3 \\ 4x_1 - x_3 &= 1 \end{aligned}$$

12. $x_1 = 4$, $x_2 = -1$, $x_3 = 2$ is a solution set of following L.I. equations

$$\begin{aligned} 4x_1 + 9x_2 + x_3 &= 9 \\ x_1 + x_2 + x_3 &= 5 \end{aligned}$$

Can we reduce the solution to a basic feasible solution? Give reasons.

13. Prove that $x_1 = 2$, $x_2 = 1$ and $x_3 = 3$ is a feasible solution of the set of equations

$$\begin{aligned} 4x_1 + 2x_2 - 3x_3 &= 1 \\ -6x_1 - 4x_2 + 5x_3 &= -1 \end{aligned}$$

Reduce the F.S. to a B.F.S. by reduction theory.

[V.B.U.(H)'83; C.U.(H)'87, 2000; C.U.(P)'95]

14. For the system of equations

$$\begin{aligned} x_1 + x_2 + 2x_3 &= 9 \\ 3x_1 + 2x_2 + 5x_3 &= 22 \end{aligned}$$

show that the feasible solution $(2, 3, 2)$ is not basic.

Reduce the feasible solution to one or more basic feasible solutions by using reduction theory.

15. Solve the L.P.P. by direct method [Optimal solution exists]

Maximize, $z = -x_1 - x_2 - x_3$

$$\text{subject to} \quad \begin{aligned} x_1 - x_2 + 2x_3 &= 2 \\ -x_1 + 2x_2 - x_3 &= 1 \end{aligned} \quad x_1 \geq 0, x_2 \geq 0, x_3 \geq 0.$$

16. Prove that $x_1 = 2$, $x_2 = 1$. $x_3 = 0$ is a F.S. to the following set of equations

$$\begin{aligned} 2x_1 - x_2 + 3x_3 &= 3 \\ -6x_1 + 3x_2 + 7x_3 &= -9 \end{aligned}$$

Is the solution basic? Justify your answer. If the solution be not basic, reduce the F.S. to B.F.S. one.

Answers

1. $\min z = -6$ at $[0, 2, 11, 0]$. **2.** (a) $\max z = 7$ at $[0, 1, 3, 0]$; (b) $\max z = 155/7$ at $[0, 23/7, 40/7]$. **3.** (a) $\max z = 40$ at $[0, 0, 8, 6]$; (b) $\max z = 116/7$ at $[18/7, 0, 2/7, 0]$. **4.** $[3, 5, 0]$, $[1/2, 0, 5/2]$. **5.** $[0, 5, 5]$, $[5/3, 5/3, 0]$. **6.** (a) $[x_1 = 7, x_3 = 0, x_2 = 0$, (non-basic)]; $[x_2 = 4\frac{1}{5}, x_3 = 0, x_1 = 0$ (non-basic)]; (b) $[26/9, 34/9, 0]$, $[0, 9/2, 13/4]$. **7.** $[0, 0, 13/5, 4/5]$, $[13/3, 0, 0, 5/3]$. **8.** $[1, 3, 3, 0]$, $[4, 0, 0, 3]$. **9.** $[9/5, 0, 22/5]$, $[61/7, 22/7, 0]$. **10.** $[3, 0, 0, 4]$, $[3, 4, 0, 0,]$. **12.** No, we can only reduce a feasible solution to a basic feasible solution. **13.** $[1, 0, 1]$. **14.** $[4, 5, 0]$, $[0, 1, 4]$. **15.** $\max z = -3$ at $[0, 4/3, 5/3]$. **16.** No. $[x_1 = 3/2$ (basic), $x_2 = 0$, $x_3 = 0$ (basic)].

Chapter 7

Graphical or Geometrical Method of Solving a L.P.P.

7.1 Introduction

We know from the fundamental theorem of L.P.P. (algebrical approach) that if a L.P.P. admits an optimal solution then at least one of the B.F.S. will be an optimal solution. Again we know that there is one to one correspondence between the basic feasible solutions and extreme points of the convex set of feasible solutions (In the absence of degenaracy). There are two types of convex set of feasible solutions (i) convex polyhedron, which is strictly bounded and has finite number of extreme points (ii) Convex polytope, has finite number of extreme points but not bounded from above. In the case of convex polyhedron, each and every problem has both finite maximum and finite minimum. But this is not true for a convex polytope. Here all objective functions may have finite maximum or finite minimum but not both; there is some problem which has neither finite maximum nor finite minimum and the problem has said to have unbounded solution. We are mainly concerned with the convex polyhedron where the optimum value will be determined from the extreme points only. We are mainly interested with such problems though there are some problems where the convex set is a convex polytope. I have discussed such problems very carefully and scientifically.

► **Example 7.1.1** *Find the convex set of feasible solutions, if any, of the following problems and comment on the nature of the convex sets:*

The constraints are

(a)
$$\begin{aligned} x_1 + x_2 &\geq 2 \\ 2x_1 + 3x_2 &\leq 6 \\ x_1 - x_2 &\leq 2, \quad x_1, x_2 \geq 0. \end{aligned}$$

(b)
$$\begin{aligned} x_1 - x_2 &\leq 1 \\ x_1 + x_2 &\geq 4 \\ x_1 - 3x_2 &\leq 3, \quad x_1, x_2 \geq 0. \end{aligned}$$

(c)
$$\begin{aligned} 2x_1 + 3x_2 &\leq 30 \\ 4x_1 + 3x_2 &\geq 60, \quad x_1, x_2 \geq 0. \end{aligned}$$

(d)
$$\begin{aligned} -x_1 + x_2 &\geq 3 \\ x_1 - x_2 &\geq 2, \quad x_1, x_2 \geq 0. \end{aligned}$$

Solution: (a) AB is the straight line represented by $x_1 + x_2 = 2$. To determine the region $x_1 + x_2 \geq 2$, take any point not on the line AB, say $(0, 0)$ and for $(0, 0)$, the value of the expression $x_1 + x_2$ is 0 (zero) which is less than or equal to 2, which does not satisfy the constraint. Thus the opposite side of the region, which contains the point $(0, 0)$, will be the region, determined by $x_1 + x_2 \geq 2$. In the same manner other two regions defined by the constraints $2x_1 + 3x_2 \leq 6$, $x_1 - x_2 \leq 2$ have been determined. $x_1 \geq 0$, $x_2 \geq 0$, absolutely represents the first quadrant. Now the intersection of the regions is the convex set of feasible solutions which is a strictly bounded convex set ABC with three extreme points $A(2, 0)$, $B(0, 2)$ and $C(12/5, 2/5)$ which are finite in number [The co-ordinates of the point C is the intersection of the line $2x_1 + 3x_2 = 6$, $x_1 - x_2 = 2$]. Hence the convex set of F.S. is a convex polyhedron which can be generated by the convex combination of three extreme points $(2, 0), (0, 2)$ and $(12/5, 2/5)$, i.e., any point on the region ABC can be expressed as the convex combination of the extreme points.

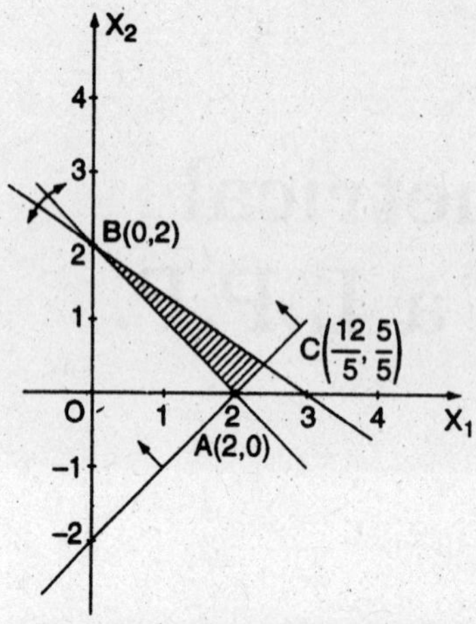

Fig. 7.1

(b) Here the convex set of feasible solutions is the shaded region of the first quadrant which is bounded from below only. It has only two extreme points $C(5/2, 3/2)$ and $D(0, 4)$. Here the convex set of F.S. is not a convex polyhedron as it is not possible to express all points of the feasible set as the convex combination of the extreme points $(5/2, 3/2)$ and $(0, 4)$. For example, $(4, 4)$ is a point belonging to the F.S. set. But the point cannot be expressed as the convex combination of the points $(5/2, 3/2)$ and $(0, 4)$. Such type of convex set of F.S. is called a convex polytope.

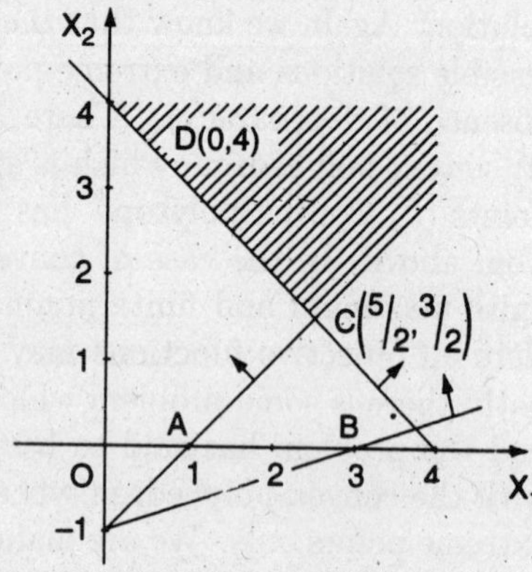

Fig. 7.2

Remark: Here the constraint $x_1 - 3x_2 \leq 3$ is redundant as the presence or absence of this constraint does not change the nature of the convex set of F.S.

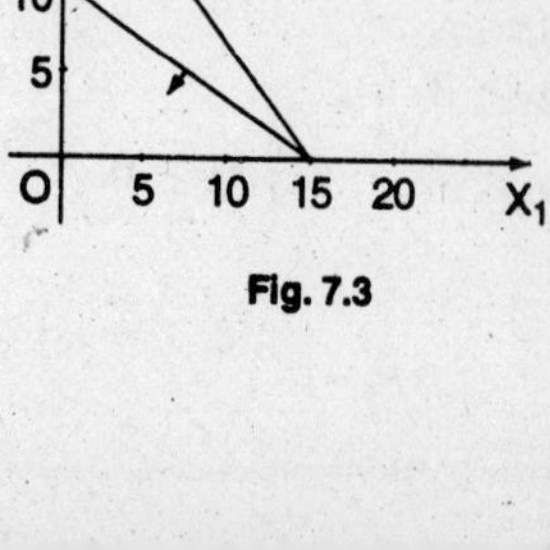

Fig. 7.3

(c) Here the convex set of F.S. is strictly a point $A(15, 0)$. This is the only extreme point of the set, as this point cannot be expressed as the convex combination of any other two points belonging to the set and this question does not arise at all as there are no other points belonging to the set. This set is a convex polyhedron with only one extreme point.

(d) Here the convex set of F.S. is a null set, i.e., there is no feasible solution set defined by the constraints.

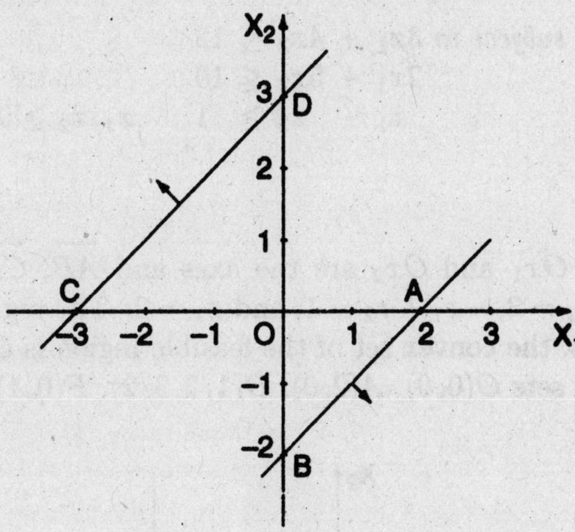

Fig. 7.4

▶ **Example 7.1.2** *Draw graphically the feasible space, if any, given by the following L.P.P. and find out the extreme points of the feasible region:*

(a) *Maximize,* $z = 2x_1 + x_2$

$$\text{subject to} \quad x_1 + x_2 \leq 2$$
$$-x_1 + x_2 \leq 1$$
$$x_1 \leq 2, \quad x_1, x_2 \geq 0.$$

[C.U.(P)'89]

(b) *Maximize,* $z = 9x_1 + 7x_2$

$$\text{subject to} \quad x_1 + 2x_2 \leq 7$$
$$x_1 - x_2 \leq 4, \quad x_1, x_2 \geq 0.$$

[C.U.(P)'87,'92]

(c) *Maximize,* $z = 2x_1 + 5x_2$

$$\text{subject to} \quad 5x_1 + 6x_2 \geq 30$$
$$3x_1 + 2x_2 \leq 21$$
$$x_1 + x_2 \leq 12, \quad x_1, x_2 \geq 0.$$

(d) *Minimize,* $z = 4x_1 + 5x_2$

$$\text{subject to} \quad x_1 + 2x_2 \geq 2$$
$$5x_1 + 3x_2 \leq 15, \quad x_1, x_2 \geq 0.$$

(e) *Maximize,* $3x_1 + 5x_2$

$$\text{subject to} \quad x_1 + x_2 \leq 5$$
$$x_1 - x_2 \leq 4, \quad x_1, x_2 \geq 0.$$

(f) *Minimize,* $z = 3x_1 - x_2$

$$\text{subject to} \quad x_1 + x_2 \leq 2$$
$$2x_1 + 3x_2 \geq 6, \quad x_1, x_2 \geq 0.$$

[C.U.(P)'90]

(g) *Maximize,* $4x_1 + 3x_2$

$$\begin{aligned} \textit{subject to } 3x_1 + 4x_2 &\leq 12 \\ 2x_1 + 5x_2 &\leq 10 \\ x_1 + x_2 &\geq 1, \quad x_1, x_2 \geq 0. \end{aligned}$$

Solution: (a) Ox_1 and Ox_2 are the axes and $\overleftrightarrow{AB}$, $\overleftrightarrow{CD}$, $\overleftrightarrow{AE}$ represent the straight lines $x_1 + x_2 = 2$, $-x_1 + x_2 = 1$, and $x_1 = 2$. The region of the inequations are shown by arrows, the convex set of the feasible region is $OADFO$, the extreme points of the convex sets $O(0,0)$, $A(2,0)$, $D(1/2, 3/2)$, $F(0,1)$.

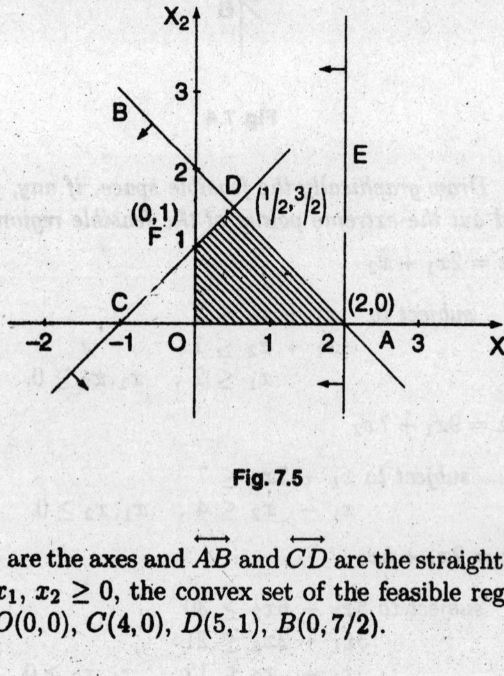

Fig. 7.5

(b) Ox_1 and Ox_2 are the axes and $\overleftrightarrow{AB}$ and $\overleftrightarrow{CD}$ are the straight lines $x_1 + 2x_2 = 7$, $x_1 - x_2 = 4$, since x_1, $x_2 \geq 0$, the convex set of the feasible region $OCDBO$; the extreme points are $O(0,0)$, $C(4,0)$, $D(5,1)$, $B(0, 7/2)$.

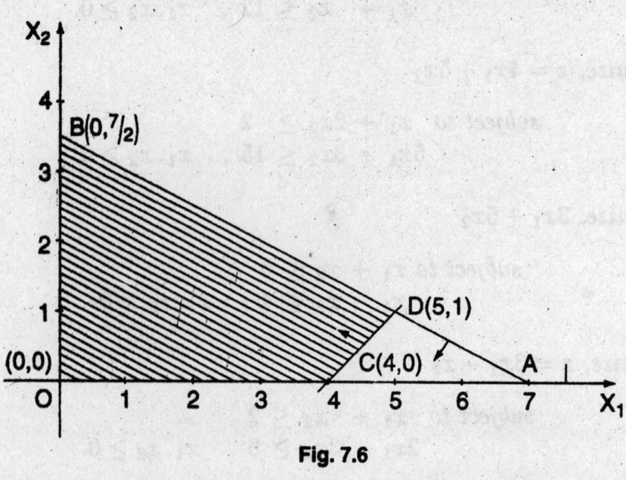

Fig. 7.6

(c) Ox_1 and Ox_2 are the axes and $\overleftrightarrow{AB}$, $\overleftrightarrow{CD}$ and $\overleftrightarrow{EF}$ are the straight lines $5x_1 + 6x_2 = 30$, $3x_1 + 2x_2 = 21$ and $x_1 + x_2 = 12$; since $x_1 \geq 0$, $x_2 \geq 0$ the convex set of the feasible region is $ACDBA$ and the extreme points are, $A(6,0), C(7,0), D(0,21/2), B(0,5)$. Here the constraint $x_1 + x_2 \leq 12$ is redundant.

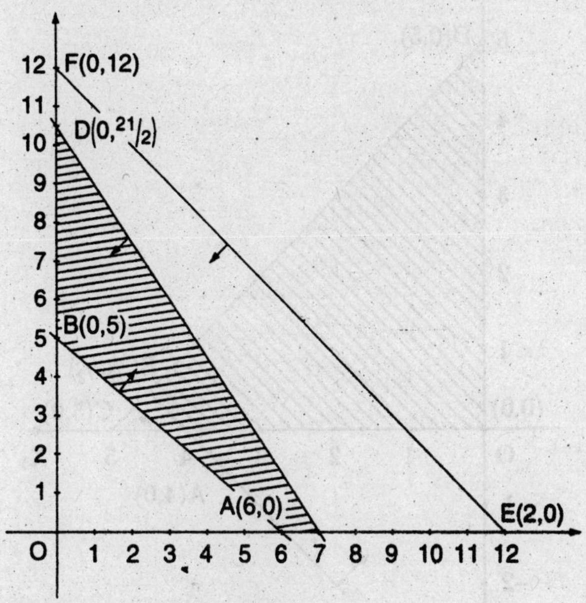

Fig. 7.7

(d) Ox_1 and Ox_2 be the axes and $\overleftrightarrow{AB}$, $\overleftrightarrow{CD}$ are the equations of the straight lines $x_1 + 2x_2 = 2$ and $5x_1 + 3x_2 = 15$ respectively. Since x_1 and $x_2 \geq 0$, the convex set of feasible region is $ACDB$ and the extreme points one $A(2,0), C(3,0), D(0,5), B(0,1)$.

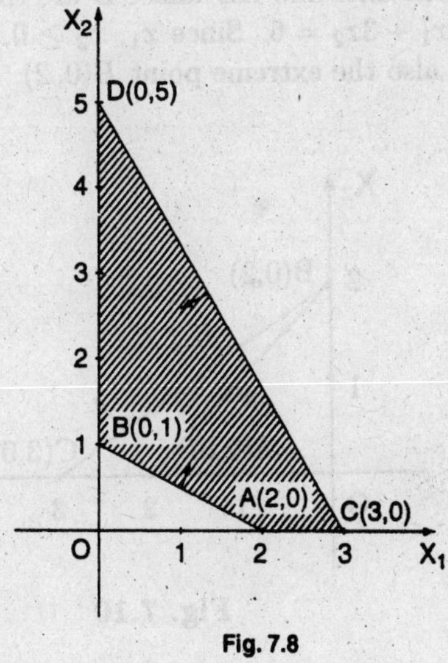

Fig. 7.8

(e) Ox_1 and Ox_2 be the axes and $\overleftrightarrow{AB}$ and $\overleftrightarrow{CD}$ are the straight lines $x_1 - x_2 = 4$, and $x_1 + x_2 = 5$ respectively. Since x_1 and $x_2 \geq 0$ the convex set of the feasible region is $OAEDQ$, and the extreme points are $A(4,0)$, $E(9/2, 1/2)$, $D(0,5)$, $O(0,0)$.

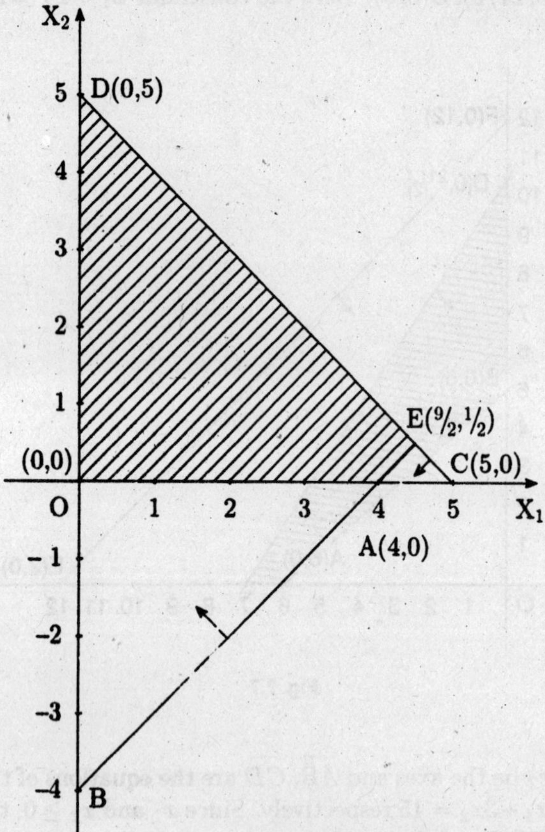

Fig. 7.9

(f) Ox_1 and Ox_2 be the axes and $\overleftrightarrow{AB}$ and $\overleftrightarrow{CB}$ are the equations of the straight lines $x_1 + x_2 = 2$ and $2x_1 + 3x_2 = 6$. Since x_1, $x_2 \geq 0$, the only feasible region is the point $(0, 2)$ which is also the extreme point $B(0, 2)$.

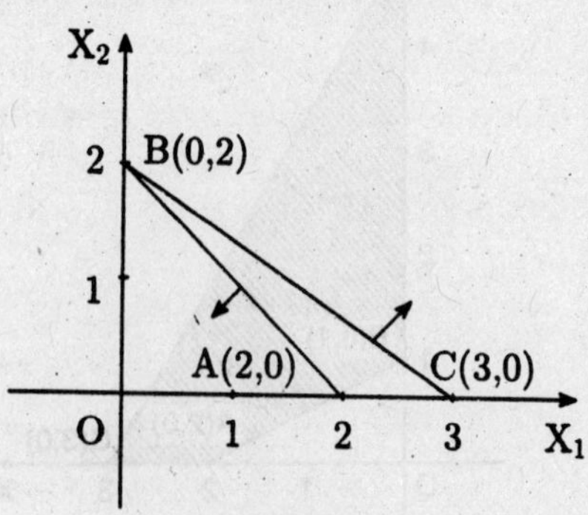

Fig. 7.10

(g) Ox_1 and Ox_2 be the axes and $\overleftrightarrow{AB}$, $\overleftrightarrow{CD}$ and $\overleftrightarrow{EF}$ are the equations of the straight lines $3x_1 + 4x_2 = 12$, $2x_1 + 5x_2 = 10$ and $x_1 + x_2 = 1$, since $x_1, x_2 \geq 0$, the convex set of the feasible region is $EAKDFE$ and the extreme points are $E(1,0), A(4,0), K(20/7, 6/7), D(0,2), F(0,1)$.

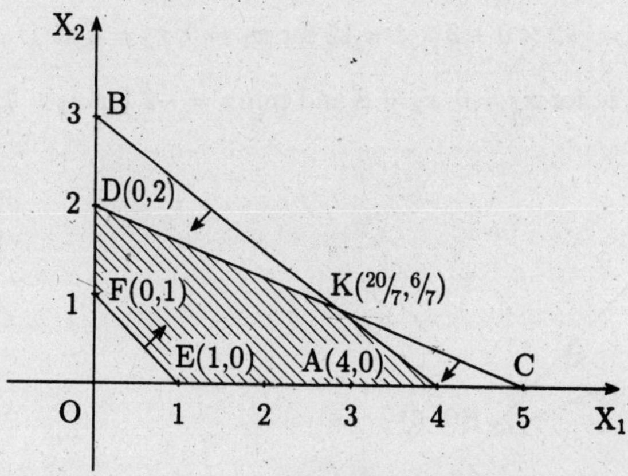

Fig. 7.11

▶ **Example 7.1.3** *Solve geometrically the L.P.P.*

$$\text{Maximize,} \quad z = 4x_1 + 7x_2$$

subject to

$$\begin{aligned} 2x_1 + 5x_2 &\leq 40 \\ x_1 + x_2 &\leq 11 \\ x_2 &\geq 4, \quad x_1 \geq 0, x_2 \geq 0. \end{aligned}$$ **[C.U.(P)'97 (H) 2000]**

Solution: As $x_2 \geq 4$, then the condition $x_2 \geq 0$ is redundant.

Let $\overleftrightarrow{AB}$, $\overleftrightarrow{BC}$, $\overleftrightarrow{DA}$ and $\overleftrightarrow{DC}$ denote the straight lines given by the equations $x_1 + x_2 = 11$, $2x_1 + 5x_2 = 40$, $x_2 = 4$ and $x_1 = 0$ respectively.

The convex set of feasible solutions of L.P.P. is the convex region $ABCD$. This region is also known as an admissible region. It is a strictly bounded region. The four extreme points are $A(7,4)$, $B(5,6)$, $C(0,8)$ and $D(0,4)$. The region is a convex polyhedron with finite number of extreme points.

The value of the objective function z is

$$\begin{aligned} z_1 &= 4 \times 7 + 7 \times 4 = 56 \text{ for } x_1 = 7, x_2 = 4 \text{ at } A \\ z_2 &= 4 \times 5 + 7 \times 6 = 62 \text{ for } x_1 = 5, x_2 = 6 \text{ at } B \\ z_3 &= 4 \times 0 + 7 \times 8 = 56 \text{ for } x_1 = 0, x_2 = 8 \text{ at } C \\ z_4 &= 4 \times 0 + 7 \times 4 = 28 \text{ for } x_1 = 0, x_2 = 4 \text{ at } D \end{aligned}$$

Hence the maximum value of $z = 62$ at $x_1 = 5, x_2 = 6$.

Note: (1) With the same constraints, the optimum value of the different objective functions can be determined. For example, if the objective function is $z = -2x_1 + 3x_2$ then

$$\begin{aligned} z_1 &= -2 \times 7 + 3 \times 4 = -2 \text{ for } x_1 = 7, x_2 = 4 \text{ at } A \\ z_2 &= -2 \times 5 + 3 \times 6 = 8 \text{ for } x_1 = 5, x_2 = 6 \text{ at } B \\ z_3 &= -2 \times 0 + 3 \times 8 = 24 \text{ for } x_1 = 0, x_2 = 8 \text{ at } C \\ z_4 &= -2 \times 0 + 3 \times 4 = 12 \text{ for } x_1 = 0, x_2 = 4 \text{ at } D \end{aligned}$$

Hence $\max z = 24$ for $x_1 = 0, x_2 = 8$ and $\min z = -2$ for $x_1 = 7, x_2 = 4$

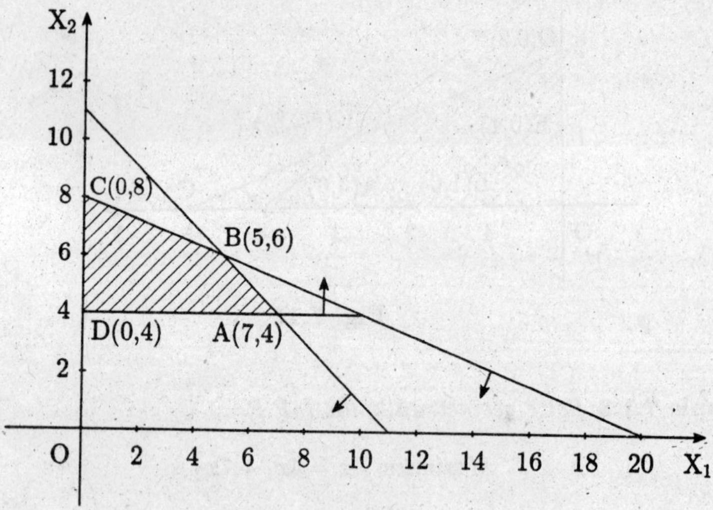

Fig. 7.12

(ii) Objective function $z = -x_1 - 4x_2$. Then

$$\begin{aligned} z_1 &= -1 \times 7 - 4 \times 4 = -23 \text{ for } x_1 = 7, x_2 = 4 \text{ at } A \\ z_2 &= -1 \times 5 - 4 \times 6 = -29 \text{ for } x_1 = 5, x_2 = 6 \text{ at } B \\ z_3 &= -1 \times 0 - 4 \times 8 = -32 \text{ for } x_1 = 0, x_2 = 8 \text{ at } C \\ z_4 &= -1 \times 0 - 4 \times 4 = -16 \text{ for } x_1 = 0, x_2 = 4 \text{ at } D \end{aligned}$$

Hence $\max z = -16$ for $x_1 = 0, x_2 = 4$ and $\min z = -32$ for $x_1 = 0, x_2 = 8$.

(iii) Objective function $z = 2x_1 + 5x_2$. Then

$$\begin{aligned} z_1 &= 34 \text{ at } x_1 = 7, x_2 = 4; \\ z_2 &= 40 \text{ at } x_1 = 5, x_2 = 6; \\ z_3 &= 40 \text{ at } x_1 = 0, x_2 = 8; \\ z_4 &= 20 \text{ at } x_1 = 4, x_2 = 0. \end{aligned}$$

Thus $\max z = 40$ at $x_1 = 5$, $x_2 = 6$ and $x_1 = 0$, $x_2 = 8$. Hence alternative optimal solutions exist.

2. Here the convex set of feasible solutions is strictly bounded, i.e., convex polyhedron. That is why any objective function has both finite maximum and minimum.

▶ **Example 7.1.4** *Solve graphically the L.P.P.*

$$\text{Maximize,} \quad z = 5x_1 - 2x_2$$

subject to

$$5x_1 + 6x_2 \geq 30$$
$$9x_1 - 2x_2 = 72$$
$$x_2 \leq 9, \quad x_1, x_2 \geq 0.$$

Solution: Let $\overleftrightarrow{AD}$, $\overleftrightarrow{CD}$ and $\overleftrightarrow{DE}$ denote the straight lines given by the equations, $5x_1 + 6x_2 = 30$, $x_2 = 9$ and $9x_1 - 2x_2 = 72$ respectively.

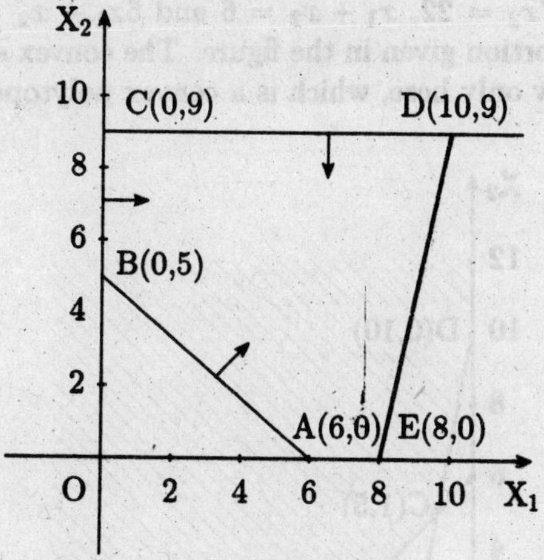

Fig. 7.13

The line AB, and the right hand portion of it, is the region represented by $5x_1 + 6x_2 \geq 30$; $x_1 \geq 0$, is the region RHS of the line BC (including the line)

x_1 axis and upward portion of it, is the region, $x_2 \geq 0$.

CD line and the downward portion of it is the region $x_2 \leq 9$. **And** DE **line is the line represented by the equation** $9x_1 - 2x_2 = 72$. Hence the **convex region** satisfied by all the constraints, is the segment DE of the line DE. Therefore, the extreme points of the convex region are only $D(10, 9)$ and $E(8, 0)$. Then

$$z_1 = 5 \times 10 - 2 \times 9 = 32 \text{ for } x_1 = 10, x_2 = 9 \text{ at } D$$
$$z_2 = 5 \times 8 - 2 \times 0 = 40 \text{ for } x_1 = 8, x_2 = 0 \text{ at } E$$

Hence max $z = z_2 = 40$ for $x_1 = 8$, $x_2 = 0$ at E.

Note: If the objective function $z = -x_1 + x_2$ then

$$z_1 = -10 + 9 = -1 \text{ for } x_1 = 10, x_2 = 9 \text{ at } D$$
$$z_2 = -8 + 0 = -8 \text{ for } x_1 = 8, x_2 = 0 \text{ at } E$$

Hence max $z = z_1 = -1$ for $x_1 = 10, x_2 = 9$ at D.

Note. The constraint $5x_1 + 6x_2 \geq 30$ has no contribution in determining the convex set of feasible solutions. This constraint is known as a **redundant constraint.**

▶ **Example 7.1.5** *Solve graphically the L.P.P.*

$$\text{Minimize,} \quad z = 2x_1 + 3x_2$$

subject to

$$\begin{aligned} 2x_1 + 7x_2 &\geq 22 \\ x_1 + x_2 &\geq 6 \\ 5x_1 + x_2 &\geq 10, \quad x_1, x_2 \geq 0. \end{aligned}$$

[C.U.(P)'94;'99]

Solution: Let $\overleftrightarrow{AB}$, $\overleftrightarrow{BC}$, $\overleftrightarrow{CD}$ denote respectively the straight lines given by the equations $2x_1 + 7x_2 = 22$, $x_1 + x_2 = 6$ and $5x_1 + x_2 = 10$. The admissible region is the dotted portion given in the figure. The convex set of feasible solutions is bounded from below only here, which is a convex polytope.

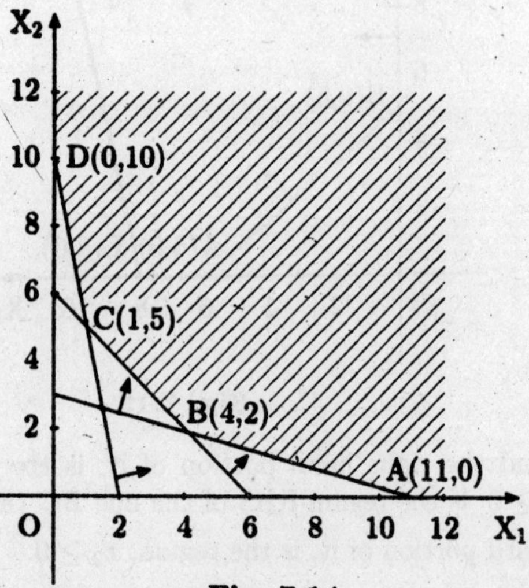

Fig. 7.14

The four extreme points are $A(11, 0), B(4, 2), C(1, 5)$ and $D(0, 10)$. Then

$$\begin{aligned} z_1 &= 2 \times 11 + 3 \times 0 = 22 \text{ for } x_1 = 11, x_2 = 0 \text{ at } A \\ z_2 &= 2 \times 4 + 3 \times 2 = 14 \text{ for } x_1 = 4, x_2 = 2 \text{ at } B \\ z_3 &= 2 \times 1 + 3 \times 5 = 17 \text{ for } x_1 = 1, x_2 = 5 \text{ at } C \\ z_4 &= 2 \times 0 + 3 \times 10 = 30 \text{ for } x_1 = 0, x_2 = 10 \text{ at } D \end{aligned}$$

Hence $\min(z_1, z_2, z_3, z_4) = \min(22, 14, 17, 30) = 14$ for $x_1 = 4, x_2 = 2$ at B.

To verify that the minimum is finite take any two points $(4.7, 1.8)$ and $(3.9, 2.1)$ very close to the extreme point B [extreme point at which the objective function attains its lowest] on the line segments BA and BC respectively and the value of the objective function corresponding to the points $(4.7, 1.8)$ and $(3.9, 2.1)$ are $2 \times 4.7 + 3 \times 1.8 = 14.8$ and $2 \times 3.9 + 3 \times 2.1 = 14.1$ both of which are greater than the lowest value 14 at B. Thus the objective function attains its minimum value 14 at $B(x_1 = 4, x_2 = 2)$.

Note. (a) If $z = 2x_1 + 3x_2$ and the problem is to be maximized then no finite value of z will be obtained. Here $\max(z_1, z_2, z_3, z_4) = 30$ at $D(0, 10)$. But for points on Dx_2, eg, at $(0, 11)$ the value of the objective function is 33 which is greater than 30 and in the same way it can be established that the objective function has no finite maximum. In that case, the problem is said to have an **unbounded solution.**

(b) Any objective function $lx_1 - mx_2$ (both l, m +ve or both −ve) has no finite maximum or minimum subject to the constraints given in the problem (7.1.5).

▶ **Example 7.1.6** *Solve graphically the L.P.P.*

$$\text{Minimize,} \quad z = -2x_1 + x_2$$

subject to

$$\begin{aligned} x_1 + \ x_2 &\geq 6 \\ 3x_1 + 2x_2 &\geq 16 \\ x_2 &\leq 9, \quad x_1 \geq 0, x_2 \geq 0. \end{aligned}$$

Solution: The feasible region is given in the figure (shaded region). The region is bounded from below only. The four extreme points are $A(6, 0)$, $P(4, 2)$, $B(0, 8)$ and $C(0, 9)$.

$$\begin{aligned} z_1 &= -12 \text{ at } A, \\ z_2 &= -6 \text{ at } P, \\ z_3 &= 8 \text{ at } B \text{ and} \\ z_4 &= 9 \text{ at } C. \end{aligned}$$

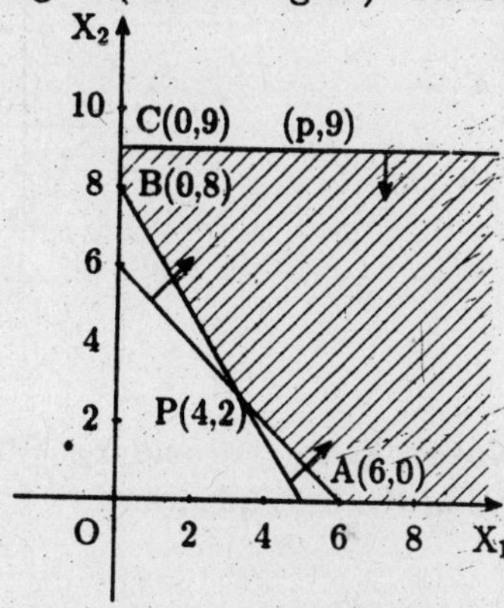

Fig. 7.15

Min. $(z_1, z_2, z_3, z_4) = -12$ at $A(6, 0)$. But for points very close to A on the line Ax_1, eg, at $(7, 0)$ the value of the objective function is -14 which is less than -12 and in the same way it can be established that the objective function has no finite minimum. In that case the problem is said to have an **unbounded solution.**

Note. (a) If $z = -2x_1 + x_2$ and the problem is to maximize it, then finite value of z will be obtained. Here $\max(z_1, z_2, z_3, z_4) = 9$ at $C(0, 9)$. To verify that maximum is finite, take any two points $(0, 8)$ and $(p, 9)$ in the feasible region very close to C on CB segment and on line $x_2 = 9$. The values of z are 8 and $9 - 2p (p > 0)$ both of which are less than 9. Hence $\max z = 9$ at $C(0, 9)$.

(b) This objective function has finite maximum but no finite minimum (in this convex set of F.S. which is bounded from below only). Again if the objective function be $2x_1 - x_2$ then it has finite minimum at $C(0, 9)$ but no finite maximum.

▶ **Example 7.1.7** *Solve graphically the L.P.P.*

$$\text{Minimize,} \quad z = -2x_1 + 5x_2$$

subject to

$$\begin{aligned} x_1 + x_2 &\geq 7 \\ 3x_1 + x_2 &\leq 6 \quad x_1 \geq 0, x_2 \geq 0. \end{aligned}$$

[**Hints.** The region restricted by the constraints $x_1 + x_2 \geq 7$ and $3x_1 + x_2 \leq 6$ is the shaded portion given in the figure 7.16 and the region restricted by the constraints $x_1 \geq 0$, $x_2 \geq 0$ is the first quadrant. Therefore in two dimensional plane, there exists no region satisfying all the four restrictions or constraints. Hence there is no **admissible** or **feasible** region. Therefore the problem has no feasible solution and no question of finding out minimum value of the objective function arises then.]

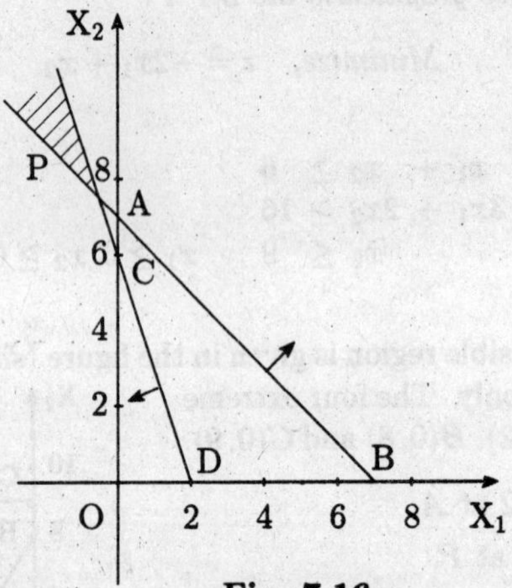

Fig. 7.16

Note. As $x_1 \geq 0$ and $x_2 \geq 0$, then the admissible or feasible region must be always in the first quadrant.

7.1.1 Moving Hyperplane Method:

There is another geometrical method of solving a L.P.P. which is called the moving hyperplane method. But judging its difficulty this method may be considered as suitable only when the convex set is a polytope. Below given is an example of that type.

▶ **Example 7.1.8** *Minimize,* $z = 3x_1 - 2x_2$ *subject to*

$$\begin{aligned} 3x_1 + 4x_2 &\geq 12 \\ x_1 - 3x_2 &\leq 6 \\ x_1 - 2x_2 &\leq -4\,, \quad x_1 \geq 0, x_2 \geq 0. \end{aligned}$$

Solution: The convex set of the feasible region is given by the shaded region which is a convex polytope. The three extreme points of the feasible region are $A(6,0), B(4,0)$ and $C(\frac{4}{5}, \frac{12}{5})$.

For $z = k = 3x_1 - 2x_2$, the objective function represents a straight line and for different values of k, we get a system of parallel straight lines.

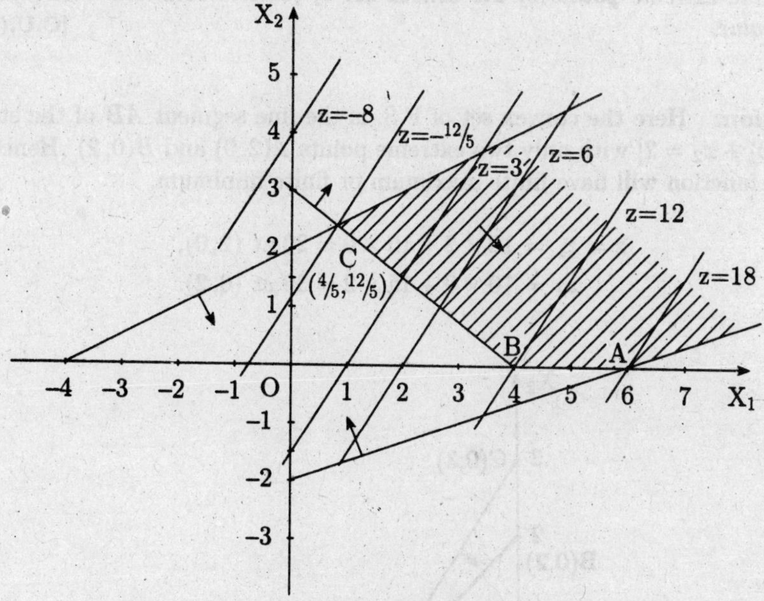

Fig. 7.17

Now on the feasible region, the value of the objective function increases as it moves parallel to itself in the direction given in the figure (by arrow). On the other hand if it moves in the opposite direction parallel to itself, the value of the objective function gradually decreases and the minimum value of the objective function is $\min z = -\frac{12}{5}$ when it passes through the extreme point $C(\frac{4}{5}, \frac{12}{5})$ of the convex set of F.S. The value of the objective function further decreases as given in the fig. 7.17. But now no part of the straight line is in the feasible region. Thus the minimum value of the objective function is $-\frac{12}{5}$ at $x_1 = \frac{4}{5}, x_2 = \frac{12}{5}$.

Note. (i) But if the problem be maximize, $z = 3x_1 - 2x_2$ with the same set of constraints, then within the feasible region, the value of the objective function gradually increases as the line of the objective function moves parallel to itself (as given by arrow) and the objective function has no finite maximum and the problem is said to have an unbounded solution.

(ii) If the problem has a bounded optimal solution, then one of the extreme points gives the optimal solution which we have seen in the above problem, but if the problem has no finite value of the objective function, then mere presence of one or more extreme points do not give any assurance of having a finite optimal solution which we have seen in the above case.

► **Example 7.1.9** *Represent geometrically the set of constraints of following L.P.P.*

$$\textit{Maximize,} \quad z = 10x_1 + 15x_2$$

subject to

$$x_1 + x_2 = 2$$
$$3x_1 + x_2 \leq 6\,, \quad x_1, x_2 \geq 0,$$

and find the extreme points of the convex set of feasible solutions. Also find the optimal value. **[C.U.(P)'83]**

Solution: Here the convex set of F.S. is the line segment AB of the straight line $AB[x_1 + x_2 = 2]$ with only two extreme points $A(2,0)$ and $B(0,2)$. Hence each objective function will have finite maximum or finite minimum.

$$z = z_1 = 10 \times 2 + 15 \times 0 = 20 \text{ at } (2,0),$$
$$z_2 = 10 \times 0 + 15 \times 2 = 30 \text{ at } (0,2).$$

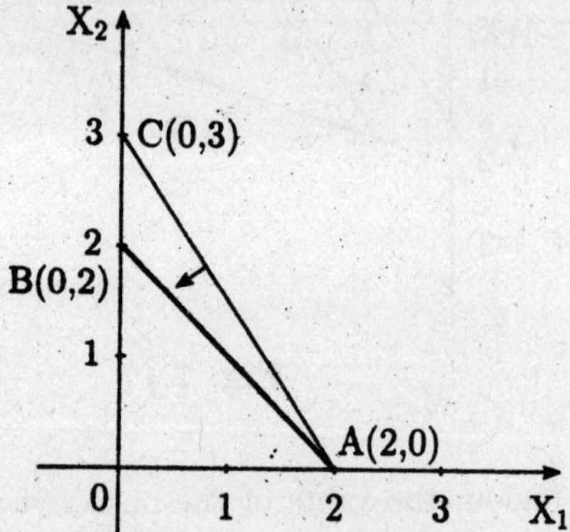

Fig. 7.18

Hence $\max z = 30$ at $x_1 = 0, x_2 = 2$.

▶ **Example 7.1.10** *Solve the following L.P.P. by graphical method.*

$$\textit{Minimize,} \quad z = 4x_1 + x_2$$

subject to

$$x_1 + 2x_2 \leq 3$$
$$4x_1 + 3x_2 = 6$$
$$3x_1 + x_2 \geq 3, \quad x_1, x_2 \geq 0.$$

[C.U.(P)'97]

Solution: Here the convex set of F.S. is the line segment CD, the extreme points are $C(\frac{3}{2},0), D(\frac{3}{5},\frac{6}{5})$. Thus

$$z_1 = 4 \times \frac{3}{2} + 1 \times 0 = 6 \text{ at } C\left(\frac{3}{2},0\right)$$

$$z_2 = 4 \times \frac{3}{5} + 1 \times \frac{6}{5} = \frac{18}{5} \text{ at } D\left(\frac{3}{5},\frac{6}{5}\right)$$

Hence min $z = \frac{18}{5}$ at $x_1 = \frac{3}{5}, x_2 = \frac{6}{5}$.

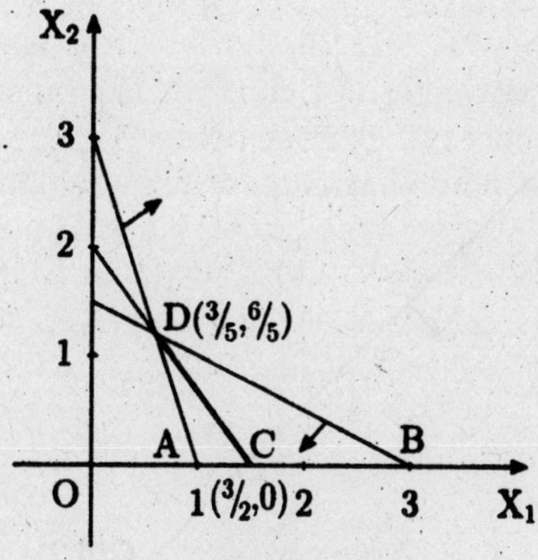

Fig. 7.19

▶ **Example 7.1.11** *Solve the following Linear Programming Problems after finding the extreme points of the convex set of the feasible solutions.*

(a) *Maximize,* $z = 3x_1 + 5x_2$

$$\text{subject to } \begin{aligned} x_1 - 3x_2 &\leq 4 \\ -x_1 + x_2 &\geq -4 \\ x_1 + x_2 &\leq 8, \quad x_1, x_2 \geq 0. \end{aligned}$$

(b) *Maximize,* $z = 4x_1 + 7x_2$

$$\text{subject to } \begin{aligned} 12x_1 + 7x_2 &\leq 42 \\ 5x_1 + 4x_2 &\leq 20 \\ 2x_1 + 3x_2 &\geq 6, \quad x_1, x_2 \geq 0. \end{aligned}$$

(c) *Maximize,* $z = x_1 + 3x_2$

$$\text{subject to } \begin{aligned} 3x_1 + x_2 &\leq 11 \\ x_1 - x_2 &\leq 1 \\ 2x_1 - x_2 &\geq -1, \quad x_1, x_2 \geq 0. \end{aligned}$$

(d) *Maximize,* $z = 2x_1 - x_2$

$$\text{subject to } \begin{aligned} x_1 - x_2 &\leq 1 \\ x_1 &\leq 3, \quad x_1, x_2 \geq 0. \end{aligned}$$

[C.U.(P)'95]

Solution: (a) Ox_1 and Ox_2 be the axes. $\overleftrightarrow{AB}$, $\overleftrightarrow{AC}$ and $\overleftrightarrow{LN}$ represent the straight lines $x_1 - 3x_2 = 4$, $-x_1 + x_2 = -4$ and $x_1 + x_2 = 8$ respectively. The convex set of the feasible region is $ABCA$ where the extreme points are $A(4,0)$, $B(7,1)$, $C(6,2)$ and the feasible region is a convex polyhedron. Therefore any objective function has finite maximum and minimum.

$$\begin{aligned} z_1 &= 3 \times 4 + 5 \times 0 = 12 \text{ at } (4,0) \\ z_2 &= 3 \times 7 + 5 \times 1 = 26 \text{ at } (7,1) \\ z_3 &= 3 \times 6 + 5 \times 2 = 28 \text{ at } (6,2) \end{aligned}$$

Then max $z = 28$ at $C(6, 2)$.

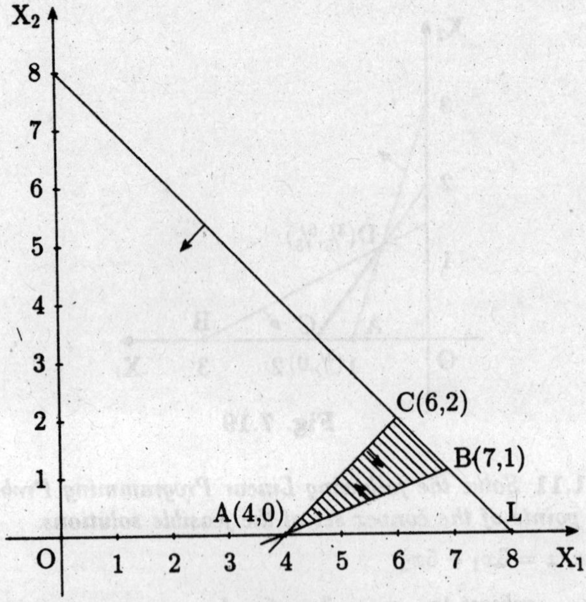

Fig. 7.20

Remark: Similarly min $z = 12$ at $A(4, 0)$.

(b) Ox_1 and Ox_2 be the axes. $\overleftrightarrow{AB}$, $\overleftrightarrow{CD}$, $\overleftrightarrow{EF}$ represent the straight lines $12x_1 + 7x_2 = 42$, $5x_1 + 4x_2 = 20$ and $2x_1 + 3x_2 = 6$ respectively. And the convex set of the feasible region is $AKDFEA$ and extreme points are $A(\frac{7}{2}, 0)$, $K(\frac{28}{13}, \frac{30}{13})$, $D(0, 5)$, $F(0, 2)$, $E(3, 0)$ which is a convex polyhedron.

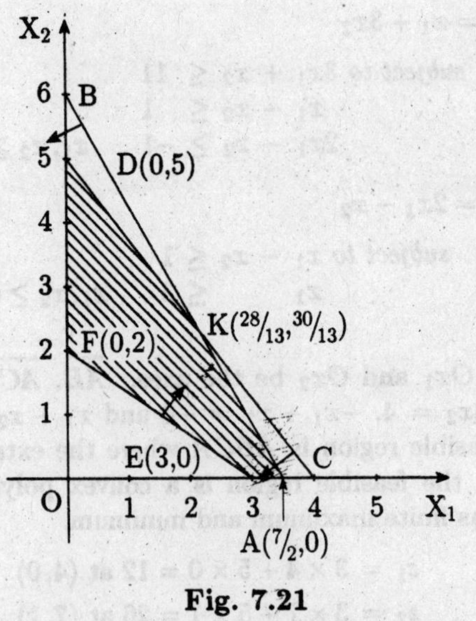

Fig. 7.21

$$z_1 = 4 \times \frac{7}{2} + 7 \times 0 = 14 \text{ at } \left(\frac{7}{2}, 0\right)$$

$$z_2 = 4 \times \frac{28}{13} + 7 \times \frac{30}{13} = \frac{322}{13} \text{ at } \left(\frac{28}{13}, \frac{30}{13}\right)$$
$$z_3 = 4 \times 0 + 7 \times 5 = 35 \text{ at } (0, 5)$$
$$z_4 = 4 \times 0 + 7 \times 2 = 14 \text{ at } (0, 2)$$
$$z_5 = 4 \times 3 + 7 \times 0 = 12 \text{ at } (3, 0)$$

Max $z = 35$ at $D(0,5)$.

Remark: Min $z = 12$ at $E(3,0)$.

(c) Ox_1 and Ox_2 be the axes. $\overleftrightarrow{AB}$, $\overleftrightarrow{BC}$, $\overleftrightarrow{DA}$ represent the straight lines $3x_1 + x_2 = 11$, $x_1 - x_2 = 1$ and $2x_1 - x_2 = -1$ respectively. And the convex set of the feasible region is $ABCODA$ which is a convex polyhedron and extreme points are $A(3,2)$, $B(2,5)$, $C(0,1)$, $O(0,0)$, $D(1,0)$.

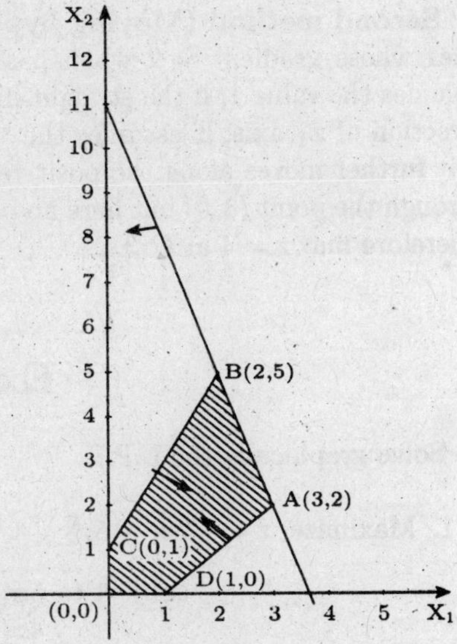

Fig. 7.22

$$z_1 = 1 \times 3 + 3 \times 2 = 9 \text{ at } (3, 2)$$
$$z_2 = 1 \times 2 + 3 \times 5 = 17 \text{ at } (2, 5)$$
$$z_3 = 1 \times 0 + 3 \times 1 = 3 \text{ at } (0, 1)$$
$$z_4 = 1 \times 0 + 3 \times 0 = 0 \text{ at } (0, 0)$$
$$z_5 = 1 \times 1 + 3 \times 0 = 1 \text{ at } (1, 0)$$

$\max z = 17$ at $B(2,5)$.

Remark: Min $z = 0$ at $O(0,0)$.

(d) **First method:** Ox_1 and Ox_2 be the set of axes. AC and BC are the straight lines $x_1 - x_2 = 1$ and $x_1 = 3$ respectively. Here the convex set of the feasible region is not a convex polyhedron, but a convex polytope which is unbounded above; the extreme points are $O(0,0), A(1,0), C(3,2)$.

$$z_1 = 2 \times 0 - 1 \times 0 = 0 \text{ at } (0, 0)$$
$$z_2 = 2 \times 1 - 1 \times 0 = 2 \text{ at } (1, 0)$$
$$z_3 = 2 \times 3 - 1 \times 2 = 4 \text{ at } (3, 2)$$

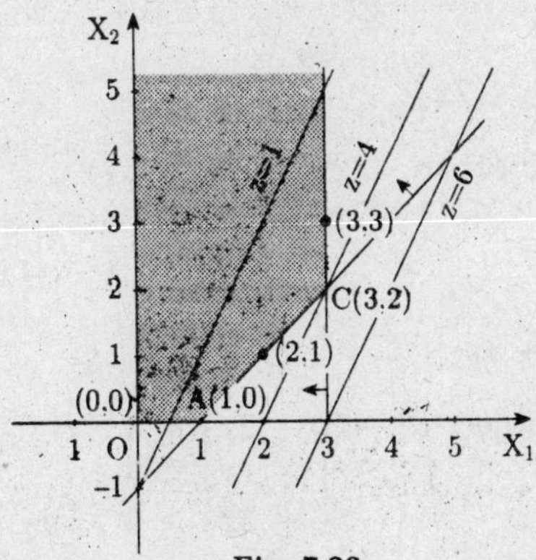

Fig. 7.23

Maximum value of z attains at $C(3,2)$. Now to test that maximum value is finite at $C(3,2)$, we take two neighbouring points $(2,1)$ and $(3,3)$ on the straight lines AC and BC on the feasible region.

$$z_4 = 2\times 2 - 1\times 1 = 3 \text{ at } (2,1)$$
$$z_5 = 2\times 3 - 1\times 3 = 3 \text{ at } (3,3)$$

both of which are less then 4, then $\max z = 4$ at $C(3,2)$.

Second method (Moving hyperplane method): We draw a set of straight lines whose gradient $= 2$ when passes through the point $(0,-1)$, $z = 2x_1 - x_2$ assumes the value 1, if the straight line moves parallel to itself towards the positive direction of x_1 axis, it assumes the value $z = 4$ when passes through the point C. If it further moves along the positive direction it assumes the value 6 when passes through the point $(3,0)$ but here no part of the straight line is on the feasible region. Therefore $\max z = 4$ at $C(3,2)$.

Exercise 7

Solve graphically the L.P.P.

1. Maximize, $z = -2x_1 + 5x_2$

$$\begin{aligned} \text{subject to } 5x_1 + 2x_2 &\le 45 \\ 4x_1 + 5x_2 &\le 53 \\ x_1 &\ge 2\,, \quad x_1, x_2 \ge 0. \end{aligned}$$

Find also the minimum value of z.

2. Minimize, $z = 4x_1 - 3x_2$

$$\begin{aligned} \text{subject to } 2x_1 - x_2 &\ge 4 \\ 4x_1 + 3x_2 &\le 28\,, \quad x_1, x_2 \ge 0. \end{aligned}$$

3. Maximize, $z = x_1 + 3x_2$

$$\begin{aligned} \text{subject to } x_1 + x_2 &\le 5 \\ 6x_1 - x_2 &\le 30 \\ 9x_1 + 2x_2 &\ge 24\,, \quad x_1, x_2 \ge 0. \end{aligned}$$

4. Minimize, $z = -2x_1 + 7x_2$

$$\begin{aligned} \text{subject to } 3x_1 + 2x_2 &\le 17 \\ -2x_1 + 3x_2 &\le 6 \\ x_2 &\ge 1\,, \quad x_1, x_2 \ge 0. \end{aligned}$$

5. Minimize, $z = 2x_1 + 3x_2$

$$\begin{aligned} \text{subject to } -x_1 + 2x_2 &\le 4 \\ x_1 + x_2 &\le 6 \\ x_1 + 3x_2 &\ge 9\,, \quad x_1, x_2 \ge 0. \end{aligned}$$

6. **Maximize,** $z = -7x_1 + 4x_2$

$$\text{subject to } 5x_1 - 2x_2 \geq 0,\quad x_1 - 6x_2 \leq 0,\quad x_1 + x_2 \leq 7,\quad x_1, x_2 \geq 0.$$

7. **Minimize,** $z = x_1 + 5x_2$

$$\text{subject to } x_1 + x_2 \leq 4,\quad 2x_1 + x_2 \leq 6,\quad 4x_1 + 3x_2 \geq 28,\quad x_1, x_2 \geq 0.$$

8. **Maximize,** $z = 3x_1 + 4x_2$

$$\text{subject to } x_1 - x_2 \leq -1,\quad -x_1 + x_2 \leq 0,\quad x_1, x_2 \geq 0.$$

9. **Maximize,** $z = 7x_1 + 3x_2$

$$\text{subject to } 4x_1 + 5x_2 \leq 40,\quad x_1 \geq 3,\quad x_2 \geq 4,\quad x_1 \geq 0, x_2 \geq 0.$$

10. **Maximize,** $z = 2x_1 - 6x_2$

$$\text{subject to } 3x_1 + 2x_2 \leq 6,\quad x_1 - x_2 \geq -1,\quad -x_1 - 2x_2 \geq 1,\quad x_1, x_2 \geq 0.$$

11. **Maximize,** $z = x_1 + 3x_2$

$$\text{subject to } x_1 - x_2 \geq -1,\quad x_1 + x_2 \geq 3,\quad x_1, x_2 \geq 0.$$

12 **Maximize,** $z = 3x_1 + 2x_2$

$$\text{subject to } 3x_1 + 2x_2 \leq 10,\quad 3x_1 + 2x_2 \geq 20,\quad x_1 \geq 0, x_2 \geq 0.$$

13. **Minimize,** $z = 4x_1 - x_2$

$$\text{subject to } x_1 + 2x_2 \geq 10,\quad x_1 \leq 12,\quad x_1 \geq 0, x_2 \geq 0.$$

14. (a) **Maximize,** $z = \frac{1}{7}x_1 + 2x_2$

$$\text{subject to } 3x_1 + 4x_2 \geq 12,\quad x_2 \leq 10,\quad x_1 \geq 0, x_2 \geq 0.$$

(b) **Maximize,** $z = x_1 + 5x_2$

$$\text{subject to } 3x_1 + 4x_2 \leq 6,\quad x_1 + 3x_2 \geq 3,\quad x_1 \geq 0, x_2 \geq 0.$$ **[C.U.(H)'80]**

(c) Maximize, $z = 3x_1 + 2x_2$

$$\begin{aligned} \text{subject to } -2x_1 + x_2 &\le 1 \\ x_1 \qquad &\le 2 \\ x_1 + x_2 &\le 3, \quad x_1, x_2 \ge 0. \end{aligned}$$

[C.U.(P)'89]

15. Maximize, $z = 2x_1 + x_2$

$$\begin{aligned} \text{subject to } 4x_1 + 3x_2 &\le 12 \\ 4x_1 + x_2 &\le 8 \\ 4x_1 - x_2 &\le 8, \quad x_1 \ge 0, x_2 \ge 0. \end{aligned}$$

Find out the extreme points of the convex set of feasible solutions and hence find out the maximum value of the objective function. **[C.U.(P)'81]**

16. Represent geometrically the set of constraints of the following L.P.P.

$$\text{Maximize,} \quad z = 10x_1 + 15x_2$$

$$\begin{aligned} \text{subject to } x_1 + x_2 &= 2 \\ 3x_1 + 2x_2 &\le 6, \quad x_1, x_2 \ge 0. \end{aligned}$$

Find the extreme points of the convex set of feasible solutions. Also find the optimal value. **[C.U.(P)'83]**

17. Minimize, $z = x_1 + 7x_2$

$$\begin{aligned} \text{subject to } -x_1 + 2x_2 &\le 8 \\ x_1 - x_2 &\le 4, \quad x_1, x_2 \ge 0. \end{aligned}$$

[C.U.(P)'92]

18. Maximize, $z = 2x_1 + 4x_2$

$$\begin{aligned} \text{subject to } 2x_1 + 3x_2 &\le 48 \\ x_1 + 3x_2 &\le 42 \\ x_1 + x_2 &\le 21, \quad x_1, x_2 \ge 0. \end{aligned}$$

[C.U.(P)'93]

19. Maximize, $z = 500x_1 + 400x_2$

$$\begin{aligned} \text{subject to } 10x_1 + 8x_2 &\le 800 \\ x_1 \qquad &\le 60 \\ x_2 &\le 75, \quad x_1, x_2 \ge 0. \end{aligned}$$

Prove that alternative optimal solutions exist. Find at least two optimal solutions where the value of the variables are all integers (Solved example 2.3.4). **[C.U.(H)'91,'94,'96]**

Solution: Here the straight line AB, CA and DB represent the straight lines $10x_1 + 8x_2 = 800$, $x_1 = 60$, $x_2 = 75$, and the convex set of the feasible region is $OCABDO$ and the extreme points are $O(0,0)$, $C(60,0)$, $A(60,25)$, $B(20,75)$, $D(0,75)$, which is a convex polyhedron.

$$\begin{aligned} z_1 &= 500 \times 0 + 400 \times 0 = 0 \text{ at } (0,0) \\ z_2 &= 500 \times 60 + 400 \times 0 = 30,000 \text{ at } (60,0) \\ z_3 &= 500 \times 60 + 400 \times 25 = 40,000 \text{ at } (60,25) \\ z_4 &= 500 \times 20 + 400 \times 75 = 40,000 \text{ at } (20,75) \text{ and} \\ z_5 &= 500 \times 0 + 400 \times 75 = 30,000 \text{ at } (0,75) \end{aligned}$$

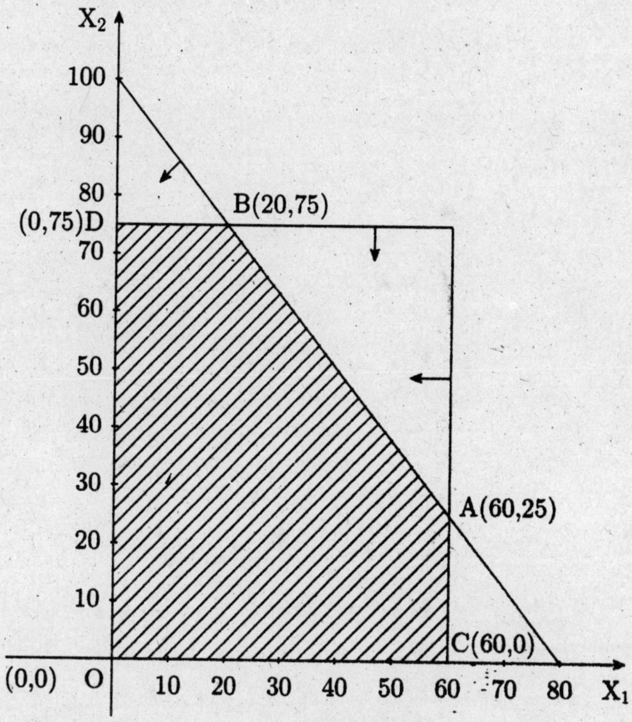

Fig. 7.24

Thus max z = Rs. 40,000 at ($x_1 = 60$, $x_2 = 25$) and ($x_1 = 20$, $x_2 = 75$). Therefore multiple optimal solutions exist and the two values are ($x_1 = 60$, $x_2 = 25$) and ($x_1 = 20$, $x_2 = 75$), where all values of the solutions are integers.

Remark. Since there are more than one optimal solutions then any convex combination of them are also solutions but they may or may not be integers. For example, one solution $x_1 = \frac{1}{2}60+\frac{1}{2}\times20 = 40$ and $x_2 = \frac{1}{2}\times25+\frac{1}{2}\times75 = 50$, which is also an integer solution.

Answers

1. max $z = 41$ at ($x_1 = 2$, $x_2 = 9$), min $z = -18$ at ($x_1 = 9$, $x_2 = 0$) 2. min $z = 4$ at ($x_1 = 4$, $x_2 = 4$) 3. max $z = 11$ at ($x_1 = 2$, $x_2 = 3$) 4. min $z = -3$ at ($x_1 = 5$, $x_2 = 1$) 5. min $z = 10\frac{1}{5}$ at ($x_1 = \frac{6}{5}$, $x_2 = \frac{13}{5}$) 6. max $z = 6$ at ($x_1 = 2$, $x_2 = 5$) 7. No. F.S. 8. No F.S. 9. max $z = 47$ at ($x_1 = 5$, $x_2 = 4$) 10. No F.S. 11. Unbounded solution 12. No. F.S. 13. Unbounded solution. 14. (a) Unbounded solution. (b) max $z = 7\frac{1}{2}$ at ($x_1 = 0$, $x_2 = \frac{3}{2}$). (c) max $z = 8$ at ($x_1 = 2$, $x_2 = 1$) 15. Extreme points are $(0,0)$, $(2,0)$, $(\frac{3}{2},2)$ and $(0,4)$; max $z = 5$ at ($x_1 = \frac{3}{2}$, $x_2 = 2$). The constraint $4x_1 - x_2 \leq 8$ is redundant. 16. Extreme points are $(2,0)$, $(0,2)$ and max $z = 30$ at $(0,2)$. 17. min $z = 0$ at ($x_1 = 0$, $x_2 = 0$) 18. max $z = 60$ at ($x_1 = 6$, $x_2 = 12$).

Chapter 8

Simplex Method or Simplex Algorithm (I)

8.1 Introduction

All numerical techniques, applied to solve a L.P.P., are known as *algorithm*. There are various numerical methods of solving a L.P.P. Simplex algorithm or simplex method is one of them and it is a suitable method. We know from the fundamental theorem of L.P.P. that if a problem has an optimal solution then at least one B.F.S. must be optimal. In this method initially a B.F.S. and the corresponding value of the objective function is obtained. Then by a suitable transformation (which will be discussed now) a new B.F.S. and the corresponding value of the objective function is obtained which is greater or less (at least equal) than the value of the objective function for the preceeding solution in the problem of maximization or minimization respectively. The process is continued until an optimal value of the objective function is attained (if it exists at all). The extra advantage of this method is that, it gives correct information about the existence of the optimal solution.

8.2 Procedure

After the introduction of the slack and surplus variables and by proper adjustment of z, let a L.P.P. be, optimize $z = \mathbf{cx}$

$$\text{subject to} \quad A\mathbf{x} = \mathbf{b}, \quad \mathbf{x} \geq 0, \cdots, [A]_{m\times n}, \tag{8.2.1}$$

where $\mathbf{c} = (c_1, c_2, \cdots, c_r, \overbrace{0, 0, \cdots, 0}^{n-r})$ an n component row vector.

$\mathbf{x} = [x_1, x_2, \cdots, x_r, x_{r+1}, \cdots, x_n]$ an n component column vector.

The components $x_{r+1}, x_{r+2}, \cdots, x_n$ are either slack or surplus variables.

We make an assumption ($m < n$) [m, the number of constraints] and further make an assumption that all components of $\mathbf{b}$ are non-negative by proper adjustment. Of course, the *second assumption is non-restrictive.*

The coefficient matrix A will be an $m \times n$ matrix and $A = (\mathbf{a}_1, \mathbf{a}_2, \cdots, \mathbf{a}_n)$, where $\mathbf{a}_j$ is the jth column vector of the coefficient matrix A, which are called the *activity vectors* associated with the variable x_j $[j = 1, 2, \cdots, n]$. As none of the m converted equations is redundant then there exists at least one set of m-column vectors of the coefficient matrix A, which are linearly independent. Let β_1, β_2, $\cdots$, β_m be a set of m linearly independent column vectors taken from $\mathbf{a}_1$, $\mathbf{a}_2$, $\cdots$, $\mathbf{a}_n \in A$. Then one basis matrix B is given by

$$B = B(\beta_1, \beta_2, \cdots, \beta_m).$$

Let $x_{B1}, x_{B2}, \cdots, x_{Bm}$ be the basic variables associated with the column vectors $\beta_1, \beta_2, \cdots, \beta_m$ respectively. Then the basic variable vector is

$$\mathbf{x}_B = [x_{B1}, x_{B2}, \cdots, x_{Bm}]. \tag{8.2.2}$$

The solution set corresponding to the basic variables is

$$\hat{\mathbf{x}}_B \text{ or simply } \mathbf{x}_B = B^{-1}\mathbf{b} \tag{8.2.3}$$

We assume that $\mathbf{x}_B \geq \mathbf{0}$, i.e., the solution is a B.F S. and in that case, the basis B is called an *admissible basis* in the simplex theory.

Let $c_{B1}, c_{B2}, \cdots, c_{Bm}$ be the coefficients of $x_{B1}, x_{B2}, \cdots, x_{Bm}$, respectively in the objective function $z = \mathbf{cx}$ then

$$\mathbf{c}_B = (c_{B1}, c_{B2}, \cdots, c_{Bm}) \tag{8.2.4}$$

is an m-component row vector which is known as the *associated cost vector*.

Now a value z_B is defined as

$$z_B = c_{B1}x_{B1} + c_{B2}x_{B2} + \cdots + c_{Bm}x_{Bm} = \mathbf{c}_B\mathbf{x}_B \tag{8.2.5}$$

z_B, the inner product of $\mathbf{c}_B$ (row vector) and $\mathbf{x}_B$ (column vector), is the value of the objective function, corresponding to the B.F.S., where the basis matrix is B. z_B does not depend on non-basic variables (as each of them of zero).

Now as $(\beta_1, \beta_2, \cdots, \beta_m)$ are linearly independent, then all the column vectors $\mathbf{a}_j$ can be expressed as the linear combination of $\beta_1, \beta_2, \cdots, \beta_m$.

Let

$$\mathbf{a}_j = \beta_1 y_{1j} + \beta_2 y_{2j} + \cdots + \beta_m y_{mj} = B\mathbf{y}_j \tag{8.2.6}$$

the inner product of B and $\mathbf{y}_j$, where $\mathbf{y}_j$ is an m-component column vector given by

$$\mathbf{y}_j = [y_{1j}, y_{2j}, \cdots, y_{mj}]$$

From (8.2.6) we get

$$\mathbf{y}_j = B^{-1}\mathbf{a}_j. \tag{8.2.7}$$

Net evaluation: Evaluation is the inner product of the row vector $\mathbf{c}_B$ and the column vector $\mathbf{y}_j$ which is usually denoted by z_j and z_j is given by

$$z_j = \mathbf{c}_B\mathbf{y}_j = \mathbf{c}_B B^{-1}\mathbf{a}_j = c_{B1}y_{1j} + c_{B2}y_{2j} + \cdots + c_{Bm}y_{mj} \tag{8.2.8}$$

and $z_j - c_j$ is called as *net evaluation*.

The function $z_j - c_j$ $[j = 1, 2, \cdots, n]$ (net evaluation) plays a very important role in determining the optimal stage, in the case of solving the L.P.P. by the simplex method.

If the coefficient matrix A contains m unit column vectors which are linearly independent, then this set of vectors constitute a basis matrix. Let $\mathbf{e}_1, \mathbf{e}_2, \cdots, \mathbf{e}_i, \cdots, \mathbf{e}_m$ be m independent unit vectors of the coefficient matrix, all of which may not be placed consecutively in the ascending order of $i (i = 1, 2, \cdots, m)$. For example, $\mathbf{e}_1, \mathbf{e}_2$ and $\mathbf{e}_3$ may occur at 5th, 7th and 3rd column of the matrix A, respectively. But still if we consider a basis matrix B in the form given by $B = B(\mathbf{e}_1, \mathbf{e}_2, \cdots, \mathbf{e}_m)$ then B is an *identity matrix*. Hence the components of the solution set corresponding to the basic variables are $x_{Bi} = b_i$ $(i = 1, 2, \cdots, m)$. But if the unit vector $\mathbf{e}_i$ of the basis matrix B (identity matrix) occur at the jth column of the coefficient matrix A, then x_j, the jth component of the B.F.S. is equal to x_{Bi} and c_j, the coefficient of x_j in the obejctive function is equal to c_{Bi} which is nothing but the coefficient of x_{Bi}. For example, if $\mathbf{e}_1$ occurs at 5th column, then $x_{B1} = x_5 = b_1$ and $c_{B1} = c_5$. Similarly, if $\mathbf{e}_2$ occurs at 3rd column then $x_{B2} = x_3 = b_2$ and $c_{B2} = c_3$ and so on.

If B is an identity matrix then from (8.2.7)

$$\mathbf{y}_j = B^{-1}\mathbf{a}_j = I_m^{-1}\mathbf{a}_j = \mathbf{a}_j \cdots, \tag{8.2.9}$$

i.e., the vector $\mathbf{y}_j$ is nothing but the column vector $\mathbf{a}_j$ due to this transformation

N.B. (1) In the simplex method, generally all equations are to be adjusted in such a way that there exist m independent unit vectors $\mathbf{e}_1, \mathbf{e}_2, \cdots, \mathbf{e}_m$ in the coefficient matrix which together constitute the initial basis matrix and $\mathbf{b} \geq \mathbf{0}$ so that $B^{-1}\mathbf{b} = I_m^{-1}\mathbf{b} = \mathbf{b} \geq \mathbf{0}$. So the identity matrix is taken to be initial basis to start the simplex procedure. But this is not strictly necessary from the theoretical point of view which has been discussed in the solved Example 8.2.1.

(2) For a basis matrix B, $\mathbf{x}_B = B^{-1}\mathbf{b}$ and the value of the objective function corresponding to $\mathbf{x}_B$ is $z_B = \mathbf{c}_B\mathbf{x}_B$. The value of the objective function does not depend on the values of the non-basic variables (each of them is zero). Hence for calculation of the value of the objective function z, $\mathbf{x}_B = B^{-1}\mathbf{b}$ may be considered as a B.F.S.

► **Example 8.2.1** *Find a basic feasible solution, if there be any, of the following set of linearly independent equations and if such solution exists, taking that basis as an admissible basis, calculate all* $\mathbf{y}_j$, $z_j - c_j$ $[j = 1, 2, \cdots, 4]$ *and the value of the objective function corresponding to that B.F.S.*

Maximize, $z = 2x_1 - 4x_2 - x_3 + 4x_4$ *subject to*

$$\begin{aligned} 3x_1 - 5x_2 + x_3 - 2x_4 &= 7 \\ 6x_1 - 10x_2 - x_3 + 5x_4 &= 11, \quad x_j \geq 0,\ j = 1, 2, \cdots, 4. \end{aligned}$$

Solution: $\mathbf{a}_1 = \begin{bmatrix} 3 \\ 6 \end{bmatrix}$, $\mathbf{a}_2 = \begin{bmatrix} -5 \\ -10 \end{bmatrix}$, $\mathbf{a}_3 = \begin{bmatrix} 1 \\ -1 \end{bmatrix}$, $\mathbf{a}_4 = \begin{bmatrix} -2 \\ 5 \end{bmatrix}$ and $\mathbf{b} = \begin{bmatrix} 7 \\ 11 \end{bmatrix}$. One square matrix $(\mathbf{a}_1, \mathbf{a}_2) = \begin{bmatrix} 3 & -5 \\ 6 & -10 \end{bmatrix}$, taking two at a time from the four vectors $\mathbf{a}_1, \cdots, \mathbf{a}_4$ and det of the matrix is zero. Hence the above square matrix cannot be considered as a basis matrix. Now another square matrix $B = (\mathbf{a}_1, \mathbf{a}_3) = \begin{bmatrix} 3 & 1 \\ 6 & -1 \end{bmatrix}$ and det $B = -9 \neq 0$. Hence B can be considered as a basis matrix and basic solution corresponding to the basis B, is $\mathbf{x}_B = B^{-1}\mathbf{b} = \frac{1}{-9}\begin{bmatrix} -1 & -1 \\ -6 & 3 \end{bmatrix}\begin{bmatrix} 7 \\ 11 \end{bmatrix} = \begin{bmatrix} 2 \\ 1 \end{bmatrix} \geq \mathbf{0}$. The solution is feasible and hence B can be taken as an *admissible basis*.

Thus $\mathbf{x}_B = [x_{B1}, x_{B2}] = [x_1, x_3] = [2, 1]$.

$\mathbf{c} = (c_1, c_2, c_3, c_4) = (2, -4, -1, 4), \mathbf{c}_B = (c_{B1}, c_{B2}) = (c_1, c_3) = (2, -1)$, the cost components corresponding to the basic variables x_1 and x_3.

Now

$$\begin{aligned}
[\mathbf{y}_1, \mathbf{y}_2, \mathbf{y}_3, \mathbf{y}_4] &= B^{-1}[\mathbf{a}_1, \mathbf{a}_2, \mathbf{a}_3, \mathbf{a}_4] \\
&= \frac{1}{-9}\begin{bmatrix} -1 & -1 \\ -6 & 3 \end{bmatrix}\begin{bmatrix} 3 & -5 & 1 & -2 \\ 6 & -10 & -1 & 5 \end{bmatrix} \\
&= \begin{bmatrix} 1 & -5/3 & 0 & 1/3 \\ 0 & 0 & 1 & -3 \end{bmatrix}.
\end{aligned}$$

Hence

$$\mathbf{y}_1 = \begin{bmatrix} 1 \\ 0 \end{bmatrix}, \mathbf{y}_2 = \begin{bmatrix} -5/3 \\ 0 \end{bmatrix}, \mathbf{y}_3 = \begin{bmatrix} 0 \\ 1 \end{bmatrix} \text{ and } \mathbf{y}_4 = \begin{bmatrix} 1/3 \\ -3 \end{bmatrix},$$

Then

$$\mathbf{y}_1 = \begin{bmatrix} y_{11} \\ y_{21} \end{bmatrix} = \begin{bmatrix} 1 \\ 0 \end{bmatrix}, \quad \mathbf{y}_2 = \begin{bmatrix} y_{12} \\ y_{22} \end{bmatrix} = \begin{bmatrix} -5/3 \\ 0 \end{bmatrix},$$

$$\mathbf{y}_3 = \begin{bmatrix} y_{13} \\ y_{23} \end{bmatrix} = \begin{bmatrix} 0 \\ 1 \end{bmatrix} \text{ and } \mathbf{y}_4 = \begin{bmatrix} y_{14} \\ y_{24} \end{bmatrix} = \begin{bmatrix} 1/3 \\ -3 \end{bmatrix}.$$

$$\begin{aligned}
z_1 - c_1 &= \mathbf{c}_B\mathbf{y}_1 - c_1 = c_{B1}y_{11} + c_{B2}y_{21} - c_1 = 2 \times 1 + (-1) \times 0 - 2 = 0 \\
z_2 - c_2 &= \mathbf{c}_B\mathbf{y}_2 - c_2 = c_{B1}y_{12} + c_{B2}y_{22} - c_2 = 2 \times (-5/3) + (-1) \times 0 + 4 = 2/3 \\
z_3 - c_3 &= \mathbf{c}_B\mathbf{y}_3 - c_3 = c_{B1}y_{13} + c_{B2}y_{23} - c_3 = 2 \times 0 + (-1) \times 1 - (-1) = 0 \\
z_4 - c_4 &= \mathbf{c}_B\mathbf{y}_4 - c_4 = c_{B1}y_{14} + c_{B2}y_{24} - c_4 = 2 \times 1/3 + (-1) \times (-3) - 4 = -1/3 \\
z_0 &= \mathbf{c}_B\mathbf{x}_B = c_{B1}x_{B1} + c_{B2}x_{B2} = 2 \times 2 + (-1) \times 1 = 3
\end{aligned}$$

Note: It is important to note that $\mathbf{y}_1$ and $\mathbf{y}_3$ corresponding to the basis vectors $\mathbf{a}_1$ and $\mathbf{a}_3$ are unit vectors and they can be arranged to form a unit matrix and $z_1 - c_1 = z_3 - c_3 = 0$ corresponding to the basis vectors $\mathbf{a}_1$ and $\mathbf{a}_3$. [See theorem 8.5.1 and the Appendix]

8.3 The Theoretical Development of the Simplex Method or Simplex Algorithm

The theoretical development of the method is very difficult to understand for the students at this stage. We shall just discuss the working principle of the method. Detailed discussion is given in the Appendix. Interested students will go through it very carefully.

Let a given maximization problem be

$$\text{Maximize,}\quad z = c_1x_1 + c_2x_2 + \cdots + c_jx_j + \cdots + c_nx_n$$

subject to

$$\begin{array}{ccccccccccc} a_{11}x_1 &+& a_{12}x_2 &+\cdots+& a_{1j}x_j &+\cdots+& a_{1n}x_n &=& b_1 \\ a_{21}x_1 &+& a_{22}x_2 &+\cdots+& a_{2j}x_j &+\cdots+& a_{2n}x_n &=& b_2 \\ \cdots & & \cdots & & \cdots & & \cdots & & \cdots \\ a_{i1}x_1 &+& a_{i2}x_2 &+\cdots+& a_{ij}x_j &+\cdots+& a_{in}x_n &=& b_i \\ \cdots & & \cdots & & \cdots & & \cdots & & \cdots \\ a_{m1}x_1 &+& a_{m2}x_2 &+\cdots+& a_{mj}x_j &+\cdots+& a_{mn}x_n &=& b_m \end{array}$$

where $x_j \geq 0$, $j = 1, 2, \cdots, n$ $(n > m)$.

Some of the variables may be slack or surplus variables.

In matrix notation, the problem is

$$\text{Maximize,}\quad z = \mathbf{cx}$$

subject to $\mathbf{a}_1x_1 + \mathbf{a}_2x_2 + \cdots + \mathbf{a}_jx_j + \cdots + \mathbf{a}_nx_n = \mathbf{b}$.

[The vector $\mathbf{a}_j$; $j = 1, 2, \cdots, n$ are called *activity vectors*], where

$$\begin{aligned} \mathbf{c} &= (c_1, c_2, \cdots, c_j, \cdots, c_n) \rightarrow \text{ a row vector} \\ \mathbf{x} &= [x_1, x_2, \cdots, x_j, \cdots, x_m] \rightarrow \text{ a column vector} \\ \mathbf{a}_j &= [a_{1j}, a_{2j}, \cdots, a_{ij}, \cdots, a_{mj}] \rightarrow \text{ a column vector} \\ \mathbf{b} &= [b_1, b_2, \cdots, b_j, \cdots, b_m] \rightarrow \text{ a column vector.} \end{aligned} \tag{8.3.1}$$

8.3.1 Search for a basis which will give a B.F.S.

Out of the vectors $\mathbf{a}_1$, $\mathbf{a}_2$, $\cdots$, $\mathbf{a}_j$, $\cdots$, $\mathbf{a}_n$, we shall have to select arbitrarily m vectors which are linearly independent (there exists always at least one set of such vectors, since the equations are linearly independent) which form a basis matrix B. With that basis, find out the basic solution and let us assume that the B.S., $\mathbf{x}_B = B^{-1}\mathbf{b} \geq \mathbf{0}$, i.e., the solution is a B.F.S. Such basis is called an *admissible basis* to start the simplex method.

In all practical problems, there always exists an identity matrix I_m and $\mathbf{b} \geq \mathbf{0}$ and

$$\mathbf{x}_B = I_m^{-1}\mathbf{b} = I_m\mathbf{b} = \mathbf{b} \geq \mathbf{0}. \tag{8.3.2}$$

So it will not be difficult to find an initial B.F.S.

Let the B.F.S.

$$\mathbf{x}_B = [x_{B1}, x_{B2}, \cdots, x_{Bm}]$$

which are some m variables corresponding to the m activity vectors belongs to the basis B. Let the cost components of the variables $[x_{B1}, x_{B2}, \cdots, x_{Bm}]$ be $(c_{B1}, c_{B2}, \cdots, c_{Bm})$ which are some m components belongs to **c**. Thus corresponding to the basic variables.

$$\mathbf{c}_B = (c_{B1}, c_{B2}, \cdots, c_{Bm}),$$

an m component row vector.

Now we calculate the column vectors $\mathbf{y}_1, \mathbf{y}_2, \cdots, \mathbf{y}_j, \cdots, \mathbf{y}_n$, where

$$\mathbf{y}_j = [y_{1j}, y_{2j}, \cdots, y_{ij}, \cdots, y_{mj}]$$

by using the formula

$$\mathbf{y}_j = B^{-1}\mathbf{a}_j, \quad [j = 1, 2, \cdots, n] \tag{8.3.3}$$

Since in all practical purposes, the initial basis is I_m, thus

$$\mathbf{y}_j = I_m^{-1}\mathbf{a}_j = I_m\mathbf{a}_j = \mathbf{a}_j.$$

Thus initially,

$$y_{ij} = a_{ij} \tag{8.3.4}$$

But this is not essential, which has been shown in a problem (8.5.1) to understand the real theory lies behind this method.

Now we find out

$$z_j = \mathbf{c}_B\mathbf{y}_j = \mathbf{c}_B B^{-1}\mathbf{a}_j.$$

Thus,

$$z_j = c_{B1}y_{1j} + c_{B2}y_{2j} + \cdots + c_{Bi}y_{ij} + \cdots + c_{Bm}y_{mj} \tag{8.3.5}$$

and the value of the objective function corresponding to the basis B which is denoted by $z_B = z_\circ$ given by

$$z_\circ = c_{B1}x_{B1} + c_{B2}x_{B2} + \cdots + c_{Bi}x_{Bi} + \cdots + c_{Bm}x_{Bm} \tag{8.3.6}$$

Now we calculate all

$$z_j - c_j, \quad [j = 1, 2, \cdots, n] \quad \text{by using (8.3.5).}$$

8.3.2 Optimality test

For a maximization problem, if at any stage, all $z_j - c_j \geq 0$ $[j = 1, 2, \cdots, n]$ the problem is at the optimal stage. If at least one $z_j - c_j < 0$, then the problem is not at the optimal stage and we shall have to proceed further. If at least one $z_j - c_j < 0$ and at least one $y_{ij} > 0$, then the value of the objective function can be improved further or at least remains same. If any $z_j - c_j < 0$ and all $y_{ij} \leq 0$ $[i = 1, 2, \cdots, m]$ the problem has no finite optimal value and the problem is said to have an unbounded solution. [Actually all the data are to be placed in a table and the format of the simplex table will be placed in some later stage.]

8.3.3 Selection of vector to enter in the next basis

If the above condition is satisfied, i.e., at least one $z_j - c_j < 0$ and at least one $y_{ij} > 0$ $[i = 1, 2, \cdots, m]$ then we shall have to select a new basis. Thus one new vector is to be selected from $\mathbf{a}_j$ which is not in the previous basis to replace a vector from the previous basis to form a new basis. The following method is to be followed to select a vector to enter in the new basis.

If

$$z_k - c_k = \min_j (z_j - c_j, z_j - c_j < 0), \tag{8.3.7}$$

then $\mathbf{a}_k$ is the vector is to enter in the new basis and the kth column of the simplex table is called the *key column* or *pivot column*. If the selection is not unique, we can select any vector out of them arbitrarily and the column corresponding to that vector is the key column, the vector $\mathbf{a}_k$ will replace one vector from the previous basis.

8.3.4 Selection of vector which will leave the previous basis and be replaced by $\mathbf{a}_k$

If $\mathbf{a}_k$ is the vector to enter in the new basis, then the rule of selection of the vector to leave the current basis is given below.

If

$$\min_i \left(\frac{x_{Bi}}{y_{ik}}, y_{ik} > 0 \right) = y_{rk}, \tag{8.3.8}$$

then the vector in the rth position of the basis will be replaced by $\mathbf{a}_k$. rth row of the table is called the *key row* and y_{rk} which is called the *key element,* always positive, lies on the intersection of the kth column and the rth row. If the value of r is not unique, i.e.,

$$\min_i \left(\frac{x_{Bi}}{y_{ik}}, y_{ik} > 0 \right) \tag{8.3.9}$$

occur for more than one value of i, then select any value of such i as r and in that case next solution will be definitely degenerate and the vector in the rth position of the previous basis B will be replaced by $\mathbf{a}_k$ to form a new basis B_1 which will give a new basic feasible solution and here we get more than one basis all of which give basic feasible solutions. All these can be done from the table very easily.

8.3.5 Future Procedure

For the basis B_1, the basic solution $\mathbf{x}'_B = B_1^{-1}\mathbf{b}$ will be also feasible, i.e., $\mathbf{x}'_B \geq \mathbf{0}$.

Now, for B_1, we require to calculate $\mathbf{y}'_j = B_1^{-1}\mathbf{a}_j$.

$z'_B = \mathbf{c}'_B\mathbf{x}'_B$ [c'_B will be formed by replacing c_r by c_k] and finally $z'_j - c_j$ corresponding to the basis B_1 and all these can be done by using the formula given below.

From the simplex table, the quantities of interest can be calculated easily.

If we take $\mathbf{x}_B = \mathbf{y}_0$, i.e., $x_{Bi} = y_{i0}$ $[i = 1, 2, \cdots, m]$ and $z_0 = \mathbf{c}_B\mathbf{x}_B = y_{m+1,0}$ and $z_j - c_j = y_{m+1,j}$ $[j = 1, 2, \cdots, n]$.

then the rule of transformation from the present table to the next is given below in a compact form.

$z_j - c_j$ can also be calculated by using the formula

$$z_j - c_j = \mathbf{c}_B\mathbf{y}_j - c_j = c_{B1}y_{1j} + c_{B2}y_{2j} + \cdots + c_{Bm}y_{mj} - c_j. \tag{8.3.10}$$

But in the initial table the value of $z_j - c_j$ can only be calculated by using the formula given in (8.3.10).

8.4 Rule of Construction of a new table from the previous table

Rule of transformation: Divide the elements y_{rj} $[j = 0, 1, \cdots, n]$ of the rth row of the previous table by y_{rk} $[y^*_{rk} =$ key element of the previous table$]$ to get the rth row of next or current table. Thus

$$y'_{rj} = \frac{y_{rj}}{y^*_{rk}}, \quad i = r, \quad j = [0, 1, 2, \cdots, n]$$

and to get the elements of the other rows of the next table, we have

$$y'_{rj} = y_{ij} - y_{ik}\frac{y_{rj}}{y^*_{rk}} = \frac{y_{ij}y^*_{rk} - y_{rj}y_{ik}}{y^*_{rk}}$$

$$= \left|\begin{array}{cc} y_{ij} & y_{rj} \\ (1) & (2) \\ y_{ik} & y^*_{rk} \end{array}\right| \Big/ y^*_{rk},$$

$$i = 1, 2, \cdots, m, m+1, i \neq r. \tag{8.4.1}$$

8.5 Nature of the Problems

In solving a L.P.P. by any method (here we consider the simplex method) generally following cases arise.

1. *The problem will be solved and we get the finite value of the objective function with finite solution set.*

2. *Unbounded value of the objective function (No finite value of the objective function exists).*

   To solve a maximization problem, if at any stage, at least one $z_j - c_j < 0$ and all $y_{ij} \leq 0$, $[i = 1, 2, \cdots, m]$, the finite value of the objective function does not exist and the problem is said to have an unbounded solution.

3. *Multiple optimal solutions:* If for a problem, optimal value remains same but there exist more than one solution sets, then we say that the multiple optimal solutions or alternative optimal solutions exist. The condition for

the existence of multiple optimal solutions is that, at the optimal stage, at least one $z_j - c_j = 0$ corresponding to a non-basis vector. We know that all $z_j - c_j = 0$ corresponding to the basis vectors at each stage. The proof is given in the Appendix.

4. The problem will have no feasible solutions. This case will be discussed in the next chapter (Big M method).

5. *Solution of a minimization problem* will be discussed later on.

Table 8.5.1 : Simplex Table (at any stage)

		$\mathbf{c}$	c_1	c_2	$\cdots$	c_k		c_j	c_n	
Basis Vectors	$\mathbf{c}_B$	$\mathbf{b}$	$\mathbf{a}_1$	$\mathbf{a}_2$	$\cdots$	$\mathbf{a}_k$	$\cdots$	$\mathbf{a}_j$	$\mathbf{a}_n$	$\frac{x_{Bi}}{y_{ik}}, y_{ik} > 0$
β_1	c_{B1}	$x_{B1} = y_{10}$	y_{11}	y_{12}		y_{1k}		y_{1j}	y_{1n}	
β_2	c_{B2}	$x_{B2} = y_{20}$	y_{21}	y_{22}		y_{2k}		y_{2j}	y_{2n}	
$\vdots$	$\vdots$	$\vdots$	$\vdots$	$\vdots$		$\vdots$		$\vdots$	$\vdots$	
β_r^*	c_{Br}	$x_{Br} = y_{r0}$	y_{r1}	y_{r2}		y_{rk}	$\cdots$	y_{rj}	y_{rn}	
$\vdots$	$\vdots$	$\vdots$	$\vdots$	$\vdots$		$\vdots$		$\vdots$	$\vdots$	
β_m	c_{Bm}	$x_{Bm} = y_{m0}$	y_{m1}	y_{m2}		y_{mk}		y_{mj}	y_{mn}	
$z_j - c_j$		$z = y_{m+1 \cdot 0}$	$z_1 - c_1 = y_{m+1 \cdot 1}$	$z_2 - c_2 = y_{m+1 \cdot 2}$		$z_k - c_k = y_{m+1 \cdot k}$		$z_j - c_j = y_{m+1 \cdot j}$	$z_n - c_n = y_{m+1 \cdot n}$	

Explanation: In the first column, lists the vectors of the current basis, in the 2nd column is the current associated cost vector. In the third column under $\mathbf{b}$, are the values of the current basic variables. Column vectors under $\mathbf{a}_j$ list the current values of $\mathbf{y}_j$. Last row $\{(m+1)\text{th row}\}$ gives the current value of z and $z_j - c_j$ $[j = 1, 2, \cdots, n]$. If y_{rk} be the key element then in the next table $\mathbf{a}_k$ will replace β_r, the vector in the rth position of the basis, to form a new basis and c_k will replace c_{Br} to form new associated cost vector. The vector, which will be replaced by $\mathbf{a}_k$ is marked with asterisk. rth row is the key row and kth column under $\mathbf{a}_k$ is the key column.

From the format of the simplex table, one pertinent question must arises in mind. If $\mathbf{x}_B$ and $\mathbf{y}_j$ $[j = 1, \cdots, n]$ are placed under $\mathbf{b}$ and $\mathbf{a}_j$, respectively, then why shall we write $\mathbf{b}$ and $\mathbf{a}_j$ on the heading line of the format table instead of $\mathbf{x}_B$ and $\mathbf{y}_j$? The answer is very simple. After the proper choice of the *admissible basis* matrix in each stage, $\mathbf{x}_B$ and $\mathbf{y}_j$ at that stage can be obtained pre-multiplying the basis inverse with $\mathbf{b}$ and $\mathbf{a}_j$ $[j = 1, \cdots, n]$, where $\mathbf{b}$ and $\mathbf{a}_j$ remain unchanged in each and every iteration. That is why, on the heading line, notations $\mathbf{b}$ and $\mathbf{a}_j$ vectors have been kept intact instead of $\mathbf{x}_B$ and $\mathbf{y}_j$ $[j = 1, \cdots, n]$ which will generally change in each iteration. From the above table we can easily understand.

(i) Which vectors are in the current basis? (ii) Which vector will leave the basis? (iii) Which vector will enter in the next basis? The unit $\mathbf{y}_j$ vectors corresponding to the basis vectors are not the vectors of the basis at all.

Rule of construction of the next table from the initial table: If the initial B.F.S. is not an optimal solution, we shall have to proceed further to get a new B.F.S. which may optimize the objective function. To get the new B.F.S. we shall have to construct a new table from the initial simplex table. From (8.4.1) we get the new B.F.S. and the changed values y'_{ij}. If we transform the initial simplex table following the transformation (8.4.1), we at once get $\mathbf{x}'_B, \mathbf{y}'_j$, value of the objective function and $z'_j - c_j$ from the next table. Once again we re-state the formula of transformation. The rule of transformation is given below.

Rule of transformation: Divide the elements y_{rj} $[j = 0, 1, \cdots, n]$ of rth row (key row) of the initial table by $\mathbf{y}_{rk}$ = key element] to get the rth row of the next table. Thus

$$y'_{rj} = \frac{y_{rj}}{y^*_{rk}}, \quad i = r, j = (0, 1, 2, \cdots, n)$$

and to get the elements of the other rows of the next table, we have

$$y'_{ij} = y_{ij} - y_{ik}\frac{y_{rj}}{y^*_{rk}} = \frac{y_{ij}y^*_{rk} - y_{rj}y_{ik}}{y^*_{rk}}$$

$$= \left|\begin{array}{cc} y_{ij} & y_{rj} \\ 1 & 2 \\ \nwarrow & \nearrow \\ y_{ik} & y^*_{rk} \end{array}\right| \Big/ y^*_{rk} \quad \text{obtained from (8.4.1)}$$

$$i = 1, 2, \cdots, m+1, i \neq r.$$

The elements of $(m+1)$th row represents the value of z, which are z_0 or z_B and $z_j - c_j$.

Since computers use a very large + integer instead of M, but here M, in Big–M method thus here it is better to use the formula

$z_j - c_j = \mathbf{c}_B\mathbf{y}_j - c_j$ to calculate $z_j - c_j$ in each table in Big–M method.

General theoretical procedure in solving a L.P.P. using simplex theory. (Note that this method will not be used in practice in future but has been solved to give an exact idea behind the simplex theory).

▶ **Example 8.5.1** *Solve the L.P.P. (by using general theory).*

$$\textit{Maximize, } z = x_1 - x_2 + 2x_3 + 3x_4$$

subject to

$$\begin{aligned} 2x_1 + x_2 + 3x_3 + 2x_4 &= 11 \\ 3x_1 - 3x_2 + 5x_3 + x_4 &= 17, \quad x_j \geq 0, \ j = 1, 2, 3, 4. \end{aligned}$$

Solution: **Step 1:** Search for a basis which will produce a feasible solution.

$$A = \begin{bmatrix} 2 & 1 & 3 & 2 \\ 3 & -3 & 5 & 1 \end{bmatrix}, \quad R(A) = 2.$$

Thus the two equations are linearly independent and consistent.

$$\mathbf{a}_1 = \begin{bmatrix} 2 \\ 3 \end{bmatrix}, \quad \mathbf{a}_2 = \begin{bmatrix} 1 \\ -3 \end{bmatrix}, \quad \mathbf{a}_3 = \begin{bmatrix} 3 \\ 5 \end{bmatrix}, \quad \mathbf{a}_4 = \begin{bmatrix} 2 \\ 1 \end{bmatrix} \quad \text{and} \quad \mathbf{b} = \begin{bmatrix} 11 \\ 17 \end{bmatrix}$$

$$\mathbf{c} = (c_1, c_2, c_3, c_4) = (1, -1, 2, 3).$$

Two vectors are required to form a basis.

First consider a square matrix

$$B = (\mathbf{a}_1, \mathbf{a}_2) = \begin{bmatrix} 2 & 1 \\ 3 & -3 \end{bmatrix}, \quad \det B = -9 \neq 0.$$

Hence B can be considered as a basis.

$$B^{-1} = -\frac{1}{9} \begin{bmatrix} -3 & -1 \\ -3 & 2 \end{bmatrix}$$

The basic solution corresponding to the basis B is

$$\mathbf{x}_B = B^{-1}\mathbf{b} = -\frac{1}{9} \begin{bmatrix} -3 & -1 \\ -3 & 2 \end{bmatrix} \begin{bmatrix} 11 \\ 17 \end{bmatrix} = \begin{bmatrix} 50/9 \\ -1/9 \end{bmatrix}$$

Here the solution is not feasible. Thus B cannot be considered as an *admissible basis* to start with the simplex procedure.

Now consider the square matrix

$$B = (\mathbf{a}_1, \mathbf{a}_3) = \begin{bmatrix} 2 & 3 \\ 3 & 5 \end{bmatrix}, \quad \det B = 1 \neq 0.$$

Hence $B = \begin{bmatrix} 2 & 3 \\ 3 & 5 \end{bmatrix}$ is also a basis. The B.S. corresponding to the new basis B is

$$\mathbf{x}_B = B^{-1}\mathbf{b} = \frac{1}{1} \begin{bmatrix} 5 & -3 \\ -3 & 2 \end{bmatrix} \begin{bmatrix} 11 \\ 17 \end{bmatrix} = \begin{bmatrix} 4 \\ 1 \end{bmatrix}$$

which is feasible.

Then $\mathbf{x}_B = [x_{B1}, x_{B2}] = [x_1, x_3] = [4, 1]$ and $B = \begin{bmatrix} 2 & 3 \\ 3 & 5 \end{bmatrix}$ can be considered as an *admissible basis* to start with the simplex procedure.

Step 2: Calculation of $\mathbf{y}_j$ and $z_j - c_j$ $[j = 1, 2, 3, 4]$ corresponding to the basis $B = \begin{bmatrix} 2 & 3 \\ 3 & 5 \end{bmatrix}$. We have

$$\begin{aligned} [\mathbf{y}_1, \mathbf{y}_2, \mathbf{y}_3, \mathbf{y}_4] &= B^{-1}(\mathbf{a}_1, \mathbf{a}_2, \mathbf{a}_3, \mathbf{a}_4) \\ &= \begin{bmatrix} 5 & -3 \\ -3 & 2 \end{bmatrix} \begin{bmatrix} 2 & 1 & 3 & 2 \\ 3 & -3 & 5 & 1 \end{bmatrix} \quad \begin{bmatrix} 1 & 14 & 0 & 7 \\ 0 & -9 & 1 & -4 \end{bmatrix} \end{aligned}$$

Thus,

$$\mathbf{y}_1 = \begin{bmatrix} y_{11} \\ y_{21} \end{bmatrix} = \begin{bmatrix} 1 \\ 0 \end{bmatrix}, \quad \mathbf{y}_2 = \begin{bmatrix} y_{12} \\ y_{22} \end{bmatrix} = \begin{bmatrix} 14 \\ -9 \end{bmatrix},$$

$$\mathbf{y}_3 = \begin{bmatrix} y_{13} \\ y_{23} \end{bmatrix} = \begin{bmatrix} 0 \\ 1 \end{bmatrix} \text{ and } \mathbf{y}_4 = \begin{bmatrix} y_{14} \\ y_{24} \end{bmatrix} = \begin{bmatrix} 7 \\ -4 \end{bmatrix}$$

Here $B = (\mathbf{a}_1, \mathbf{a}_3) = (\beta_1, \beta_2)$ say. Thus $\mathbf{c}_B = (c_{B1}, c_{B2}) = (c_1, c_3) = (1, 2)$ the cost components corresponding to the basic variables x_1 and x_3.

$$\begin{aligned}
z_1 - c_1 &= \mathbf{c}_B\mathbf{y}_1 - c_1 = c_{B1}y_{11} + c_{B2}y_{21} = 1 \times 1 + 2 \times 0 - 1 = 0 \\
z_2 - c_2 &= \mathbf{c}_B\mathbf{y}_2 - c_2 = c_{B1}y_{12} + c_{B2}y_{22} = 1 \times 14 + 2 \times (-9) + 1 = -3 \\
z_3 - c_3 &= \mathbf{c}_B\mathbf{y}_3 - c_3 = c_{B1}y_{13} + c_{B2}y_{23} = 1 \times 0 + 2 \times 1 - 2 = 0 \\
z_4 - c_4 &= \mathbf{c}_B\mathbf{y}_4 - c_4 = c_{B1}y_{14} + c_{B2}y_{24} = 1 \times 7 + 2 \times (-4) - 3 = -4 \\
z_0 &= \mathbf{c}_B\mathbf{x}_B = c_{B1}x_{B1} + c_{B2}x_{B2} = 1 \times 4 + 2 \times 1 = 6.
\end{aligned}$$

Conclusion: Here $z_2 - c_2$ and $z_4 - c_4$ are both negative with at least one of y_{i2} and $y_{i4} > 0$ $(i = 1, 2)$. Thus the B.F.S. $\mathbf{x}_B = [x_{B1}, x_{B2}] = [x_1, x_3] = [4, 1]$ corresponding to the basis

$$B = (\mathbf{a}_1, \mathbf{a}_3) = \begin{bmatrix} 2 & 3 \\ 3 & 5 \end{bmatrix}$$

is not an optimal solution of the problem and we shall have to proceed further to obtain an optimal solution [if it exists at all].

Step 3: Search for a new vector to enter in the next basis and the vector which will leave the current basis $(\mathbf{a}_1, \mathbf{a}_3)$.

$$\begin{aligned}
\min_j(z_j - c_j, z_j - c_j < 0) &= \min(z_2 - c_2, z_4 - c_4) \\
&= \min(-3, -4) = -4
\end{aligned}$$

which occurs for $j = 4$.

Then the fourth vector $\mathbf{a}_4$ will enter in the next basis. Thus, the fourth column is the key column. Again,

$$\begin{aligned}
\min_i\left(\frac{x_{Bi}}{y_{i4}},\ y_{i4} > 0\right) &= \min\left(\frac{x_{B1}}{y_{14}},\ \frac{x_{B2}}{y_{24}},\ y_{i4} > 0\right) \\
&= \min\left(\frac{4}{7}, \cdots\right),
\end{aligned}$$

$$\left[\frac{x_{B2}}{y_{24}} = \frac{1}{-4} \text{ is not considered as } y_{24} = -4 < 0\right]$$

which occurs for $i = 1$. Then first row is the key row and $y_{14} = 7$ is the key element. Then the first vector $\beta_1 = \mathbf{a}_1$ of the previous basis will leave the basis and replaced by $\mathbf{a}_4 = \beta_1^*$ (say) to form a new basis and the new basis will be

$$B = (\mathbf{a}_4, \mathbf{a}_3) = (\beta_1^*, \beta_2) \text{ say } = \begin{bmatrix} 2 & 3 \\ 1 & 5 \end{bmatrix}$$

and $y_{14} = 7$ which is the element at the point of intersection of key row [first row] and the key column [fourth column] is the key element.

Now we solve the problem by using simplex tables.

Simplex tables

		c	c_1=1	c_2=−1	c_3=2	c_4=3	
Basis	c_B	b	$\mathbf{a}_1$	$\mathbf{a}_2$	$\mathbf{a}_3$	$\mathbf{a}_4$	x_{Bi}/y_{i4}, y_{i4}>0
$\mathbf{a}_1$*	c_{B1}=1	x_{B1}=4	y_{11}=1	y_{12}=14	y_{13}=0	y^*_{14}=7*	x_{B1}/y_{14}=4*/7
$\mathbf{a}_3$	c_{B2}=2	x_{B2}=1	y_{21}=0	y_{22}=−9	y_{23}=1	y_{24}=−4	...
	z_j−c_j	z_0=6	z_1−c_1=0	z_2−c_2=−3	z_3−c_3=0	z_4−c_4=−4*	
$\mathbf{a}_4$	c_{B1}=3	x_{B1}=4/7	y_{11}=1/7	y_{12}=2	y_{13}=0	y_{14}=1	
$\mathbf{a}_3$	c_{B2}=2	x_{B2}=23/7	y_{21}=4/7	y_{22}=−1	y_{23}=1	y_{24}=0	
	z_j−c_j	z_0=58/7	z_1−c_1=4/7	z_2−c_2=5	z_3−c_3=0	z_4−c_4=0	

All the quantities of interest are marked with asterisk.

Construction of the second table from the first table. First row is the key row and $y_{14} = 7$ is the key element and $\mathbf{a}_4$ will remove $\mathbf{a}_1$ to form a new basis $B = (\mathbf{a}_4, \mathbf{a}_3)$

8.5.1 Calculation of third row, $(z_j - c_j)$ row

$$\begin{aligned}
z_0 &= y_{30} = \frac{7 \times 6 - (-4) \times 4}{7} = \frac{58}{7},\\
z_1 - c_1 &= y_{31} = \frac{7 \times 0 - (-4) \times 1}{7} = \frac{4}{7},\\
z_2 - c_2 &= y_{32} = \frac{7 \times (-3) - (-4) \times 14}{7} = 5,\\
z_3 - c_3 &= y_{33} = 0,\\
z_4 - c_4 &= y_{34} = 0,
\end{aligned}$$

all $z_j - c_j \geq 0$, $j = 1, 2, 3, 4$ thus we reach at the optimal stage. We need not calculate the second row. But we calculate the first and the second row to verify the problem in another method.

8.5.2 Calculation of the first row

Devide all the elements of the key row including x_{B1} by 7 (key element) to get the first row of the second table.

8.5.3 Calculation of the second row

$$\begin{aligned}
y_{20} &= x_{B2} = \frac{7\times 1-(-4)\times 4}{7} = \frac{23}{7},\\
y_{21} &= \frac{7\times 0-(-4)\times 1}{7} = \frac{4}{7},\\
y_{22} &= \frac{7\times(-9)-(-4)\times 14}{7} = -1,\\
y_{23} &= 1,\\
y_{24} &= 0.
\end{aligned}$$

Of course, we can calculate the $z_j - c_j$ row after completing the body of the second table by using the formula $z_j - c_j = \mathbf{c}_B\mathbf{y}_j - c_j$ and the value of the objective function needs not to be calculated at each stage.

$$\begin{aligned}
z_1 - c_1 &= 3\times\frac{1}{7}+2\times\frac{4}{7}-1 = \frac{4}{7},\\
z_2 - c_2 &= 3\times 2+2\times(-1)+1 = 5,\\
z_3 - c_3 &= 0,\\
z_4 - c_4 &= 0,\\
z_0 &= 3\times\frac{4}{7}+2\times\frac{23}{7} = \frac{58}{7} = \max z,
\end{aligned}$$

Thus max $z = 58/7$ at $x_1 = 0$, $x_2 = 0$, $x_3 = 23/7$, $x_4 = 4/7$.

Remark: (1) For B.F.S. $\mathbf{x}_B = [x_1, x_3] = [4, 1]$, the value of the objective function is

$$z_B = \mathbf{c}_B\mathbf{x}_B = (1,2)[4,1] = 6 \le \frac{58}{7} = \max z.$$

(2) $B = (\mathbf{a}_1, \mathbf{a}_3)$ is accepted as an *initial admissible basis* to start the simplex procedure. Of course, other basis which can produce a feasible solution can also be taken as an initial admissible basis. For example, $B = (\mathbf{a}_4, \mathbf{a}_3)$ which produces a feasible solution can be taken as an initial admissible basis and in that case *step 2* will give the optimal value of the problem.

(3) In each stage, the *vectors* $\mathbf{a}_1, \mathbf{a}_2, \cdots, \mathbf{a}_4, \mathbf{b}$ *do not change.* The changes those take place are; replacement of one vector of the basis by another vector which is not in the basis to form a new basis, the corresponding changes of $\mathbf{y}_j$ and B.F.S. $\mathbf{x}_B = B^{-1}\mathbf{b}$ etc. To get $\mathbf{y}_j$ and $\mathbf{x}_B$, we only require to pre-multiply B^{-1} with $\mathbf{a}_j$ and $\mathbf{b}$, where B is the basis at that stage. In this problem, the first admissible basis is $B_1 = (\mathbf{a}_1, \mathbf{a}_3)$. The second basis is $B_2 = (\mathbf{a}_4, \mathbf{a}_3)$ which is obtained by replacing $\mathbf{a}_1$ of the previous basis by a vector $\mathbf{a}_4 \in A$, which is not in the previous basis. Thus to form a new basis from the old one, replacement takes place but no numerical change of the vectors in the basis takes place.

(4) It is important to note that $\mathbf{y}_j$ corresponding to the basis vectors in each stage are unit vectors and they are such that, they can be arranged to form a unit matrix and $z_j - c_j = 0$ corresponding to all basis vectors. In **step 2.** $\mathbf{y}_1$ and

$\mathbf{y}_3$ are unit vectors and $[\mathbf{y}_1, \mathbf{y}_3] = \begin{bmatrix} 1 & 0 \\ 0 & 1 \end{bmatrix} = I_2$ and $z_1 - c_1 = z_3 - c_3 = 0$.

Similarly in **step 3,** $\mathbf{y}_3$ and $\mathbf{y}_4$ are unit vectors and $[\mathbf{y}_4, \mathbf{y}_3] = \begin{bmatrix} 1 & 0 \\ 0 & 1 \end{bmatrix} = I_2$ and $z_4 - c_4 = z_3 - c_3 = 0$ etc.

(5) This problem has been solved in Example (6.1.2) by using the *fundamental theorem of L.P.P.* making an assumption that an optimal solution exists. But here we need not make any such assumption.

8.6 Relation between of the maximum value and minimum value of the objective function

Theorem 8.6.1 *Minimum value of z is the negative of the maximum* $(-z)$ *with the same solution set, i.e.,* $\min z = -\max(-z)$ *with the same solution set.*

Proof: $z = \mathbf{cx}$. Let z attain its minimum at $\mathbf{x} = \mathbf{x}_0$ then $\min z = \mathbf{cx}_0$.

Hence

$$\mathbf{cx} \geq \mathbf{cx}_0, \quad \text{or,} \quad -\mathbf{cx} \leq -\mathbf{cx}_0.$$

Therefore,

$$\max(-\mathbf{cx}) = -\mathbf{cx}_0, \quad \text{or,} \quad \mathbf{cx}_0 = -\max(-\mathbf{cx}),$$

$$\text{or,} \quad \min(z) = -\max(-z) \tag{8.6.1}$$

with the same solution set. Similarly,

$$\max(z) = -\min(-z). \tag{8.6.2}$$

Theorem 8.6.2 *The column vectors* $\mathbf{y}_j$ *corresponding to* $\mathbf{a}_j$ *(which are the basis vectors at that stage) are unit vectors and the* $\mathbf{y}_j$ *under the initial unit basis vectors gives the basis inverse in each table.*

Proof: Let at any iteration the basis vectors be $\mathbf{a}_1, \mathbf{a}_2, \cdots, \mathbf{a}_m$. Then the basis at that stage is $B = (\mathbf{a}_1, \mathbf{a}_2, \cdots, \mathbf{a}_m)$. We have $\mathbf{y}_j = B^{-1}\mathbf{a}_j$. Therefore, $\mathbf{y}_j$ corresponding to the vectors $\mathbf{a}_1, \mathbf{a}_2, \cdots, \mathbf{a}_m$ are given by

$$(\mathbf{y}_1, \mathbf{y}_2, \cdots, \mathbf{y}_m) = B^{-1}(\mathbf{a}_1, \mathbf{a}_2, \cdots, \mathbf{a}_m),$$

$$\text{or,} \quad (\mathbf{y}_1, \mathbf{y}_2, \cdots, \mathbf{y}_m) = B^{-1}B = I_m. \tag{8.6.3}$$

Hence the column vectors $\mathbf{y}_j$ corresponding to the basis vectors $\mathbf{a}_j$ are always unit vectors and they can be arranged so as to form a unit matrix.

And finally, if the initial basis be unit basis I_m then $\mathbf{y}_j$ vectors under the initial unit basis vectors gives the basis inverse in each iteration

$$\text{since} \quad \mathbf{y}_j (j = 1, 2, \cdots, m) = B^{-1}I_m = B^{-1}. \tag{8.6.4}$$

8.7 Determination of the initial B.F.S. to a L.P.P.

Case 1: When all constraints are connected with sign "$\leq$" and all $b_i \geq 0$. First of all convert all constraint inequations into equations by adding slack variables one to each of the inequations. If there are m constraint inequations with initial n variables $x_1, x_2, \cdots, x_n$ the converted equations contain $(n+m)$ variables of which m variables are slack variables.

The converted equations are

$$\begin{aligned} a_{11}x_1 + a_{12}x_2 + \cdots + a_{1n}x_n + x_{n+1} \qquad\qquad &= b_1 \\ a_{21}x_1 + a_{22}x_2 + \cdots + a_{2n}x_n + \quad \cdots + x_{n+2} \qquad &= b_2 \\ \cdots\cdots\cdots\cdots\cdots\cdots\cdots\cdots\cdots\cdots\cdots\cdots\cdots\cdots & \\ a_{m1}x_1 + a_{m2}x_2 + \cdots + a_{mn}x_n + \quad \cdots + \cdots + x_{n+m} &= b_m \end{aligned}$$

where $x_{n+1}, x_{n+2}, \cdots, x_{n+m}$ are m-slack variables.

The coefficient matrix A of the above set of equations is

$$\left[\begin{array}{cccc|ccc} a_{11} & a_{12} & \cdots & a_{1n} & 1 & 0\cdots 0 \\ a_{21} & a_{22} & \cdots & a_{2n} & 0 & 1\cdots 0 \\ \cdots & \cdots & \cdots & \cdots & \cdots & \cdots \\ a_{m1} & a_{m2} & \cdots & a_{mn} & 0 & 0\cdots 1 \end{array}\right] \quad (\text{last } m \text{ columns: } m \text{ slack vectors})$$

It is a $m \times (n+m)$ matrix.

The column vectors associated with slack variables are known as *slack vectors*. All slack vectors are unit vectors which are linearly independent. Hence the m slack vectors constitute a basis matrix B which is always true for this type of problem. $B = (\mathbf{e}_1, \mathbf{e}_2, \cdots, \mathbf{e}_m)$. The B.F.S. $\mathbf{x}_B = B^{-1}\mathbf{b} = \mathbf{b} \geq \mathbf{0}$. But as the unit vector $\mathbf{e}_i$ occurs at the $(n+i)$th position, then basic variable $x_{Bi} = x_{n+i} = b_i$ and hence the B.F.S. is given by

$$\mathbf{x}_B = [x_{n+1}, x_{n+2}, \cdots, x_{n+m}] = [b_1, b_2, \cdots, b_m] \geq \mathbf{0}. \qquad (8.7.1)$$

The B.F.S. $\mathbf{x}_B$ is taken as initial B.F.S. in solving the problem by simplex method.

Case 2: When the constraints are of mixed type connected with signs "$\leq$"·"$\geq$" and "$=$".

Initially make all b_i's $(i = 1, 2, \cdots, m)$ positive by suitable adjustment. Then convert all the inequations into equations by introducing slack or surplus variables one in each constraint (whenever necessary). The column vectors associated with the surplus variables are known as *surplus vectors*. It is important to note that in this type of problem initially coefficient matrix does not necessarily contain a basis matrix which is a unit matrix. Let us consider a numerical example. Let the constraints be

$$\begin{aligned} 2x_1 + 3x_2 - \ x_3 &\leq 4 \\ x_1 + \ x_2 + 4x_3 &\geq 7 \\ x_1 - 3x_2 + 2x_3 &= 8 \end{aligned}$$

Here all b_i's are already positive. After the addition of slack variable x_4 to first constraint and surplus variable x_5 to the second constraint, the converted equations are

$$\begin{aligned} 2x_1 + 3x_2 - x_3 + x_4 \quad &= 4 \\ x_1 + x_2 + 4x_3 + \quad - x_5 &= 7 \\ x_1 - 3x_2 + 2x_3 + \quad &= 8 \end{aligned}$$

The coefficient matrix A is given by

$$A = \begin{bmatrix} 2 & 3 & -1 & 1 & 0 \\ 1 & 1 & 4 & 0 & -1 \\ 1 & -3 & 2 & 0 & 0 \end{bmatrix} \quad \text{and} \quad \mathbf{b} = \begin{bmatrix} 4 \\ 7 \\ 8 \end{bmatrix} \geq \mathbf{0}.$$

There is no unit basis matrix in A. A new technique has been adopted which is known as *artificial variable technique* to get an unit basis matrix from the coefficient matrix. A set of variables are to be added again to the left hand side of the equations (in each equation whenever necessary) to get a unit matrix out of the coefficient matrix. These variables are known as the *artificial variables* and the column vectors associated with the artificial variables are known as *artificial vectors.* But care should be taken to ensure that a minimum number of artificial variables be inserted to get a unit basis matrix. In the above example, the constraint equations are

$$\begin{aligned} 2x_1 + 3x_2 - x_3 + x_4 \qquad\qquad &= 4 \\ x_1 + x_2 + 4x_3 \quad - x_5 + x_6 \quad &= 7 \\ x_1 - 3x_2 + 2x_3 \qquad\qquad + x_7 &= 8 \end{aligned}$$

after the introduction of the artificial variables x_6 and x_7. The coefficient matrix is then

$$\begin{bmatrix} 2 & 3 & -1 & 1 & 0 & 0 & 0 \\ 1 & 1 & 4 & 0 & -1 & 1 & 0 \\ 1 & -3 & 2 & 0 & 0 & 0 & 1 \end{bmatrix}$$

which contains a unit basis matrix $B = (\mathbf{e}_1, \mathbf{e}_2, \mathbf{e}_3)$. The fourth column vector is a slack vector and sixth and seventh column vectors are artificial vectors. Now the basic feasible solution (artificial) is given by

$$\mathbf{x}_B = B^{-1}\mathbf{b} = \mathbf{b} = [x_{B1}, x_{B2}, x_{B3}] = [x_4, x_6, x_7] = [4, 7, 8] \tag{8.7.2}$$

as $\mathbf{e}_1, \mathbf{e}_2$ and $\mathbf{e}_3$ occur at fourth, sixth and seventh column respectively. Applying this method to numerical examples, the initial B.F.S. can be determined. The artificial variables are essentially non-negative quantities. The method of solving this type of problems will be discussed later [by Big. M-method].

Note: No artificial variable is added to the first equation.

8.8 Advantages and disadvantages of using unit matrix as the initial basis matrix

We can find the initial B.F.S. at once due to the presence of unit basis provided $\mathbf{b} \geq \mathbf{0}$. Moreover, we can determine the value of y_{ij} for all i and j easily for the initial simplex table because $\mathbf{y}_j = B^{-1}\mathbf{a}_j = \mathbf{a}_j$, i.e., $y_{ij} = a_{ij}$ in the initial simplex table. But the artificial vectors included in the unit basis create some difficulties in solving the problem which we shall discuss later.

Note: In simplex method, the initial B.F.S. can be determined by other method when initial basis is not a unit basis which we have discussed previously in Example 4.3.1 and Example 4.3.2. But this method is too difficult and after determining the initial B.F.S. and the vectors $\mathbf{y}_j$, using the relation $\mathbf{y}_j = B^{-1}\mathbf{a}_j$, the starting simplex table can be constructed and the problem can be solved. But this method is rarely used in practice [See Example 8.5.1].

8.9 Construction of the initial simplex table

We have stated earlier, that to get an initial B.F.S. easily, the constraints are to be converted in such a way that a unit basis matrix B is present in the coefficient matrix with all $b_i \geq 0$. Then, initial B.F.S.

$$\mathbf{x}_B = B^{-1}\mathbf{b} = \mathbf{b} \geq \mathbf{0} \quad \text{and} \quad z = \mathbf{c}_B\mathbf{x}_B = \mathbf{c}_B\mathbf{b}$$

(initial value of the objective function) and

$$\mathbf{y}_j = B^{-1}\mathbf{a}_j = \mathbf{a}_j.$$

Then $y_{ij} = a_{ij}$ for all values of i and j. Hence the elements y_{ij} can be determined easily from the coefficient matrix. Now

$$z_j - c_j = \mathbf{c}_B B^{-1}\mathbf{a}_j - c_j = \mathbf{c}_B\mathbf{y}_j - c_j = \mathbf{c}_B\mathbf{a}_j - c_j.$$

With these data an initial simplex table (chart) can be constructed according to table given in table (8.5.1).

8.10 Computational procedure (when m slack vectors constitute the initial basis $B = I_m$ (problem of maximization) and $b \geq 0$

1. Convert all constraints into equations by adding m slack vectors one to each constraint.

2. Readjust the objective function accordingly.

3. Calculate the I.B.F.S. $\mathbf{x}_B = B^{-1}\mathbf{b} = \mathbf{b}$, associated cost vector $\mathbf{c}_B$, value of the objective function $z_B = \mathbf{c}_B\mathbf{x}_B = \mathbf{c}_B\mathbf{b}$, $\mathbf{y}_j = B^{-1}\mathbf{a}_j = \mathbf{a}_j$ and $z_j - c_j = \mathbf{c}_B\mathbf{y}_j - c_j = \mathbf{c}_B\mathbf{a}_j - c_j$ for all j.

4. With these data construct the initial simplex table as given in table (8.5.1).

5. If all $z_j - c_j \geq 0$, the initial B.F.S. is optimal and thus the value of the objective function will be the last element of the column headed by **b**. If at least one $z_j - c_j < 0$, the solution is not optimal, then proceed further to construct the next table to get the next B.F.S. which may optimize the objective function. For this, the selection of key element y_{rk} is essential. [If $z_k - c_k$ is minimum (negative), the vector $\mathbf{a}_k$ will enter in the next basis, replacing the vector in the rth position under the column *vector in the basis* and c_k will replace c_{Br}. If minimum is not unique select any one of them. The value of r is to be determined by min ratio rule which has been discussed previously in (8.3.8) and (8.3.9).]

6. Construct the next table following the rule of transformation given in (8.4.1). The value of z and $z_j - c_j$ are given in the last row of the table. If $z_j - c_j \geq 0$ for all j the B.F.S. is optimal. If at least one $z_j - c_j < 0$, then proceed further to construct a new table to get an optimal solution and so on until the conditions $z_j - c_j \geq 0$ are satisfied for all columns. At the optimal stage, the optimal value of the objective function is the last element under the column headed by **b**. Of course, we may calculate the optimal value of the objective function by using the relation $z = \mathbf{c}_B\mathbf{x}_B$. [All values are to be calculated from the final table]. The value of the objective function obtained from these two methods will be same. But this does not ensure the correctness of the solution of the problem. But by using duality theory we may verify the correctness of the solution with a fair degree of accuracy.

7. At any stage if $z_j - c_j < 0$ for at least one column with $y_{ij} \leq 0$ for all i, no finite value of the objective function exists.

   The problem is said to have an unbounded solution. We need not calculate the value of the objective function.

8. If none of $z_j - c_j < 0$ but some $z_j - c_j = 0$ corresponding to non-basis vectors, multiple solutions exist.

9. All discussions are related to the problem maximization of the objective function z. For a minimization problem find out the maximum value of $(-z = z')$ and from mini-max theorem min $z = -\max(-z) = -\max z'$ with the same solution set. Hence the problem can be solved by the method above. Of course, minimization problem can be solved as a minimization problem and in that case the vector to be entered in the basis corresponding to positive maximum of $z_j - c_j$ and at the optimal stage all $z_j - c_j \leq 0$ which has been shown in the worked out Example (8.11.1).

8.11 Solution to a L.P.P. when m slack vectors constitute a unit basis and $b_i \geq 0$ for all i

▶ **Example 8.11.1** *Solve the L.P.P.*

$$\textit{maximize}, \quad z = 5x_1 + 2x_2 + 2x_3$$

subject to

$$\begin{aligned} x_1 + 2x_2 - 2x_3 &\leq 30 \\ x_1 + 3x_2 + x_3 &\leq 36, \quad x_1, x_2, x_3 \geq 0. \end{aligned}$$

Solution: This is a maximization problem. $b_i \geq 0$ for $i = 1, 2$ and the constraints involved with sign '$\leq$'.

Introducing two slack variables x_4 and x_5 one to each constraint, we get the following converted equations

$$\begin{aligned} x_1 + 2x_2 - 2x_3 + x_4 \quad &= 30 \\ x_1 + 3x_2 + x_3 \quad + x_5 &= 36, \end{aligned}$$

$x_1, x_2, x_3 \geq 0$, $x_4, x_5 \geq 0$ (slack variables).

The adjusted objective function $z = 5x_1 + 2x_2 + 2x_3 + 0x_4 + 0x_5$

$$\mathbf{c} = (5, 2, 2, 0, 0).$$

The coefficient of variables constitute the column vectors,

$$\mathbf{a}_1 = \begin{bmatrix} 1 \\ 1 \end{bmatrix}, \quad \mathbf{a}_2 = \begin{bmatrix} 2 \\ 3 \end{bmatrix}, \quad \mathbf{a}_3 = \begin{bmatrix} -2 \\ 1 \end{bmatrix},$$

$$\mathbf{a}_4 = \begin{bmatrix} 1 \\ 0 \end{bmatrix}, \quad \mathbf{a}_5 = \begin{bmatrix} 0 \\ 1 \end{bmatrix} \quad \text{and} \quad \mathbf{b} = \begin{bmatrix} 30 \\ 36 \end{bmatrix} \geq \mathbf{0}.$$

Here $\mathbf{a}_4(\mathbf{e}_1)$ and $\mathbf{a}_5(\mathbf{e}_2)$, the slack vectors are unit vectors which together constitute a unit basis matrix $B = I_2$ and taking this as initial basis, initial B.S. $\mathbf{x}_B = B^{-1}\mathbf{b} = I_2^{-1}\mathbf{b} = I_2\mathbf{b} = \mathbf{b} \geq \mathbf{0}$ which is feasible. Thus $B = I_2$ is the initial admissible basis to start the problem.

$$\begin{aligned} \mathbf{x}_B &= [x_{B1}, x_{B2}] = [x_4, x_5] = [b_1, b_2] = [30, 36] \\ \mathbf{c}_B &= (c_{B1}, c_{B2}) = (c_4, c_5) = (0, 0) = \mathbf{0} \\ z_B &= \mathbf{c}_B\mathbf{x}_B = c_{B1}x_{B1} + c_{B2}x_{B2} = 0 \\ \mathbf{y}_j &= \mathbf{B}^{-1}\mathbf{a}_j = I_2\mathbf{a}_j = \mathbf{a}_j, \text{ i.e.,} \\ y_{ij} &= a_{ij} \end{aligned}$$

in the initial table $[j = 1, 2, 3, 4, 5]$ but they are no longer the coefficient of variable x_j; they are now quite different quantities.

$$z_j - c_j = \mathbf{c}_B\mathbf{y}_j - c_j = \mathbf{c}_B\mathbf{a}_j - c_j = \mathbf{0}\mathbf{a}_j - c_j = -c_j \quad [j = 1, 2, \ldots, 5].$$

[**Note:** It is obvious that $z_4 - c_4$ and $z_5 - c_5$ corresponding to the basis vectors $\mathbf{a}_4(\mathbf{e}_1)$ and $\mathbf{a}_5(\mathbf{e}_2)$ are zero].

With the help of the data given above, we can construct the initial simplex table as given in the table (8.5.1) putting column vector $\mathbf{x}_B = \mathbf{y}_0$ under b and $\mathbf{y}_j$ under $\mathbf{a}_j$ $[j = 1, 2, \cdots, 5]$.

Initial simplex table

		c	5	2	2	0	0	
Basis	c_B	b	$\mathbf{a}_1$	$\mathbf{a}_2$	$\mathbf{a}_3$	$\mathbf{a}_4(\mathbf{e}_1)$	$\mathbf{a}_5(\mathbf{e}_2)$	x_{Bi}/y_{i1}, $y_{i1}>0$
$\mathbf{a}_4$*	0	30	1*	2	–2	1	0	30/1=30*
$\mathbf{a}_5$	0	36	1	3	1	0	1	36/1=36
$z_j–c_j$		z_0=0	–5*	–2	–2	0	0	

As $z_1 - c_1$, $z_2 - c_2$ and $z_3 - c_3$ are all negative and at least one component of $\mathbf{y}_1, \mathbf{y}_2$ and $\mathbf{y}_3$ are positive, therefore the present solution is not optimal. Min $(z_j - c_j, z_j - c_j < 0) = \min(z_1 - c_1 = -5,\ z_2 - c_2 = -2,\ z_3 - c_3 = -2) = -5$ which is $z_1 - c_1$ which occurs for $j = 1$. Then the first column is the key column and the first vector $\mathbf{a}_1$ is the vector to enter in the next basis by replacing one vector from $\mathbf{a}_4$ and $\mathbf{a}_5$ which form the initial basis. $z_1 - c_1 = -5$ is marked with asterisk.

$$\min_{i=1,2}\left(\frac{x_{Bi}}{y_{i1}}, y_{i1} > 0\right) = \min\left(\frac{x_{B1}}{y_{11}}, \frac{x_{B2}}{y_{21}}\right) = \min\left(\frac{30}{1}, \frac{36}{1}\right)$$

$$= 30 \text{ which occurs at } i = 1.$$

Therefore, $r = 1$ and the first row is the key row and $y_{11} = 1$ is the key element and since $r = 1$, therefore the first vector $\mathbf{a}_4$ under the column (basis) will leave the current basis and will be replaced by $\mathbf{a}_1$ and $\mathbf{a}_1, \mathbf{a}_5$ will constitute the basis in the next iteration. Vector $\mathbf{a}_4$, the key element $y_{11} = 1$ and min $\left(\frac{x_{B1}}{y_{11}}\right) = 30$ are marked with asterisk to see the quantities of interest at a glance.

Rule of construction of the second table from the first table: Initially we can calculate $z_0 = y_{30}$ and $z_j - c_j = y_{3j} (j = 1, 2, \cdots, 5)$ elements of $z_j - c_j$ row without computing all other rows. If all $z_j - c_j \geq 0$, we need not calculate all other rows. We only require to calculate all the components of the column under b which are the components of B.F.S. But if at least one $z_j - c_j < 0$, we shall have to complete the whole table.

Calculation

Third row:

$$z_0 = y_{30} = \frac{1 \times 0 - (30 \times -5)}{1} = 150,$$

$$z_1 - c_1 = y_{31} = 0,$$

$$z_2 - c_2 = y_{32} = \frac{1 \times (-2) - 2 \times (-5)}{1} = 8,$$

$$z_3 - c_3 = y_{33} = \frac{1 \times (-2) - (-5 \times -2)}{1} = -12,$$

$$z_4 - c_4 = y_{34} = \frac{1 \times 0 - (-5 \times 1)}{1} = 5,$$

$$z_5 - c_5 = y_{35} = 0.$$

Here at least one $z_j - c_j < 0$. Thus we shall have to complete the table. Of course, after completing the first and second row of the table, we can complete the third row ($z_j - c_j$ row) by using the formula $z_j - c_j = \mathbf{c}_B\mathbf{y}_j - c_j = c_{B1}y_{1j} + c_{B2}y_{2j} - c_j$ $j = 1, 2, \cdots, 5$ and for $j = 0$, $z_0 = c_{B1}x_{B1} + c_{B2}x_{B2}$ etc. But the initial rule is followed by all modern computers.

First row (the row corresponding to the key row of the first table): The elements of the first row of the second table will be obtained by dividing all elements of the key row of the first table by $y_{11} = 1$ (key element), starting from the element $x_{B1} = 30$ under the column **b**.

Second row:

$$y_{20} = x_{B2} = \frac{1 \times 36 - 1 \times 30}{1} = 6,$$

$$y_{21} = 0,$$

$$y_{22} = \frac{1 \times 3 - 1 \times 2}{1} = 1,$$

$$y_{23} = \frac{1 \times 1 - 1 \times (-2)}{1} = 3,$$

$$y_{24} = \frac{1 \times 0 - 1 \times 1}{1} = -1,$$

$$y_{25} = \frac{1 \times 1 - 0 \times 1}{1} = 1.$$

The same notation x_{Bi} and y_{ij} have been used in each table. But they differ from table to table. With these data, the second table is to be made complete.

2nd Simplex table: (1st iteration)

		c	5	2	2	0	0	
Basis	c_B	b	a_1	a_2	a_3	$a_4(e_1)$	$a_5(e_2)$	Min ratio x_{Bi}/y_{i3}, y_{i3}>0
a_1	5	30	1	2	–2	1	0	...
a_5*	0	6	0	1	3*	–1	1	6/3=2*
$z_j–c_j$		150	0	8	–12*	5	0	

As $\mathbf{a}_1$ is the first basis vector, the $c_1 = 5$ will replace $c_{B1} = c_4 = 0$ to form a new $\mathbf{c}_B$ in the second table.

Note: Of course we can first complete the table and finally the $z_j - c_j$ row or the third row can be calculated by using the formula.

$z_0 = \mathbf{c}_B\mathbf{x}_B$ and $z_j - c_j = \mathbf{c}_B\mathbf{y}_j - c_j$ (verify it). But using the first rule of calculation of z_0 and $z_j - c_j$ we get some advantage and there lies the sweetness of the formula and this formula will be extremely helpful if there are three or more than three constraints and this transform formula is being used in all modern computers.

In the second table, $z_3 - c_3 = -12$ with at least one $y_{i3} > 0$. Hence the solution is not optimal. $\mathbf{a}_3$ is the vector to enter in the next basis. The third column under $\mathbf{a}_3$ is the key column.

$$\min_{i=1,2}\left(\frac{x_{Bi}}{y_{i3}}, y_{i3} > 0\right) = \min\left(\cdots, \frac{x_{B2}}{y_{23}}\right) = \left(\cdots, \frac{6}{3}\right)$$

$$= 2 \text{ which occurs at } i = 2.$$

Therefore, $r = 2$ and the second row is the key row and $y_{23} = 3$ which is marked with asterisk is the key element. Key row and key column are blocked by dotted lines and all quantities of interest are marked with asterisks. Since $r = 2$, the second vector $\mathbf{a}_5$ under the column 'basis' will leave the basis will be replaced by $\mathbf{a}_3$ and the next basis will be $(\mathbf{a}_1, \mathbf{a}_3)$. As the solution is not optimal, the next table is to be made to find the optimal solution.

Rule of construction of the third table from the second table

Third row ($z_j - c_j$ row):

$$y_{30} = z_0 = \frac{3 \times 150 - 6 \times (-12)}{3} = 174,$$

$$z_1 - c_1 = y_{31} = 0,$$

$$z_2 - c_2 = y_{32} = \frac{3 \times 8 - 1 \times (-12)}{3} = 12,$$

$$z_3 - c_3 = y_{33} = 0,$$

$$z_4 - c_4 = y_{34} = \frac{3 \times 5 - (-12 \times -1)}{3} = 1,$$

$$z_5 - c_5 = y_{35} = \frac{3 \times 0 - 1 \times (-12)}{3} = 4.$$

Here all $z_j - c_j \geq 0$, $(j = 1, 2, \cdots, 5)$. Hence the third table is the optimal table and the optimal value of the objective function, $\max z = 174$ which can also be found by making the table complete.

Second row (row corresponding to the key row of the second table): Divide all elements of the second row of the second table by the key element 3 to get the second row of the third table.

First row:

$$y_{10} = x_{B1} = \frac{3 \times 30 - 6 \times (-2)}{3} = 34,$$

$$y_{11} = 1,$$

$$y_{12} = \frac{3 \times 2 - 1 \times (-2)}{3} = \frac{8}{3},$$

$$y_{13} = 0,$$

$$y_{14} = \frac{3 \times 1 - (-2 \times -1)}{3} = \frac{1}{3},$$
$$y_{15} = \frac{3 \times 0 - (1 \times -2)}{3} = \frac{2}{3},$$

and the table is complete:

Third simplex table (2nd iteration)

		$\mathbf{c}$	5	2	2	0	0
Basis	$\mathbf{c}_B$	$\mathbf{b}$	$\mathbf{a}_1$	$\mathbf{a}_2$	$\mathbf{a}_3$	$\mathbf{a}_4(\mathbf{e}_1)$	$\mathbf{a}_5(\mathbf{e}_2)$
$\mathbf{a}_1$	5	34	1	$\frac{8}{3}$	0	$\frac{1}{3}$	$\frac{2}{3}$
$\mathbf{a}_3$	2	2	0	$\frac{1}{3}$	1	$-\frac{1}{3}$	$\frac{1}{3}$
$z_j - c_j$		174	0	12	0	1	4

As $\mathbf{a}_3$ is the 2nd basis vector, $c_3 = 3$ will replace $c_{B2} = c_4 = 0$ to form new $\mathbf{c}_B$ in the third table. Now verify the result of $z_j - c_j$ row using the formula $z_j - c_j = \mathbf{c}_B \mathbf{y}_j - c_j$ $[i = 1, 2, \cdots, 5]$ and $z_0 = \mathbf{c}_B \mathbf{x}_B$. If we use the first method of calculating the $z_j - c_j$ row or the third row we even need not write $\mathbf{c}_B$ from the second table and onwards.

We have already traced that all $z_i - c_j \geq 0$ and hence the solution in optimal. Final basis is $B = (\mathbf{a}_1, \mathbf{a}_3)$. Hence the optimal value of z, $\max z = z_0 = 174$ corresponding to the B.F.S. $\mathbf{x}_B = [x_{B1}, x_{B2}] = [x_1, x_3] = [34, 2]$ be at $x_1 = 34$, $x_3 = 2$, $x_2 = 0$ (non-basic). The slack variables are not present in the B.F.S. But they may present at the optimal stage.

Check formula at each stage: If the initial basis be a unit basis and initial unit vectors be slack vectors, then there exists a check formula at each stage which is almost perfect but not with hundred percent certainty.

Procedure: The components of $\mathbf{b}$ are b_1 and b_2, where $b_1 = 30$, $b_2 = 36$. $\mathbf{a}_4(\mathbf{e}_1)$ and $\mathbf{a}_5(\mathbf{e}_2)$ form the initial unit basis which are slack vectors; $z_4 - c_4 = 0$, $z_5 - c_5 = 0$. $(z_4 - c_4)b_1 + (z_5 - c_5)b_2 = 0 \times 30 + 0 \times 36 = 0$ which is equal to z_0 obtained from the initial table which verifies the correctness of the procedure. In the second table, $z_4 - c_4 = 5$, $z_5 - c_5 = 0$, and $(z_4 - c_4)b_1 + (z_5 - c_5)b_2 = 5 \times 30 + 0 \times 36 = 150$ which is equal to z_0 obtained from the second table and in the final table $z_4 - c_4 = 1$, $z_5 - c_5 = 4$, $(z_4 - c_4)b_1 + (z_5 - c_5)b_2 = 1 \times 30 + 4 \times 36 = 174$ which is equal to the optimal value of the objective function. Thus we may assume that the problem solved, is correct.

Note: (1) As the final basis $B = (\mathbf{a}_1, \mathbf{a}_3)$ the basic variables at the final stage are $x_{B1} = x_1 = 34$, $x_{B2} = x_3 = 2$.

(2) The vectors $\mathbf{a}_1, \mathbf{a}_2, \cdots, \mathbf{a}_5$ and $\mathbf{b}$ do not change in any iteration; only the column vectors $\mathbf{y}_j$ under the heading $\mathbf{a}_j$ and $\mathbf{x}_B = \mathbf{y}_0$ under the heading $\mathbf{b}$ change in each iteration due to the change of basis in each stage.

(3) During computation in each table, the same notation x_{Bi} is used to denote the ith component of the basic variable. Similarly the same notation y_{ij} is used to

represent the element corresponding to ith row and jth column. Practically they differ from table to table.

(4) It is important to note that during each iteration, the column vectors $\mathbf{y}_j$ under the heading $\mathbf{a}_j$ (basis vectors) are always unit vectors and they can be arranged to form a unit basis and $z_j - c_j$ under the heading $\mathbf{a}_j$ (basis vectors) are zero.

(5) It is interesting to note that at each stage $\mathbf{y}_4$ and $\mathbf{y}_5$ under the slack vectors $\mathbf{a}_4(\mathbf{e}_1), \mathbf{a}_5(\mathbf{e}_2)$ constitute the basis inverse in each iteration. (Proof of which is given in theorem 8.5.1).

For example, the final basis

$$B = (\mathbf{a}_1, \mathbf{a}_3) = \begin{bmatrix} 1 & -2 \\ 1 & 1 \end{bmatrix} \quad \text{and} \quad B^{-1} = \begin{bmatrix} 1/3 & 2/3 \\ -1/3 & 1/3 \end{bmatrix}$$

which has been reflected under the vectors $\mathbf{a}_4$ and $\mathbf{a}_5$ in the third or final table.

(6) During computation of the second table and onwards, $\mathbf{c}$ and $\mathbf{c}_B$ are not essential.

▶ **Example 8.11.2** *Solve the following L.P.P. by simplex method.*

$$\textit{Maximize}, \quad z = 4x_1 + 7x_2$$

subject to

$$\begin{aligned} 2x_1 + \ x_2 &\le 1000 \\ x_1 + \ x_2 &\le 600 \\ -x_1 - 2x_2 &\ge -1000, \quad x_1, x_2 \ge 0. \end{aligned}$$

[C.U.(P)'80]

Solution. This is a maximization problem.

Multiplying the third constraint by (-1) we get $x_1 + 2x_2 \le 1000$.

Hence all $b_i \ge 0$ and all constraints are attached with sign "$\le$" type. Introducing three slack variables x_3, x_4, x_5 one to each constraint we get the following converted equations

$$\begin{aligned} 2x_1 + x_2 + x_3 \qquad\qquad &= 1000 \\ x_1 + x_2 \qquad + x_4 \qquad &= 600 \\ x_1 + 2x_2 \qquad\qquad + x_5 &= 1000 \end{aligned}$$

The adjusted objective function z is given by

$$z = 4x_1 + 7x_2 + 0 \cdot x_3 + 0 \cdot x_4 + 0 \cdot x_5, \quad \mathbf{c} = (4, 7, 0, 0, 0).$$

$$\mathbf{a}_1 = \begin{bmatrix} 2 \\ 1 \\ 1 \end{bmatrix}, \quad \mathbf{a}_2 = \begin{bmatrix} 1 \\ 1 \\ 2 \end{bmatrix}, \quad \mathbf{a}_3 = \begin{bmatrix} 1 \\ 0 \\ 0 \end{bmatrix},$$

$$\mathbf{a}_4 = \begin{bmatrix} 0 \\ 1 \\ 0 \end{bmatrix}, \quad \mathbf{a}_5 = \begin{bmatrix} 0 \\ 0 \\ 1 \end{bmatrix} \quad \text{and} \quad \mathbf{b} = \begin{bmatrix} 1000 \\ 600 \\ 1000 \end{bmatrix} \ge \mathbf{0}.$$

Here the slack vectors $\mathbf{a}_3, \mathbf{a}_4, \mathbf{a}_5$ are unit vectors $\mathbf{e}_1, \mathbf{e}_2$ and $\mathbf{e}_3$ which together constitute the unit basis $B = (\mathbf{a}_3, \mathbf{a}_4, \mathbf{a}_5) = I_3$ and taking this as initial basis we have

Initial B.S. $\mathbf{x}_B = B^{-1}\mathbf{b} = I_3^{-1}\mathbf{b} = I_3\mathbf{b} = \mathbf{b} \geq \mathbf{0}$ which is feasible.

Then B is the admissible basis and

$$\begin{aligned}
\mathbf{x}_B &= [x_{B1}, x_{B2}, x_{B3}] = [x_3, x_4, x_5] = [b_1, b_2, b_3] = [1000, 600, 1000],\\
\mathbf{c}_B &= (c_{B1}, c_{B2}, c_{B3}) = (c_3, c_4, c_5) = (0, 0, 0) = \mathbf{0},\\
z_0 &= \mathbf{c}_B\mathbf{x}_B = \mathbf{0}\mathbf{x}_B = 0,\\
\mathbf{y}_j &= B^{-1}\mathbf{a}_j = I_3\mathbf{a}_j = \mathbf{a}_j = [j = 1, 2, 3, 4, 5],\\
z_j - c_j &= \mathbf{c}_B\mathbf{y}_j - c_j = \mathbf{0}\mathbf{y}_j - c_j = -c_j[j = 1, 2, 3, 4, 5].
\end{aligned}$$

With the data given above, we can construct the initial simplex table.

Initial simplex table:

		$\mathbf{c}$	4	7	0	0	0	
Basis	$\mathbf{c}_B$	$\mathbf{b}$	$\mathbf{a}_1$	$\mathbf{a}_2$	$\mathbf{a}_3(\mathbf{e}_1)$	$\mathbf{a}_4(\mathbf{e}_2)$	$\mathbf{a}_5(\mathbf{e}_3)$	Min. ratio
$\mathbf{a}_3$	0	1000	2	1	1	0	0	$\frac{1000}{1} = 1000$
$\mathbf{a}_4$	0	600	1	1	0	1	0	$\frac{600}{1} = 600$
$\mathbf{a}_5^*$	0	1000	1	2*	0	0	1	$\frac{1000}{2} = 500$*
$z_j - c_j$		$z_0 = 0$	−4	−7*	0	0	0	

Construction of the 2nd table from the first table:

Fourth row ($z_j - c_j$ row):

$$z_0 = \frac{2 \times 0 - (-7 \times 1000)}{2} = 3500,$$

$$\begin{aligned}
y_{41} &= z_1 - c_1 = \frac{2 \times (-4) - (-7 \times 1)}{2} = -\frac{1}{2},\\
y_{42} &= z_2 - c_2 = 0,\\
y_{43} &= z_3 - c_3 = 0,\\
y_{44} &= z_4 - c_4 = 0,\\
y_{45} &= z_5 - c_5 = \frac{2 \times 0 - (-7 \times 1)}{2} = \frac{7}{2}
\end{aligned}$$

Third row: Divide all elements of the key row (third row) of the first table by 2 to get all elements of the third row of the second table.

First row:

$$\begin{aligned}
x_{B1} = y_{10} &= \frac{2 \times 1000 - 1 \times 1000}{2} = 500,\\
y_{11} &= \frac{2 \times 2 - 1 \times 1}{2} = \frac{3}{2},
\end{aligned}$$

$$y_{12} = 0,$$
$$y_{13} = \frac{2\times 1 - 1\times 0}{2} = 1,$$
$$y_{14} = 0,$$
$$y_{15} = \frac{2\times 0 - 1\times 1}{2} = -\frac{1}{2}.$$

Second row:

$$x_{B2} = y_{20} = \frac{2\times 600 - 1\times 1000}{2} = 100,$$
$$y_{21} = \frac{2\times 1 - 1\times 1}{2} = \frac{1}{2},$$
$$y_{22} = 0,$$
$$y_{23} = 0,$$
$$y_{24} = \frac{2\times 1 - 1\times 0}{2} = 1,$$
$$y_{25} = \frac{2\times 0 - 1\times 1}{2} = -\frac{1}{2}.$$

All $z_j - c_j (j = 1, 2, \cdots, 5)$ are not greater than or equal to zero. $z_1 - c_1 = -\frac{1}{2} < 0$ and all at least one of $y_{i1} \geq 0 (i = 1, 2, 3)$. Then we shall have to construct the next table.

Second simplex table:

$\mathbf{a}_2$ is the vector to enter in the next basis and $\mathbf{a}_5$ will leave the basis and will be replaced by $\mathbf{a}_2$ and $c_{B3} = c_5$ will be replaced by $c_2 = 7$.

		c	4	7	0	0	0	
Basis	$\mathbf{c}_B$	b	$\mathbf{a}_1$	$\mathbf{a}_2$	$\mathbf{a}_3(\mathbf{e}_1)$	$\mathbf{a}_4(\mathbf{e}_2)$	$\mathbf{a}_5(\mathbf{e}_3)$	Min. ratio
$\mathbf{a}_3$	0	500	$\frac{3}{2}$	0	1	0	$-\frac{1}{2}$	$500/\frac{3}{2} = \frac{1000}{3}$
$\mathbf{a}_4^*$	0	100	$\frac{1}{2}^*$	0	0	1	$-\frac{1}{2}$	$100/\frac{1}{2} = 200^*$
$\mathbf{a}_2$	7	500	$\frac{1}{2}$	1	0	0	$\frac{1}{2}$	$500/\frac{1}{2} = 1000$
$z_j - c_j$		3500	$-\frac{1}{2}^*$	0	0	0	$\frac{7}{2}$	

Construction of the third table from the second table:

Fourth row ($z_j - c_j$ row),

$$z_0 = \frac{\frac{1}{2}\times 3500 - (-\frac{1}{2}\times 100)}{\frac{1}{2}} = 3600, \quad \left[\frac{1}{2} = \text{key element}\right]$$
$$y_{41} = z_1 - c_1 = 0,$$
$$y_{42} = z_2 - c_2 = 0,$$
$$y_{43} = z_3 - c_3 = 0,$$
$$y_{44} = z_4 - c_4 = \frac{\frac{1}{2}\times 0 - (-\frac{1}{2}\times 1)}{\frac{1}{2}} = 1,$$

$$y_{45} = z_5 - c_5 = \frac{\frac{1}{2} \times \frac{7}{2} - (-\frac{1}{2} \times -\frac{1}{2})}{\frac{1}{2}} = 3.$$

Here all $z_j - c_j \geq 0$. Thus we reach at the optimal stage. We need not complete the table except to calculate x_{B1}, x_{B2} and x_{B3}. But here also we make the table complete.

Second row: Divide all elements of the second row (key row) of the second table by the key element 1/2 to get the second row of the third table.

First row:

$$x_{B1} = y_{10} = \frac{\frac{1}{2} \times 500 - 100 \times \frac{3}{2}}{\frac{1}{2}} = 200,$$

$$y_{11} = 0,$$

$$y_{12} = 0,$$

$$y_{13} = \frac{\frac{1}{2} \times 1 - \frac{3}{2} \times 0}{\frac{1}{2}} = 1,$$

$$y_{14} = \frac{\frac{1}{2} \times 0 - \frac{3}{2} \times 1}{\frac{1}{2}} = -3,$$

$$y_{15} = \frac{\frac{1}{2} \times (-\frac{1}{2}) - \frac{3}{2} \times (-\frac{1}{2})}{\frac{1}{2}} = 1.$$

Third row:

$$x_{B3} = y_{30} = \frac{\frac{1}{2} \times 500 - 1 \times 100}{\frac{1}{2}} = 400,$$

$$y_{31} = 0,$$

$$y_{32} = 1,$$

$$y_{33} = 0,$$

$$y_{34} = \frac{\frac{1}{2} \times 0 - \frac{1}{2} \times 1}{\frac{1}{2}} = -1,$$

$$y_{35} = \frac{\frac{1}{2} \times \frac{1}{2} - \frac{1}{2} \times (-\frac{1}{2})}{\frac{1}{2}} = 1.$$

Third simplex table:

$\mathbf{a}_1$ is the vector to enter in the next basis by replacing $\mathbf{a}_4$ of the previous basis and $c_1 = 4$ will replace $c_{B2} = c_4 = 0$ from the previous $\mathbf{c}_B$.

Basis	c_B	b	$\mathbf{a}_1$	$\mathbf{a}_2$	$\mathbf{a}_3(\mathbf{e}_1)$	$\mathbf{a}_4(\mathbf{e}_2)$	$\mathbf{a}_5(\mathbf{e}_3)$
$\mathbf{a}_3$	0	200	0	0	1	−3	1
$\mathbf{a}_1$	4	200	1	0	0	2	−1
$\mathbf{a}_2$	7	400	0	1	0	−1	1
$z_j - c_j$		3600	0	0	0	1	3

All $z_j - c_j \geq 0$, then the solution is optimal. Hence the optimum value of the objective function z corresponding to the final basis $(\mathbf{a}_3, \mathbf{a}_1, \mathbf{a}_2)$, is given by $\max z = 3600$ at B.F.S.

$$\mathbf{x}_B = [x_3, x_1, x_2] = [200, 200, 400],$$

i.e., for $x_1 = 200$, $x_2 = 400$, the objective function of the original problem attains it maximum [$x_3 = 200$ is a slack variable, we need not consider it].

Check formula in each stage:

The initial basis is a unit basis and basis vectors are slack vectors.

Procedure: The components of $\mathbf{b}$ are $b_1 = 1000$, $b_2 = 600$, $b_3 = 1000$.

In the first table $z_3 - c_3 = 0$, $z_4 - c_4 = 0$, $z_5 - c_5 = 0$,

$$(z_3 - c_3)b_1 + (z_4 - c_4)b_2 + (z_5 - c_5)b_3 = 0$$

which is equal to z_0 in the first table.

In the second table $z_3 - c_3 = 0$, $z_4 - c_4 = 0$, $z_5 - c_5 = \frac{7}{2}$,

$$(z_3 - c_3)b_1 + (z_4 - c_4)b_2 + (z_5 - c_5)b_3 = \frac{7}{2} \times 1000 = 3500$$

which is equal to z_0 in the second table.

In the third table $z_3 - c_3 = 0$, $z_4 - c_4 = 1$, $z_5 - c_5 = 3$,

$$(z_3 - c_3)b_1 + (z_4 - c_4)b_2 + (z_5 - c_5)b_3 = 1 \times 600 + 3 \times 1000 = 3600$$

which is z_0 of third table. We may assume that the problem is solved correctly.

► **Example 8.11.3** *Solve by simplex method.*

$$\textit{Maximize,} \quad z = x_1 - x_2 + 3x_3$$

subject to

$$\begin{aligned} x_1 + x_2 + x_3 &\leq 10 \\ 2x_1 \quad\;\; - x_3 &\leq 2 \\ 2x_1 - 2x_2 + 3x_3 &\leq 0, \quad x_1, x_2, x_3 \geq 0. \end{aligned}$$

[C.U.(P)'88]

Solution. This is a maximization problem.

$b_i \geq 0$ for all i and the constraints are involved with the sign "$\leq$". Introducing three slack variables x_4, x_5, x_6 one in each constraint, we get the following converted equations

$$\begin{aligned} x_1 + x_2 + x_3 + x_4 \qquad\qquad &= 10 \\ 2x_1 \quad\;\; - x_3 \quad + x_5 \qquad &= 2 \\ 2x_1 - 2x_2 + 3x_3 \qquad\quad + x_6 &= 0 \end{aligned}$$

The adjusted objective function z is given by

$$z = x_1 - x_2 + 3x_3 + 0 \cdot x_4 + 0 \cdot x_5 + 0 \cdot x_6.$$

Here all slack vectors together constitute a unit basis matrix $B = I_3$ and then as

$$\mathbf{b} = [10, 2, 0], \quad \mathbf{x}_B = B^{-1}\mathbf{b} \geq \mathbf{0}$$

which gives a feasible solution.

Thus initial B.F.S. $= \mathbf{x}_B = B^{-1}\mathbf{b} = \mathbf{b} = [x_{B1}, x_{B2}, x_{B3}] = [x_4, x_5, x_6]$
$= [b_1, b_2, b_3] = [10, 2, 0]$.

Here the solution is degenerate.

[This problem can be solved by usual method; though Degeneracy occurs at the initial stage].

$$\mathbf{c} = (1, -1, 3, 0, 0, 0), \quad \mathbf{c}_B = (c_4, c_5, c_6) = (0, 0, 0) = \mathbf{0}$$

$$\mathbf{y}_j = B^{-1}\mathbf{a}_j = I_3^{-1}\mathbf{a}_j = \mathbf{a}_j [j = 1, 2, \cdots, 6]$$

$$z_B = \text{Value of the objective function} = \mathbf{c}_B\mathbf{x}_B = 0$$

$$z_j - c_j = \mathbf{c}_B\mathbf{y}_j - c_j = \mathbf{0}\mathbf{y}_j - c_j = -c_j.$$

With these data we shall construct the initial table.

Now without going details we shall solve the problem in a compact form.

Simplex tables:

		c	1	−1	3	0	0	0	
Basis	$\mathbf{c}_B$	b	$\mathbf{a}_1$	$\mathbf{a}_2$	$\mathbf{a}_3$	$\mathbf{a}_4(\mathbf{e}_1)$	$\mathbf{a}_5(\mathbf{e}_2)$	$\mathbf{a}_6(\mathbf{e}_3)$	Min. ratio
$\mathbf{a}_4$	0	10	1	1	1	1	0	0	$\frac{10}{1} = 10$
$\mathbf{a}_5$	0	2	2	0	−1	0	1	0	......
$\mathbf{a}_6^*$	0	0	2	−2	3*	0	0	1	$\frac{0}{3} = 0^*$
$z_j - c_j$		0	−1	1	−3*	0	0	0	
$\mathbf{a}_4^*$	0	10	$\frac{1}{3}$	$\frac{5}{3}^*$	0	1	0	$-\frac{1}{3}$	$10/\frac{5}{3} = 6^*$
$\mathbf{a}_5$	0	2	$\frac{8}{3}$	$-\frac{2}{3}$	0	0	1	$\frac{1}{3}$	......
$\mathbf{a}_3$	3	0	$\frac{2}{3}$	$-\frac{2}{3}$	1	0	0	$\frac{1}{3}$	......
$z_j - c_j$		0	1	−1*	0	0	0	1	
$\mathbf{a}_2$		6							
$\mathbf{a}_5$		6							
$\mathbf{a}_3$		4							
$z_j - c_j$		6	$\frac{6}{5}$	0	0	$\frac{3}{5}$	0	$\frac{4}{5}$	

As all $z_j - c_j \geq 0$ $[j = 1, 2, \cdots, 6]$ in the third table, then the third table is the optimal table and we need not complete the third table. We require to calculate only the column, under the vector **b** which gives the optimal B.F.S. and the optimal value of z. Thus $\max z = 6$ at $x_2 = 6$, $x_5 = 6$, $x_3 = 4$, i.e., for $x_1 = 0$ (non-basic), $x_2 = 6$, $x_3 = 4$ the original problem attains its maximum.

Note. (1) In the first table $\frac{x_{B2}}{y_{23}}$ is not calculated as $y_{23} < 0$. Similarly in the second table $\frac{x_{B2}}{y_{22}}$ and $\frac{x_{B3}}{y_{32}}$ are not calculated as $y_{22}, y_{32} \leq 0$.

Calculation of z_B, elements of $z_j - c_j$ row and $\mathbf{x}_B$

$$
\begin{aligned}
y_{40} &= z_B = \frac{\frac{5}{3} \times 0 - 10 \times (-1)}{\frac{5}{3}} = 6, \\
y_{41} &= z_1 - c_1 = \frac{\frac{5}{3} \times 1 - \frac{1}{3} \times (-1)}{\frac{5}{3}} = \frac{6}{5}, \\
y_{42} &= z_2 - c_2 = 0, \\
y_{43} &= z_3 - c_3 = 0, \\
y_{44} &= z_4 - c_4 = \frac{\frac{5}{3} \times 0 - 1 \times (-1)}{\frac{5}{3}} = \frac{3}{5}, \\
y_{45} &= z_5 - c_5 = 0, \\
y_{46} &= z_6 - c_6 = \frac{\frac{5}{3} \times 1 - (-\frac{1}{3}) \times (-1)}{\frac{5}{3}} = \frac{4}{5}, \\
x_{B1} &= \frac{10}{\frac{5}{3}} = 6, \\
x_{B2} &= \frac{\frac{5}{3} \times 2 - 10 \times (-\frac{2}{3})}{\frac{5}{3}} = 6, \\
x_{B3} &= \frac{\frac{5}{3} \times 0 - 10 \times (-\frac{2}{3})}{\frac{5}{3}} = 4.
\end{aligned}
$$

We now solve a problem in a compact form.

▶ **Example 8.11.4** *Solve the L.P.P. by simplex method.*

$$\text{Maximize,} \quad z = 4x_1 + 3x_2$$

subject to

$$
\begin{aligned}
3x_1 + x_2 &\leq 15 \\
3x_1 + 4x_2 &\leq 24, \quad x_1, x_2 \geq 0,
\end{aligned}
$$

Adding slack variables one to each constraint, the converted equations are

$$
\begin{aligned}
3x_1 + x_2 + x_3 \qquad &= 15 \\
3x_1 + 4x_2 \qquad + x_4 &= 24, \quad x_1, x_2, x_3, x_4 \geq 0.
\end{aligned}
$$

Simplex tables:

Basic	$\mathbf{c}_B$	c / $\mathbf{b}$	4 / $\mathbf{a}_1$	3 / $\mathbf{a}_2$	0 / $\mathbf{a}_3(e_1)$	0 / $\mathbf{a}_4(e_2)$	Min. ratio
$\mathbf{a}_3^*$	0	15	3*	1	1	0	$\frac{15}{3} = 5^*$
$\mathbf{a}_4$	0	24	3	4	0	1	$\frac{24}{3} = 8$
$z_j - c_j$		0	−4*	−3	0	0	
$\mathbf{a}_1$	4	5	1	$\frac{1}{3}$	$\frac{1}{3}$	0	$\frac{5}{1/3} = 15$
$\mathbf{a}_4^*$	0	9	0	3*	−1	1	$\frac{9}{3} = 3^*$
$z_j - c_j$		20	0	$-\frac{5}{3}^*$	$\frac{4}{3}$	0	
$\mathbf{a}_1$	4	4					
$\mathbf{a}_2$	3	3					
$z_j - c_j$		25	0	0	$\frac{7}{9}$	$\frac{5}{9}$	

Final basis $B = (\mathbf{a}_1, \mathbf{a}_2)$ optimal value of $z = \max z = 25$ at B.F.S. $\mathbf{x}_B = [x_{B1}, x_{B2}] = [x_1, x_2] = [4, 3]$, i.e., at $x_1 = 4$, and $x_2 = 3$.

► **Example 8.11.5** *Solve the L.P.P.*

$$\textit{Minimize,} \quad z = -2x_1 + 3x_2$$

subject to

$$\begin{aligned} 2x_1 - 5x_2 &\leq 7 \\ 4x_1 + x_2 &\leq 8 \\ 7x_1 + 2x_2 &\leq 16, \quad x_1 \geq 0, x_2 \geq 0. \end{aligned}$$

Solution. The problem is a problem of minimization.

Let $z' = -z$; then $\min z = -\max(-z) = -\max z'$. Hence the problem is a problem of maximization of $z' = -z = -(-2x_1 + 3x_2) = 2x_1 - 3x_2$ and finally $\min z = -\max(z')$ with the same solution set. $b_i \geq 0$ for all i and constraints are associated with the sign "$\leq$".

Introducing three slack variables x_3, x_4 and x_5 (one in each inequation) we get the following converted equations

$$\begin{aligned} 2x_1 - 5x_2 + x_3 \qquad\qquad &= 7 \\ 4x_1 + x_2 \qquad + x_4 \qquad &= 8 \\ 7x_1 + 2x_2 \qquad\qquad + x_5 &= 16. \end{aligned}$$

The adjusted objective function is $z' = 2x_1 - 3x_2 + 0.x_3 + 0.x_4 + 0.x_5$.

Here all the slack vectors are unit vectors which produce a unit basis. Initial

$$\text{B.F.S.} = \mathbf{x}_B = [x_{B1}, x_{B2}, x_{B3}] = [x_3, x_4, x_5] = [7, 8, 16]$$

$$\mathbf{c}_B = (c_{B1}, c_{B2}, c_{B3}) = (c_3, c_4; c_5) = (0, 0, 0)$$

$$z = \mathbf{c}_B \mathbf{x}_B = 0 \text{ and } \mathbf{y}_j = B^{-1}\mathbf{a}_j = \mathbf{a}_j$$

$$z_j - c_j = \mathbf{c}_B \mathbf{y}_j - c_j = \mathbf{0}\mathbf{y}_j - c_j = -c_j$$

Now with the values of $z_j - c_j$ etc. we construct the initial table and solve accordingly.

Simplex tables:

		c	2	−3	0	0	0	
Basis	$\mathbf{c}_B$	**b**	$\mathbf{a}_1$	$\mathbf{a}_2$	$\mathbf{a}_3(\mathbf{e}_1)$	$\mathbf{a}_4(\mathbf{e}_2)$	$\mathbf{a}_5(\mathbf{e}_3)$	Min. ratio
$\mathbf{a}_3$	0	7	2	−5	1	0	0	$\frac{7}{2}$
$\mathbf{a}_4^*$	0	8	4*	1	0	1	0	$\frac{8}{4} = 2^*$
$\mathbf{a}_5$	0	16	7	2	0	0	1	$\frac{16}{7}$
$z_j - c_j$		0	−2*	3	0	0	0	
$\mathbf{a}_3$	0	3	0	$-\frac{11}{2}$	1	$-\frac{1}{2}$	0	
$\mathbf{a}_1$	2	2	1	$\frac{1}{4}$	0	$\frac{1}{4}$	0	
$\mathbf{a}_5$	0	2	0	$\frac{1}{4}$	0	$-\frac{7}{4}$	1	
$z_j - c_j$		4	0	$\frac{7}{2}$	0	$\frac{1}{2}$	0	

As none of $z_j - c_j < 0$, therefore the solution is optimal.

Hence $\max z' = 4$.

Now $\min z = -\max z' = -4$. Hence the minimum value of z is -4 corresponding to the optimal basic feasible solution.

$\mathbf{x}_B = [x_3, x_1, x_5] = [3, 2, 2]$, i.e., for $x_1 = 2$, $x_2 = 0$, the objective function of the original problem attains its minimum [x_2 is a non-basic variable.].

▶ **Example 8.11.6** *Solve the L.P.P.*

$$\textit{Maximize,} \quad z = x_1 + x_2 + 3x_3$$

subject to

$$\begin{aligned} x_1 + 2x_2 - x_3 &\le 10 \\ 3x_2 + 2x_3 &\le 8 \\ x_2 + 3x_3 &\le 15, \quad x_1 \ge 0, x_2 \ge 0 \textit{ and } x_3 \ge 0. \end{aligned}$$

Solution. $b_i \ge 0$ for all i.

Introducing three slack variables x_4, x_5 and x_6, one to each constraint we get the following equations

$$\begin{aligned} x_1 + 2x_2 - x_3 + x_4 &= 10 \\ 0 \cdot x_1 + 3x_2 + 2x_3 + x_5 &= 8 \\ 0 \cdot x_1 + x_2 + 3x_3 + x_6 &= 15 \end{aligned}$$

The adjusted objective function z is given by

$$z = x_1 + x_2 + 3x_3 + 0 \cdot x_4 + 0 \cdot x_5 + 0 \cdot x_6.$$

The slack vectors constitute a basis matrix which is a unit matrix. But in this problem the column vector $\mathbf{a}_1$, associated with the variable x_1 is also a unit vector $(\mathbf{e}_1)$. Hence in this problem unit basis matrix is not unique. But as the problem is a problem of maximization and the coefficient of x_1 in the objective function is a positive quantity, the initial basis matrix may be selected in such a way, that x_1 is a basic variable, i.e., the column vector $\mathbf{a}_1(\mathbf{e}_1)$ associated with x_1 be included in the initial unit basis. And due to this selection the problem may be solved quickly. Unit vector $\mathbf{e}_1$, associated with the variable x_4 be kept outside the basis matrix, i.e., x_4 is to be considered as a non-basic variable.

Therefore, initial B.F.S.

$$\mathbf{x}_B = [x_1, x_5, x_6] = [10, 8, 15]$$

$$\mathbf{c}_B = (c_1, c_5, c_6) = (1, 0, 0), \mathbf{y}_j = \mathbf{a}_j$$

$$z = \mathbf{c}_B \mathbf{x}_B = 1 \times 10 + 0 \times 8 + 0 \times 15 = 10$$

$$z_1 - c_1 = z_5 - c_5 = z_6 - c_6 = 0$$

$$z_2 - c_2 = (1, 0, 0)[2, 3, 1] - 1 = 1$$

$$z_3 - c_3 = (1, 0, 0)[-1, 2, 3] - 3 = -4$$

$$z_4 - c_4 = (1, 0, 0)[1, 0, 0,] - 0 = 1.$$

Simplex tables:

		c	1	1	3	0	0	0	
Basis	$\mathbf{c}_B$	b	$\mathbf{a}_1(\mathbf{e}_1)$	$\mathbf{a}_2$	$\mathbf{a}_3$	$\mathbf{a}_4(\mathbf{e}_1)$	$\mathbf{a}_5(\mathbf{e}_2)$	$\mathbf{a}_6(\mathbf{e}_3)$	Min. ratio
$\mathbf{a}_1$	1	10	1	2	−1	1	0	0	......
$\mathbf{a}_5^*$	0	8	0	3	2*	0	1	0	$\frac{8}{2} = 4^*$
$\mathbf{a}_6$	0	15	0	1	3	0	0	1	$\frac{15}{3} = 5$
$z_j - c_j$		10	0	1	−4*	1	0	0	
$\mathbf{a}_1$	1	14	1	$\frac{7}{2}$	0	1	$\frac{1}{2}$	0	
$\mathbf{a}_3$	3	4	0	$\frac{3}{2}$	1	0	$\frac{1}{2}$	0	
$\mathbf{a}_6$	0	3	0	$-\frac{7}{2}$	0	0	$-\frac{3}{2}$	1	
$z_j - c_j$		26	0	7	0	1	2	0	

As none of $z_j - c_j < 0$, therefore the solution set is optimal and the optimal value of z is 26 for the B.F.S. $\mathbf{x}_B = [x_1, x_3, x_6] = [14, 4, 3]$, i.e., for $x_1 = 14$, $x_2 = 0$ and $x_3 = 4$ the original objective function attains its maximum [x_2 is a non-basic variable.].

Now we solve a problem and observe how much the method be able to save time and labour.

► **Example 8.11.7** *Solve, the L.P. problem by simplex method.*

$$\text{Maximize,} \quad z = 3x_1 + 5x_2 + 4x_3$$

subject to

$$2x_1 + 3x_2 \leq 8$$
$$2x_2 + 5x_3 \leq 10$$
$$3x_1 + 2x_2 + 4x_3 \leq 15, \quad x_1, x_2, x_3 \geq 0.$$

[Meerut M.Sc.(Math)'84]

Solution. $\mathbf{b} = [8, 10, 15] \geq \mathbf{0}$. Thus introducing three slack variables, x_4, x_5 and x_6, one to each constraint and taking initial basis $B = (\mathbf{a}_4, \mathbf{a}_5, \mathbf{a}_6) = I_3$, we can start the initial simplex table and then solve in a compact table as shown below.

		c	3	5	4	0	0	0	
Basis	$\mathbf{c}_B$	**b**	$\mathbf{a}_1$	$\mathbf{a}_2$	$\mathbf{a}_3$	$\mathbf{a}_4(\mathbf{e}_1)$	$\mathbf{a}_5(\mathbf{e}_2)$	$\mathbf{a}_6(\mathbf{e}_3)$	Min. ratio
$\mathbf{a}_4^*$	0	8	2	3*	0	1	0	0	$\frac{8}{3}$*
$\mathbf{a}_5$	0	10	0	2	5	0	1	0	$\frac{10}{2} = 5$
$\mathbf{a}_6$	0	15	3	2	4	0	0	1	$\frac{15}{2}$
$z_j - c_j$		0	−3	−5*	−4	0	0	0	
$\mathbf{a}_2$		$\frac{8}{3}$	$\frac{2}{3}$	1	0	$\frac{1}{3}$	0	0	......
$\mathbf{a}_5^*$		$\frac{14}{3}$	$-\frac{4}{3}$	0	5*	$-\frac{2}{3}$	1	0	$\frac{14}{3}/5 = \frac{14}{15}$*
$\mathbf{a}_6$		$\frac{29}{3}$	$\frac{5}{3}$	0	4	$-\frac{2}{3}$	0	1	$\frac{29}{3}/4 = \frac{29}{12}$
$z_j - c_j$		$\frac{40}{3}$	$\frac{1}{3}$	0	−4*	$\frac{5}{3}$	0	0	
$\mathbf{a}_2$		$\frac{8}{3}$	$\frac{2}{3}$	1	0	$\frac{1}{3}$	0	0	$\frac{8}{3}/\frac{2}{3} = 4$
$\mathbf{a}_3$		$\frac{14}{15}$	$-\frac{4}{15}$	0	1	$-\frac{2}{15}$	$\frac{1}{5}$	0	......
$\mathbf{a}_6^*$		$\frac{89}{15}$	$\frac{41}{15}$*	0	0	$-\frac{2}{25}$	$-\frac{4}{5}$	1	$\frac{89}{15}/\frac{41}{15} = \frac{89}{41}$*
$z_j - c_j$		$\frac{256}{15}$	$-\frac{11}{15}$*	0	0	$\frac{17}{15}$	$\frac{4}{5}$	0	
$\mathbf{a}_2$		$\frac{50}{41}$							
$\mathbf{a}_3$		$\frac{62}{41}$							
$\mathbf{a}_1$		$\frac{89}{41}$							
$z_j - c_j$		$\frac{765}{41}$	0	0	0	$\frac{45}{41}$	$\frac{24}{41}$	$\frac{11}{41}$	

In the fourth table, all $z_j - c_j \geq 0$. Thus the fourth table is the optimal table We now only calculate the elements under the column vector **b** which gives B.F.S.

and the value of the objective function. We need not complete the table.

$$\max z = \frac{765}{41} \quad \text{at} \quad x_1 = \frac{89}{41}, \quad x_2 = \frac{50}{41} \quad \text{and} \quad x_3 = \frac{62}{41}.$$

This method is extremely helpful for three or more than three constraints and for fractional cost coefficients. But here the final basis inverse will not be available. Of course, the value of the objective function needs not to be calculated in each table. It may be calculated at the optimal table only by using the formula $z_B = \mathbf{c}_B\mathbf{x}_B$.

Problem having Multiple Optimal Solutions

► **Example 8.11.8** *Use simplex method to solve the following L.P.P.*

$$\textit{Maximize,} \quad z = 5x_1 + 2x_2$$

subject to

$$\begin{aligned} 6x_1 + 10x_2 &\le 30 \\ 10x_1 + 4x_2 &\le 20, \quad x_1, x_2 \ge 0. \end{aligned}$$

[C.U. M.Com.'85]

Is the solution unique? If not, write down the convex combination of the alternative optima.

Solution: The constraints, after the addition of slack variables x_3 and x_4, one to each, are

$$\begin{aligned} 6x_1 + 10x_2 + x_3 \quad &= 30 \\ 10x_1 + 4x_2 \quad + x_4 &= 20, \quad x_j \ge 0, j = 1, 2, \cdots, 4. \end{aligned}$$

The adjusted objective function $z = 5x_1 + 2x_2 + 0 \cdot x_3 + 0 \cdot x_4$.

$$\mathbf{c} = (5, 2, 0, 0)$$

$$\mathbf{b} = \begin{bmatrix} 30 \\ 20 \end{bmatrix} \ge \mathbf{0} \quad \text{and} \quad B = (\mathbf{a}_3, \mathbf{a}_4) = I_2 \text{ is a unit matrix.}$$

$$\mathbf{x}_B = B^{-1}\mathbf{b} = \mathbf{b} \ge \mathbf{0}.$$

Thus with the initial basis B we can start the problem

$$\mathbf{y}_j = B^{-1}\mathbf{a}_j = \mathbf{a}_j, \mathbf{c}_B = (0, 0).$$

		c	5	2	0	0	
c_B	Basis	b	a_1	a_2	$a_3(e_1)$	$a_4(e_2)$	Min. ratio
0	a_3	30	6	10	1	0	$\frac{30}{6}=5$
0	a_4^*	20	10^*	4	0	1	$\frac{20}{10}=2^*$
	z_j-c_j	0	-5^*	-2	0	0	
0	a_3^*	18	0	$\frac{38}{5}^*$	1	$-\frac{3}{5}$	$\frac{18}{38/5}=\frac{90}{38}^*$
5	a_1	2	1	$\frac{2}{5}$	0	$\frac{1}{10}$	$\frac{2}{2/5}=5$
	z_j-c_j	10	0	0^*	0	$\frac{1}{2}$	
2	a_2	$\frac{45}{19}$	0	1	$\frac{5}{38}$	$-\frac{3}{38}$	
5	a_1	$\frac{20}{19}$	1	0	$-\frac{2}{19}$	$\frac{5}{38}$	
	z_j-c_j	10	0	0	0	$\frac{1}{2}$	

Here in the second table all $z_j - c_j \geq 0$. Hence the optimal solution has been obtained $\max z = 10$ at $x_3 = 18, x_1 = 2$, i.e., for $x_1 = 2, x_2 = 0$ (non-basic), the problem attains its maximum. But here $z_2 - c_2 = 0$ corresponding to a non-basic vector $\mathbf{a}_2$. Thus the solution is not unique. Using $\mathbf{a}_2$ as a vector to enter in the next basis we have $\max z = 10$ remains same but the optimal solution will change which has been shown from the table 3. Other optimal basic solution is $x_1 = \frac{20}{19}, x_2 = \frac{45}{19}$. We know that if there are more than one optimal solution then there exist infinite optimal solutions which will be obtained from the convex combination of the optimal solutions $\mathbf{x}_1 = [2, 0]$, $\mathbf{x}_2 = [\frac{20}{19}, \frac{45}{19}]$. Any optimal solution $\mathbf{x}$ is given by [Alternative optima]

$$\begin{aligned}\mathbf{x} &= \lambda\mathbf{x}_1 + (1-\lambda)\mathbf{x}_2, 0 \leq \lambda \leq 1 \\ &= \lambda[2, 0] + (1-\lambda)\left[\frac{20}{19}, \frac{45}{19}\right]\end{aligned}$$

For example, if we take $\lambda = \frac{1}{2}$ then $\mathbf{x} = [1\frac{10}{19}, \frac{45}{38}]$ which is also an alternative optimal solution.

Note. $\mathbf{x}_1 = [x_1 = 2, x_2 = 0]$

Problem having an Unbounded Solution

► **Example 8.11.9** *Use the simplex method to solve the L.P.P.*

$$\textit{Maximize,} \quad 2x_2 + x_3$$

subject to

$$\begin{aligned} x_1 + x_2 - 2x_3 &\leq 7 \\ -3x_1 + x_2 + 2x_3 &\leq 3, \quad x_1, x_2 \text{ and } x_3 \geq 0 \end{aligned}$$ [C.U.(H)'89]

Solution: Adding two slack variables x_4 and x_5, one to each constraint, the constraints are

$$\begin{aligned} x_1 + x_2 - 2x_3 + x_4 \quad &= 7 \\ -3x_1 + x_2 + 2x_3 \quad + x_5 &= 3 , \quad x_1 \geq 0, j = 1, \cdots, 5. \end{aligned}$$

and the objective function is $0x_1 + 2x_2 + x_3 + 0 \cdot x_4 + 0x_5$, $\mathbf{b} = [7, 3] \geq \mathbf{0}$ and $B = [\mathbf{a}_4, \mathbf{a}_5] = I_2$ will be the initial unit basis.

Simplex tables

		c	0	2	1	0	0	
Basis	$\mathbf{c}_B$	**b**	$\mathbf{a}_1$	$\mathbf{a}_2$	$\mathbf{a}_3$	$\mathbf{a}_4(\mathbf{e}_1)$	$\mathbf{a}_5(\mathbf{e}_2)$	**Min. ratio**
$\mathbf{a}_4$	0	7	1	1	−2	1	0	$\frac{7}{1} = 7$
$\mathbf{a}_5^*$	0	3	−3	1*	2	0	1	$\frac{3}{1} = 3^*$
$z_j - c_j$		0	0	−2*	−1	0	0	
$\mathbf{a}_4^*$	0	4	4*	0	−4	1	−1	$\frac{4}{4} = 1^*$
$\mathbf{a}_2$	2	3	−3	1	2	0	1	……
$z_j - c_j$		6	−6*	0	3	0	2	
$\mathbf{a}_1$	0	1	1	0	−1	$\frac{1}{4}$	$-\frac{1}{4}$	
$\mathbf{a}_2$	2	6	0	1	−1	$\frac{3}{4}$	$\frac{1}{4}$	
$z_j - c_j$		12	0	0	−3	$\frac{3}{2}$	$\frac{1}{2}$	

Note. (1) After completing the $z_j - c_j$ of the third table, we have seen that $z_3 - c_3 = -3 < 0$. Thus we require to complete the table. After completing, we have seen that $y_{i3} \leq 0$ for $i = 1, 2$ for which $z_3 - c_3$ is negative. Then the only conclusion is that the problem has no finite optimal value and the problem is said to have unbounded solution.

(2) For a problem having unbounded solution we cannot trace it before completing the final table.

► **Example 8.11.10** *Solve the L.P.P. by simplex method and prove that alternative optimal solutions exist. Find two optimal solutions.*

$$\textit{Maximize,} \quad z = 2x_1 - x_2 + 3x_3 + x_4$$

subject to

$$\begin{aligned} 2x_1 + \ x_2 + 3x_3 + 5x_4 &\leq 12 \\ 3x_1 + 2x_2 + \ x_3 + 4x_4 &\leq 15 , \quad x_1, x_2, x_3 \textit{ and } x_4 \geq 0. \end{aligned}$$

Simplex table

		c	2	−1	3	1	0	0	
Basis	c_B	**b**	a_1	a_2	a_3	a_4	$a_5(e_1)$	$a_6(e_2)$	Min. ratio
a_5^*	0	12	2	1	3*	5	1	0	$\frac{12}{3}=4^*$
a_6	0	15	3	2	1	4	0	1	$\frac{15}{1}=15$
z_j-c_j		0	−2	1	−3*	−1	0	0	......
a_3	3	4	$\frac{2}{3}$	$\frac{1}{3}$	1	$\frac{5}{3}$	$\frac{1}{3}$	0	$\frac{4}{2/3}=6$
a_6^*	0	11	$\frac{7}{3}^*$	$\frac{5}{3}$	0	$\frac{7}{3}$	$-\frac{1}{3}$	1	$\frac{11}{7/3}=\frac{33}{7}^*$
z_j-c_j		12	0*	2	0	4	1	0	
a_3	3	$\frac{6}{7}$							
a_1	2	$\frac{33}{7}$							
z_j-c_j		12	0	2	0	4	1	0	

In the second table, all $z_j - c_j \geq 0$. Therefore, we reach at the optimal stage. Then $\max z = 12$ at $x_1 = 0$, $x_2 = 0$, $x_3 = 4$ and $x_4 = 0$. Now in the table $z_1 - c_1 = 0$ corresponding to a non-basis vector $\mathbf{a}_1$. Hence alternative optimal solutions exist. Thus taking $\mathbf{a}_1$ to vector enter in the basis we get another optimal solution which is $x_1 = \frac{33}{7}$, $x_2 = 0$, $x_3 = \frac{6}{7}$, $x_4 = 0$ and $\max z = 12$.

▶ **Example 8.11.11** *Solve the L.P.P. by simplex method*

$$\text{Maximize,} \quad z = 2x_1 - 3x_2 - 2x_3 + 6x_4$$

subject to

$$\begin{aligned} 5x_1 - x_2 + 2x_3 + 6x_4 &\leq 20 \\ 2x_1 + 3x_2 + 4x_3 - 5x_4 &\leq 16 \\ x_1 + 2x_2 - 3x_3 + x_4 &\leq 2, \quad x_1, x_2, x_3 \text{ and } x_4 \geq 0. \end{aligned}$$

Simplex tables

		c	2	−3	−2	6	0	0	0	
Basis	$\mathbf{c}_B$	**b**	$\mathbf{a}_1$	$\mathbf{a}_2$	$\mathbf{a}_3$	$\mathbf{a}_4$	$\mathbf{a}_5(\mathbf{e}_1)$	$\mathbf{a}_6(\mathbf{e}_2)$	$\mathbf{a}_7(\mathbf{e}_3)$	Min. ratio
$\mathbf{a}_5$	0	20	5	−1	2	6	1	0	0	$\frac{20}{6}=\frac{10}{3}$
$\mathbf{a}_6$	0	16	2	3	4	−5	0	1	0	......
$\mathbf{a}_7^*$	0	2	1	2	−3	1*	0	0	1	$\frac{2}{1}=2^*$
z_j-c_j		0	−2	3	2	−6*	0	0	0	
$\mathbf{a}_5^*$	0	8	−1	−13	20*	0	1	0	−6	$\frac{8}{20}=\frac{2}{5}^*$
$\mathbf{a}_6$	0	26	7	13	−11	0	0	1	5	......
$\mathbf{a}_4$	6	2	1	2	−3	1	0	0	1	......
z_j-c_j		12	4	15	−16*	0	0	0	6	
$\mathbf{a}_3$	−2	$\frac{2}{5}$								
$\mathbf{a}_6$	0	$\frac{152}{5}$								
$\mathbf{a}_4$	6	$\frac{16}{5}$								
z_j-c_j		$\frac{92}{5}$	$\frac{16}{5}$	$\frac{23}{5}$	0	0	$\frac{4}{5}$	0	$\frac{6}{5}$	

All $z_j - c_j \geq 0$ in the third table. Thus we reach at the optimal stage. Then

$$\max z = \frac{92}{5} \quad \text{at} \quad x_1 = 0, \quad x_2 = 0, \quad x_3 = \frac{2}{5} \text{ and } \quad x_4 = \frac{16}{5},$$

Verification of the result.

$$\max z = -2 \times \frac{2}{5} + 0 \times \frac{152}{5} + 6 \times \frac{16}{5} = \frac{92}{5}.$$

By using Duality theory

$$\max z = 20 \times \frac{4}{5} + 0 \times 16 + 2 \times \frac{6}{5} = \frac{92}{5}$$

Thus the correctness of the solution has been verified.

8.12 Minimization Problems, Solved as Minimization Problems

► **Example 8.12.1** *Solve the L.P.P. by simplex method*

$$\textit{Minimize,} \quad z = x_1 - 3x_2 + 2x_3$$

subject to

$$\begin{aligned} 3x_1 - x_2 + 2x_3 &\leq 7 \\ -2x_1 + 4x_2 &\leq 12 \\ -4x_1 + 3x_2 + 8x_3 &\leq 10, \quad x_1, x_2, x_3 \geq 0. \end{aligned}$$ **[C.U.(P)'81,'96,'99]**

Simplex table

		c	1	−3	2	0	0	0	
Basis	c_B	**b**	$\mathbf{a}_1$	$\mathbf{a}_2$	$\mathbf{a}_3$	$\mathbf{a}_4(\mathbf{e}_1)$	$\mathbf{a}_5(\mathbf{e}_2)$	$\mathbf{a}_6(\mathbf{e}_3)$	**Min. ratio**
$\mathbf{a}_4$	0	7	3	−1	2	1	0	0	……
$\mathbf{a}_5^*$	0	12	−2	4*	0	0	1	0	$\frac{12}{4} = 3^*$
$\mathbf{a}_6$	0	10	−4	3	8	0	0	1	$\frac{10}{3}$
$z_j - c_j$		0	−1	3*	−2	0	0	0	
$\mathbf{a}_4^*$	0	10	$\frac{5}{2}^*$	0	2	1	$\frac{1}{4}$	0	$\frac{10}{5/2} = 4^*$
$\mathbf{a}_2$	−3	3	$-\frac{1}{2}$	1	0	0	$\frac{1}{4}$	0	……
$\mathbf{a}_6$	0	1	$-\frac{5}{2}$	0	8	0	$-\frac{3}{4}$	1	……
$z_j - c_j$		−9	$\frac{1}{2}^*$	0	−2	0	$-\frac{3}{4}$	0	
$\mathbf{a}_1$	1	4							
$\mathbf{a}_2$	−3	5							
$\mathbf{a}_5$	0	11							
$z_j - c_j$		−11	0	0	$-\frac{12}{5}$	$-\frac{1}{5}$	$-\frac{4}{5}$	0	

All $z_j - c_j \leq 0$. Thus we reach at the optimal stage.

$$\min(z) = -11 \text{ at } x_1 = 4, \quad x_2 = 5, \quad x_3 = 0 \quad \text{(non-basic)}.$$

Remark. Here, to select the vector to enter in the next basis, we shall have to select kth vector in the manner,

$$z_k - c_k = \max_{j=1,\cdots,6} (z_j - c_j, z_j - c_j > 0).$$

Here, that is why we select the second vector to enter in the next basis and at the optimal stage all $z_j - c_j \leq 0$.

Exercise 8

1. Solve the L.P.P. by simplex method:

(a) Maximize, $z = 3x_1 + 2x_2$

$$\text{subject to } 2x_1 + x_2 \leq 12$$
$$6x_1 + 5x_2 \leq 40, \quad x_1 \geq 0, x_2 \geq 0.$$

(b) Maximize, $z = 4x_1 + x_2$

$$\text{subject to } x_1 + 2x_2 \leq 10$$
$$4x_1 + 3x_2 \leq 24, \quad x_1 \geq 0, x_2 \geq 0.$$

(c) Maximize, $z = 5x_1 + 3x_2$

$$\text{subject to } 3x_1 + 5x_2 \leq 15$$
$$5x_1 + 2x_2 \leq 10, \quad x_1 \geq 0, x_2 \geq 0.$$ **[C.U.(P)'93]**

(d) Maximize, $z = 2x_1 + 3x_2$

$$\text{subject to } 7x_1 + 4x_2 \leq 28$$
$$7x_1 + 12x_2 \leq 52, \quad x_1 \geq 0, x_2 \geq 0.$$

(e) Maximize, $z = 5x_1 + 4x_2 + x_3$

$$\text{subject to } 6x_1 + x_2 + 2x_3 \leq 12$$
$$8x_1 + 2x_2 + x_3 \leq 30$$
$$4x_1 + x_2 - 2x_3 \leq 16, \quad x_1 \geq 0, x_2 \geq 0 \text{ and } x_3 \geq 0.$$

(f) Maximize, $z = 10x_1 + x_2 + 2x_3$

$$\text{subject to } x_1 + x_2 - 2x_3 \leq 10$$
$$4x_1 + x_2 + x_3 \leq 20, \quad x_1, x_2, \text{ and } x_3 \geq 0.$$

[C.U.(P)'87]

2. (a) Maximize, $z = 3x_1 + 2x_2$

$$\text{subject to } 5x_1 + x_2 \leq 10$$
$$4x_1 + 5x_2 \leq 60, \quad x_1 \geq 0, x_2 \geq 0.$$

(b) Solve the following L.P.P. with the help of simplex method:
Maximize, $z = 2x_1 + 2x_2$

$$\text{subject to } x_1 - x_2 \geq -1$$
$$-0.5x_1 + x_2 \leq 2, \quad x_1 \geq 0, x_2 \geq 0.$$

[C.U.M.Com.'86]

(c) Given the following linear programme:
Maximize, $z = 3x_1 + 4x_2$

$$\text{subject to } 2x_1 + 3x_2 \leq 120$$
$$3x_1 + 2x_2 \leq 105, \quad x_1 \geq 0, x_2 \geq 0.$$

[C.U.M.Com.'88]

(i) Find an optimum solution to this programme by applying simplex technique.

(ii) Is the solution unique? Give reason for your answer.

(iii) Is the solution degenerate? Give reason for your answer.

(d) Solve the L.P.P. by using simplex algorithm.
Maximize, $z = 2x_1 - x_2 + \frac{8}{3}x_3$

$$\text{subject to } 3x_1 + x_2 - 2x_3 \leq 6$$
$$2x_1 + 5x_2 + x_3 \leq 14$$
$$x_1 + 4x_2 + 2x_3 \leq 8, \quad x_1, x_2, \text{ and } x_3 \geq 0.$$

Mention the optimal basis. From the optimal simplex table write down the optimal basis inverse. Verify that, by using ordinary matrix inverse method.

(e) Solve the problem:
Maximize, $z = 2x_1 + x_2 - 4x_3 + x_4$

$$\text{subject to } 2x_1 - 3x_2 + 4x_3 + x_4 \leq 6$$
$$x_1 + 2x_2 + 3x_3 + 4x_4 \leq 8$$
$$2x_1 + x_2 + 4x_3 + 5x_4 \leq 10, \quad x_1, x_2, x_3 \text{ and } x_4 \geq 0.$$

Prove that the problem has multiple optimal solutions and prove that two of them are $x_1 = \frac{9}{2}, x_2 = 1, x_3 = 0, x_4 = 0$ and $x_1 = 4$, $x_2 = 2$, $x_3 = 0$, $x_4 = 0$ and $\max z = 10$.

3. Maximize, $z = 20x_1 + 6x_2 + 8x_3$

$$\text{subject to } 8x_1 + 2x_2 + 3x_3 \leq 250$$
$$4x_1 + 3x_2 \leq 150$$
$$2x_1 + x_3 \leq 50, \quad x_1 \geq 0, x_2 \geq 0, \text{ and } x_3 \geq 0.$$

4. Maximize, $z = 5x_1 + 3x_2$

$$\text{subject to } x_1 + x_2 \leq 2$$
$$5x_1 + 2x_2 \leq 15$$
$$3x_1 + 8x_2 \leq 12, \quad x_1 \geq 0, x_2 \geq 0.$$

5. (a) Maximize, $z = 3x_1 + 2x_2 + x_3$

$$\text{subject to } x_1 + 2x_2 - 3x_3 \leq 8$$
$$3x_2 + x_3 \leq 6, \quad x_1 \geq 0, x_2 \geq 0 \text{ and } x_3 \geq 0.$$

(b) Maximize, $z = 3x_1 + 6x_2 + 2x_3$

$$\text{subject to } 3x_1 + 4x_2 + x_3 \leq 2$$
$$x_1 + 2x_2 + 3x_3 \leq 1, \quad x_1, x_2 \text{ and } x_3 \geq 0.$$

[C.U.(P)'87,'90]

6. Maximize, $z = 3x_1 + 5x_2 + 4x_3$

$$\text{subject to } 2x_1 \quad 3x_3 \leq 18$$
$$2x_2 + 5x_3 \leq 18$$
$$3x_1 + 2x_2 + 4x_3 \leq 25 \quad x_1, x_2 \text{ and } x_3 \geq 0.$$

7. (a) Maximize, $z = x_1 + 3x_2 + 2x_3$

$$\text{subject to } x_1 + 2x_2 \leq 10$$
$$2x_1 + x_3 \leq 8$$
$$2x_2 + x_3 \leq 6, \quad x_1, x_2 \text{ and } x_3 \geq 0$$

Also prove that the problem has multiple optimal solutions. Find two optimal B.F.S.

(b) Maximize, $z = 4x_1 + 10x_2$

$$\text{subject to } 2x_1 + x_2 \leq 50$$
$$2x_1 + 5x_2 \leq 100$$
$$2x_1 + 3x_2 \leq 90, \quad x_1, x_2 \geq 0.$$

[Madras B.Sc.'84]

(c) Given the following L.P.P. Maximize $z = 2.5x_1 + x_2$

$$\text{subject to } 3x_1 + 5x_2 \leq 15$$
$$5x_1 + 2x_2 \leq 10, \quad x_1, x_2 \geq 0.$$

[C.U.M.Com.'87]

(i) Show that there exist multiple optimal solutions to the problem.

(ii) Suppose the first constraint is changed to $1.5x_1 + 2.5x_2 \leq 7.5$ and the remaining part of the problem remain unchanged. Will that give us a unique solution? Give reasons for your answer.

[*Hints:* The change of the constraint will not give unique solution as the constraint $1.5x_1 + 2.5x_2 \leq 7.5$ is same as the constraint $3x_1 + 5x_2 \leq 15$. Only the scale is being changed which does not change the nature of the solution etc.]

8. (a) Minimize, $z = 3x_1 - 2x_2$

$$\text{subject to } 4x_1 + x_2 \leq 8$$
$$2x_1 + 4x_2 \leq 20, \quad x_1, x_2 \geq 0.$$

(b) Minimize, $z = -3x_1 + 4x_2$

$$\text{subject to } x_1 + x_2 \leq 8$$
$$2x_1 + 5x_2 \leq 22$$
$$x_2 \leq 4, \quad x_1, x_2 \geq 0.$$

9. Minimize, $z = x_1 - 3x_2 + 2x_3$

$$\text{subject to } 3x_1 - x_2 + 2x_3 \leq 7$$
$$-2x_1 + 4x_2 \leq 12$$
$$-4x_1 + 3x_2 + 8x_3 \leq 10, \quad x_1, x_2 \text{ and } x_3 \geq 0.$$

[C.U.(P)'81,'96,'99]

10. Maximize, $z = 10x_1 + x_2 + 2x_3$

$$\text{subject to } x_1 + x_2 - 2x_3 \leq 10$$
$$4x_1 + x_2 + x_3 \leq 20, \quad x_j \geq 0, j = 1, 2, 3.$$

[C.U.(P)'86, M.Sc.(Appl.Math.)'77]

11. Maximize, $z = 2x_1 + 4x_2 + x_3 + x_4$

$$\text{subject to } x_1 + 3x_2 + x_4 \leq 4$$
$$2x_1 + x_2 \leq 3$$
$$x_2 + x_3 + x_4 \leq 3, \quad x_1, x_2, x_3, x_4 \geq 0.$$

12. Maximize, $z = x_1 + x_2$

$$\text{subject to } 2x_1 + 3x_2 \leq 22$$
$$2x_1 + x_2 \leq 14$$
$$x_1 - x_2 \leq 4$$
$$3x_1 - 2x_2 \geq -6, \quad x_1, x_2 \geq 0.$$

[C.U.(P)'94]

Verify the results by using graphical method.

13. Maximize, $z = 4x_1 + 3x_2 - 14x_3$

$$\text{subject to } \begin{aligned} x_1 + 2x_2 - 3x_3 &\le 4 \\ 4x_1 + x_2 + 2x_3 &\le 10 \\ 3x_1 + 5x_2 + 2x_3 &\le 12, \quad x_1, x_2 \text{ and } x_3 \ge 0. \end{aligned}$$

14. (a) Maximize, $z = 3x_1 + x_2 + 3x_3$

$$\text{subject to } \begin{aligned} 2x_1 + x_2 + x_3 &\le 2 \\ x_1 + 2x_2 + 3x_3 &\le 5 \\ 2x_1 + 2x_2 + x_3 &\le 6, \quad x_1, x_2 \text{ and } x_3 \ge 0. \end{aligned}$$

[B.U.(H)'82]

(b) Maximize, $z = 3x_1 + 2x_2 + 5x_3$

$$\text{subject to } \begin{aligned} x_1 + 2x_2 + x_3 &\le 430 \\ 3x_1 + 2x_3 &\le 460 \\ x_1 + 4x_2 &\le 420, \quad x_1, x_2 \text{ and } x_3 \ge 0. \end{aligned}$$

[C.U.(P)'83, V.U.(H)'93]

(c) Maximize, $z = 3x_1 + 2x_2 + 5x_3$

$$\text{subject to } \begin{aligned} x_1 + 2x_2 + x_3 &\le 430 \\ 3x_1 + 2x_2 &\le 460 \\ x_2 + 4x_3 &\le 420, \quad x_1, x_2 \text{ and } x_3 \ge 0. \end{aligned}$$

15. Solve the problem 4 given in (2.6).

16. (a) Solve the problem 5 given in (2.6).

(b) Solve the problem 13 by simplex method given in (2.6).

[C.U.M.Com.'89]

17. A factory is engaged in manufacturing three products A, B and C which involve lathe work, grinding and assembling. The cutting, grinding and assembling times required for one unit of A, are $2, 1$ and 1 hours, respectively. Similarly they are $3, 1, 3$ hours for one unit B and $1, 3, 1$ hours for one unit of C. The profits on A, B and C are Rs. 2, Rs. 2 and Rs. 4 per unit, respectively. Assuming that there are available 300 hours of lathe time, 300 hours of grinding time and 240 hours of assemble time, how many units of each product should be produced to maximize profits [solve by simplex method].

18. A company produces two types of leather belts, say type A and B. Belt A is superior quality and belt B is of lower quality. Profits on the two types belts are 40 are 30 paise per belt, respectively. Each belt of A type requires twice as much time as required by a belt of B type. If all belts were type B, the company could produce 1,000 belts per day. But the supply of leather is sufficient only for 800 belts per day. Belt A requires a fancy buckle and only 400 fancy buckles are available for this per day. For belt of type B, only 700 buckles are available per day. How should the company manufacture the two types of belt in order to have a maximum overall profit? [C.U.(H)'95]

Answers

1. (a) Max $z = 19$ at $x_1 = 5, x_2 = 2$.
   (b) Max $z = 24$ at $x_1 = 6, x_2 = 0$.
   (c) Max $z = \frac{235}{19}$ at $x_1 = \frac{20}{19}, x_2 = \frac{45}{19}$.
   (d) Max $z = 13\frac{4}{7}$ at $x_1 = \frac{16}{7}, x_2 = 3$.
   (e) Max $z = 48$ for $x_1 = 0, x_2 = 12$ and $x_3 = 0$.
   (f) Max $z = 50$ for $x_1 = 5, x_2 = 0, x_3 = 0$.

2. (a) Max $z = 20$ for $x_1 = 0, x_2 = 10$.
   (b) The problem is said to have an unbounded solution.
   (c) Max $z = 165$ at $x_1 = 15$ and $x_2 = 30$. The optimal solution is unique as $z_3 - c_3 = \frac{6}{5} \neq 0$ and $z_4 - c_4 = \frac{1}{5} \neq 0$ corresponding to non-basis vectors. The optimal solution is not degenerate as the optimal basic variable is $x_1 = 15 \neq 0$ and $x_2 = 30 \neq 0$.
   (d) Max $z = 13$ at $x_1 = \frac{7}{2}$, $x_2 = 0$ and $x_3 = \frac{9}{4}$, optimal basis

$$B(\mathbf{a}_1, \mathbf{a}_5, \mathbf{a}_3) = \begin{bmatrix} 3 & 0 & -2 \\ 2 & 1 & 1 \\ 1 & 0 & 2 \end{bmatrix}, \quad B^{-1} = \begin{bmatrix} 1/4 & 0 & 1/4 \\ -3/8 & 1 & -7/8 \\ -1/8 & 0 & 3/8 \end{bmatrix}$$

   which can be obtained from $(\mathbf{y}_3, \mathbf{y}_4, \mathbf{y}_5)$ under the initial unit basis vectors $\mathbf{a}_3$, $\mathbf{a}_4$ and $\mathbf{a}_5$ (in the final table).

3. Max $z = 700$ for $x_1 = 0, x_2 = 50$ and $x_3 = 50$.

4. Max $z = 10$ for $x_1 = 2, x_2 = 0$.

5. (a) Max $z = 84$ for $x_1 = 26, x_2 = 0$ and $x_3 = 6$.
   (b) Max $z = 3$ at $x_1 = 0$ (basic), $x_2 = \frac{1}{2}$ and $x_3 = 0$, degeneracy occurs.

6. Max $z = 52$ for $x_1 = \frac{7}{3}, x_2 = 9$ and $x_3 = 0$.

7. (a) Max $z = 13$; [$x_1 = 4, x_2 = 3, x_3 = 0$ (basic)], [$x_1 = 4$, $x_2 = 3$, $x_4 = 0$ (basic)]. The solutions are degenerate.
   (b) Max $z = 200$ at $x_1 = \frac{75}{4}, x_2 = \frac{25}{2}$, alternative optimal solutions exist.

8. (a) Min $z = -10$ at $x_1 = 0, x_2 = 5$;
   (b) Min $z = -24$ for $x_1 = 8, x_2 = 0$.

9. Min $z = -11$ for $x_1 = 4, x_2 = 5, x_3 = 0$.

10. Max $z = 50$ for $x_1 = 5, x_2 = 0, x_3 = 0$.

11. Max $z = 8$ for $x_1 = 1, x_2 = 1, x_3 = 2, x_4 = 0$.

12. Max $z = 9$ for $x_1 = 5, x_2 = 4$.

13. Max $z = \frac{82}{7}$ for $x_1 = \frac{16}{7}, x_2 = \frac{6}{7}, x_3 = 0$.

14. (a) Max $z = \frac{27}{5}$ for $x_1 = \frac{1}{5}, x_2 = 0, x_3 = \frac{8}{5}$.
    (b) Max $z = 1350$ for $x_1 = 0, x_2 = 100, x_3 = 230$.
    (c) Max $z = 985$ at $x_1 = \frac{460}{3}$, $x_2 = 0$, $x_3 = 105$.

15. Min cost = Rs. 13.00, liquid product 1 unit, dry product 5 units.

16. (a) Min cost = Rs. 2.05 for 15 units of X food and $\frac{5}{4}$ units of Y food.
    (b) Max. rev. = Rs. 2,550 for selling 150 red and 250 blue pen.

17. Product A, 120 product B, 0 and product C, 60. Maximum profit = Rs. 480.

18. Max profit = Rs. 260.00 at A type 200 and B type 600.

Chapter 9

Simplex Method or Simplex Algorithm (II)

9.1 Solution to the problem when some artificial variables are added to get a unit basis in the coefficient matrix

9.1.1 Artificial variable technique

Before using the simplex method, all inequations are to be converted into equations by introducing either slack or surplus variables. Next problem is to get an initial B.F.S. which will be obtained smoothly and easily if the coefficient matrix contains a unit basis and $\mathbf{b} \geq \mathbf{0}$. We have explained previously how some artificial variables are added to the left hand side of the converted equations to get a unit basis matrix. This technique is used only to get the initial B.F.S. of the problem easily. But it is interesting to note that an equation is being kept an equation even after the addition of a new variable (artificial variable) only in one side of an equation which is not correct from the mathematical point of view. It is only true if the values of all artificial variables be equal to zero. So in solving the problem of this type by using simplex method, one must be sure at the optimal stage, that all artificial variables are at zero level. If it is not possible to bring all artificial variables at zero level at the optimal stage, we conclude that the problem has no feasible solution. In an attempt to solve the problem, by using simplex method, the following three cases may arise when the optimality conditions are satisfied (at the optimal stage).

1. All artificial vectors are not present in the basis which indicates that all artificial variables are at zero level at the optimal stage. Thus the solution obtained is a B.F.S.

2. Some artificial vectors are present in the basis and some artificial variables are at positive level at the optimal stage. In that case there exists no feasible solution to the problem.

3. All artificial variables are at zero level but at least one artificial vector is present in the basis at the optimal stage. Here the solution under test is an optimal solution. Here the converted equations are consistent but some of the constraints may be redundant. By redundancy we mean that the system has more than enough constraints. There are two methods of solving problems of this type:

   (a) *Charnes* method of penalties or Big M-method.

   (b) Two phase method due to Dantzig, Orden and Wolfe.

We consider initially the Big-M method.

9.2 Charnes method of penalties or Big M-method

In this method, initially minimum number of artificial variables are to be inserted in the equations to get a unit basis matrix from the coefficient matrix. To each of the artificial variable, a very high negative price or cost, say — M(M is +ve, very large) is attached in the objective function of the problem. M is so large that the sign of α determines the sign of the expression $\alpha M + \beta$. Due to very high negative prices, the objective function cannot be improved in the presence of artificial variables. With the negative high prices, we proceed to solve the problem in usual method and following cases may arise.

1. At any stage, all artificial vectors may be driven out from the basis. In that case, all artificial variables are at zero level at that stage. Now if the optimality conditions are satisfied at that stage, the problem is solved and the problem has an optimal solution. If the optimality conditions are not satisfied we may proceed further to get an optimal solution omitting all column vectors corresponding to the artificial variables. The artificial vectors are used only as agents to get a unit basis.

2. Some artificial variables are at positive level, though the optimality conditions are satisfied. In that case the problem has no feasible solution at all. Here some artificial vectors must be present in the basis at the final stage.

3. All artificial variables are at zero level and at least one artificial vector is present in the basis. If at this stage, the optimality conditions are satisfied the solution obtained is an optimal solution. Here one or more constraints may be redundant. If the optimality conditions are not satisfied, proceed further as in the previous cases to get an optimal solution.

In this connection it is important to note that once an artificial vector leaves the basis, we forget all about it for ever and never consider it as a vector to enter into the basis at any next iteration. The rule of construction of the tables are same as given in the previous cases.

▶ **Example 9.2.1** *Solve the L.P.P.*

$$\textit{Minimize,} \quad z = 4x_1 + 3x_2$$

subject to

$$\begin{aligned} x_1 + 2x_2 &\geq 8 \\ 3x_1 + 2x_2 &\geq 12, \quad x_1 \geq 0, x_2 \geq 0. \end{aligned}$$

by Charnes Big M-method. **[C.U.(P)'89,'93; V.U.(P)'90]**

Solution: The problem here is to minimize z. The components of **b** are already +ve.

Let $z' = -z$; then $\min z = -\max(-z) = -\max z'$.

Hence the problem is to maximize $z' = -4x_1 - 3x_2$.

Introducing two surplus variables x_3 and x_4 we get the following converted equations

$$\begin{aligned} x_1 + 2x_2 - x_3 \quad\quad &= 8 \\ 3x_1 + 2x_2 \quad\quad - x_4 &= 12 \end{aligned}$$

The coefficient matrix does not contain a unit basis matrix. To get a unit basis matrix, two artificial variables x_5 and x_6 are added, one in each equation and the equations are

$$\begin{aligned} x_1 + 2x_2 - x_3 \quad\quad + x_5 \quad\quad &= 8 \\ 3x_1 + 2x_2 \quad\quad - x_4 \quad\quad + x_6 &= 12 \end{aligned}$$

Now the coefficient matrix does contain a unit basis.

The adjusted objective function z' is given by

$$z' = -4x_1 - 3x_2 + 0.x_3 + 0.x_4 - Mx_5 - Mx_6$$

[assigning a very large negative price to each of the artificial variables x_5 and x_6, M is positive].

Here vectors

$$\mathbf{a}_5 = \begin{bmatrix} 1 \\ 0 \end{bmatrix} \quad \text{and} \quad \mathbf{a}_6 = \begin{bmatrix} 0 \\ 1 \end{bmatrix}$$

are artificial vectors and the initial basis

$$B = (\mathbf{a}_5, \mathbf{a}_6) = \begin{bmatrix} 1 & 0 \\ 0 & 1 \end{bmatrix}$$

and hence initial (artificial) B.F.S.

$$\begin{aligned} \mathbf{x}_B &= [x_5, x_6] = [8, 12] \\ \mathbf{c}_B &= (c_5, c_6) = (-M, -M), \\ \mathbf{y}_j &= B^{-1}\mathbf{a}_j = \mathbf{a}_j. \end{aligned}$$

$$\begin{aligned} z &= \mathbf{c}_B\mathbf{x}_B = -20M \\ z_1 - c_1 &= (-M,-M)[1,3] + 4 = -4M + 4 \\ z_2 - c_2 &= (-M,-M)[2,2] + 3 = -4M + 3 \\ z_3 - c_3 &= (-M,-M)[-1,0] = M \\ z_4 - c_4 &= (-M,-M)[0,-1] = M \quad \text{and} \\ z_5 - c_5 &= z_6 - c_6 = 0 \end{aligned}$$

Initial Simplex table

		c	−4	−3	0	0	−M	−M	
Basis	$\mathbf{c}_B$	**b**	$\mathbf{a}_1$	$\mathbf{a}_2$	$\mathbf{a}_3$	$\mathbf{a}_4$	$\mathbf{a}_5(\mathbf{e}_1)$	$\mathbf{a}_6(\mathbf{e}_2)$	**Min. ratio**
$\mathbf{a}_5^*$	$-M$	8	1	2*	−1	0	1	0	$\frac{8}{2} = 4^*$
$\mathbf{a}_6$	$-M$	12	3	2	0	−1	0	1	$\frac{12}{2} = 6$
$z_j - c_j$			$-4M+4$	$-4M+3^*$	M	M	0	0	

$\mathbf{a}_2$ is the incoming vector as $z_2 - c_2$ is the −ve minimum and $\mathbf{a}_5$ is the outgoing vector. We shall leave the column corresponding to the artificial vector $\mathbf{a}_5$ for ever.

Second Simplex table

		c	−4	−3	0	0	−M	
Basis	$\mathbf{c}_B$	**b**	$\mathbf{a}_1$	$\mathbf{a}_2$	$\mathbf{a}_3$	$\mathbf{a}_4$	$\mathbf{a}_6$	**Min. ratio**
$\mathbf{a}_2$	−3	4	$\frac{1}{2}$	1	$-\frac{1}{2}$	0	0	$4/\frac{1}{2} = 8$
$\mathbf{a}_6^*$	$-M$	4	2*	0	1	−1	1	$\frac{4}{2} = 2^*$
$z_j - c_j$			$-2M + \frac{5}{2}^*$	0	$-M + \frac{3}{2}$	M	0	

$\mathbf{a}_1$ is incoming vector and $\mathbf{a}_6$ is the outgoing vector. Remove the column corresponding to $\mathbf{a}_6$ for ever.

Third simplex table

		c	−4	−3	0	0
Basis	$\mathbf{c}_B$	**b**	$\mathbf{a}_1$	$\mathbf{a}_2$	$\mathbf{a}_3$	$\mathbf{a}_4$
$\mathbf{a}_2$	−3	3	0	1	$-\frac{3}{4}$	$\frac{1}{4}$
$\mathbf{a}_1$	−4	2	1	0	$\frac{1}{2}$	$-\frac{1}{2}$
$z_j - c_j$		−17	0	0	$\frac{1}{4}$	$\frac{5}{4}$

None of $z_j - c_j < 0$; the solution obtained is an optimal solution. No artificial vector is present in the final basis. Therefore, all artificial variables are at zero level at the final stage. Hence the optimal solution obtained is a B.F.S. and the maximum value of z' is −17 for $x_1 = 2$ and $x_2 = 3$.

Therefore, $\min z = -\max(z') = -(-17) = 17$ for $x_1 = 2$ and $x_2 = 3$.

▶ **Example 9.2.2** *Solve the L.P.P.*

$$\text{Maximize,} \quad z = 2x_1 - 3x_2$$

subject to

$$\begin{aligned} -x_1 + x_2 &\geq -2 \\ 5x_1 + 4x_2 &\leq 46 \\ 7x_1 + 2x_2 &\geq 32, \quad x_1 \geq 0, x_2 \geq 0. \end{aligned}$$

[C.U.(H)'85,'87,'92; (P)'95, 99]

Solution: 1st constraint is $x_1 - x_2 \leq 2$ [making component of **b**, +ve].

Introducing slack variables x_3, x_4 and surplus variable x_5 one to each of the constraints respectively we get the following converted equations

$$\begin{aligned} x_1 - x_2 + x_3 \quad\quad &= 2 \\ 5x_1 + 4x_2 \quad + x_4 \quad &= 46 \\ 7x_1 + 2x_2 \quad\quad - x_5 &= 32 \end{aligned}$$

The coefficient matrix does not contain a unit basis matrix. To get a unit basis, one artificial variable x_6 is added to the left hand side of the third equation and the set of equations are

$$\begin{aligned} x_1 - x_2 + x_3 \quad\quad &= 2 \\ 5x_1 + 4x_2 \quad + x_4 \quad &= 46 \\ 7x_1 + 2x_2 \quad\quad - x_5 + x_6 &= 32, x_j \geq 0, \; j = (1, 2, \cdots, 6) \end{aligned}$$

The adjusted objective function z is given by

$$z = 2x_1 - 3x_2 + 0.x_3 + 0.x_4 + 0.x_5 - Mx_6$$

[assigning very large negative price to the artificial variable x_6.]

Here

$$\mathbf{a}_3 = \begin{bmatrix} 1 \\ 0 \\ 0 \end{bmatrix}, \quad \mathbf{a}_4 = \begin{bmatrix} 0 \\ 1 \\ 0 \end{bmatrix}, \quad \mathbf{a}_6 = \begin{bmatrix} 0 \\ 0 \\ 1 \end{bmatrix}, \quad \mathbf{b} = \begin{bmatrix} 2 \\ 46 \\ 32 \end{bmatrix} \geq 0$$

and the initial unit basis $B = (\mathbf{a}_3, \mathbf{a}_4, \mathbf{a}_6)$. Hence initial B.F.S. (artificial)

$$\begin{aligned} \mathbf{x}_B &= [x_3, x_4, x_6] = [2, 46, 32] \\ \mathbf{c}_B &= (c_3, c_4, c_6) = (0, 0, -M) \\ z &= -32M \\ \mathbf{y}_j &= B^{-1}\mathbf{a}_j = \mathbf{a}_j \end{aligned}$$

$\mathbf{a}_6$ is an artificial vector.

Simplex tables

Basis	c_B	c / b	2 / a_1	−3 / a_2	0 / $a_3(e_1)$	0 / $a_4(e_2)$	0 / a_5	−M / $a_6(e_3)$	Min. ratio
a_3^*	0	2	1^*	−1	1	0	0	0	$\frac{2}{1} = 2^*$
a_4	0	46	5	4	0	1	0	0	$\frac{46}{5} = 9\frac{1}{5}$
a_6	$-M$	32	7	2	0	0	−1	1	$\frac{32}{7} = 4\frac{4}{7}$
$z_j - c_j$			$-7M - 2^*$	$-2M + 3$	0	0	M	0	
a_1	2	2	1	−1	1	0	0	0	...
a_4	0	36	0	9	−5	1	0	0	$\frac{36}{9} = 4$
a_6^*	$-M$	18	0	9^*	−7	0	−1	1	$\frac{18}{9} = 2^*$
$z_j - c_j$			0	$-9M + 1^*$	$7M + 2$	0	M	0	
a_1	2	4	1	0	$\frac{2}{9}$	0	$-\frac{1}{9}$		
a_4	0	18	0	0	2	1	1		
a_2	−3	2	0	1	$-\frac{7}{9}$	0	$-\frac{1}{9}$		
$z_j - c_j$		2	0	0	$\frac{25}{9}$	0	$\frac{1}{9}$		

As none of $z_j - c_j < 0$, therefore, the optimality conditions are satisfied. The artificial vector $\mathbf{a}_6$ is not present in the final basis. Therefore, the artificial variable x_6 is zero at the final stage. Hence the optimal solution obtained is a B.F.S. and the maximum value of z is 2 for $x_1 = 4$ and $x_2 = 2$.

Note: We even need not compute the mid-body of the third table except the column under **b**.

► **Example 9.2.3** *Solve the L.P.P.*

$$\textit{Maximize,} \quad z = x_1 + 2x_2$$

subject to

$$\begin{aligned} x_1 - 5x_2 &\leq 10 \\ 2x_1 - x_2 &\geq 2 \\ x_1 + x_2 &= 10, \quad x_1 \geq 0, x_2 \geq 0. \end{aligned}$$

[C.U.(P)'91,96]

Solution: Introducing slack variable x_3 and surplus variable x_4 to the first and second constraints respectively, we get the following equations

$$\begin{aligned} x_1 - 5x_2 + x_3 &= 10 \\ 2x_1 - x_2 - x_4 &= 2 \\ x_1 + x_2 &= 10 \end{aligned}$$

The coefficient matrix does not contain a unit basis. In order to get a unit basis two artificial variables x_5 and x_6 are added one to each of second and third equation respectively and the new equations are

$$\begin{aligned} x_1 - 5x_2 + x_3 \qquad\qquad\qquad\qquad &= 10 \\ 2x_1 - \ x_2 \qquad - x_4 + x_5 \qquad\quad &= 2 \\ x_1 + \ x_2 \qquad\qquad\qquad + x_6 &= 10 \ , x_j \geq 0, [j = 1, 2, \cdots, 6] \end{aligned}$$

One slack vector, corresponding to the slack variable x_3 and the two artificial vectors corresponding to the artificial variables x_5 and x_6 constitute a unit basis.

The adjusted objective function z is given by

$$z = x_1 + 2x_2 + 0.x_3 + 0.x_4 - Mx_5 - Mx_6$$

[assigning very large negative price $-M$ to each of the artificial variables x_5 and x_6].

Initial (artificial) B.F.S.

$$\mathbf{x} = [x_3, x_5, x_6] = [10, 2, 10]$$

$$z = \mathbf{c}_B\mathbf{x}_B = 0 \times 10 - M \times 2 - M \times 10 = -12M$$

$$\mathbf{y}_j = \mathbf{a}_j$$

Simplex tables

		c	1	2	0	0	$-M$	$-M$	
Basis	$\mathbf{c}_B$	**b**	$\mathbf{a}_1$	$\mathbf{a}_2$	$\mathbf{a}_3(\mathbf{e}_1)$	$\mathbf{a}_4$	$\mathbf{a}_5(\mathbf{e}_2)$	$\mathbf{a}_6(\mathbf{e}_3)$	**Min. ratio**
$\mathbf{a}_3$	0	10	1	-5	1	0	0	0	$\frac{10}{1} = 10$
$\mathbf{a}_5^*$	$-M$	2	2^*	-1	0	-1	1	0	$\frac{2}{2} = 1^*$
$\mathbf{a}_6$	$-M$	10	1	1	0	0	0	1	$\frac{10}{1} = 10$
$z_j - c_j$			$-3M - 1^*$	-2	0	M	0	0	...
$\mathbf{a}_3$	0	9	0	$-\frac{9}{2}$	1	$\frac{1}{2}$		0	...
$\mathbf{a}_1$	1	1	1	$-\frac{1}{2}$	0	$-\frac{1}{2}$		0	...
$\mathbf{a}_6^*$	$-M$	9	0	$\frac{3}{2}^*$	0	$\frac{1}{2}$		1	$\frac{9}{3/2} = 6^*$
$z_j - c_j$			0	$-\frac{3}{2}M - \frac{5}{2}^*$	0	$-M/2 - \frac{1}{2}$		0	
$\mathbf{a}_3$	0	36	0	0	1	2			
$\mathbf{a}_1$	1	4	1	0	0	$-\frac{1}{3}$			
$\mathbf{a}_2$	2	6	0	1	0	$\frac{1}{3}$			
$z_j - c_j$		16	0	0	0	$\frac{1}{3}$			..

None of $z_j - c_j < 0$. Therefore, the optimality conditions are satisfied. The artificial vectors $\mathbf{a}_5$ and $\mathbf{a}_6$ are driven out from the final basis. Hence the artificial variables are at zero level. Thus the solution obtained is a B.F.S. The maximum value of the objective function of the original problem is 16 for $x_1 = 4$ and $x_2 = 6$.

► **Example 9.2.4** *Solve the L.P.P.*

$$\textit{Minimize,} \quad z = 4x_1 + 8x_2 + 3x_3$$

subject to

$$\begin{aligned} x_1 + x_2 \quad &\geq 2 \\ 2x_1 \quad + x_3 &\geq 0\,, \quad x_1 \geq 0, x_2 \geq 0, x_3 \geq 0. \end{aligned}$$

[C.U.(H)'90, 2000]

Hints. No artificial vector is required to insert in the coefficient matrix to get a unit basis matrix. A unit matrix $\mathbf{a}_2 = \mathbf{e}_1$ and $\mathbf{a}_3 = \mathbf{e}_2$ is already present there.

Let $z' = -z = -4x_1 - 8x_2 - 3x_3$

Now $\min z = -\max(z')$. Solve the problem without using artificial vector.

[*Ans.* $\min z = -\max z' = -(-10) = 10$ for $x_1 = \frac{5}{2}$, $x_2 = 0$, $x_3 = 0$. [x_2 and x_3 are non-basic variables.]

Problem having an unbounded solution

▶ **Example 9.2.5** *Solve the L.P.P.*

$$\text{Maximize,} \quad z = 3x_1 - x_2$$

subject to

$$\begin{aligned} -x_1 + \ x_2 &\geq 2 \\ 5x_1 - 2x_2 &\geq 2\,, \quad x_1 \geq 0, x_2 \geq 0. \end{aligned}$$

Introducing surplus variables x_3 and x_4, one to each of the constraints, we get the following equations

$$\begin{aligned} -x_1 + \ x_2 - x_3 \qquad &\geq 2 \\ 5x_1 - 2x_2 \qquad - x_4 &\geq 2 \end{aligned}$$

The coefficient matrix does not contain a unit basis.

To get a unit basis, two artificial variables x_5 and x_6 are added to the left hand side of the equations and then

$$\begin{aligned} -x_1 + \ x_2 - x_3 \qquad + x_5 \qquad &\geq 2 \\ 5x_1 - 2x_2 \qquad - x_4 \qquad + x_6 &\geq 2 \end{aligned}$$

The adjusted objective function z is given by

$$z = 3x_1 - x_2 + 0.x_3 + 0.x_4 - Mx_5 - Mx_6$$

Initial (artificial) B.F.S. is

$$\begin{aligned} \mathbf{x}_B &= [x_5, x_6] = [2, 2] \\ \mathbf{c}_B &= [c_5, c_6] = [-M, -M] \\ z &= c_B \mathbf{x}_B = -4M, \mathbf{y}_j = \mathbf{a}_j \end{aligned}$$

Simplex tables

	c	3	−1	0	0	−M	−M	
Basis $\mathbf{c}_B$	**b**	$\mathbf{a}_1$	$\mathbf{a}_2$	$\mathbf{a}_3$	$\mathbf{a}_4$	$\mathbf{a}_5(\mathbf{e}_1)$	$\mathbf{a}_6(\mathbf{e}_2)$	**Min. ratio**
$\mathbf{a}_5$ $-M$	2	−1	1	−1	0	1	0	......
$\mathbf{a}_6^*$ $-M$	2	5*	−2	0	−1	0	1	$\frac{2}{5}^*$
$z_j - c_j$		$-4M-3^*$	$M+1$	M	M	0	0	......
$\mathbf{a}_5^*$ $-M$	$\frac{12}{5}$	0	$\frac{3}{5}^*$	−1	$-\frac{1}{5}$	1		$\frac{12}{5}/\frac{3}{5} = 4^*$
$\mathbf{a}_1$ 3	$\frac{2}{5}$	1	$-\frac{2}{5}$	0	$-\frac{1}{5}$	0		......
$z_j - c_j$		0	$-\frac{3}{5}M - \frac{1}{5}^*$	M	$\frac{M}{5} - \frac{3}{5}$	0		
$\mathbf{a}_2$ −1	4	0	1	$-\frac{5}{3}$	$-\frac{1}{3}$			
$\mathbf{a}_1$ 3	2	1	0	$-\frac{2}{3}$	$-\frac{1}{3}$			
$z_j - c_j$		0	0	$-\frac{1}{3}$	$-\frac{2}{3}$			

All artificial vectors are driven out from the basis. Now in the third column $z_3 - c_3 = -\frac{1}{3} < 0$ with $y_{i3} \leq 0$ for $i = 1$ and 2. Hence the problem has no finite optimal value of the objective function, i.e., the problem has said to have an unbounded solution.

9.2.1 Problem having no F.S.

► **Example 9.2.6** *Solve the L.P.P.*

$$\textit{Maximize,} \quad z = 2x_1 - 3x_2$$

subject to

$$\begin{aligned} 2x_1 + x_2 &\leq 8 \\ 10x_1 + 11x_2 &\geq 100\,, \quad x_1 \geq 0, x_2 \geq 0. \end{aligned}$$

Introducing slack variable x_3 and surplus variable x_4 one to each of the constraints respectively we get the following converted equations

$$\begin{aligned} 2x_1 + x_2 + x_3 &= 8 \\ 10x_1 + 11x_2 - x_4 &= 100 \end{aligned}$$

The coefficient matrix does not contain a unit basis. To get a unit basis one artificial variable x_5 is to be added to the L.H.S of the 2nd equation and the set of equations are

$$\begin{aligned} 2x_1 + x_2 + x_3 &= 8 \\ 10x_1 + 11x_2 - x_4 + x_5 &= 100 \end{aligned}$$

$z = 2x_1 - 3x_2 + 0.x_3 + 0.x_4 - Mx_5$ [Assigning very large −ve price to the artificial variable x_5]

The vectors $\mathbf{a}_3$ and $\mathbf{a}_5$ constitute a unit basis. Initial solution

$$\mathbf{x}_B = [x_3, x_5] = [8, 100]$$
$$\mathbf{c}_B = (c_3, c_5) = (0, -M) \text{ and } \mathbf{y}_j = B^{-1}\mathbf{a}_j = \mathbf{a}_j$$
$$z = \mathbf{c}_B\mathbf{x}_B = -100M$$

Simplex tables

		c	2	−3	0	0	$-M$	
Basis	$\mathbf{c}_B$	**b**	$\mathbf{a}_1$	$\mathbf{a}_2$	$\mathbf{a}_3(\mathbf{e}_1)$	$\mathbf{a}_4$	$\mathbf{a}_5(\mathbf{e}_2)$	**Min. ratio**
$\mathbf{a}_3^*$	0	8	2	1*	1	0	0	$\frac{8}{1} = 8^*$
$\mathbf{a}_5$	$-M$	100	10	11	0	−1	1	$\frac{100}{11} = 9\frac{1}{11}$
$z_j - c_j$			$-2 - 10M$	$-3 - 11M^*$	0	M	0	
$\mathbf{a}_2$	−3	8	2	1	1	0	0	
$\mathbf{a}_5$	$-M$	12	−12	0	−11	−1	1	
$z_j - c_j$			$12M - 8$	0	$11M - 3$	M	0	

The optimality conditions have been satisfied. But the artificial vector $\mathbf{a}_5$ is in the basis at positive level. Hence the only conclusion is that the problem has no F.S. in this case. There is no need to calculate the value of the objective function at the final stage.

N.B. Verify the result by geometrical method.

Example 9.2.7 *Solving by Big M-method prove that the following L.P.P. has no feasible solution.*

$$\textit{Maximize,} \quad z = 2x_1 - x_2 + 5x_3$$

subject to

$$\begin{aligned} x_1 + 2x_2 + 2x_3 &\leq 2 \\ \tfrac{5}{2}x_1 + 3x_2 + 4x_3 &= 12 \\ 4x_1 + 3x_2 + 2x_3 &\geq 24, \quad x_1, x_2, x_3 \geq 0. \end{aligned}$$

$\mathbf{b} = [2, 12, 24] \geq \mathbf{0}$. In order to get a unit basis we require to add one slack variable x_4, one artificial variable x_5 and one surplus and other artificial variable x_6 and x_7 in the three constraints respectively. Thus $z = 2x_1 - x_2 + 5x_3 + 0.x_4 - Mx_5 + 0.x_6 - Mx_7$ and the constraints are

$$\begin{aligned} x_1 + 2x_2 + 2x_3 + x_4 &= 2 \\ \tfrac{5}{2}x_1 + 3x_2 + 4x_3 + x_5 &= 12 \\ 4x_1 + 3x_2 + 2x_3 - x_6 + x_7 &= 24, \quad x_j \geq 0[j = 1, 2, \cdots, 7]. \end{aligned}$$

Simplex tables

	c	2	−1	5	0	−M	0	−M	
Basis c_B	**b**	$\mathbf{a}_1$	$\mathbf{a}_2$	$\mathbf{a}_3$	$\mathbf{a}_4(\mathbf{e}_1)$	$\mathbf{a}_5(\mathbf{e}_2)$	$\mathbf{a}_6$	$\mathbf{a}_7(\mathbf{e}_3)$	**Min. ratio**
$\mathbf{a}_4^*$ 0	2	1^*	2	2	1	0	0	0	$\frac{2}{1}=2^*$
$\mathbf{a}_5$ $-M$	12	$\frac{5}{2}$	3	4	0	1	0	0	$12/\frac{5}{2}=\frac{24}{5}$
$\mathbf{a}_7$ $-M$	24	4	3	2	0	0	−1	1	$\frac{24}{4}=6$
z_j-c_j		$-\frac{13}{2}M^*-2$	$-6M+1$	$-6M-5$	0	0	M	0	
$\mathbf{a}_1$	2								
$\mathbf{a}_5$	7								
$\mathbf{a}_7$	16								
z_j-c_j		0	$7M+5$	$7M-1$	$\frac{13}{2}M+2$	0	M	0	

All $z_j - c_j \geq 0$, $j = 1, 2, \cdots, 7$ in the second table. Thus we need not complete the second table. Two artificial variables x_5 and x_7 are present at the positive level in the optimal solution. Then the only conclusion is that the problem has no feasible solution.

Note: $z_j - c_j$ of the second table is calculated using the formula as shown in the examples solved previously.

▶ **Example 9.2.8** *Solve the L.P.P. using artificial variables*

$$\textit{Minimize,}\quad z = -3x_1 + 2x_2$$

subject to

$$\begin{aligned} x_1 - 4x_2 &\leq -14 \\ -3x_1 + 2x_2 &\leq \quad 6, \quad x_1, x_2 \geq 0. \end{aligned}$$

[C.U.(P)'88]

The problem can be written in the maximization standard form

$$\text{maximize,}\quad z' = 3x_1 - 2x_2 + 0.x_3 - M.x_4 + 0.x_5$$

subject to

$$\begin{aligned} -x_1 + 4x_2 - x_3 + x_4 \quad &= 14 \\ -3x_1 + 2x_2 \qquad\qquad + x_5 &= \ 6 \end{aligned}$$

where $x_1, x_2 \geq 0$ are decision variables, $x_3 \geq 0$ surplus variable, $x_4 \geq 0$ artificial variable, and $x_5 \geq 0$ slack variable.

Simplex tables

		c	3	−2	0	$-M$	0	
Basis	c_B	**b**	$\mathbf{a}_1$	$\mathbf{a}_2$	$\mathbf{a}_3$	$\mathbf{a}_4(\mathbf{e}_1)$	$\mathbf{a}_5(\mathbf{e}_2)$	**Min. ratio**
$\mathbf{a}_4$	$-M$	14	−1	4	−1	1	0	$\frac{14}{4}=\frac{7}{2}$
$\mathbf{a}_5^*$	0	6	−3	2*	0	0	1	$\frac{6}{2}=3^*$
$z_j - c_j$			$M-3$	$-4M+2^*$	M	0	0	
$\mathbf{a}_4^*$	$-M$	2	5*	0	−1	1	−2	
$\mathbf{a}_2$	−2	3	$-\frac{3}{2}$	1	0	0	$\frac{1}{2}$	
$z_j - c_j$			$-5M^*$	0	M	0	$2M-1$	
$\mathbf{a}_1$	3	$\frac{2}{5}$	1	0	$-\frac{1}{5}$		$-\frac{2}{5}$	
$\mathbf{a}_2$	−2	$\frac{18}{5}$	0	1	$-\frac{3}{10}$		$-\frac{1}{10}$	
$z_j - c_j$			0	0	0		−1	

In the third table $z_5 - c_5 = -1$ (negative), and both y_{15}, y_{25} are negative. Thus we stop here. The problem has no finite optimal solution and the problem is said to have an unbounded solution.

► **Example 9.2.9** *Solve the following problem by Big M-method*

$$\textit{Minimize,} \quad z = 2x_1 + 3x_2$$

subject to

$$\begin{aligned} 2x_1 + 7x_2 &\geq 22 \\ x_1 + x_2 &\geq 6 \\ 5x_1 + x_2 &\geq 10, \quad x_1, x_2 \geq 0. \end{aligned}$$

Verify it by using graphical method. **[C.U.(P)'94,99]**

Solution: After the introduction of the surplus and artificial variables to get a unit matrix, the constraints are

$$\begin{aligned} 2x_1 + 7x_2 - x_3 \qquad\qquad + x_6 \qquad\qquad &= 22 \\ x_1 + x_2 \qquad - x_4 \qquad\qquad + x_7 \qquad &= 6 \\ 5x_1 + x_2 \qquad\qquad - x_5 \qquad\qquad + x_8 &= 10, \quad x_j \geq 0[j = 1, 2, \cdots, 8.] \end{aligned}$$

x_6, x_7, x_8 are artificial variables and the objective function,

$$\text{maximize} \quad -2x_1 - 3x_2 + 0.x_3 + 0 \cdot x_4 + 0 \cdot x_5 - Mx_6 - Mx_7 - Mx_8,$$

where $\min z = -\max(-z)$

		c	−2	−3	0	0	0	$-M$	$-M$	$-M$	
Basis	$\mathbf{c}_B$	**b**	$\mathbf{a}_1$	$\mathbf{a}_2$	$\mathbf{a}_3$	$\mathbf{a}_4$	$\mathbf{a}_5$	$\mathbf{a}_6$	$\mathbf{a}_7$	$\mathbf{a}_8$	Min. ratio
$\mathbf{a}_6^*$	$-M$	22	2	7^*	−1	0	0	1	0	0	$\frac{22}{7}=\frac{22}{7}^*$
$\mathbf{a}_7$	$-M$	6	1	1	0	−1	0	0	1	0	$\frac{6}{1}=6$
$\mathbf{a}_8$	$-M$	10	5	1	0	0	−1	0	0	1	$\frac{10}{1}=10$
z_j-c_j			$-8M+2$	$-9M+3^*$	M	M	M	0	0	0	
$\mathbf{a}_2$	−3	$\frac{22}{7}$	$\frac{2}{7}$	1	$-\frac{1}{7}$	0	0		0	0	$\frac{22}{7}/\frac{2}{7}=11$
$\mathbf{a}_7$	$-M$	$\frac{20}{7}$	$\frac{5}{7}$	0	$\frac{1}{7}$	−1	0		1	0	$\frac{20}{7}/\frac{5}{7}=4$
$\mathbf{a}_8^*$	$-M$	$\frac{48}{7}$	$\frac{33}{7}^*$	0	$\frac{1}{7}$	0	−1		0	1	$\frac{48}{7}/\frac{33}{7}=\frac{16}{11}^*$
z_j-c_j			$-\frac{38M}{7}^*+\frac{8}{7}$	0	$-\frac{2M}{7}+\frac{3}{7}$	M	M		0	0	
$\mathbf{a}_2$	−3	$\frac{30}{11}$	0	1	$-\frac{5}{33}$	0	$\frac{2}{33}$		0		$\frac{30}{11}/\frac{2}{33}=45$
$\mathbf{a}_7^*$	$-M$	$\frac{20}{11}$	0	0	$\frac{4}{33}$	−1	$\frac{5}{33}^*$		1		$\frac{20}{11}/\frac{5}{33}=12^*$
$\mathbf{a}_1$	−2	$\frac{16}{11}$	1	0	$\frac{1}{33}$	0	$-\frac{7}{33}$		0		......
z_j-c_j			0	0	$-\frac{4M}{33}+\frac{13}{33}$	M	$-\frac{5M}{33}+\frac{8}{33}^*$		0		
$\mathbf{a}_2$	−3	2	0	1	$-\frac{1}{5}$	$\frac{2}{5}$	0				
$\mathbf{a}_5$	0	12	0	0	$\frac{4}{5}$	$-\frac{33}{5}$	1				
$\mathbf{a}_1$	−2	4	1	0	$\frac{1}{5}$	$-\frac{7}{5}$	0				
z_j-c_j		−14	0	0	$\frac{1}{5}$	$\frac{8}{5}$	0				

Max $(-z) = -14$ at $x_1 = 4$, $x_2 = 2$, $x_5 = 12$, i.e., Min $z = -(-14) = 14$ at $x_1 = 4, x_2 = 2$. Graphically the problem has been solved in Example 7.1.5.

▶ **Example 9.2.10** *Solve the following L.P. problem by Big M-method and prove that the problem has finite optimal solution and finite value of the objective function.*

$$\textit{Minimize,} \quad z = 3x_1 + 5x_2$$

subject to

$$x_1 + 2x_2 \geq 8$$
$$3x_1 + 2x_2 \geq 12$$
$$5x_1 + 6x_2 \leq 60, \quad x_1, x_2 \geq 0.$$

[B.U.(H)'87]

Solution: After the introduction of slack, surplus and artificial variables so that there exists a unit matrix which will from initial unit basis; the constraints are

$$\begin{aligned} x_1 + 2x_2 - x_3 \qquad\qquad + x_6 \quad &= 8 \\ 3x_1 + 2x_2 \qquad - x_4 \qquad\qquad + x_7 &= 12 \\ 5x_1 + 6x_2 \qquad\qquad + x_5 \qquad\qquad &= 60, \quad x_1, x_2 \geq 0; x_3, x_4 \geq 0 \end{aligned}$$

surplus variables $x_5 \geq 0$, slack variable; x_6, $x_7 \geq 0$, artificial variables.

The objective function can be converted as

$$\text{Maximize } (-z) = -3x_1 - 5x_2 + 0 \cdot x_3 + 0 \cdot x_4 + 0 \cdot x_5 - Mx_6 - Mx_7,$$

where $\min z = -\max(-z)$.

We now solve the maximization problem.

Simplex tables

		c	−3	−5	0	0	0	−M	−M	
Basis	$\mathbf{c}_B$	**b**	$\mathbf{a}_1$	$\mathbf{a}_2$	$\mathbf{a}_3$	$\mathbf{a}_4$	$\mathbf{a}_5(\mathbf{e}_3)$	$\mathbf{a}_6(\mathbf{e}_1)$	$\mathbf{a}_7(\mathbf{e}_2)$	**Min. ratio**
$\mathbf{a}_6$	$-M$	8	1	2	−1	0	0	1	0	$\frac{8}{1}=8$
$\mathbf{a}_7^*$	$-M$	12	3^*	2	0	−1	0	0	1	$\frac{12}{3}=4^*$
$\mathbf{a}_5$	0	60	5	6	0	0	1	0	0	$\frac{60}{5}=12$
z_j-c_j			$-4M+3^*$	$-4M+5$	M	M	0	0	0	
$\mathbf{a}_6^*$	$-M$	4	0	$\frac{4}{3}^*$	−1	$\frac{1}{3}$	0	1		$\frac{4}{4/3}=3^*$
$\mathbf{a}_1$	−3	4	1	$\frac{2}{3}$	0	$-\frac{1}{3}$	0	0		$\frac{4}{2/3}=6$
$\mathbf{a}_5$	0	40	0	$\frac{8}{3}$	0	$\frac{5}{3}$	1	0		$\frac{40}{8/3}=15$
z_j-c_j			0	$-\frac{4}{3}M+3^*$	M	$-\frac{1}{3}M+1$	0	0		
$\mathbf{a}_2$	−5	3	0	1	$-\frac{3}{4}$	$\frac{1}{4}$	0			
$\mathbf{a}_1$	−3	2	1	0	$\frac{1}{2}$	$-\frac{1}{2}$	0			
$\mathbf{a}_5$	0	32	0	0	2	1	1			
z_j-c_j		−21	0	0	$\frac{9}{4}$	$\frac{1}{4}$	0			

All $z_j - c_j \geq 0$ and all artificial variables are driven out from the basis. Then the finite optimal value with finite optimal solution exists. $\max(-z) = -21$ at $x_1 = 2, x_2 = 3$. Therefore, $\min z = -(-21) = 21$ at $x_1 = 2, x_2 = 3$.

Remark. Verify the result by graphical method.

▶ **Example 9.2.11** *Solve the following L.P. Problem by Big M-method and prove that the problem has no feasible solution.*

$$\textit{Maximize,} \quad z = 5x_1 + 11x_2$$

subject to

$$\begin{aligned} 2x_1 + \ x_2 &\leq 4 \\ 3x_1 + 4x_2 &\geq 24 \\ 2x_1 - 3x_2 &\geq 6, \quad x_1 \geq 0, x_2 \geq 0. \end{aligned}$$

Solution: After the introduction of slack, surplus and artificial variables so that there exists a unit matrix which will form an initial unit basis, the constraints are

$$\begin{aligned} 2x_1 + \ x_2 + x_3 \qquad\qquad\qquad &= 4 \\ 3x_1 + 4x_2 \qquad - x_4 \qquad + x_6 \qquad &= 24 \\ 2x_1 - 3x_2 \qquad\qquad - x_5 \qquad + x_7 &= 6 \end{aligned}$$

$x_1, x_2 \geq 0, x_3 \geq 0$ slack variable, $x_4, x_5 \geq 0$, surplus variable, $x_6, x_7 \geq 0$ are artificial variables and the adjusted objective function is

$$z = 5x_1 + 11x_2 + 0 \cdot x_3 + 0 \cdot x_4 + 0 \cdot x_5 - Mx_6 - Mx_7$$

Simplex tables

Basis	c_B	c b	5 a_1	11 a_2	0 $a_3(e_1)$	0 a_4	0 a_5	$-M$ $a_6(e_2)$	$-M$ $a_7(e_3)$	Min. ratio
a_3^*	0	4	2^*	1	1	0	0	0	0	$\frac{4}{2} = 2^*$
a_6	$-M$	24	3	4	0	-1	0	1	0	$\frac{24}{4} = 6$
a_7	$-M$	6	2	-3	0	0	-1	0	1	......
$z_j - c_j$			$-5M - 5^*$	$-M - 11$	0	M	M	0	0	
a_1	5	2	1	$\frac{1}{2}$	$\frac{1}{2}$	0	0	0	0	
a_6	$-M$	18	0	$\frac{5}{2}$	$-\frac{3}{2}$	-1	0	1	0	
a_7	$-M$	2	0	-4	-1	0	-1	0	1	
$z_j - c_j$			0	$\frac{3}{2}M - \frac{17}{2}$	$\frac{5}{2}M + \frac{5}{2}$	M	M	0	0	

All $z_j - c_j \geq 0$ [M is so large that $\frac{3}{2}M - \frac{17}{2} \geq 0$]. But the artificial variables are present in the basis at positive level. The only conclusion is that the problem has no feasible solution.

Exercise 9

Solve the following L.P.P. by Big-M method:

1. (a) Minimize, $z = x_1 + x_2$

$$\text{subject to } 2x_1 + x_2 \geq 4$$
$$x_1 + 7x_2 \geq 7, \quad x_1, x_2 \geq 0.$$ **[C.U.(H)'85]**

(b) Minimize, $z = 2x_1 + x_2$

$$\text{subject to } 8x_1 + 5x_2 \geq 50$$
$$x_1 + x_2 \geq 7, \quad x_1, x_2 \geq 0.$$

(c) Minimize, $z = 15x_1 + 24x_2$

$$\text{subject to } 3x_1 + 3x_2 \geq 4$$
$$x_1 + 4x_2 \geq 3, \quad x_1, x_2 \geq 0.$$

(d) Maximize, $z = 10x_1 + 15x_2$

$$\text{subject to } x_1 + x_2 = 2$$
$$3x_1 + 2x_2 \leq 6, \quad x_1, x_2 \geq 0.$$ **[C.U.(P)'85]**

(e) Maximize, $z = x_1 + 5x_2$

$$\text{subject to } 3x_1 + 4x_2 \leq 6$$
$$x_1 + 3x_2 \geq 3, \quad x_1, x_2 \geq 0.$$ **[C.U.(H)'80]**

2. (a) Maximize, $z = 5x_1 + 8x_2$

$$\text{subject to } \begin{aligned} 3x_1 + 2x_2 &\geq 3 \\ x_1 + 4x_2 &\geq 4 \\ x_1 + x_2 &\leq 5, \quad x_1, x_2 \geq 0. \end{aligned}$$

(b) Minimize, $z = 4x_1 + 2x_2$

$$\text{subject to } \begin{aligned} 3x_1 + x_2 &\geq 27 \\ x_1 + x_2 &\geq 21 \\ x_1 + 2x_2 &\geq 30 \end{aligned}$$

[C.U.(P)'86,'96; M.Sc.(Appl.Math.)'77]

3. Maximize, $z = x_1 + 2x_2 + 3x_3 - x_4$

$$\text{subject to } \begin{aligned} x_1 + 2x_2 + 3x_3 &= 15 \\ 2x_1 + x_2 + 5x_3 &= 20 \\ x_1 + 2x_2 + x_3 + x_4 &= 10, \quad x_1, x_2, x_3 \text{ and } x_4 \geq 0. \end{aligned}$$

[C.U.(P)'90; V.U.(H)'95]

4. Maximize, $z = 2x_1 + x_2 + 3x_3$

$$\text{subject to } \begin{aligned} x_1 + x_2 + 2x_3 &\leq 5 \\ 2x_1 + 3x_2 + 4x_3 &= 12. \end{aligned}$$

[C.U.(P)'90; J.U.(M.Sc)'81]

5. Minimize, $z = 30x_1 + 36x_2$

$$\text{subject to } \begin{aligned} x_1 + x_2 &\geq 5 \\ 2x_1 + 3x_2 &\geq 2 \\ -2x_1 + x_2 &\geq 2, \quad x_1, x_2 \geq 0. \end{aligned}$$

6. Maximize, $z = x_1 - 2x_2 + 3x_3$

$$\text{subject to } \begin{aligned} -2x_1 + x_2 + 3x_3 &= 2 \\ 2x_1 + 3x_2 + 4x_3 &= 1, \quad x_1, x_2 \text{ and } x_3 \geq 0. \end{aligned}$$

7. Find the B.F.S or solutions of the L.P.P. Maximize, $z = x_1 + 2x_2 + 4x_3$ subject to

$$\begin{aligned} -x_1 + x_2 \quad &= 5 \\ -2x_1 \quad + x_3 &= 2, \quad x_1, x_2 \text{ and } x_3 \geq 0. \end{aligned}$$

From the B.F.S. obtained from the problem can we compute directly the optimal value of the of the objective function z? Justify it with proper reason. Verify your statement by using simplex algorithm.

[*Hints:* $[x_1 = 0, x_2 = 5, x_3 = 2]$. No, we cannot compute directly the optimal value of z from the B.F.S. This can only be done if we have previous information that the problem has finite optimal value. That is why it may give erroneous result, if we compute max z directly from here. It can be shown by simplex algorithm that the problem has no finite optimal value of the objective function.]

8. (a) Maximize, $z = 3x_1 - x_2$

$$\text{subject to } 2x_1 + x_2 \geq 2$$
$$x_1 + 3x_2 \geq 3$$
$$x_2 \leq 4, \quad x_1, x_2 \geq 0.$$

[C.U.(P)'92]

(b) Maximize, $z = -2x_1 + 5x_2$

$$\text{subject to } 4x_1 - 5x_2 \geq -20$$
$$x_1 + x_2 \geq 10$$
$$x_2 \geq 2, \quad x_1, x_2 \geq 0.$$

9. (a) Maximize, $z = 3x_1 + 6x_2 + 2x_3$

$$\text{subject to } 3x_1 + 4x_2 + x_3 = 2$$
$$x_1 + 3x_2 + 2x_3 = 1, \quad x_1, x_2 \text{ and} x_3 \geq 0.$$

[C.U.(P)'90]

(b) Minimize, $z = 2x_1 + x_2$

$$\text{subject to } 3x_1 + x_2 \geq 3$$
$$4x_1 + 3x_2 \geq 6$$
$$x_1 + 2x_2 \geq 2.$$

10. Solve the following L.P.P. using artificial variables. Minimize, $z = -3x_1 + 2x_2$ subject to

$$x_1 - 4x_2 \leq -14$$
$$-3x_1 + 2x_2 \leq 6, \quad x_1, x_2 \geq 0.$$

[C.U.(P)'88]

11. Solve the following problem graphically and satisfy that the system is infeasible. Maximize, $z = 2x + 3y$ subject to

$$x + y \leq 1$$
$$x + 2y \geq 3.$$

[C.U.(M.Com)'89]

Solve it again by simplex method and comment on the identification of the criterion of the infeasibility of the system in the simplex method.

12. Minimize, $z = 2x_1 - 3x_2 + 6x_3$

$$\text{subject to } 3x_1 - 4x_2 - 6x_3 \leq 2$$
$$2x_1 + x_2 + 2x_3 \geq 11$$
$$x_1 + 3x_2 - 2x_3 \leq 5, \quad x_j \geq 0, [j = 1, 2, 3.]$$

13. (a) Minimize, $z = 2x_1 + x_2$

$$\text{subject to } 3x_1 + x_2 \geq 3$$
$$4x_1 + 3x_2 \geq 6$$
$$x_1 + 2x_2 \geq 2, \quad x_j \geq 0, [j = 1, 2.]$$

(b) Maximize, $z = 2x_1 + 3x_2$

$$\begin{aligned} \text{subject to } x_1 + x_2 &\le 8 \\ x_1 + 2x_2 &= 5 \\ 2x_1 + x_2 &\le 8, \quad x_1 \ge 0, x_2 \ge 0. \end{aligned}$$

[C.U.(P)'84,'91]

14. Minimize, $z = 4x_1 + x_2$

$$\begin{aligned} \text{subject to } 3x_1 + x_2 &= 3 \\ 4x_1 + 3x_2 &\ge 6 \\ x_1 + 2x_2 &\le 3, \quad x_1, x_2 \ge 0. \end{aligned}$$

[C.U.(H)'83,'91,'99]

15. (a) Maximize, $z = 2x_1 + 3x_2 - x_3$

$$\begin{aligned} \text{subject to } 2x_1 + 5x_2 - x_3 &\le 5 \\ x_1 + x_2 + 2x_3 &= 6 \\ 2x_1 - x_2 + 3x_3 &= 7, \quad x_1, x_2 \text{ and } x_3 \ge 0. \end{aligned}$$

(b) Maximize, $z = 5x_1 - 2x_2 + 3x_3$

$$\begin{aligned} \text{subject to } 2x_1 + 2x_2 - x_3 &\ge 2 \\ 3x_1 - 4x_2 \quad &\le 3 \\ x_2 + 3x_3 &\le 5, \quad x_1, x_2, x_3 \ge 0. \end{aligned}$$

[C.U.(H)'86, 2000; Kalyani (H)'82]

16. Prove that, Maximize, $z = 2x_1 + x_2 - x_3 + 4x_4$

$$\begin{aligned} \text{subject to } 3x_1 - 5x_2 + x_3 - 2x_4 &= 7 \\ 6x_1 - 10x_2 - x_3 + 5x_4 &= 11, \quad x_j \ge 0, [j = 1, \cdots, 4.] \end{aligned}$$

has an unbounded solution.

17. Prove that, Minimize, $z = 3x_1 - x_2 + 10x_3$

$$\begin{aligned} \text{subject to } x_1 + 2x_2 + 3x_3 &\le 6 \\ 4x_1 + x_2 + x_3 &= 32 \\ 2x_1 + x_2 + 2x_3 &\ge 72, \quad x_j \ge 0, [j = 1, 2, 3.] \end{aligned}$$

has no feasible solution.

18. Maximize, $z = x_1 - x_2 + 2x_3 + 3x_4$

$$\begin{aligned} \text{subject to } 2x_1 + x_2 + 3x_3 + 2x_4 &= 11 \\ 3x_1 - 3x_2 + 5x_3 + x_4 &= 17, \quad x_j \ge 0, [j = 1, \cdots, 4.] \end{aligned}$$

19. Minimize, $3x_1 - 2x_2$

$$\begin{aligned} \text{subject to } 3x_1 + 4x_2 &\ge 12 \\ x_1 - 3x_2 &\le 6 \\ x_1 - 2x_2 &\le -4, \quad x_1, x_2 \ge 0. \end{aligned}$$

Verify the result by using graphical method.

20. Maximize, $6x_1 - 4x_2 + 4x_3 + x_4$ subject to

$$\begin{aligned} 2x_1 - x_2 + 3x_3 + x_4 &= 6 \\ 4x_1 - 2x_2 - x_3 + 2x_4 &= 10, \quad x_j \geq 0, [j = 1, \cdots, 4.] \end{aligned}$$

21. Solve the following L.P. problem using big-M method.

Minimize, $-3x_1 + x_2 + 3x_3 - x_4$

$$\text{subject to } \begin{aligned} x_1 + 2x_2 - x_3 + x_4 &= 0 \\ 2x_1 - 2x_2 + 3x_3 + 3x_4 &= 9 \\ x_1 - x_2 + 2x_3 - x_4 &= 6, \quad x_j \geq 0, [j = 1, 2, \cdots, 4.] \end{aligned}$$

[C.U.(H)'89,'97]

22. Prove that

$$B = (\mathbf{a}_1, \mathbf{a}_3) = \begin{bmatrix} 2 & 3 \\ 3 & 5 \end{bmatrix}$$

can be considered as an admissible basis to start the simplex procedure of the problem 18 and then solve the problem by applying simplex algorithm without using any artificial variable.

[*Hints:* $\det B = 1 \neq 0, B^{-1} = \begin{bmatrix} 5 & -3 \\ -3 & 2 \end{bmatrix}, B^{-1}\mathbf{b} = \begin{bmatrix} 5 & -3 \\ -3 & 2 \end{bmatrix}\begin{bmatrix} 11 \\ 17 \end{bmatrix} = \begin{bmatrix} 4 \\ 1 \end{bmatrix} \geq \mathbf{0}$. Hence B can be taken as an admissible basis to start the simplex procedure. $\mathbf{c}_B = (1, 2)$. Now calculate $\mathbf{y}_j = B^{-1}\mathbf{a}_j [j = 1, 2, \cdots, 4]$ and $z_j - c_j = \mathbf{c}_B\mathbf{y}_j - c_j$ and now construct the simplex table without using artificial variables and solve the problem by applying simplex algorithm technique.]

23. Prove that the L.P.P. Maximize, $z = 5x_1 + x_2 - 2x_3 + x_4$ subject to

$$\begin{aligned} x_1 + 5x_2 - 8x_3 + 3x_4 &= 6 \\ 3x_1 - x_2 + x_3 + x_4 &= 2, \quad x_j \geq 0, [j = 1, 2, \cdots, 4.] \end{aligned}$$

has an unbounded solution.

Answers

1. (a) Min $z = 31/13$ at $x_1 = 21/13$, $x_2 = 10/13$;
   (b) Min $z = 10$ at $x_1 = 0$, $x_2 = 10$;
   (c) Min $z = 25$ at $x_1 = 7/9$, $x_2 = 5/9$;
   (d) Max $z = 30$ at $x_1 = 0$, $x_2 = 2$;
   (e) Max $z = 15/2$ at $x_1 = 0$, $x_2 = 3/2$.
2. (a) Max $z = 40$ at $x_1 = 0$, $x_2 = 5$;
   (b) Min $z = 48$ at $x_1 = 3$, $x_2 = 18$.
3. Max $z = 15$ at $x_1 = 5/2$, $x_2 = 5/2$, $x_3 = 5/2$, $x_4 = 0$.
4. Max $z = 8$ at $x_1 = 3$, $x_2 = 2$, $x_3 = 0$.
5. Min $z = 174$ at $x_1 = 1$, $x_2 = 4$.
6. No feasible solution.

8. (a) Max $z = 9$ at $x_1 = 3$, $x_2 = 0$; (b) Unbounded solution.

9. (a) $x_1 = 2/5$, $x_2 = 1/5$, $x_3 = 0$ Max $z = 12/5$; (b) Min $z = 12/5$ at $x_1 = 3/5$, $x_2 = 6/5$.

10. Unbounded solution.

11. Solving by Big-M method, at the optimal level the artificial variable x_5 is present at the positive level and $x_5 = 3$ there. Hence the problem is infeasible.

12. Min $z = 9$ at $x_1 = 0$, $x_2 = 4$, $x_3 = 7/2$.

13. (a) Min $z = 12/5$ at $x_1 = 3/5$, $x_2 = 6/5$;
    (b) Max $z = 28/3$ at $x_1 = 11/3$, $x_2 = 2/3$;
    (c) Max $z = -3$ at $x_1 = 0$, $x_2 = 4/3$, $x_3 = 5/3$.

14. Min $z = 18/5$ at $x_1 = 3/5$, $x_2 = 6/5$.

15. (a) Max $z = 3$ at $x_1 = 1$, $x_2 = 1$, $x_3 = 2$.
    (b) Max $z = 85/3$ at $x_1 = 23/3$, $x_2 = 5$, $x_3 = 0$.

16. Unbounded solution.

17. No feasible solution.

18. Max $z = 58/7$ at $x_1 = 0$, $x_2 = 0$, $x_3 = 23/7$, $x_4 = 4/7$.

19. Min $z = -12/5$ at $x_1 = 4/5$, $x_2 = 12/5$.

20. Max $z = 116/7$ at $x_1 = 18/7$, $x_2 = 0$, $x_3 = 2/7$, $x_4 = 0$.

21. Min $z = 7$ at $x_1 = 1$, $x_2 = 1$, $x_3 = 3$, $x_4 = 0$.

22. Initial table

		c	1	−1	2	3
Basis	c_B	b	a_1	a_2	a_3	a_4
a_1	1	4	1	14	0	7
a_3	2	1	0	−9	1	−4
$z_j - c_j$		6	0	−3	0	−4

Chapter 10

Simplex Method or Simplex Algorithm (III)

10.1 Introduction

Two phase method due to Danzig, Orden and Wolfe

Two phase method (simplex) is only applicable when we require to introduce at least one artificial variable to get the initial unit basis. We need not apply this method in other cases. Solved problems given below will give clear idea about this method.

In this method, the problems are to be solved in two phases. In the first phase, all artificial vectors are to be removed from the basis matrix which is initially a unit basis. This can be done mathematically by the following device.

Let us consider a new objective function $U = \Sigma x_a$ [x_a is an artificial variable]. From the mathematical point of view all artificial variables must be at zero level at the optimal stage. Hence the minimum value of U should be equal to zero. But practically this is not true in all problems. To get the minimum value of U, an auxiliary L.P.P.

Minimize $U = \Sigma x_a$ subject to all constraints is to be solved first and in an attempt to solve the problem following three cases may arise.

(i) Min $U = 0$ and all artificial vectors are removed from the basis. Then removing all columns corresponding to the artificial vectors from the matrix formed with the elements y_{ij}, proceed further **to solve the original problem** by usual method (with the changed value of **c** and $\mathbf{c}_B$) in the **second phase.**

(ii) Min $U = 0$, but some artificial vectors are present at zero level in the basis matrix. In that case some constraints may be redundant. Taking the final simplex table of the auxiliary L.P.P. as the initial simplex table of the original problem (with the changed value of **c** and $\mathbf{c}_B$) we shall have to proceed further to get the optimal value of the objective function z, of the original problem.

But here the costs of artificial variables, only which are basic, are taken to be zero and drop all $\mathbf{y}_j$ corresponding to the non-basic artificial vectors.

(iii) Min $U > 0$ and the optimality conditions are satisfied in the auxiliary L.P.P. In that case the problem has no feasible solution and there is no question of proceeding further in the second phase.

► **Example 10.1.1** *Solve the L.P.P. by two phase method.*

$$\textit{Minimize,} \quad z = 3x_1 + 5x_2$$

subject to

$$\begin{aligned} x_1 + 2x_2 &\geq 8 \\ 3x_1 + 2x_2 &\geq 12 \\ 5x_1 + 6x_2 &\leq 60, \quad x_1, x_2 \geq 0. \end{aligned}$$

Initially the constraints are being converted into equations by introducing surplus variables x_3 and x_4 in the first and second constraints respectively and slack variable x_5 in the third constraint. The converted equations are

$$\begin{aligned} x_1 + 2x_2 - x_3 \qquad\qquad &= 8 \\ 3x_1 + 2x_2 \qquad - x_4 \qquad &= 12 \\ 5x_1 + 6x_2 \qquad\qquad + x_5 &= 60 \end{aligned}$$

But the co-efficient matrix does not contain a unit basis matrix. The equations are again converted in such a way, using artificial variables x_6 and x_7, that the new co-efficient matrix contains a unit basis matrix. The column vectors associated with the artificial variables are known as artificial vectors.

Remark. It should be remembered that Big-M. method and two phase method will be used only when artificial variables are required to get an initial unit basis, with all $b_i \geq 0$.

The equations are (after adding x_6 and x_7)

$$\begin{aligned} x_1 + 2x_2 - x_3 \qquad\qquad + x_6 \qquad &= 8 \\ 3x_1 + 2x_2 \qquad - x_4 \qquad\qquad + x_7 &= 12 \\ 5x_1 + 6x_2 \qquad\qquad + x_5 \qquad\qquad &= 60 \end{aligned}$$

$$x_j \geq 0; \; j = 1, 2, \cdots, 7.$$

In the **first phase**, let the objective function be $U = x_6 + x_7$ of the auxiliary L.P.P. and U is to be minimized.

Then put $U' = -x_6 - x_7$ and $\min U = -\max U'$. Hence the auxiliary L.P.P. is to maximize U'.

$$U' = 0.x_1 + 0.x_2 + 0.x_3 + 0.x_4 + 0.x_5 - x_6 - x_7.$$

$\mathbf{a}_6(\mathbf{e}_1), \mathbf{a}_7(\mathbf{e}_2)$ are two artificial vectors.

Simplex tables of the auxiliary L.P.P.

		c^*	0	0	0	0	0	−1	−1	
Basis	c_B^*	**b**	a_1	a_2	a_3	a_4	$a_5(e_3)$	$a_6(e_1)$	$a_7(e_2)$	**Min. ratio**
a_6^*	−1	8	1	2^*	−1	0	0	1	0	$\frac{8}{2} = 4^*$
a_7	−1	12	3	2	0	−1	0	0	1	$\frac{12}{2} = 6$
a_5	0	60	5	6	0	0	1	0	0	$\frac{60}{6} = 10$
$z_j^* - c_j^*$		−20	−4	-4^*	1	1	0	0	0	
a_2	0	4	$\frac{1}{2}$	1	$-\frac{1}{2}$	0	0	$\frac{1}{2}$	0	$4/\frac{1}{2} = 8$
a_7^*	−1	4	2^*	0	1	−1	0	−1	1	$\frac{4}{2} = 2^*$
a_5	0	36	2	0	3	0	1	−3	0	$\frac{36}{2} = 18$
$z_j^* - c_j^*$		−4	-2^*	0	−1	1	0	2	0	
a_2	0	3	0	1	$-\frac{3}{4}$	$\frac{1}{4}$	0	$\frac{3}{4}$	$-\frac{1}{2}$	
a_1	0	2	1	0	$\frac{1}{2}$	$-\frac{1}{2}$	0	$-\frac{1}{2}$	$\frac{1}{2}$	
a_5	0	32	0	0	2	1	1	−2	−1	
$z_j^* - c_j^*$		0	0	0	0	0	0	1	1	

None of $z_j^* - c_j^* < 0$. Hence the optimality conditions are satisfied and $\min U = -\max U' = -(0) = 0$.

Here all the artificial vectors are driven out from the basis and $\min U = 0$. Thus we proceed to solve the original L.P.P. after removing the columns corresponding to the artificial vectors.

Remark. In phase 1, all artificial variables are legitimate variables. Therefore, we cannot cancel the coloumns corresponding to the artificial vectors.

2nd phase

The objective function of the original L.P.P. is $z = 3x_1 + 5x_2$ and the problem is to minimize it.

Then put $z' = -3x_1 - 5x_2$ and $\min z = -\max z'$. Hence the problem is ultimately a problem of maximization.

$$z' = -3x_1 - 5x_2 + 0.x_3 + 0.x_4 + 0.x_5.$$

Initial simplex table of the original L.P.P.

		c	−3	−5	0	0	0
Basis	c_B	**b**	a_1	a_2	a_3	a_4	a_5
a_2	−5	3	0	1	$-\frac{3}{4}$	$\frac{1}{4}$	0
a_1	−3	2	1	0	$\frac{1}{2}$	$-\frac{1}{2}$	0
a_5	0	32	0	0	2	1	1
$z_j - c_j$		−21	0	0	$\frac{9}{4}$	$\frac{1}{4}$	0

Note. In this table the value of the associated cost vector changes and the elements of $z_j - c_j$ row change according to new $\mathbf{c}_B$.

As none of the $z_j - c_j < 0$, the optimality conditions are satisfied.

Hence $\max z' = -21$ at $x_1 = 2$, $x_2 = 3$ or $\min z = -\max z' = -(-21) = 21$ at $x_1 = 2$ and $x_2 = 3$.

▶ **Example 10.1.2** *Solve the L.P.P.*

$$\text{Maximize,} \quad z = 5x_1 + 11x_2$$

subject to

$$2x_1 + x_2 \le 4$$
$$3x_1 + 4x_2 \ge 24$$
$$2x_1 - 3x_2 \ge 6, \; x_1 \ge 0, x_2 \ge 0$$

by two phase method and prove that the problem has no feasible solution.

The constraints after the addition of slack, surplus and artificial variables (to get an initial identity matrix) are

$$\begin{aligned} 2x_1 + x_2 + x_3 \qquad\qquad\qquad\qquad &= 4 \\ 3x_1 + 4x_2 \qquad - x_4 \qquad + x_6 \qquad &= 24 \\ 2x_1 - 3x_2 \qquad\qquad - x_5 \qquad + x_7 &= 6 \end{aligned}$$

x_6 and x_7 are artificial variables.

Thus in the first phase the objective function of the auxiliary L.P.P. is $U = x_6 + x_7$ and we are to minimize it. Put $U' = -x_6 - x_7$ and $\min U = -\max(U')$

$$U' = 0.x_1 + 0.x_2 + 0.x_3 + 0.x_4 + 0.x_5 - x_6 - x_7.$$

Simplex tables of the auxiliary L.P.P. (Phase 1)

		$\mathbf{c}^*$	0	0	0	0	0	−1	−1	
Basis	$\mathbf{c}_B^*$	**b**	$\mathbf{a}_1$	$\mathbf{a}_2$	$\mathbf{a}_3(\mathbf{e}_1)$	$\mathbf{a}_4$	$\mathbf{a}_5$	$\mathbf{a}_6(\mathbf{e}_2)$	$\mathbf{a}_7(\mathbf{e}_3)$	**Min. ratio**
$\mathbf{a}_3^*$	0	4	2*	1	1	0	0	0	0	$\frac{4}{2} = 2^*$
$\mathbf{a}_6$	−1	24	3	4	0	−1	0	1	0	$\frac{24}{3} = 8$
$\mathbf{a}_7$	−1	6	2	−3	0	0	−1	0	1	$\frac{6}{2} = 3$
$z_j^* - c_j^*$		−30	−5*	−1	0	1	1	0	0	
$\mathbf{a}_1$		2								
$\mathbf{a}_6$		18								
$\mathbf{a}_7$		2								
$z_j^* - c_j^*$		−20	0	$\frac{3}{2}$	$\frac{5}{2}$	1	1	0	0	

After completing the $z_j^* - c_j^*$ row of the second table, it has been found that all $z_j^* - c_j^* \ge 0$ and $\max U' = -20$, i.e., $\min U = 20$ and one or more artificial variables are at positive level in the final basis. Then the only conclusion is that the problem has no feasible solution and we need not compute the body of the table and we need not go further to the phase II.

▶ **Example 10.1.3** *Solve the following L.P.P. by two phase method.*

$$\text{Minimize,} \quad z = 4x_1 + x_2$$

subject to

$$x_1 + 2x_2 \le 3$$
$$4x_1 + 3x_2 \ge 6$$
$$3x_1 + x_2 = 3, \text{ and } x_1, x_2 \ge 0. \qquad \text{[C.U.(Hons.) '83,'91,'94]}$$

Here all components of **b** are non-negative. Thus if we find a unit basis we can start the problem. To get a unit basis, we require to introduce two artificial variables as given below and the constraints are then

$$x_1 + 2x_2 + x_3 = 3$$
$$4x_1 + 3x_2 - x_4 + x_5 = 6$$
$$3x_1 + x_2 + x_6 = 3, \; x_j \ge 0, j = 1, 2, \cdots, 6.$$

Here x_3 is a slack, x_4 surplus x_5, x_6 are artificial variables.

Phase I

As there are two artificial variables x_5 and x_6, our problem is to find the minimum value of the auxiliary objective function

$$\mathsf{U} = x_5 + x_6 = 0.x_1 + 0.x_2 + 0.x_3 + 0.x_4 + x_5 + x_6,$$

subject to the constraints given above. Now let

$$\mathsf{U}' = -\mathsf{U} \text{ and } \min \mathsf{U} = -\max(-\mathsf{U}) = -\max \mathsf{U}'.$$

Here $\mathbf{c}^* = (0, 0, 0, 0 - 1, -1)$ and x_5 and x_6 now are considered as legitimate variables.

Phase I : Simplex tables

		$\mathbf{c}^*$	0	0	0	0	−1	−1	
$\mathbf{c}_B^*$	Basis	**b**	$\mathbf{a}_1$	$\mathbf{a}_2$	$\mathbf{a}_3(\mathbf{e}_1)$	$\mathbf{a}_4$	$\mathbf{a}_5(\mathbf{e}_2)$	$\mathbf{a}_6(\mathbf{e}_3)$	Min. ratio
0	$\mathbf{a}_3^*$	3	1	2	1	0	0	0	$\frac{3}{1} = 3$
−1	$\mathbf{a}_5$	6	4	3	0	−1	1	0	$\frac{6}{4} = \frac{3}{2}$
−1	$\mathbf{a}_6^*$	3	3^*	1	0	0	0	1	$\frac{3}{3} = 1^*$
$z_j^* - c_j^*$		−9	-7^*	−4	0	1	0	0	
0	$\mathbf{a}_3^*$	2	0	$\frac{5}{3}^*$	1	0	0	$-\frac{1}{3}$	$2/\frac{5}{3} = \frac{6}{5}^*$
−1	$\mathbf{a}_5$	2	0	$\frac{5}{3}$	0	−1	1	$-\frac{4}{3}$	$2/\frac{5}{3} = \frac{6}{5}$
0	$\mathbf{a}_1$	1	1	$\frac{1}{3}$	0	0	0	$\frac{1}{3}$	$1/\frac{1}{3} = 3$
$z_j^* - c_j^*$		−2	0	$-\frac{5}{3}^*$	0	1	0	$\frac{7}{3}$	
0	$\mathbf{a}_2$	$\frac{6}{5}$	0	1	$\frac{3}{5}$	0	0	$-\frac{1}{5}$	
−1	$\mathbf{a}_5$	0	0	0	−1	−1	1	−1	
0	$\mathbf{a}_1$	$\frac{3}{5}$	1	0	$-\frac{1}{5}$	0	0	$\frac{2}{5}$	
$z_j^* - c_j^*$		0	0	0	1	1	0	2	

Note. In the second table, min. ratio is $\frac{6}{5}$ for $i = 1$ and 2 which is minimum. Let us take first row as key row for simplicity.

All $z_j^* - c_j^* \geq 0$, thus the optimality conditon is satisfied.

Min $\cup = -\max \cup' = 0$. But here one artificial vector $\mathbf{a}_5$ is present in the basis at zero level. Then there may have redundant constraints. We shall start the phase II assuming the cost of the artificial variable x_5 be zero and the basis in the 2nd phase will be $B = (\mathbf{a}_2, \mathbf{a}_5, \mathbf{a}_1)$ which will give a feasible solution $\mathbf{x}_B \geq \mathbf{0}$ with $x_2 = \frac{6}{5}$, x_5 (Artificial) $= 0$, $x_1 = \frac{3}{5}$ [In the second phase we simply consider $x_5 = 0$ with cost component also zero.]

Phase II

Now we shall have to find the minimum value of z. Now

$$\min z = -\max(-z) = -\max(-4x_1 - x_2 + 0x_3 + 0x_4 + 0x_5)$$

		c	-4	-1	0	0	0
$\mathbf{c}_B$	**Basis**	**b**	$\mathbf{a}_1$	$\mathbf{a}_2$	$\mathbf{a}_3$	$\mathbf{a}_4$	$\mathbf{a}_5$
-1	$\mathbf{a}_2$	$\frac{6}{5}$	0	1	$\frac{3}{5}$	0	0
0	$\mathbf{a}_5$	0	0	0	-1	-1	1
-4	$\mathbf{a}_1$	$\frac{3}{5}$	1	0	$-\frac{1}{5}$	0	0
	$z_j - c_j$	$-\frac{18}{5}$	0	0	$\frac{1}{5}$	0	0

$z_j - c_j \geq 0$ for all j. Hence the optimality condition is being satisfied.

But $z_4 - c_4 = 0$ corresponding to a non-basic variable. The alternative optimal solution exists. Max $(-z) = -\frac{18}{5}$, i.e., $\min z = -(-\frac{18}{5}) = \frac{18}{5}$ at $x_1 = \frac{3}{5}$, $x_2 = \frac{6}{5}$, $x_3 = 0$ (non-basic).

Note. But in phase I, in the 2nd table, as the min. ratio is $\frac{6}{5}$ for $i = 1$ and 2 we may take $\frac{5}{3} = y_{22}$ as the key element and then $\mathbf{a}_5$ (artificial) will leave the basis and the last table of the phase I will be with the basis $(\mathbf{a}_3, \mathbf{a}_2, \mathbf{a}_1)$ and having degenerate B.F.S.

► **Example 10.1.4** *Solve the following L.P.P. by two phase method and prove that the problem is said to have an unbounded solution :*

$$\textit{Maximize,} \quad z = 5x_1 + x_2 - 2x_3 + x_4$$

subject to

$$\begin{aligned} x_1 + 5x_2 - 8x_3 + 3x_4 &= 6 \\ 3x_1 - x_2 + x_3 + x_4 &= 2, \; x_j \geq 0, j = 1, \cdots, 4. \end{aligned}$$

Solution. Two artificial variables x_5 and x_6 are to be introduced one to each equation to get an initial unit basis and the equations are then

$$\begin{aligned} x_1 + 5x_2 - 8x_3 + 3x_4 + x_5 \quad &= 6 \\ 3x_1 - x_2 + x_3 + x_4 \quad + x_6 &= 2, \; x_j \geq 0, j = 1, \cdots, 6. \end{aligned}$$

Thus in phase I our problem is to minimize an auxiliary objective function $\cup = 0.x_1 + 0.x_2 + 0.x_3 + 0.x_4 + x_5 + x_6$, i.e., to maximize $(-\cup)$.

Phase I

		$\mathbf{c}^*$	0	0	0	0	−1	−1	
Basis	c_B^*	**b**	$\mathbf{a}_1$	$\mathbf{a}_2$	$\mathbf{a}_3$	$\mathbf{a}_4$	$\mathbf{a}_5(\mathbf{e}_1)$	$\mathbf{a}_6(\mathbf{e}_2)$	**Min. ratio**
$\mathbf{a}_5$	−1	6	1	5	−8	3	1	0	$\frac{6}{1} = 6$
$\mathbf{a}_6^*$	−1	2	3^*	−1	1	1	0	1	$\frac{2}{3} = \frac{2}{3}^*$
$z_j^* - c_j^*$		−8	-4^*	−4	7	−4	0	0	
$\mathbf{a}_5^*$	−1	$\frac{16}{3}$	0	$\frac{16}{3}^*$	$-\frac{25}{3}$	$\frac{8}{3}$	1	$-\frac{1}{3}$	$\frac{16}{3}/\frac{16}{3} = 1^*$
$\mathbf{a}_1$	0	$\frac{2}{3}$	1	$-\frac{1}{3}$	$\frac{1}{3}$	$\frac{1}{3}$	0	$\frac{1}{3}$	...
$z_j^* - c_j^*$		$-\frac{16}{3}$	0	$-\frac{16}{3}^*$	$\frac{25}{3}$	$-\frac{8}{3}$	0	$\frac{4}{3}$	
$\mathbf{a}_2$	0	1	0	1	$-\frac{25}{16}$	$\frac{1}{2}$	$\frac{3}{16}$	$-\frac{1}{16}$	
$\mathbf{a}_1$	0	1	1	0	$-\frac{3}{16}$	$\frac{1}{2}$	$\frac{1}{16}$	$\frac{5}{16}$	
$z_j^* - c_j^*$		0	0	0	0	0	1	1	

All $z_j^* - c_j^* \geq 0$ in the third table and all artificial variables are driven out from the basis. Then after completing the third table we shall have to go to the phase II.

Phase II

Our problem is to maximize $5x_1 + x_2 - 2x_3 + x_4$. Then $\mathbf{c} = (5, 1, -2, 1)$.

Simplex table

		c	5	1	−2	−1
Basis	$\mathbf{c}_B$	**b**	$\mathbf{a}_1$	$\mathbf{a}_2$	$\mathbf{a}_3$	$\mathbf{a}_4$
$\mathbf{a}_2$	1	1	0	1	$-\frac{25}{16}$	$\frac{1}{2}$
$\mathbf{a}_1$	5	1	1	0	$-\frac{3}{16}$	$\frac{1}{2}$
$z_j - c_j$		6	0	0	$-\frac{1}{2}$	2

Now $z_3 - c_3 < 0$ with all $y_{i3} < 0$, $i = 1, 2$. Then the problem is said to have an unbounded solution.

► **Example 10.1.5** *Do we require to introduce artificial variable or variables to get initially a unit basis matrix? If so, solve the problem by two phase method.*

Solve the L.P.P.,

$$\text{Maximize,} \quad z = 2x_1 - 3x_2$$

subject to

$$\begin{aligned} -\ x_1 + \ x_2 &\geq -\ 2 \\ 5x_1 + 4x_2 &\leq \ 46 \\ 7x_1 + 2x_2 &\geq \ 32, \quad x_1 \geq 0, x_2 \geq 0. \end{aligned}$$

[C.U.(Hons.) '85,'87,'92] [C.U.(Pass) '95,'97]

Initially we require to adjust the constraints in such a manner that $\mathbf{b} \geq \mathbf{0}$. Here we have the inequations

$$\begin{aligned} x_1 - \ x_2 &\leq 2 \\ 5x_1 + 4x_2 &\leq 46 \\ 7x_1 + 2x_2 &\geq 32, \quad x_1 \geq 0, x_2 \geq 0. \end{aligned}$$

Here one artificial variable is required to introduce in the third constraint to get a unit basis and thus two phase method is applicable here to solve the problem. The problem is,

$$\text{Maximize,} \quad z = 2x_1 - 3x_2$$

subject to

$$\begin{aligned} x_1 - \ x_2 + x_3 \qquad\qquad &= 2 \\ 5x_1 + 4x_2 \qquad + x_4 \qquad &= 46 \\ 7x_1 + 2x_2 \qquad\qquad - x_5 + x_6 &= 32 \end{aligned}$$

Here x_6 is an artificial variable and $x_6 \geq 0$. x_4, x_5 are also ≥ 0.

Phase I. In phase 1, we require to minimize the auxiliary function

$$U = 0x_1 + 0x_2 + 0x_3 + 0x_4 + 0x_5 + x_6$$

$x_j \geq 0$. Here x_6 is the legitimate variable and we cannot remove the column $\mathbf{a}_6$ corresponding to the artificial variable x_6 though it can be removed from the basis

$$\min U = -\max -U = -\max U^*, \quad \text{where } U^* = -U.$$

Simplex tables

		$\mathbf{c}^*$	0	0	0	0	0	−1	
Basis	$\mathbf{c}_B^*$	**b**	$\mathbf{a}_1$	$\mathbf{a}_2$	$\mathbf{a}_3(\mathbf{e}_1)$	$\mathbf{a}_4(\mathbf{e}_2)$	$\mathbf{a}_5$	$\mathbf{a}_6(\mathbf{e}_2)$	**Min. ratio**
$\mathbf{a}_3^*$	0	2	1*	−1	1	0	0	0	$\frac{2}{1} = 2^*$
$\mathbf{a}_4$	0	46	5	4	0	1	0	0	$\frac{46}{5} = 9\frac{1}{5}$
$\mathbf{a}_6$	−1	32	7	2	0	0	−1	1	$\frac{32}{7} = 4\frac{4}{7}$
$z_j^* - c_j^*$		−32	−7*	−2	0	0	1	1	
$\mathbf{a}_1$	0	2	1	−1	1	0	0	0	....
$\mathbf{a}_4$	0	36	0	9	−5	1	0	0	$\frac{36}{9} = 4$
$\mathbf{a}_6^*$	−1	18	0	9*	−7	0	−1	1	$\frac{18}{2} = 2^*$
$z_j^* - c_j^*$		−18	0	−9	7	0	1	1	
$\mathbf{a}_1$	0	4	1	0	$\frac{2}{9}$	0	$-\frac{1}{9}$	$\frac{1}{9}$	
$\mathbf{a}_4$	0	18	0	0	$\frac{2}{9}$	1	0	$-\frac{1}{9}$	
$\mathbf{a}_2$	0	2	0	1	$-\frac{7}{9}$	0	$-\frac{1}{9}$	$\frac{1}{9}$	
$z_j^* - c_j^*$		0	0	0	0	0	0	0	

All $z_j^* - c_j^* \geq 0$. We reach at the optimal stage.

Min $U = -\max(-U) = -(0) = 0$ and no artificial variable is present in the basis at the optimal table. Thus we can proceed to the **Phase II**. The original problem is a maximization problem. Here the objective function is $z = 2x_1 - 3x_2$ and the vectors of the basis is $B = (\mathbf{a}_1, \mathbf{a}_4, \mathbf{a}_2)$. The matter encircled by the bold lines is to be taken for Phase II and the basis is the basis in the last table of the Phase I.

Phase II : Simplex table

		c	2	0	−3	0	0
Basis	c_B	b	$\mathbf{a}_1$	$\mathbf{a}_2$	$\mathbf{a}_3$	$\mathbf{a}_4$	$\mathbf{a}_5$
$\mathbf{a}_1$	2	4	1	0	$\frac{2}{9}$	0	$-\frac{1}{9}$
$\mathbf{a}_4$	0	18	0	0	2	1	0
$\mathbf{a}_2$	−3	2	0	1	$-\frac{7}{9}$	0	$-\frac{1}{9}$
$z_j^* - c_j^*$		2	0	0	$\frac{25}{9}$	0	$\frac{1}{9}$

In Phase II, all $z_j - c_j \geq 0$. Thus we reach at the optimal stage $\max z = 2$ at $x_1 = 4, x_2 = 2$.

Remark. This problem has also been solved in Big-M method. Compare the procedure and computed values and compare yourself which method is easier out of these two methods.

10.2 Problems of L.P. when one or more variables are unrestricted in sign

Considering from the physical point of view, we have seen that the variables are non-negative. But from the theoretical point of view, we need to discuss the case when some of the variables are unrestricted in sign (in duality theory), i.e., its value may be positive, negative or zero. An unrestricted variable x can be expressed as the difference of two non-negative variables x' and x'' and $x = x' - x''$. In case when $x' > x''$, x is positive; $x' < x''$, x is negative. $x' = x''$, x is zero. Using this technique, we may able to solve a L.P.P. when some of the variables are unrestricted in sign.

▶ **Example 10.2.1** *Solve the L.P.P.*

$$\textit{Maximize,} \quad z = 2x_1 + 5x_2$$

subject to

$$\begin{aligned} 2x_1 + x_2 &\geq 12 \\ x_1 + x_2 &\leq 4 \end{aligned}$$

$x_1 \geq 0$ *and* x_2 *is unrestricted in sign.*

Solution. As x_2 is unrestricted in sign, x_2 may be written as $x_2 = x_2' - x_2''$ where x_2' and x_2'' are non-negative.

Hence the problem can be written as

$$\text{Maximize,} \quad z = 2x_1 + 5x_2' - 5x_2''$$

subject to

$$2x_1 + x_2' - x_2'' \geq 12$$
$$x_1 + x_2' - x_2'' \leq 4, \text{ and } x_1 \geq 0, x_2' \geq 0, x_2'' \geq 0.$$

Introducing surplus variable x_3, artificial variable x_4 in the first constraint and slack variable x_5 in the second constraint, the problem reduces to

$$\text{Maximize,} \quad z = 2x_1 + 5x_2' - 5x_2'' + 0.x_3 - Mx_4 + 0.x_5$$

subject to

$$2x_1 + x_2' - x_2'' - x_3 + x_4 = 12$$
$$x_1 + x_2' - x_2'' + x_5 = 4$$

and all variables are non-negative.

$$\mathbf{a}_1 = \begin{bmatrix} 2 \\ 1 \end{bmatrix}, \mathbf{a}_2' = \begin{bmatrix} 1 \\ 1 \end{bmatrix}, \mathbf{a}_2'' = \begin{bmatrix} -1 \\ -1 \end{bmatrix}, \mathbf{a}_3 = \begin{bmatrix} -1 \\ 0 \end{bmatrix},$$

$$\mathbf{a}_4 = \begin{bmatrix} 1 \\ 0 \end{bmatrix}, \mathbf{a}_5 = \begin{bmatrix} 0 \\ 1 \end{bmatrix} \text{ and } \mathbf{b} = \begin{bmatrix} 12 \\ 4 \end{bmatrix} \geq 0.$$

The artificial vector $\mathbf{a}_4$ and slack vector $\mathbf{a}_5$ constitute the initial unit basis matrix.

Thus I.B.F.S. $\mathbf{x}_B = [x_4, x_5] = [12, 4], \mathbf{c}_B = (-M, 0)$

$$z = \mathbf{c}_B \mathbf{x}_B = -12M \text{ and } \mathbf{y}_j = B^{-1}\mathbf{a}_j = \mathbf{a}_j \text{ for all } j.$$

Simplex tables

		c	2	5	−5	0	−M	0	
Basis	$\mathbf{c}_B$	**b**	$\mathbf{a}_1$	$\mathbf{a}_2'$	$\mathbf{a}_2''$	$\mathbf{a}_3$	$\mathbf{a}_4(\mathbf{e}_1)$	$\mathbf{a}_5(\mathbf{e}_2)$	**Min. ratio**
$\mathbf{a}_4$	$-M$	12	2	1	−1	−1	1	0	$\frac{12}{2} = 6$
$\mathbf{a}_5^*$	0	4	1*	1	−1	0	0	1	$\frac{4}{1} = 4^*$
$z_j - c_j$		$-12M$	$-2M-2^*$	$-M-5$	$M+5$	M	0	0	
$\mathbf{a}_4$	$-M$	4	0	−1	1*	−1	1	−2	$\frac{4}{1} = 4^*$
$\mathbf{a}_1$	2	4	1	1	−1	0	0	0	....
$z_j - c_j$		$-4M+8$	0	$M-3$	$-M+3^*$	M	0	$2M+2$	
$\mathbf{a}_2''$	−5	4	0	−1	1	−1		−2	
$\mathbf{a}_1$	2	8	1	0	0	−1		−1	
$z_j - c_j$		−4	0	0	0	3		8	

None of $z_j - c_j < 0$. Hence the optimality conditions are satisfied and no artificial vector is present in the basis. Hence $\max z = -4$ at $x'' = 4$, $x_1 = 8$ and $x_2' = 0$ [non-basic variable], i.e., $\max z = -4$ at $x_1 = 8$ and $x_2 = x_2' - x_2'' = -4$.

Note. (1) As the value of $z_2' - c_2$ corresponding to the non-basic variable $\mathbf{a}_2'$ is zero at the final stage, and $y_{ij} \leq 0$, therefore a non-basic alternative optimal solution exists.

(2) The two vectors $\mathbf{a}_2'$ and $\mathbf{a}_2''$ are not linearly independent because $\mathbf{a}_2'+\mathbf{a}_2''=0$. Hence at any stage these two vectors cannot occur simultaneously in the basis and at least one of the two variables x_2' and x_2'' must be a non-basic variable at each stage.

► **Example 10.2.2** *Solve the following L.P.P.*

$$\textit{Maximize,}\quad 5x_1 - 2x_2 + x_3$$

subject to

$$2x_1 + 4x_2 + x_3 \le 6$$
$$2x_1 + x_2 + 3x_3 \ge 2,\ x_1, x_2 \ge 0, x_3 \textit{ unrestricted.}$$

[C.U.(Hons.) '94]

Since x_3 is unrestricted then $x_3 = x_3' - x_3''$, $x_3', x_3'' \ge 0$ and the problem in the new form is

$$\text{Maximize,}\quad 5x_1 - 2x_2 + x_3' - x_3''$$

subject to

$$2x_1 + 4x_2 + x_3' - x_3'' \le 6$$
$$2x_1 + x_2 + 3x_3' - 3x_3'' \ge 2,\ x_1, x_2, x_3', x_3'' \ge 0.$$

Solution. We can start to solve the problem in the following manner in Big-M method.

Simplex tables

		c	5	−2	1	−1	0	0	$-M$	
Basis	$\mathbf{c}_B$	**b**	$\mathbf{a}_1$	$\mathbf{a}_2$	$\mathbf{a}_3'$	$\mathbf{a}_3''$	$\mathbf{a}_4(\mathbf{e}_1)$	$\mathbf{a}_5$	$\mathbf{a}_6(\mathbf{e}_2)$	**Min. ratio**
$\mathbf{a}_4$	0	6	2	4	1	−1	1	0	0	$\frac{6}{2}=3$
$\mathbf{a}_6^*$	$-M$	2	2*	1	1	−1	0	−1	1	$\frac{2}{2}=1^*$
z_j-c_j			$-2M+5^*$	$-M+2$	$-M-1$	$M+1$	0	M	0	
$\mathbf{a}_4$	0	4	0	3	0	0	1	$\frac{1}{2}$		
$\mathbf{a}_1$	5	1	1	$\frac{1}{2}$	$\frac{1}{2}$	$-\frac{1}{2}$	0	$-\frac{1}{2}$		
z_j-c_j			0	$\frac{9}{2}$	$\frac{3}{2}$	$-\frac{3}{2}$	0	$-\frac{5}{2}$		

In the fourth column, all $y_{i4} \le 0$ for $i = 1, 2$ and $z_4 - c_4 < 0$ and hence the problem is said to have an unbounded solution.

Exercise

Solve all problems given in the exercise of the chapter 9 by two phase method.

Short Answer Type Questions with Answers

1. (a) What are slack and surplus variables?

   (b) Convert the following constraints (inequation) into equations using either slack or surplus variables.

   (i) $2x_1 + x_2 + x_3 \leq 7$;

   (ii) $x_1 - 2x_2 + x_3 \leq -14$;

   (iii) $7x_1 + x_2 + 3x_3 - x_4 \geq 15$;

   (iv) $-4x_1 + 2x_2 - x_3 \geq -10$;

   (v) $|2x_1 - 3x_2 + x_3| \leq 24$;

   (vi) $|3x_1 + x_2 - x_3| \geq 10$.

   [all variables are non-negative]

   [**Ans.** (a) A non-negative variable x_{r+1} is added to the left hand side of the inequation $a_{11}x_1 + a_{12}x_2 + \cdots + a_{1r}x_r \leq b_1$ to convert it into an equation $a_{11}x_1 + a_{12}x_2 + \cdots + a_{1r}x_r + x_{r+1} = b_1$ the variable x_{r+1} is called a *slack variable*; similarity a non-negative variable x_{r+1} is subtracted from the left hand side of the constraint (inequation) $a_{11}x_1 + a_{12}x_2 + \cdots + a_{ir}x_r \geq b_r$ to convert it into an equation $a_{11}x_1 + a_{12}x_2 + \cdots + a_{1r}x_r - x_{r+1} = b_r$. This variable x_{r+1} is called a *surplus variable*.

   (b) $x_1 - 2x_2 + x_3 + x_4 = -14$. Here x_4 is a slack variable. But if we need to convert -14 to 14 then the equation reduces to $-x_1 + 2x_2 - x_3 - x_4 = 14$ and here the slack variable x_4 changes to a surplus variable.

   (e) $|2x_1 - 3x_2 + x_3| \leq 24$ means $2x_1 - 3x_2 + x_3 \leq 24$ and $2x_1 - 3x_2 + x_2 \geq -24$. Then after adding slack and subtracting surplus variables we have $2x_1 - 3x_2 + x_3 + x_4 = 24$ and $2x_1 - 3x_2 + x_3 - x_5 = -24$. Here x_4 is a slack and x_5 is a surplus variable.]

2. Transform the L.P.P.,

$$\text{Maximize,} \quad z = 2x_1 + 3x_2$$

subject to

$$\begin{aligned} 2x_1 - 3x_2 &\leq 10 \\ -2x_1 + x_2 &\leq -14 \\ 3x_1 + 4x_2 &= 17, \ x_1 \geq 0, x_2 \geq 0. \end{aligned}$$

such that all constraints are of equality type (standard form).

[**Ans.** Maximize, $z = 2x_1 + 3x_2 + 0.x_3 + 0.x_4$

subject to

$$\begin{aligned} 2x_1 + 3x_2 + x_3 \qquad &= 10 \\ -2x_1 + x_2 \qquad + x_4 &= -14 \\ 3x_1 + 4x_2 \qquad &= 17, \ x_j \geq 0, [j = 1, 2, \cdots, 4.] \end{aligned}$$

Of course, the second constraint can be written as $2x_1 - x_2 - x_4 = 14$ and here the variable x_4 is a surplus variable.]

3. Transform the L.P.P.,

$$\text{Minimize,} \quad z = 2x_1 - x_2 + x_3$$

subject to

$$\begin{aligned} 3x_1 - 4x_2 + x_3 &\leq 14 \\ x_1 - x_2 + 2x_3 &\leq 10, \ x_j \geq 0, [j = 1, 2, 3.] \end{aligned}$$

such that all constraints are of equality type.

[**Ans.** Minimize, $z = 2x_1 - x_2 + x_3 + 0.x_4 + 0.x_5 + 0.x_6$

subject to

$$\begin{aligned} 3x_1 - 4x_2 + x_3 + x_4 &= 14 \\ x_1 - x_2 + 2x_3 + x_5 &= 10 \\ x_1 - x_2 + 2x_3 - x_6 &= -10,\ x_j \geq 0, [j = 1, 2, \cdots, 6.] \end{aligned}$$

x_6 is a surplus variable. But if we write the equation in the form $-x_1 + x_2 - 2x_3 + x_6 = 10$, then x_6 reduces to a slack variable.]

4. Transform the L.P.P.;

$$\text{Maximize,} \quad z = 4x_1 + x_2$$

subject to

$$\begin{aligned} 3x_1 - x_2 &\leq 10 \\ 2x_1 + x_2 &\geq 20 \\ 3x_1 - x_2 &= -14,\ x_j \geq 0, [j = 1, 2.] \end{aligned}$$

such that the usual simplex procedure can be applied with the presence of an initial unit basis. Write down the initial basic feasible solution also.

[**Ans.** Maximize, $z = 4x_1 + x_2 + 0.x_3 + 0.x_4 - Mx_5 - Mx_6$

subject to

$$\begin{aligned} 3x_1 - x_2 + x_3 &= 10 \\ 2x_1 + x_2 - x_4 + x_5 &= 20 \\ -3x_1 + x_2 + x_6 &= 14,\ x_j \geq 0, [j = 1, 2, \cdots, 6.] \end{aligned}$$

x_3 is a slack, x_4 is a surplus and x_5 and x_6 are artificial variables. $\mathbf{a}_3(\mathbf{e}_1), \mathbf{a}_5(\mathbf{e}_2), \mathbf{a}_6(\mathbf{e}_3)$ constitute the initial unit basis.

I.B.F.S. (artificial) $\mathbf{x}_B = [x_3, x_5, x_6] = [10, 20, 14]$. M is a very large positive quantity.]

5. (a) Transform the L.P.P.,

$$\text{Maximize,} \quad z = 3x_1 + x_2 - 3x_3 + x_4$$

subject to

$$\begin{aligned} x_1 + x_2 - 3x_3 + x_4 &= 17 \\ -3x_1 + x_3 - x_4 &\leq -27,\ x_j \geq 0, [j = 1, 2, 3, 4.] \end{aligned}$$

such that usual simplex procedure can be applied with the presence of an initial unit basis. Write down the initial B.F.S. also.

[**Ans.** Maximize, $z = 3x_1 + x_2 - 3x_3 = x_4 + 0.x_5 - Mx_6$

subject to

$$\begin{aligned} x_1 + x_2 - 3x_3 + x_4 &= 17 \\ 3x_1 - x_3 + x_4 - x_5 + x_6 &= 27,\ x_j \geq 0, [j = 1, \cdots, 6.] \end{aligned}$$

The vectors $\mathbf{a}_2(\mathbf{e}_1)$, and $\mathbf{a}_6(\mathbf{e}_2)$ constitute the initial basis. Only one artificial vector is required to get a unit basis. $\mathbf{a}_6$ is an artificial vector.

I.B.F.S (artificial) $\mathbf{x}_B = [x_2, x_6] = [17, 27]$.

(b) Introducing slack and surplus variables, write down the following L.P.P. in the standard maximizing form.

Minimize, $z = 3x_1 - 2x_2 + 4x_3$

subject to

$$\begin{aligned} x_1 - x_2 + 3x_3 &\geq 1 \\ 2x_1 + 3x_2 - 5x_3 &\geq -3 \\ 4x_1 + 2x_2 &\geq 2, \; x_1, x_2, x_3 \geq 0. \end{aligned}$$

[C.U.(Pass) '88]

Chapter 11

Resolution of Degeneracy

11.1 Introduction

In Chapters 8, 9, 10, in all problems which have been solved by simplex method, the B.F.S. at any iteration is in general non-degenerate. But this is not the fact in all problems. There are problems in which B.F.S. may be degenerate at any iteration (either at the initial stage or at any subsequent iteration). This type of problems are known as the *problems of degeneracy.* We may face some difficulties in solving the problems of this type all of which will be discussed step by step.

11.1.1 Only one component of B.F.S corresponding to the basic variables is zero

(a) Let $x_{Br} = 0$ and $\mathbf{a}_k$ be the incoming vector.

If $y_{rk} > 0$ then from the minimum ratio rule,

$$\min_i \left(\frac{x_{Bi}}{y_{ik}},\ y_{ik} > 0 \right) = \frac{x_{Br}}{y_{rk}} = 0.$$

Here y_{rk} is the key element which is unique and therefore the vector which is to be removed from the basis matrix is also unique. Therefore we may proceed further to construct the next table *without improving the value of the objective function* $\left(\text{as } \frac{x_{Br}}{y_{rk}} = 0\right)$ *and the B.F.S. in the next table is also degenerate.* But still if we proceed further step by step, following the definite rule, we shall obtain the optimal solution as well as the optimal value of the objective function (if there is an optimal solution at all).

(b) If $x_{Br} = 0$ but $y_{rk} \leq 0$; here x_{Br} does not enter in the calculation of minimum ratio in determining the value of the key element

$$\min \left(\frac{x_{Bi}}{y_{ik}},\ y_{ik} > 0 \right) > 0$$

[provided not all $y_{ik} \leq 0, (i = 1, 2, \cdots, m)$; $i \neq r$] and the next B.F.S. will be non-degenerate. Now the problem can be solved by the method adopted previously.

11.1.2 At least two components of B.F.S. corresponding to the basic variables are zero

Let $x_{Bi} = 0$ for $i = r$ and l and let $\mathbf{a}_k$ be the incoming vector. If both of y_{rk} and y_{lk} are greater than zero then

$$\min_i \left(\frac{x_{Bi}}{y_{ik}},\ y_{ik} > 0 \right) = \frac{x_{Br}}{y_{rk}} = \frac{x_{Bl}}{y_{lk}} = 0.$$

Here the minimum ratio is not unique. Therefore it is not possible to select an unique key element as well as the basis vector which is to be removed from the basis. But if we select the key element at random out of the two values y_{rk} and y_{ik} and proceed further to get an optimal solution, then at one stage, we may be surprised to note that the initial simplex table appears before us after some finite number of iterations. If we proceed further following the same method the same tables will cycle again and we are unable to find out the optimal value of the objective function (though there is an unique optimal value). These types of problems are known as *cycling problems.* By using the method discussed above, it may not be possible to solve the cycling problems. To solve the problems of these types, a mathematical technique has been developed by Charnes (Charnes perturbation technique) and using this technique, the key element and the outgoing vector at any stage can be determined uniquely and there is no chance of having the problems to cycle again and the optimal solution can be obtained.

But it has been found in practice that almost in all practical problems there is a little chance of a problem to cycle. Thousands and thousands of L.P.P. have already been solved by electronic computers. But none of them cycle at any stage. A very few theoretical problems have been constructed by Beale, Kuhn, Yudin and Galsthein to demonstrate how a L.P.P. cycles when one tries to solve the problem without following some definite rule of determining the key element uniquely at any stage. Hence the technique developed by Charnes, has practically no importance.

11.1.3 If $\min_i \left(\frac{x_{Bi}}{y_{ik}},\ y_{ik} > 0 \right) > 0$, but not unique

(a) If $\min_i \left(\frac{x_{Bi}}{y_{ik}},\ y_{ik} > 0 \right)$ are equal only for two values of i.

Let the minimum ratio be equal only for $i = r$ and l. Therefore,

$$\frac{x_{Br}}{y_{rk}} = \frac{x_{Bl}}{y_{lk}} > 0.$$

Here also the key element and the basis vector which is to be removed from the basis cannot be determined uniquely. But still if we select one, as a key element out of y_{rk} and y_{lk} then the next B.F.S. will be degenerate with *only one zero out of m values of the basic variables.* If we take y_{rk} as the key element then $x'_{Bl} = 0$. Thus the problem can be solved by using the method discussed in 1(a) or 1(b).

(b) If min ratio values are equal for more than two values of i. Let the min ratios be equal for $i = r, l$ and s. Therefore,

$$\frac{x_{Br}}{y_{rk}} = \frac{x_{Bl}}{y_{lk}} = \frac{x_{Bs}}{y_{sk}}.$$

Hence the key element and the basis vector which is to be removed from the basis is not unique. We have to select the key element out of these three values y_{rk}, y_{lk} and y_{sk}. If we select at random one, say y_{rk} as the key element then vector in the rth position in the basis will leave the basis and in the next table $x'_{Bl} = x'_{Bs} = 0$. Therefore the next B.F.S. is degenerate with two zeros of basic variables.

If in the next table all the optimality conditions are satisfied and thus the optimal value of the objective function is determined, no difficulty arises in solving the problem. But if the optimality conditions are not satisfied we may face the same difficulty as in (2) in selecting the key element as well as the basis vector which is to be removed from the basis. And if the proper selection is not done in each iteration there is a chance to cycle the problem. In order to avoid cycling (if there exists at all) here also, we shall have to use the technique, developed by Charnes (Charnes perturbation technique) and the problem will be solved ultimately without cycling. The problem in which at any iteration, the key element and the basis vector which is to be removed from the basis is not unique, is known as the *problem with ties* and all the basis vectors which have the same chance of being removed from the basis are known as the *tied vectors.*

From the discussions made above, it is clear that we have to make use of the Charnes perturbation technique (which will be discussed now) in solving the problems of the types discussed in (2) and 3(b) (the technique may not be required to use in all problems) only. All other problems (degenerate) can be solved ignoring all about the nature of degeneracy.

11.2 Charnes perturbation technique

Let the L.P.P. be, maximize, $z = \mathbf{cx}$ subject to $A\mathbf{x} = \mathbf{b}$, $\mathbf{x} \geq 0$ [degeneracy occurs either at the initial stage or at any subsequent iteration]. [$B = (\beta_1, \beta_2, \cdots, \beta_m)$ be the basis matrix at that stage.]

$$\text{B.F.S. } \mathbf{x}_B = B^{-1}\mathbf{b} \geq 0 \tag{11.2.1}$$

and the value of the objective function corresponding to $\mathbf{x}_B$ is

$$z = \mathbf{c}_B\mathbf{x}_B. \tag{11.2.2}$$

The requirement vector $\mathbf{b}$ is such that effective measure is to be taken to tackle the degeneracy problem (mainly cycling) in the coming iterations.

We consider a new requirement vector

$$\mathbf{b}(\varepsilon) = \mathbf{b} + \varepsilon\mathbf{a}_1 + \varepsilon^2\mathbf{a}_2 + \cdots + \varepsilon^n\mathbf{a}_n = \mathbf{b} + \sum_{j=1}^{n} \varepsilon^j \mathbf{a}_j, \tag{11.2.3}$$

where $\mathbf{b} \geq \mathbf{0}$, ε is a very small positive proper fraction, $\mathbf{a}_1, \mathbf{a}_2, \cdots, \mathbf{a}_n$ are the n column vectors of the co-efficient matrix. $\mathbf{b}$ is slightly perturbed to $\mathbf{b}(\epsilon)$. The new perturbed L.P.P. is maximize, $z = \mathbf{cx}$ subject to $A\mathbf{x} = \mathbf{b}(\varepsilon)$, $\mathbf{x} \geq \mathbf{0}$ with the requirement vector $\mathbf{b}(\varepsilon)$. Initially we make an assumption that the first m column vectors of the co-efficient matrix constitute the basis matrix B. [Rearrange the problem accordingly]. Hence, $B = (\mathbf{a}_1, \mathbf{a}_2, \cdots, \mathbf{a}_m)$. Since B is a basis matrix, therefore

$$\begin{aligned} I_m = B^{-1}B &= B^{-1}(\mathbf{a}_1, \mathbf{a}_2, \cdots, \mathbf{a}_m) \\ &= (B^{-1}\mathbf{a}_1, B^{-1}\mathbf{a}_2, \cdots, B^{-1}\mathbf{a}_m) \\ &= (\mathbf{y}_1, \mathbf{y}_2, \cdots, \mathbf{y}_m) \quad [\text{as } B^{-1}\mathbf{a}_j = \mathbf{y}_j]. \end{aligned}$$

Therefore $\mathbf{y}_j$ is an unit column vector $\mathbf{e}_j$ with unit element at the jth position $[j = 1, 2, \cdots, m]$

$$\begin{aligned} \mathbf{x}_B(\varepsilon) = B^{-1}\mathbf{b}(\varepsilon) &= B^{-1}\left(\mathbf{b} + \sum_{j=1}^{n} \varepsilon^j \mathbf{a}_j\right) \quad \text{from (11.2.3)} \\ &= B^{-1}\mathbf{b} + \sum_{j=1}^{n} B^{-1}\mathbf{a}_j \varepsilon^j = \mathbf{x}_B + \sum_{j=1}^{n} \varepsilon^j \mathbf{y}_j \qquad (11.2.4) \end{aligned}$$

$$\text{or} \quad \mathbf{x}_B(\varepsilon) = \mathbf{x}_B + \varepsilon\mathbf{e}_1 + \varepsilon^2\mathbf{e}_2 + \cdots + \varepsilon^m\mathbf{e}_m + \sum_{j=m+1}^{n} \varepsilon^j y_j.$$

Therefore,

$$x_{Bi}(\varepsilon) = x_{Bi} + \varepsilon^i + \sum_{j=m+1}^{n} \varepsilon^j y_{ij}, \quad i = 1, 2, \cdots, m.$$

Then it is always possible to select ε (small) in such a way that $\mathbf{x}_B(\varepsilon) > 0$ even if some $x_{Bi} = 0$.

Hence the B.F.S. of the perturbed problem is always non-degenerate and therefore it can be solved without taking any risk of cycling.

Now,

$$\begin{aligned} z_B = \mathbf{c}_B\mathbf{x}_B(\varepsilon) &= \mathbf{c}_B\left(\mathbf{x}_B + \sum_{j=1}^{n} \epsilon^j \mathbf{y}_j\right) = \mathbf{c}_B\mathbf{x}_B + \sum_{j=1}^{n} \epsilon^j \mathbf{c}_B\mathbf{y}_j \\ &= \mathbf{c}_B\mathbf{x}_B + \sum_{j=1}^{n} \epsilon^j \mathbf{c}_B\mathbf{y}_j = z + \sum_{j=1}^{n} \epsilon^j \mathbf{c}_B\mathbf{y}_j \qquad (11.2.5) \end{aligned}$$

Putting $\varepsilon = 0$, the value of the objective function of the original problem can be determined from the value of the objective function of the perturbed problem. Hence the optimal value of the objective function of the original problem can be determined from the optimal value of the objective function of the perturbed problem.

Furthermore, the cost vector $\mathbf{c}$ and the column vectors $\mathbf{y}_j [j = 1, 2, \cdots, n]$ remain unaffected due to the introduction of ε in the requirement vector $\mathbf{b}$. Thus $z_j - c_j$ are same for the perturbed and the original problem. In practice, the simplex tables of the perturbed problem will differ from the original problem only in one column, column headed by **b** in each iteration. Therefore using the method of selection of key element and the departing vector of the perturbed problem as in the original problem, we shall be in a position to solve the original problem without any risk of cycling.

11.2.1 Selection of the key element and the basis vector which is to be removed from the basis

Let $\mathbf{a}_k$ be the incoming vector (which is to be determined by the usual method). Now calculate

$$\min_i \left(\frac{x_{Bi}}{y_{ik}},\ y_{ik} > 0 \right).$$

If the minimum is unique and occurs at $i = r$, then y_{rk} is the key element and β_r is the departing vector (whether the B.F.S. is degenerate or non-degenerate). But if the minimum is not unique [minimum value is zero and occurs at least for two values of i, or minimum value is > 0 and occurs at least for three values of i] we shall have to use the perturbation technique to select the key element and the departing vector to avoid cycling. Rearrange the column vectors of the co-efficient matrix in such a way that the first m column vectors constitute the basis matrix. After arrangement let $\mathbf{a}_1, \mathbf{a}_2, \cdots, \mathbf{a}_m$ constitute the basis. Then calculate

$$\min_i \left(\frac{x_{Bi}(\varepsilon)}{y_{ik}},\ y_{ik} > 0 \right) = \min_i \left(\frac{x_{Bi}}{y_{ik}} + \sum_{j=1}^{n} \frac{y_{ij}\varepsilon^j}{y_{ik}},\ y_{ik} > 0 \right) \quad \text{by using (11.2.4).}$$

Let the minimum value of $\dfrac{x_{Bi}}{y_{ik}}$ occurs for $i = i_1, i_2, \cdots, i_p$. Then calculate

$$\min \left(\frac{y_{i1}}{y_{ik}},\ y_{ik} > 0 \right)$$

algebraically for those values of i, for which $\left(\dfrac{x_{Bi}}{y_{ik}},\ y_{ik} > 0 \right)$ is minimum. If minimum is unique then the key element and the departing vector can be selected uniquely.

If the unique minimum is $\dfrac{y_{ip}^{\,1}}{y_{ip}^{\,k}}$ (say), $y_{ip}^{\,k}$ is the key element and β_{ip} is the departing vector $\left[\text{assume } \varepsilon \text{ so small such that } \sum_{j=2}^{n} \varepsilon^j \dfrac{y_{ij}}{y_{ik}} = 0\right]$.

But if again the minimum is not unique, then calculate

$$\min \left(\frac{y_{i2}}{y_{ik}},\ y_{ik} > 0 \right)$$

for those values of i for which $\left(\dfrac{y_{i2}}{y_{ik}},\ y_{ik} > 0 \right)$ is minimum. If the minimum value is unique, then the key element is unique. If it is not unique then proceed further to calculate

$$\min\left(\frac{y_{i3}}{y_{ik}},\ y_{ik} > 0\right)$$

and so on and ultimately we get an unique value of the key element and the departing vector and the problem will be solved without cycling.

11.3 Other method of solving degenerate problems

If it is possible to select at least one column for which $z_j - c_j < 0$ and $(z_j - c_j)\dfrac{x_{Br}}{y_{rj}} < 0$, then the degenerate problems will be solved easily and more quickly on selecting that column vector as the incoming vector. If there exists more than one such column, then select that one as the incoming vector for which $z_j - c_j$ (for all columns of the non-basic variables with $\dfrac{x_{Br}}{y_{rj}} > 0$) then $\mathbf{a}_k$ is the incoming vector. All these can be done only by mere inspection. Moreover, the B.F.S. in the next iteration will be non-degenerate provided the min. ratio is unique at the current table.

▶ **Example 11.3.1** *Solve the L.P.P.;*

$$\textit{maximize},\quad z = 2x_1 + 5x_2$$

subject to

$$\begin{aligned} -2x_1 + \ x_2 &\le 0 \\ x_1 + 3x_2 &\le 14 \\ x_1 + \ x_2 &\le \ 8, \text{ and } x_1, x_2 \ge 0. \end{aligned}$$

The converted equations are

$$\begin{aligned} -2x_1 + \ x_2 + x_3 \qquad\qquad &= \ 0 \\ x_1 + 3x_2 \qquad + x_4 \qquad &= 14 \\ x_1 + \ x_2 \qquad\qquad + x_5 &= \ 8. \end{aligned}$$

After the addition of three slack variables x_3, x_4 and x_5 (one to each inequation) the adjusted objective function is

$$z = 2x_1 + 5x_2 + 0.x_3 + 0.x_4 + 0.x_5.$$

Initial B.F.S.

$$\begin{aligned} \mathbf{x}_B &= [x_3, x_4, x_5] = [0, 14, 8] \text{ which is degenerate} \\ \mathbf{c}_B &= (c_3, c_4, c_5) = (0, 0, 0), \mathbf{y}_j = \mathbf{a}_j \\ z &= 0 \times 0 + 0 \times 14 + 0 \times 8 = 0. \end{aligned}$$

Simplex tables

Basis	c_B	c / b	2 / a_1	5 / a_2	0 / $a_3(e_1)$	0 / $a_4(e_2)$	0 / $a_5(e_3)$	Min. ratio
a_3^*	0	0	-2	1^*	1	0	0	$\frac{0}{1} = 0^*$
a_4	0	14	1	3	0	1	0	$\frac{14}{3} = 4\frac{2}{3}$
a_5	0	8	1	1	0	0	1	$\frac{8}{1} = 8$
$z_j - c_j$		0	-2	-5^*	0	0	0	
a_2	5	0	-2	1	1	0	0	......
a_4^*	0	14	7^*	0	-3	1	0	$\frac{14}{7} = 2^*$
a_5	0	8	3	0	-1	0	1	$\frac{8}{3} = 2\frac{2}{3}$
$z_j - c_j$		0	-12^*	0	5	0	0	
a_2	5	4	0	1	$\frac{1}{7}$	$\frac{2}{7}$	0	$4/\frac{1}{7} = 28$
a_1	2	2	1	0	$-\frac{3}{7}$	$\frac{1}{7}$	0	......
a_5^*	0	2	0	0	$\frac{2}{7}^*$	$-\frac{3}{7}$	1	$2/\frac{2}{7} = 7^*$
$z_j - c_j$		24	0	0	$-\frac{1}{7}^*$	$\frac{12}{7}$	0	
a_2	5	3	0	1	0	$\frac{1}{2}$	$-\frac{1}{2}$	
a_1	2	5	1	0	0	$-\frac{1}{2}$	$\frac{3}{2}$	
a_3	0	7	0	0	1	$-\frac{3}{2}$	$\frac{7}{2}$	
$z_j - c_j$		25	0	0	0	$\frac{3}{2}$	$\frac{1}{2}$	

As none of $z_j - c_j < 0$, then the optimality conditions are satisfied. Hence the solution is optimal which is non-degenerate. The optimal value of z is 25 for $x_1 = 5, x_2 = 3$, i.e., $\max z = 25$ for $x_1 = 5, x_2 = 3$.

▶ **Example 11.3.2** *Solve the L.P.P.,*

$$\textit{maximize,} \quad z = 5x_1 + 3x_2$$

subject to

$$\begin{aligned} x_1 + \ x_2 &\leq 2 \\ 5x_1 + 2x_2 &\leq 10 \\ 3x_1 + 8x_2 &\leq 12, \text{ and } x_1 \geq 0, x_2 \geq 0. \end{aligned}$$

Introducing the slack variables x_3, x_4, x_5 (one to each inequation) we get the following converted equations

$$\begin{aligned} x_1 + \ x_2 + x_3 \qquad\qquad &= 2 \\ 5x_1 + 2x_2 \qquad + x_4 \qquad &= 10 \\ 3x_1 + 8x_2 \qquad\qquad + x_5 &= 12 \end{aligned}$$

The adjusted objective function z is given by

$$z = 5x_1 + 3x_2 + 0.x_3 + 0.x_4 + 0.x_5.$$

Initial B.F.S.

$$\mathbf{x}_B = [x_3, x_4, x_5] = [2, 10, 12] \text{ which is non-degeneratic}$$
$$\mathbf{c}_B = (c_3, c_4, c_5) = (0, 0, 0), \mathbf{y}_j = \mathbf{a}_j$$
$$z = 0 \times 2 + 0 \times 10 + 0 \times 12 = 0.$$

Initial simplex table

		c	5	3	0	0	0	
Basis	$\mathbf{c}_B$	**b**	$\mathbf{a}_1$	$\mathbf{a}_2$	$\mathbf{a}_3(\mathbf{e}_1)$	$\mathbf{a}_4(\mathbf{e}_2)$	$\mathbf{a}_5(\mathbf{e}_3)$	**Min. ratio**
$\mathbf{a}_3^*$	0	2	1*	1	1	0	0	$\frac{2}{1} = 2$
$\mathbf{a}_4$	0	10	5	2	0	1	0	$\frac{10}{5} = 2$
$\mathbf{a}_5$	0	12	3	8	0	0	1	$\frac{12}{3} = 4$
$z_j - c_j$		0	−5*	−3	0	0	0	

$z_1 - c_1 = -5$ and $z_2 - c_2 = -3$. Hence the solution is not optimal. $\mathbf{a}_1$ is the incoming vector

$$\min_i \left(\frac{x_{Bi}}{y_{i1}}, \ y_{i1} > 0 \right) = \min \left(\frac{2}{1}, \ \frac{10}{5}, \ \frac{12}{3} \right) = 2$$

which occurs at two values of i, $i = 1$ and 2. Hence the minimum ratio is not unique. But it occurs only for two values of i. Then we may take either y_{11} or y_{21} as the key element. If we take y_{11} as the key element, then in the next table $x'_{B_2} = 0$ and if we take y_{21} as the key element then in the next table $x'_{B_1} = 0$.

Let us take y_{11} as the key element, then $r = 1$ and therefore $\mathbf{a}_3$, the first vector under the column "basis" is the outgoing vector.

Second simplex table

		c	5	3	0	0	0
Basis	$\mathbf{c}_B$	**b**	$\mathbf{a}_1$	$\mathbf{a}_2$	$\mathbf{a}_3$	$\mathbf{a}_4$	$\mathbf{a}_5$
$\mathbf{a}_1$	5	2	1	1	1	0	0
$\mathbf{a}_4$	0	0	0	−3	−5	1	0
$\mathbf{a}_5$	0	6	0	5	−3	0	1
$z_j - c_j$		10	0	2	5	0	0

As none of $z_j - c_j < 0$, therefore optimality conditions are satisfied. The optimal value of $z = 10$ for $\mathbf{x}_B = [x_1, x_4, x_5] = [2, 0, 6]$, i.e., for the solution set $x_1 = 2$ and $x_2 = 0$ (non-basic), $\max z = 10$. In this problem the final B.F.S. is degenerate.

Note. We may solve the problem taking $y_{21} = 5$ as the key element but one more step is required to solve the problem.

▶ **Example 11.3.3** *Solve the L.P.P.,*

$$\textit{maximize,} \quad z = 6x_1 + 5x_2$$

subject to

$$\begin{aligned} 4x_1 + \ \ x_2 &\leq 8 \\ 2x_1 - 3x_2 &\leq 0 \\ 3x_1 - \ \ x_2 &\leq 0, \text{ and } x_1, x_2 \geq 0. \end{aligned}$$

Introducing the slack variables x_3, x_4 and x_5 one to each of the inequations we get the following converted equations

$$\begin{aligned} 4x_1 + x_2 + x_3 \qquad\qquad &= 8 \\ 2x_1 - 3x_2 \qquad + x_4 \qquad &= 0 \\ 3x_1 - x_2 \qquad\qquad + x_5 &= 0. \end{aligned}$$

The adjusted objective function z is given by

$$z = 6x_1 + 5x_2 + 0.x_3 + 0.x_4 + 0.x_5.$$

Initial B.F.S. $\mathbf{x}_B = [x_3, x_4, x_5] = [8, 0, 0]$.

The B.F.S. is degenerate

$$\mathbf{c}_B = (c_3, c_4, c_5) = (0, 0, 0), \mathbf{y}_j = \mathbf{a}_j$$

The value of the objective function

$$z = 0 \times 8 + 0 \times 0 + 0 \times 0 = 0.$$

Starting simplex table

		c	6	5	0	0	0
Basis	$\mathbf{c}_B$	**b**	$\mathbf{a}_1$	$\mathbf{a}_2$	$\mathbf{a}_3(\mathbf{e}_1)$	$\mathbf{a}_4(\mathbf{e}_2)$	$\mathbf{a}_5(\mathbf{e}_3)$
$\mathbf{a}_3^*$	0	8	4	1*	1	0	0
$\mathbf{a}_4$	0	0	2	−3	0	1	0
$\mathbf{a}_5$	0	0	3	−1	0	0	1
$z_j - c_j$		0	−6	−5*	0	0	0

According to the convention, we shall have to take $\mathbf{a}_1$ as the incoming vector. But if we take $\mathbf{a}_1$ as the incoming vector, then from min. ratio rule we get

$$\min_i \left(\frac{y_{Bi}}{y_{i1}},\ y_{i1} > 0 \right) = \min \left(\frac{3}{4}, \frac{0}{2}, \frac{0}{3} \right) = 0$$

which is not unique. Hence, it is difficult to select the key element and the outgoing vector. But if we simply consider $\mathbf{a}_2$ as the incoming vector then we find that not only the min. ratio is unique but also greater than zero. And this can be done only by simple observation. Now,

$$\min_i \left(\frac{x_{Bi}}{y_{i1}},\ y_{i2} > 0 \right) = \min \left(\frac{8}{1}, \cdots, \cdots \right) = 8$$

which occurs at $i = 1$. $\left[\frac{x_{B2}}{y_{22}}\right.$ and $\frac{x_{B3}}{x_{32}}$ do not enter in the calculation as $y_{22} \leq 0$ and $\left.y_{32} \leq 0\right]$.

Therefore if we take $\mathbf{a}_2$ as the incoming vector then in the next step the B.F.S. will be non-degenerate. We are in a position to solve the problem quickly. The key element is $y_{12} = 1$ and the outgoing vector is $\mathbf{a}_3(\mathbf{e}_1)$.

Second simplex table

		c	6	5	0	0	0
Basis	$\mathbf{c}_B$	**b**	$\mathbf{a}_1$	$\mathbf{a}_2$	$\mathbf{a}_3$	$\mathbf{a}_4$	$\mathbf{a}_5$
$\mathbf{a}_2$	5	8	4	1	1	0	0
$\mathbf{a}_4$	0	24	14	0	3	1	0
$\mathbf{a}_5$	0	8	7	0	1	0	1
$z_j - c_j$		40	14	0	5	0	0

As none of $z_j - c_j < 0$, then the optimality conditions are satisfied. Hence the optimal value of z is 40 for $x_1 = 0$ [non-basic], $x_2 = 8$, i.e., $\max z = 40$ for $x_1 = 0, x_2 = 8$.

Note. If we take $\mathbf{a}_1$ as the vector to enter the basis the problem of degeneracy occurs at the very initial stage. And to determine the key element and the vector to leave the basis we are to follow the rules prescribed in Charnes perturbed technique.

11.4 Cycling problem

▶ **Example 11.4.1** *Solve the L.P.P.,*

$$\textit{maximize,} \quad z = \frac{3}{4}x_1 - 20x_2 + \frac{1}{2}x_3 - 6x_4$$

subject to

$$\begin{aligned} \tfrac{1}{4}x_1 - 8x_2 - x_3 + 9x_4 &\leq 0 \\ \tfrac{1}{2}x_1 - 12x_2 - \tfrac{1}{2}x_3 + 3x_4 &\leq 0 \\ x_3 &\leq 1, \text{ and } x_1, x_2, x_3, x_4 \geq 0. \end{aligned}$$

This problem is due to **Baele.**

Introducing the slack variables x_5, x_6, x_7 one to each of the constraints, we get the following converted equations.

$$\begin{aligned} \tfrac{1}{4}x_1 - 8x_2 - x_3 + 9x_4 + x_5 \qquad &= 0 \\ \tfrac{1}{2}x_1 - 12x_2 - \tfrac{1}{2}x_3 + 3x_4 \qquad + x_6 \qquad &= 0 \\ x_3 \qquad\qquad + x_7 &= 1 \end{aligned}$$

The adjusted objective function z is given by

$$z = \frac{3}{4}x_1 - 20x_2 + \frac{1}{2}x_3 - 6x_4 + 0.x_5 + 0.x_6 + 0.x_7.$$

Initial B.F.S. $\mathbf{x}_B = [x_5, x_6, x_7] = [0, 0, 1]$ which is degenerate, $\mathbf{c}_B = (c_5, c_6, c_7) = (0, 0, 0), \mathbf{y}_i = \mathbf{a}_j,\ z = 0 \times 0 + 0 \times 0 + 0 \times 1 = 0.$

Initial simplex table

		c	$\frac{3}{4}$	-20	$\frac{1}{2}$	-6	0	0	0
Basis	$\mathbf{c}_B$	**b**	$\mathbf{a}_1$	$\mathbf{a}_2$	$\mathbf{a}_3$	$\mathbf{a}_4$	$\mathbf{a}_5(\mathbf{e}_1)$	$\mathbf{a}_6(\mathbf{e}_2)$	$\mathbf{a}_7(\mathbf{e}_3)$
$\mathbf{a}_5^*$	0	0	$\frac{1}{4}^*$	-8	-1	9	1	0	0
$\mathbf{a}_6$	0	0	$\frac{1}{2}$	-12	$-\frac{1}{2}$	3	0	1	0
$\mathbf{a}_7$	0	1	0	0	1	0	0	0	1
$z_j - c_j$		0	$-\frac{3}{4}^*$	20	$-\frac{1}{2}$	6	0	0	0

$z_1 - c_1 = -\frac{8}{4}$ and $z_2 - c_2 = -\frac{1}{2}$. Hence the solution is not optimal. Hence $\mathbf{a}_1$ is the incoming vector

$$\min_i \left(\frac{x_{Bi}}{y_{i1}}, y_{i1} > 0 \right) = \min \left(\frac{0}{\frac{1}{4}}, \frac{0}{\frac{1}{2}}, - \right) = 0$$

which is not unique.

Hence the problem of degeneracy occurs at the initial stage. Minimum ratio is equal to zero for $i = 1$ and 2. Hence the key element will be either $y_{11} = \frac{1}{4}$ or $y_{12} = \frac{1}{2}$. If we take $y_{11} = \frac{1}{4}$ as the key element, the outgoing vector will be $\mathbf{a}_5(\mathbf{e}_1)$ and if we take $y_{12} = \frac{1}{2}$ as the key element, the outgoing vector will be $\mathbf{a}_6(\mathbf{e}_2)$. Let us take y_{11} as the key element.

First iteration (second simplex table)

		c	$\frac{3}{4}$	-20	$\frac{1}{2}$	-6	0	0	0
Basis	$\mathbf{c}_B$	**b**	$\mathbf{a}_1$	$\mathbf{a}_2$	$\mathbf{a}_3$	$\mathbf{a}_4$	$\mathbf{a}_5$	$\mathbf{a}_6$	$\mathbf{a}_7$
$\mathbf{a}_1$	$\frac{3}{4}$	0	1	-32	-4	36	4	0	0
$\mathbf{a}_6^*$	0	0	0	4^*	$\frac{3}{2}$	-15	-2	1	0
$\mathbf{a}_7$	0	1	0	0	1	0	0	0	1
$z_j - c_j$		0	0	-4^*	$-\frac{7}{2}$	33	3	0	0

$z_2 - c_2 = -4$ and $z_3 - c_3 = -\frac{7}{2}$. Hence the solution is not optimal. Hence $\mathbf{a}_2$ is the incoming vector

$$\min_i \left(\frac{y_{Bi}}{y_{i2}}, y_{i2} > 0 \right) = \min \left(-, \frac{0}{4}, - \right) = 0$$

whichis unique and occurs at $i = 2$. Therefore y_{22} is the key element and $\mathbf{a}_6$ is the outgoing vector.

Second iteration

	c		$\frac{3}{4}$	-20	$\frac{1}{2}$	-6	0	0	0
Basis	$\mathbf{c}_B$	b	$\mathbf{a}_1$	$\mathbf{a}_2$	$\mathbf{a}_3$	$\mathbf{a}_4$	$\mathbf{a}_5$	$\mathbf{a}_6$	$\mathbf{a}_7$
$\mathbf{a}_1^*$	$\frac{3}{4}$	0	1	0	8^*	-84	-12	8	0
$\mathbf{a}_2$	-20	0	0	1	$\frac{8}{9}$	$-\frac{15}{4}$	$-\frac{1}{2}$	$\frac{1}{4}$	0
$\mathbf{a}_7$	0	1	0	0	1	0	0	0	1
$z_j - c_j$		0	0	0	-2^*	18	1	1	0

Third iteration. $\mathbf{a}_3$ is the incoming vector. Key element is not unique. Let us take $y_{13} = 8$ as the key element and $\mathbf{a}_1$ is the outgoing vector.

	c		$\frac{3}{4}$	-20	$\frac{1}{2}$	-6	0	0	0
Basis	$\mathbf{c}_B$	b	$\mathbf{a}_1$	$\mathbf{a}_2$	$\mathbf{a}_3$	$\mathbf{a}_4$	$\mathbf{a}_5$	$\mathbf{a}_6$	$\mathbf{a}_7$
$\mathbf{a}_3$	$\frac{1}{2}$	0	$\frac{1}{8}$	0	1	$-\frac{21}{2}$	$-\frac{3}{2}$	1	0
$\mathbf{a}_2^*$	-20	0	$-\frac{3}{64}$	1	0	$\frac{3}{16}^*$	$\frac{1}{16}$	$-\frac{1}{8}$	0
$\mathbf{a}_7$	0	1	$-\frac{1}{6}$	0	0	$\frac{21}{2}$	-1	$\frac{3}{2}$	1
$z_j - c_j$		0	$\frac{1}{4}$	0	0	-3^*	-2	3	0

Fourth iteration. $\mathbf{a}_4$ is the incoming vector. Key element $y_{24} = \frac{3}{16}$ is unique. Then $\mathbf{a}_2$ is the outgoing vector.

	c		$\frac{3}{4}$	-20	$\frac{1}{2}$	-6	0	0	0
Basis	$\mathbf{c}_B$	b	$\mathbf{a}_1$	$\mathbf{a}_2$	$\mathbf{a}_3$	$\mathbf{a}_4$	$\mathbf{a}_5$	$\mathbf{a}_6$	$\mathbf{a}_7$
$\mathbf{a}_3^*$	$\frac{1}{2}$	0	$-\frac{5}{2}$	56	1	0	2^*	-6	0
$\mathbf{a}_4$	-6	0	$-\frac{1}{4}$	$\frac{16}{3}$	0	1	$\frac{1}{3}$	$-\frac{2}{3}$	0
$\mathbf{a}_7$	0	1	$\frac{5}{2}$	-56	0	0	6	-2	1
$z_j - c_j$		0	$-\frac{1}{2}$	16	0	0	-1^*	1	0

Fifth iteration. $\mathbf{a}_5$ is the incoming vector. Key element is not unique. Let us take $y_{15} = 2$ as the key element. Therefore $\mathbf{a}_3$ is the outgoing vector.

	c		$\frac{3}{4}$	-20	$\frac{1}{2}$	-6	0	0	0
Basis	$\mathbf{c}_B$	b	$\mathbf{a}_1$	$\mathbf{a}_2$	$\mathbf{a}_3$	$\mathbf{a}_4$	$\mathbf{a}_5$	$\mathbf{a}_6$	$\mathbf{a}_7$
$\mathbf{a}_5$	0	0	$-\frac{5}{4}$	28	$\frac{1}{2}$	0	1	-3	0
$\mathbf{a}_4^*$	-6	0	$\frac{1}{6}$	-4	$-\frac{1}{6}$	1	0	$\frac{1}{3}^*$	0
$\mathbf{a}_7$	0	1	0	0	1	0	0	0	1
$z_j - c_j$		0	$-\frac{7}{4}$	44	$\frac{1}{2}$	0	0	-2^*	0

Sixth iteration. $\mathbf{a}_6$ is the incoming vector. Key element is unique and it is $y_{26} = \frac{1}{3}$. $\mathbf{a}_4$ is the outgoing vector.

Basis	c_B	b	$\mathbf{a}_1$	$\mathbf{a}_2$	$\mathbf{a}_3$	$\mathbf{a}_4$	$\mathbf{a}_5$	$\mathbf{a}_6$	$\mathbf{a}_7$
$\mathbf{a}_5$	0	0	$\frac{1}{4}$	-8	-1	9	1	0	0
$\mathbf{a}_6$	0	0	$\frac{1}{2}$	-12	$-\frac{1}{2}$	3	0	1	0
$\mathbf{a}_7$	0	1	0	0	1	0	0	0	1
$z_j - c_j$		0	$-\frac{3}{4}$	20	$-\frac{1}{2}$	6	0	0	0

It is interesting to note that after six iterations, tbe initial simplex table appears before us. The problem is known as a **cycling problem**. If we proceed further to solve the problem obeying the rule used before, we shall get the same iterated tables as before and the problem cannot be solved. Hence in order to solve the problem, to avoid cycling, Charnes perturbation technique is to be used.

11.4.1 Solution by Charnes perturbation technique

We have already seen that degeneracy occurs at the initial stage. Hence in order to avoid cycling we shall use Charnes perturbation technique. To use the technique we shall have to rearrange the column vectors of the co-efficient matrix such that the column vectors which are included in the basis must be placed in the first $m(=3)$ columns of the matrix. The objective function z can be re-written as

$$z = 0.x_5 + 0.x_6 + 0.x_7 + \frac{3}{4}x_1 - 20x_2 + \frac{1}{2}x_3 - 6x_4$$

Initial B.F.S. $\mathbf{x}_B = [x_5, x_6, x_7] = [0, 0, 1]$

Starting simplex table

		c	0	0	0	$\frac{3}{4}$	-20	$\frac{1}{2}$	-6
Basis	c_B	b	$\mathbf{a}'_1$	$\mathbf{a}'_2$	$\mathbf{a}'_3$	$\mathbf{a}'_4$	$\mathbf{a}'_5$	$\mathbf{a}'_6$	$\mathbf{a}'_7$
$\mathbf{a}'_1$	0	0	1	0	0	$\frac{1}{4}$	-8	-1	9
$\mathbf{a}'^{*}_2$	0	0	0	1	0	$\frac{1}{2}^*$	-12	$-\frac{1}{2}$	3
$\mathbf{a}'_3$	0	1	0	0	1	0	0	1	0
$z_j - c_j$		0	0	0	0	$-\frac{3}{4}^*$	2	$-\frac{1}{2}$	6

$\mathbf{a}'_4, \mathbf{a}'_5, \mathbf{a}'_6$ and $\mathbf{a}'_7$ are the column vectors corresponding to the variables x_1, x_2, x_3 and x_4 respectively.

$z_4 - c_4 = -\frac{3}{4}$ and $z_6 - c_6 = -\frac{1}{2}$. Hence the solution is not optimal.

Hence $\mathbf{a}'_4$ is the incoming vector.

$$\min_i \left(\frac{x_{Bi}}{y'_{ir}}, y'_{i4} > 0\right) = \min\left(\frac{x_{B1}}{y'_{14}}, \frac{x_{B2}}{y'_{24}}, \frac{x_{B3}}{y'34}\right) = \min\left(\frac{0}{1/4}, \frac{0}{1/2}, -\right) = 0$$

which is not unique but occurs at $i = 1$ and 2.

Therefore we have to calculate

$$\min_{i=1,2}\left(\frac{y'_{i1}}{y'_{i4}}\right) = \min\left(\frac{y'_{11}}{y'_{14}}, \frac{y'_{21}}{y'_{24}}\right) = \min\left(\frac{1}{1/4}, \frac{0}{1/2}\right) = 0,$$

which is unique and occurs at $i = 2$. Therefore $y'_{24} = \frac{1}{2}$ is the key element and $\mathbf{a}'_2$ is the outgoing vectors.

Second simplex table

		c	0	0	0	$\frac{3}{4}$	−20	$\frac{1}{2}$	−6
Basis	$\mathbf{c}_B$	**b**	$\mathbf{a}'_1$	$\mathbf{a}'_2$	$\mathbf{a}'_3$	$\mathbf{a}'_4$	$\mathbf{a}'_5$	$\mathbf{a}'_6$	$\mathbf{a}'_7$
$\mathbf{a}'_1$	0	0	1	$-\frac{1}{2}$	0	0	−2	$-\frac{3}{4}$	$\frac{15}{2}$
$\mathbf{a}'_4$	$\frac{3}{4}$	0	0	2	0	1	−24	−1	6
$\mathbf{a}'^{*}_3$	0	1	0	0	1	0	0	1*	0
$z_j - c_j$		0	0	$\frac{3}{2}$	0	0	2	$-\frac{5}{4}$*	$\frac{21}{2}$

$z_6 - c_6 = -\frac{5}{4} < 0$. Hence the solution is not optimal. $\mathbf{a}'_6$ is the incoming vector

$$\min\left(\frac{x_{Bi}}{y'_{i6}}, y'_{i6} > 0\right) = \min\left(-, -, \frac{1}{1}\right) = 1$$

which occurs at $i = 3$. Therefore $y'_{36} = 1$ is the key element and $\mathbf{a}'_3$ is the leaving vector.

Third simplex table

		c	0	0	0	$\frac{3}{4}$	−20	$\frac{1}{2}$	−6
Basis	$\mathbf{c}_B$	**b**	$\mathbf{a}'_1$	$\mathbf{a}'_2$	$\mathbf{a}'_3$	$\mathbf{a}'_4$	$\mathbf{a}'_5$	$\mathbf{a}'_6$	$\mathbf{a}'_7$
$\mathbf{a}'_1$		$\frac{3}{4}$							
$\mathbf{a}'_4$		1							
$\mathbf{a}'_6$		1							
$z_j - c_j$		$\frac{5}{4}$	0	$\frac{3}{2}$	$\frac{5}{4}$	0	2	0	$\frac{21}{2}$

None of $z_j - c_j < 0$. Hence the solution is optimal and non-degenerate. The maximum value of $z = \frac{5}{4}$ for $x_1 = 1$ and $x_3 = 1$ and $x_2 = 0$ [non-basic]

[as x_1 and x_2 are associated with the column vectors $\mathbf{a}'_4$ and $\mathbf{a}'_6$].

Note. If we place the first constraint in the second position and vice versa, then it requires a larger number of iterations in solving the problem.

The above cycling problem can be solved in an orthodex manner which is shown below (solved by the author).

► **Example 11.4.2** *Solve the L.P.P.,*

$$\textit{maximize,} \quad z = \frac{3}{4}x_1 - 20x_2 + \frac{1}{2}x_3 - 6x_4$$

subject to

$$\begin{aligned} \tfrac{1}{4}x_1 - \ 8x_2 - \ x_3 + 9x_4 &\leq 0 \\ \tfrac{1}{2}x_1 - 12x_2 - \tfrac{1}{2}x_3 + 3x_4 &\leq 0 \\ x_3 \qquad\qquad &\leq 1, \ x_1, x_2, x_3, x_4 \geq 0. \end{aligned}$$

Basis	c_B	c / b	$\frac{3}{4}$ / $\mathbf{a}_1$	-20 / $\mathbf{a}_2$	$\frac{1}{2}$ / $\mathbf{a}_3$	-6 / $\mathbf{a}_4$	0 / $\mathbf{a}_5$	0 / $\mathbf{a}_6$	0 / $\mathbf{a}_7$	Min. ratio
$\mathbf{a}_5$	0	0	$\frac{1}{4}$	-8	-1	9	1	0	0	....
$\mathbf{a}_6$	0	0	$\frac{1}{2}$	-12	$-\frac{1}{2}$	3	0	1	0	....
$\mathbf{a}_7^*$	0	1	0	0	1^*	0	0	0	1	$\frac{1}{1} = 1^*$
$z_j - c_j$		0	$-\frac{3}{4}$	20	$-\frac{1}{2}^*$	6	0	0	0	

$-\frac{3}{4} < -\frac{1}{2}$. Thus according to the conventional method first column will be the key column. But the degeneracy cannot be avoided. But if we take the third column as the key column the degeneracy can be avoided, then from the the first table $y_{33} = 1$ is the key element and the third row is the key row and vector $\mathbf{a}_7$ of the basis will be replaced by $\mathbf{a}_3$ to form the new basis and the next table is

Basis	c_B	c / b	$\frac{3}{4}$ / $\mathbf{a}_1$	-20 / $\mathbf{a}_2$	$\frac{1}{2}$ / $\mathbf{a}_3$	-6 / $\mathbf{a}_4$	0 / $\mathbf{a}_5$	0 / $\mathbf{a}_6$	0 / $\mathbf{a}_7$	Min. ratio
$\mathbf{a}_5$	0	1	$\frac{1}{4}$	-8	0	9	1	0	1	$\frac{1}{1/4} = 4$
$\mathbf{a}_6^*$	0	$\frac{1}{2}$	$\frac{1}{2}^*$	-12	0	3	0	1	$\frac{1}{2}$	$\frac{1/2}{1/2} = 1^*$
$\mathbf{a}_3$	$\frac{1}{2}$	1	0	0	1	0	0	0	1	....
$z_j - c_j$		$\frac{1}{2}$	$-\frac{3}{4}^*$	20	0	6	0	0	$\frac{1}{2}$	
$\mathbf{a}_5$	0	$\frac{3}{4}$								
$\mathbf{a}_1$	$\frac{3}{4}$	1								
$\mathbf{a}_3$	$\frac{1}{2}$	1								
$z_j - c_j$		$\frac{5}{4}$	0	2	0	$\frac{21}{2}$	0	$\frac{3}{2}$	$\frac{5}{4}$	

All $z_j - c_j \geq 0$. Then the solution is optimal. Max $z = 5/4$ at $x_1 = 1$, $x_2 = 0$, $x_3 = 1$.

Remark. All cycling problems cannot be solved by this method. So long, I failed to solve only one problem.

Exercise 11

1. What is degeneracy? Discuss a method to resolve it in a L.P.P.

   Solve the L.P.P.

   Maximise, $z = 3x_1 + 4x_2$

   $$\text{subject to } 2x_1 + x_2 \leq 9, \quad 5x_1 - 2x_2 \leq 0, \text{ and } x_1, x_2 \geq 0.$$

2. Maximise, $z = 3x_1 + 9x_2$

$$\text{subject to } \begin{aligned} x_1 + 4x_2 &\le 8 \\ x_1 + 2x_2 &\le 4, \text{ and } x_1, x_2 \ge 0. \end{aligned}$$

3. Maximise, $z = 3x_1 + 2x_2$

$$\text{subject to } \begin{aligned} x_1 + 2x_2 &\le 15 \\ 2x_1 - x_2 &\le 0 \\ 4x_1 - 7x_2 &\le 0, \text{ and } x_1, x_2 \ge 0. \end{aligned}$$

4. Maximise, $z = 5x_1 - 2x_2 + 3x_3$

$$\text{subject to } \begin{aligned} 2x_1 + 2x_2 - x_3 &\ge 2 \\ 3x_1 - 4x_2 &\le 3 \\ x_2 + 3x_3 &\le 5, \text{ and } x_1, x_2, x_3 \ge 0. \end{aligned}$$

5. Maximise, $z = 22x_1 + 30x_2 + 25x_3$

$$\text{subject to } \begin{aligned} 2x_1 + 3x_2 &\le 100 \\ 2x_1 + x_2 + x_3 &\le 100 \\ x_1 + 2x_2 + 2x_3 &\le 100, \text{ and } x_1, x_2, x_3 \ge 0. \end{aligned}$$

6. Maximise, $z = 3x_1 + 2x_2 + x_3$

$$\text{subject to } \begin{aligned} 2x_1 + x_2 + 2x_3 &\le 4 \\ 3x_1 - 2x_2 &\le 6 \\ 4x_1 + x_2 + x_3 &\le 8, \text{ and } x_1, x_2, x_3 \ge 0. \end{aligned}$$

7. Maximise, $z = \frac{3}{4}x_1 - 150x_2 + \frac{1}{50}x_3 - 6x_4$

$$\text{subject to } \begin{aligned} \tfrac{1}{4}x_1 - 60x_2 - \tfrac{1}{25}x_3 + 9x_4 &\le 0 \\ \tfrac{1}{2}x_1 - 90x_2 - \tfrac{1}{50}x_3 + 3x_4 &\le 0 \\ x_3 &\le 1, \text{ and } x_1, x_2, x_3, x_4 \ge 0. \end{aligned}$$

Answers

1. $\max z = 36$ for $x_1 = 0$, $x_2 = 9$.
2. $\max z = 18$ for $x_1 = 0$, $x_2 = 2$.
3. $\max z = 21$ for $x_1 = 3$, $x_2 = 6$.
4. $\max z = 19$ for $x_1 = 5$, $x_2 = 3$, $x_3 = 0$.
5. $\max z = 1650$ for $x_1 = \frac{100}{3}$, $x_2 = \frac{50}{3}$, $x_3 = \frac{50}{3}$.
6. $\max z = 8$ for $x_1 = 0$, $x_2 = 4$, $x_3 = 0$.
7. $\max z = \frac{1}{20}$ for $x_1 = \frac{1}{25}$, $x_2 = 0$, $x_3 = 1$, $x_4 = 0$.

Chapter 12

Duality Theory

Associated with every L.P.P. there exists a corresponding L.P.P. The original L.P.P. is known as the *primal* problem and the corresponding problem is known as the ***dual*** problem.

12.1 Concept of Duality

Let us consider the following diet problem.

The daily requirement of a patient is 20 and 30 units of vitamins v_1 and v_2 respectively. The food F_1 contains 3 units of v_1 and 4 units of v_2, the food F_2 contains 2 units of v_1 and 3 units of v_2. Find the minimum cost of buying the vitamins if the cost per unit of F_1 and F_2 be Rs.7 and Rs.5 respectively.

12.2 Mathematical Formulation

Chart

Food	$F_1(x_1)$	$F_2(x_2)$	Requirement
v_1	3	2	20 units
v_2	4	3	30 units
Cost	Rs.7	Rs.5	per unit

Let x_1 units of F_1 and x_2 units of F_2 be required to get the minimum amount of vitamins. This is a problem of minimization. The L.P.P. is

$$\text{Minimize, } z = 7x_1 + 5x_2$$

subject to

$$\begin{aligned} 3x_1 + 2x_2 &\geq 20 \\ 4x_1 + 3x_2 &\geq 30\,, \quad x_1 \geq 0, x_2 \geq 0. \end{aligned}$$

Let us now consider the corresponding problem.

A dealer sells the above mentioned vitamins v_1 and v_2 separately. His problem is to fix the cost per unit of v_1 and v_2 in such a way that the price of F_1 and F_2 do not exceed the amount mentioned above. His problem is also to get a maximum amount in selling the vitamins.

Let w_1 and w_2 be the price per unit of v_1 and v_2 respectively.

Therefore the problem is

$$\text{Maximize,} \quad z^* = 20w_1 + 30w_2$$

subject to

$$3w_1 + 4w_2 \leq 7$$
$$2w_1 + 3w_2 \leq 5, \quad w_1 \geq 0, w_2 \geq 0.$$

Now if we take

$$A = \begin{bmatrix} 3 & 2 \\ 4 & 3 \end{bmatrix}, \quad \mathbf{b} = \begin{bmatrix} 20 \\ 30 \end{bmatrix}, \quad \mathbf{x} = \begin{bmatrix} x_1 \\ x_2 \end{bmatrix} \quad \text{and} \quad \mathbf{c} = (7, 5)$$

the initial problem can be written as

$$\text{Minimize,} \quad z_x = \mathbf{cx}$$

subject to

$$A\mathbf{x} \geq \mathbf{b}, \quad \mathbf{x} \geq \mathbf{0}.$$

Now the corresponding problem is

$$\text{Maximize,} \quad z_w = \mathbf{b}'\mathbf{w}$$

subject to

$$A'\mathbf{w} \leq \mathbf{c}', \quad \mathbf{w} \geq \mathbf{0}.$$

$A', \mathbf{b}'$ and $\mathbf{c}'$ are the transposes of the matrix A, vector $\mathbf{b}$ and $\mathbf{c}$ respectively.

The above is an example of primal-dual problem. In the above problem all constraints are inequations. Generally the initial problem is known as the primal and the corresponding problem is known as the dual problem.

There are three types of primal-dual problems.

1. **Symmetric primal-dual problems :** Here all constraints of the both primal and dual problems are inequations and the variables are non-negative.

2. **Unsymmetric primal-dual problems :** Here all constraints of primal are equations and all primal variables non-negative.

3. **Mixed type problems :** Here some constraints of primal are equations and some variables are unrestricted.

Note: Each and every unsymmetric and mixed type problem can be converted into symmetric type problems by proper adjustment which has been shown later.

12.3 Standard Form of Primal

A L.P.P. is said to be in standard form if

1. All constraints involve the sign '$\leq$' in a problem of maximization, or
2. All constraints involve the sign '$\geq$' in a problem of minimization.

12.4 Standard symmetric primal problem

The following L.P.P.

$$\text{Maximize,} \quad z_x = c_1x_1 + c_2x_2 + \cdots + c_nx_n$$

subject to

$$\left.\begin{array}{l} a_{11}x_1 + a_{12}x_2 + \cdots + a_{1n}x_n \leq b_1 \\ a_{21}x_1 + a_{22}x_2 + \cdots + a_{2n}x_n \leq b_2 \\ \cdots\cdots\cdots\cdots\cdots\cdots\cdots\cdots\cdots \\ a_{m1}x_1 + a_{m2}x_2 + \cdots + a_{mn}x_n \leq b_m \end{array}\right\} \quad (12.4.1)$$

$x_1 \geq 0, x_2 \geq 0, \cdots, x_n \geq 0.$

is known as a standard symmetric primal.

Here the constants

$$b_i[i = 1, 2, \cdots, m]$$
$$\text{and} \quad c_j[j = 1, 2, \cdots, n]$$

are unrestricted in sign, i.e., b_i and c_j may be positive, negative or zero.

The corresponding dual problem is

$$\text{Minimize,} \quad z_w = b_1w_1 + b_2w_2 + \cdots + b_mw_m$$

subject to

$$\left.\begin{array}{l} a_{11}w_1 + a_{21}w_2 + \cdots + a_{m1}w_m \geq c_1 \\ a_{12}w_1 + a_{22}w_2 + \cdots + a_{m2}w_m \geq c_2 \\ \cdots\cdots\cdots\cdots\cdots\cdots\cdots\cdots\cdots \\ a_{1n}w_1 + a_{2n}w_2 + \cdots + a_{mn}w_m \geq c_n \end{array}\right\} \quad (12.4.2)$$

$w_1 \geq 0, w_2 \geq 0, \cdots, w_m \geq 0.$

$$\mathbf{w} = \begin{bmatrix} w_1 \\ w_2 \\ \vdots \\ w_m \end{bmatrix}$$

is an m component dual variable vector.

There are n constraints in the dual problem. Putting

$$A = \begin{bmatrix} a_{11} & a_{12} & \cdots & a_{1n} \\ a_{21} & a_{22} & \cdots & a_{2n} \\ \cdots & \cdots & \cdots & \cdots \\ a_{m1} & a_{m2} & \cdots & a_{mn} \end{bmatrix}, \; b = \begin{bmatrix} b_1 \\ b_2 \\ \vdots \\ b_m \end{bmatrix}, \; \mathbf{x} = \begin{bmatrix} x_1 \\ x_2 \\ \vdots \\ x_n \end{bmatrix} \text{ and } \mathbf{c} = (c_1, c_2, \cdots, c_n)$$

(12.4.1) can be written as

$$\text{Maximize,} \quad z_x = \mathbf{cx}$$

subject to

$$A\mathbf{x} \leq \mathbf{b}, \quad \mathbf{x} \geq \mathbf{0}. \tag{12.4.3}$$

and the corresponding dual problem (12.4.2) can be written as

$$\text{Minimize,} \quad z_w = \mathbf{b}'\mathbf{w}$$

subject to

$$A'\mathbf{w} \geq \mathbf{c}', \quad \mathbf{w} \geq \mathbf{0}. \tag{12.4.4}$$

The variable vector $\mathbf{w}$ is an m-component column vector. $A', \mathbf{b}'$ and $\mathbf{c}'$ are the transposes of the matrix A, vector $\mathbf{b}$ and $\mathbf{c}$ respectively. $\mathbf{b}$ is unrestricted in sign here.

The rule of transformation is

1. Co-efficient matrix of the dual is the transpose of the co-efficient matrix of the primal.
2. Interchanging of the cost vector and requirement vector.
3. Changing of the direction of the sign of inequalities.
4. Minimization of the objective function instead of maximization.

12.5 Important Remark

Initially, we cannot consider both maximization and minimization problem as primal problems. We shall have to stick to one of the problems as primal and after establishing the theorem (12.5.1), we may consider both types of problem as primal problems and the corresponding problem as the dual problem. But we cannot do this before establishing that most vital theorem. Before establishing that theorem, Hadley considered maximization problem as primal but Saul I. Gauss considered minimization problem as primal and they never confused it any where. In this book, except the problem, given in the concept of duality, we shall initially start taking maximization problem as the primal problem.

Theorem 12.5.1 *Dual of the dual is the primal itself.*

Proof. Let the primal problem be,

$$\text{Maximize,} \quad (z_x = \mathbf{cx})$$

subject to

$$A\mathbf{x} \leq \mathbf{b}, \quad \mathbf{x} \geq \mathbf{0}. \tag{12.5.1}$$

The dual of which is, Minimize, $(z_w = \mathbf{b}'\mathbf{w})$ subject to

$$A'\mathbf{w} \geq \mathbf{c}', \quad \mathbf{w} \geq \mathbf{0}. \tag{12.5.2}$$

The problem (12.5.2) is equivalent to the problem,

$$\text{Maximize,} \quad (-\mathbf{b}'\mathbf{w})$$

subject to

$$-A'\mathbf{w} \le -\mathbf{c}', \quad \mathbf{w} \ge \mathbf{0} \tag{12.5.3}$$

[multiplying the constraints of (12.5.2) by (-1)]

$$\text{where} \quad \min(\mathbf{b}'\mathbf{w}) = -\max(-\mathbf{b}'\mathbf{w}). \tag{12.5.4}$$

The dual problem (12.5.3) now "looks like" a primal of type (12.5.1) and hence considering it as a primal, the dual of it can be written as,

$$\text{Minimize,} \quad (-\mathbf{cx})$$

subject to

$$-A\mathbf{x} \ge -\mathbf{b}, \quad \mathbf{x} \ge \mathbf{0}. \tag{12.5.5}$$

Now, (12.5.5) is equivalent to the form,

$$\text{Maximize,} \quad (\mathbf{cx})$$

subject to

$$A\mathbf{x} \le \mathbf{b}, \quad \mathbf{x} \ge \mathbf{0} \tag{12.5.6}$$

[multiplying the constraints of (12.5.5) by -1]

$$\text{where} \quad \min(-\mathbf{cx}) = -\max(\mathbf{cx}). \tag{12.5.7}$$

Now, due to the presence of two negative signs, one before $\max(-\mathbf{b}'\mathbf{w})$ of (12.5.4) and one before $\max(\mathbf{cx})$ of (12.5.7), the maximum value of [if it exists at all] (12.5.6) will also be same as that of (12.5.1).

Hence (12.5.6) is exactly the original problem (12.5.1) and thus proves that 'the dual of the dual is the primal itself.'

From this, we conclude now, that if either problem (12.4.3) or (12.4.4) is considered as a primal then the other will be its dual.

i.e., if we consider the primal be,

$$\text{Minimize,} \quad z = \mathbf{cx}$$

subject to

$$A\mathbf{x} \ge \mathbf{b}, \quad \mathbf{x} \ge \mathbf{0} \tag{12.5.8}$$

then its dual will be,

$$\text{Maximize,} \quad z_w = \mathbf{b}'\mathbf{w}$$

subject to

$$A'\mathbf{w} \le \mathbf{c}', \quad \mathbf{w} \ge \mathbf{0} \tag{12.5.9}$$

and vice-versa.

Note: (1) $\max f(x)$ = Maximum value of $f(x)$.

(2) For verification of the proof, see page 9, *last paragraph*, page 131, 132 and page 222-224 of 'Linear Programming' by Hadley.

► **Example 12.5.1** *Write down the dual of the symmetric primal*

$$\text{maximize,} \quad z = 2x_1 - 3x_2$$

subject to

$$\begin{aligned} x_1 - 4x_2 &\leq 10 \\ -x_1 + x_2 &\leq 3 \\ -x_1 - 3x_2 &\geq 4, \quad x_1 \geq 0, x_2 \geq 0. \end{aligned}$$

The third constraint can be written as

$$x_1 + 3x_2 \leq -4.$$

Hence the problem in the standard form is,

$$\text{maximize,} \quad z_x = (2, -3)[x_1, x_2]$$

subject to

$$\begin{bmatrix} 1 & -4 \\ -1 & 1 \\ 1 & 3 \end{bmatrix} \begin{bmatrix} x_1 \\ x_2 \end{bmatrix} \leq \begin{bmatrix} 10 \\ 3 \\ -4 \end{bmatrix} \text{ and } x_1 \geq 0, x_2 \geq 0.$$

Therefore the dual problem is,

$$\text{minimize,} \quad z_w = (10, 3, -4)[w_1, w_2, w_3]$$

subject to

$$\begin{bmatrix} 1 & -1 & 1 \\ -4 & 1 & 3 \end{bmatrix} \begin{bmatrix} w_1 \\ w_2 \\ w_3 \end{bmatrix} \geq \begin{bmatrix} 2 \\ -3 \end{bmatrix} \text{ and } w_1 \geq 0, w_2 \geq 0, w_3 \geq 0.$$

or the dual problem is,

$$\text{minimize,} \quad z_w = 10w_1 + 3w_2 - 4w_3$$

subject to

$$\begin{aligned} w_1 - w_2 + w_3 &\geq 2 \\ -4w_1 + w_2 + 3w_3 &\geq -3, \quad w_1 \geq 0, w_2 \geq 0, w_3 \geq 0. \end{aligned}$$

Note: In this problem,

$$A = \begin{bmatrix} 1 & -4 \\ -1 & 1 \\ 1 & 3 \end{bmatrix}, \quad \mathbf{b} = \begin{bmatrix} 10 \\ 3 \\ -4 \end{bmatrix} \text{ and } \mathbf{c} = (2, -3)$$

and **b** is unrestricted in sign.

▶ **Example 12.5.2** *Write down the dual of the following problem,*

$$\textit{minimize}, \quad z = 2x_1 + 3x_2 + 4x_3$$

subject to

$$\begin{aligned} x_1 + x_2 + x_3 &\geq 4 \\ -2x_1 + x_2 &\geq 1 \\ 2x_2 - 3x_3 &\leq 2\,, \quad x_1 \geq 0, x_2 \geq 0, x_3 \geq 0. \end{aligned}$$

Expressing the problem in the standard form, it is given by,

$$\text{minimize}, \quad z_x = (2,3,4)[x_1, x_2, x_3]$$

subject to

$$\begin{bmatrix} 1 & 1 & 1 \\ -2 & 1 & 0 \\ 0 & -2 & 3 \end{bmatrix} \begin{bmatrix} x_1 \\ x_2 \\ x_3 \end{bmatrix} \geq \begin{bmatrix} 4 \\ 1 \\ -2 \end{bmatrix}$$

[multiply both sides of the third constraint by −1, to have "≥" as all constraints must have '≥']

Let us consider this problem as the primal problem. Hence the dual problem is,

$$\text{maximize}, \quad z_w = (4,1,-2)[w_1, w_2, w_3]$$

subject to

$$\begin{bmatrix} 1 & -2 & 0 \\ 1 & 1 & -2 \\ 1 & 0 & 3 \end{bmatrix} \begin{bmatrix} w_1 \\ w_2 \\ w_3 \end{bmatrix} \leq \begin{bmatrix} 2 \\ 3 \\ 4 \end{bmatrix} \text{ and } w_1 \geq 0, w_2 \geq 0, w_3 \geq 0$$

or the problem is,

$$\text{maximize}, \quad z_w = 4w_1 + w_2 - 2w_3$$

subject to

$$\begin{aligned} w_1 - 2w_2 &\leq 2 \\ w_1 + w_2 - 2w_3 &\leq 3 \\ w_1 + 3w_3 &\leq 4\,, \quad w_1 \geq 0, w_2 \geq 0, w_3 \geq 0. \end{aligned}$$

Note: This transformation is made directly by using the property, established in the theorem (12.5.1) and the result stated in (12.5.8) and (12.5.9).

▶ **Example 12.5.3** *Using the theory "dual of the dual is the primal" verify this in the following problems.*

(i) Maximize $z = 2x_1 + x_2 - x_3$ subject to

$$\begin{aligned} 4x_1 - x_2 + x_3 &\leq 4 \\ x_1 + 3x_2 + 4x_3 &\leq 8\,, \quad x_1, x_2, x_3 \geq 0. \end{aligned}$$

This is a maximization problem which can be written in the manner,

$$\text{maximize,} \quad z_x = (2, 1, -1)[x_1, x_2, x_3]$$

subject to

$$\begin{bmatrix} 4 & -1 & 1 \\ 1 & 3 & 4 \end{bmatrix} \begin{bmatrix} x_1 \\ x_2 \\ x_3 \end{bmatrix} \leq \begin{bmatrix} 4 \\ 8 \end{bmatrix}, \quad x_1, x_2, x_3 \geq 0.$$

The dual of it is,

$$\text{minimize,} \quad z_w = 4w_1 + 8w_2$$

subject to

$$\begin{aligned} 4w_1 + w_2 &\geq 2 \\ -w_1 + 3w_2 &\geq 1 \\ w_1 + 4w_2 &\geq -1, \quad w_1, w_2 \geq 0. \end{aligned}$$

Now considering the minimization problem as primal; the dual of it,

$$\text{maximize,} \quad 2v_1 + v_2 - v_3$$

subject to

$$\begin{aligned} 4v_1 - v_2 + v_3 &\leq 4 \\ v_1 + 3v_2 + 4v_3 &\leq 8, \quad v_1, v_2, v_3 \geq 0 \end{aligned}$$

which is nothing but the primal problem. Hence the theorem 'dual of the dual is the primal' is verified for the above problem.

(ii) Minimize, $z = 3x_1 + 4x_3$ subject to

$$\begin{aligned} 4x_1 + 2x_2 - x_3 &\geq 12 \\ x_1 + 5x_2 + x_3 &\geq 18 \\ x_1 - x_2 + 7x_3 &\geq 2, \quad x_1, x_2, x_3 \geq 0 \end{aligned}$$

Considering the minimization problem as primal, the dual of it is.

$$\text{maximize,} \quad z_w = 12w_1 + 18w_2 + 2w_3$$

subject to

$$\begin{aligned} 4w_1 + w_2 + w_3 &\leq 3 \\ 2w_1 + 5w_2 - w_3 &\leq 0 \\ -w_1 + w_2 + 7w_3 &\leq 4, \quad w_1, w_2, w_3 \geq 0. \end{aligned}$$

Now considering the maximization problem as primal, the dual of it is,

$$\text{minimize,} \quad 3v_1 + 4v_3$$

subject to

$$\begin{aligned} 4v_1 + 2v_2 - v_3 &\geq 12 \\ v_1 + 5v_2 + v_3 &\geq 18 \\ v_1 - v_2 + 7v_3 &\geq 2, \quad v_1, v_2, v_3 \geq 0. \end{aligned}$$

Thus the theorem has been verified.

(iii) Maximize, $z = 3x_1 + 2x_2$ subject to

$$2x_1 + x_2 \leq 5$$
$$x_1 + x_2 \leq 3\,, \quad x_1, x_2 \geq 0.$$

[C.U.(H)'86]

Considering the maximization problem as the primal, the dual of it is,

$$\text{minimize,} \quad 5w_1 + 3w_2$$

subject to

$$2w_1 + w_2 \geq 3$$
$$w_1 + w_2 \geq 2\,, \quad w_1, w_2 \geq 0.$$

Now considering the minimization problem as primal, the dual of it is,

$$\text{maximize,} \quad 3v_1 + 2v_2$$

subject to

$$2v_1 + v_2 \leq 5$$
$$v_1 + v_2 \leq 3\,, \quad v_1, v_2 \geq 0.$$

Hence the theorem is verified.

12.6 Properties of Symmetric Primal-Dual Problems

12.6.1 Weak Duality Theorem

Theorem 12.6.1 *If* $\mathbf{x_0}$ *be any F.S. to the primal, maximize,* $z_x = \mathbf{cx}$ *subject to* $A\mathbf{x} \leq \mathbf{b}$, $\mathbf{x} \geq \mathbf{0}$ *and* $\mathbf{w_0}$ *be any F.S. to its dual problem, minimize,* $z_w = \mathbf{b'w}$ *subject to* $A'\mathbf{w} \geq \mathbf{c'}$, $\mathbf{w} \geq \mathbf{0}$, *then* $\mathbf{cx_0} \leq \mathbf{b'w_0}$.

Proof. We have for any feasible solution $\mathbf{w_0}$ of the dual,

$$A'\mathbf{w_0} \geq \mathbf{c'}, \quad \text{or,} \quad (A'\mathbf{w_0})' \geq (\mathbf{c'})' \Rightarrow \mathbf{w_0'}A \geq \mathbf{c} \qquad (12.6.1)$$

$\mathbf{x_0}$ be any feasible solution of the primal [n-component column vector.]

Post multiplying (12.6.1) by $\mathbf{x_0}$, we have

$$(\mathbf{w_0'}A)\mathbf{x_0} \geq \mathbf{cx_0} \quad [\because \quad \mathbf{x_0} \geq \mathbf{0}]$$
$$\text{or,} \quad \mathbf{w_0'}(A\mathbf{x_0}) \geq \mathbf{cx_0}$$
$$\text{or,} \quad \mathbf{w_0'b} \geq \mathbf{cx_0} \quad [\because \quad A\mathbf{x_0} \leq \mathbf{b}]$$
$$\text{or,} \quad \mathbf{b'w_0} \geq \mathbf{cx_0} \Rightarrow \mathbf{cx_0} \leq \mathbf{b'w_0} \quad [\because \quad \mathbf{w_0'b} \text{ is a scalar}] \qquad (12.6.2)$$

Hence the theorem is proved.

Note: (1) If $\mathbf{x_0}$ and $\mathbf{w_0}$ be the optimal feasible solution of the primal and dual respectively then Max $z_x \leq$ Min z_w.

(2) The relation (12.6.2) holds good even if $\mathbf{w}_0$ be not feasible, i.e., if $\mathbf{x}_0$ be a feasible solution of primal and $\mathbf{w}_0$ be any solution of the dual (feasible or infeasible) $\mathbf{c}\mathbf{x}_0 \leq \mathbf{b}'\mathbf{w}_0$ which is obvious from the deduction.

(3) This relation is also true for unsymmetric primal problem which will be proved later.

Theorem 12.6.2 *If $\mathbf{x}^*$ and $\mathbf{w}^*$ be any two feasible solutions of the primal, maximize, $z = \mathbf{c}\mathbf{x}$, subject to $A\mathbf{x} \leq \mathbf{b}$, $\mathbf{x} \geq \mathbf{0}$ and the corresponding dual, minimize, $z_w = \mathbf{b}'\mathbf{w}$ subject to $A'\mathbf{w} \geq \mathbf{c}'$, $\mathbf{w} \geq \mathbf{0}$ respectively and $\mathbf{c}\mathbf{x}^* = \mathbf{b}'\mathbf{w}^*$, then $\mathbf{x}^*$ and $\mathbf{w}^*$ are the optimal feasible solutions of the primal and dual respectively.*

Proof. From theorem 10.6.1, for any two F.S. $\mathbf{x}_0$ and $\mathbf{w}_0$ of the primal and dual

$$\mathbf{c}\mathbf{x}_0 \leq \mathbf{b}'\mathbf{w}_0.$$

Then

$$\mathbf{c}\mathbf{x}_0 \leq \mathbf{b}'\mathbf{w}^* \quad [\text{As } \mathbf{w}^* \text{ is a F.S. of the dual}]$$
$$\text{or,} \quad \mathbf{c}\mathbf{x}_0 \leq \mathbf{b}'\mathbf{w}^* = \mathbf{c}\mathbf{x}^* \Rightarrow \mathbf{c}\mathbf{x}_0 \leq \mathbf{c}\mathbf{x}^*$$

from which we get,

$$\max(\mathbf{c}\mathbf{x}) = \mathbf{c}\mathbf{x}^*, \text{ or, } \mathbf{x}^*$$

is an optimal feasible solution of the primal. In the same way we can prove that

$$\min z_w = \min(\mathbf{b}'\mathbf{w}) = \mathbf{b}'\mathbf{w}^*,$$

i.e., $\mathbf{w}^*$ is an optimal feasible solution of the dual.

12.7 Fundamental Duality Theorem (Symmetric Primal-Dual Problem)

Theorem 12.7.1 (a) *If either the primal, maximize, $z_x = \mathbf{c}\mathbf{x}$, subject to $A\mathbf{x} \leq \mathbf{b}$, $\mathbf{x} \geq \mathbf{0}$ or the dual, minimize $z_w = \mathbf{b}'\mathbf{w}$ subject to $A'\mathbf{w} \geq \mathbf{0}$ has a finite optimal solution, then the other problem will also have a finite optimal solution.*

Furthermore, the optimal values of the objective functions in both the problems will be same; i.e., $\max z_x = \min z_w$.

Proof. We first assume that primal has an optimal feasible solution which has been obtained by simplex method.

Let us convert the constraints of the primal in the following form

$$A\mathbf{x} + I_m\mathbf{x}_s = \mathbf{b}, \quad \mathbf{x} \geq \mathbf{0},\ \mathbf{x}_s \geq \mathbf{0}, \tag{12.7.1}$$

where $\mathbf{x}_s$ is a set of m slack variables and I_m is a unit matrix of order m. It is interesting to note that $\mathbf{b}$ is unrestricted in sign as in the original primal problem. Now imagine that an optimal solution has been found without making each component of the requirement vector non-negative (it is possible to solve the problem even if some components of **b** are negative).

Let $\mathbf{x}_B$ be the optimal feasible solution of the primal problem corresponding to the final basis B and let $\mathbf{c}_B$ be the associated cost vector. Therefore,

$$\mathbf{x}_B = B^{-1}\mathbf{b} \tag{12.7.2}$$

and the corresponding optimal value of the objective function

$$\max z_x = \mathbf{c}_B\mathbf{x}_B = \mathbf{c}_B(B^{-1}\mathbf{b}). \tag{12.7.3}$$

Since $\mathbf{x}_B$ is optimal, we have

$$z_j - c_j \geq 0$$

in a maximization problem for all j in the final table. Thus

$$\begin{aligned}
&\mathbf{c}_B\mathbf{y}_j - c_j \geq 0 \quad [\mathbf{y}_j \text{ is the } j\text{-th column vector of the final table}]\\
\text{or,}\ \ &\mathbf{c}_B B^{-1}\mathbf{a}_j \geq \mathbf{c}_j\\
\text{or,}\ \ &\mathbf{c}_B B^{-1}(\mathbf{a}_1, \mathbf{a}_2, \cdots, \mathbf{a}_n, \mathbf{e}_1, \mathbf{e}_2, \cdots, \mathbf{e}_m) \geq (c_1, c_2, \cdots, c_n, \overbrace{0, 0, \cdots, 0}^{m \text{ times}})\\
&[\text{as } \mathbf{e}_1, \mathbf{e}_2 \text{ etc. are slack vectors and 0, 0 etc. are the cost}\\
&\text{components corresponding to the slack vectors}]\\
\text{or,}\ \ &\mathbf{c}_B B^{-1}(A, I_m) \geq (\mathbf{c}, \mathbf{0})
\end{aligned}$$

Equating we get,

$$\mathbf{c}_B B^{-1}A \geq \mathbf{c} \quad \text{and} \quad \mathbf{c}_B B^{-1}I_m \geq \mathbf{0}. \tag{12.7.4}$$

Putting

$$\mathbf{c}_B B^{-1} = \mathbf{w}_0' \geq \mathbf{0},$$

where $\mathbf{w}_0' = (w_1, w_2, \cdots, w_m)$, an m-component row vector, we get from (12.7.4)

$$\begin{aligned}
&\mathbf{w}_0'A \geq \mathbf{c}, && (12.7.5)\\
\text{or,}\ \ &[\mathbf{w}_0'A]' \geq \mathbf{c}' \quad [\text{taking transpose of both sides}] && (12.7.6)\\
\text{or,}\ \ &A'\mathbf{w}_0 \geq \mathbf{c}' && (12.7.7)
\end{aligned}$$

which indicates that $\mathbf{w}_0$ is a feasible solution to the dual problem.

Now we are to prove that $\mathbf{w}_0$ is also an optimal solution to the dual problem.

$$\begin{aligned}
\hat{z}_w &= \mathbf{b}'\mathbf{w}_0 = \mathbf{w}_0'\mathbf{b} \quad [\text{as } \mathbf{b}'\mathbf{w}_0 \text{ is a scalar}] && (12.7.8)\\
&= (\mathbf{c}_B B^{-1})\mathbf{b} = \mathbf{c}_B(B^{-1}\mathbf{b}) && (12.7.9)\\
&= \mathbf{c}_B\mathbf{x}_B = \mathbf{c}\mathbf{x}_B = \max z_x \quad [\because \mathbf{c}_B\mathbf{x}_B = \mathbf{c}\mathbf{x}_B, \mathbf{0}]] && (12.7.10)
\end{aligned}$$

Hence from the theorem (12.6.2) we can conclude that $\mathbf{w}_0$ is the optimal solution to the dual problem and

$$\hat{z}_w = \min z_w = \max z_x \tag{12.7.11}$$

Now similarly starting with the finite optimal value to the dual problem, if it exists, we can prove that primal has also an optimal value of the objective function and

$$\max z_x = \min z_w. \tag{12.7.12}$$

Note: (1) In fact, in solving the problem (primal) by usual simplex method, the slack vectors change to surplus vectors, for which the components of the requirement vector are negative initially.

The theorem can be re-stated in the following manner.

Statement: A feasible solution $\mathbf{x}^*$ to a primal maximization problem with objective function $\mathbf{cx}$ will be optimal, if and only if, there exists feasible solution $\mathbf{w}^*$ to the dual minimization problem with the objective function $\mathbf{b'w}$ such that $\mathbf{cx}^* = \mathbf{b'w}^*$.

Hints of the proof is given. The proof is almost exactly same as given in the above theorem.

Part I: Let $\mathbf{x}^*$ be an optimal feasible solution of the primal. We are to prove that there exists at least one feasible solution set $\mathbf{w}^*$ of the dual such that $\mathbf{cx}^* = \mathbf{b'w}^*$.

As $\mathbf{x}^*$ is an optimal feasible solution of the primal, then from the fundamental theorem of L.P.P. we can say that there exists at least one B.F.S. $\mathbf{x}_B$ corresponding to a basis (which is optimal basis) B which will make $\mathbf{cx}$ maximum and $\max(\mathbf{cx}) = \mathbf{cx}^* = \mathbf{c}_B\mathbf{x}_B$ where $\mathbf{c}_B$ is the associated cost vector corresponding to the basis B.

Now using the above theorem we can say that there exists at least one feasible solution $\mathbf{w}^*$ to the dual problem such that

$$\mathbf{c}_B\mathbf{x}_B = \mathbf{b'w}^* = \mathbf{cx}^*. \tag{12.7.13}$$

Part II. Here we make an assumption that there are two F.S. $\mathbf{x}^*$ and $\mathbf{w}^*$ of the primal and dual problems respectively such that $\mathbf{cx}^* = \mathbf{b'w}^*$ relation holds. We are to prove that $\mathbf{x}^*$ and $\mathbf{w}^*$ are the optimal feasible solutions of the primal and dual respectively. The theorem (12.7.1) is the exact proof of this part.

From the above two theorems we can derive a very important conclusion.

In the duality theory, both the primal and dual will have finite optimal solutions if and only if both problems have at least one feasible solution set.

(2) This theorem is also valid for unsymmetric primal dual problem. But here the dual variables are unrestricted in sign. The proof of the general theorem is too difficult to understand for the students at this stage. Hence the proof is omitted.

or **Part I** is just the theorem 12.7.1 and for **Part II** since the dual of the dual is the primal, then the condition is also sufficient condition.

Theorem 12.7.2 (b) *If either of the primal or dual has unbounded solution then the other will have no feasible solution.*

Proof. The primal problem is said to have an unbounded solution. If the dual problem has finite optimal solution, then the primal will also have finite optimal solution which is a contradiction. We now prove that dual will have no feasible solution.

For unboundedness of the primal objective function, $\max z_x = \max \mathbf{cx} \to \infty$ and since for any feasible solution $\mathbf{w}$ of the dual, $\mathbf{b'w} \geq \max \mathbf{cx}$ which implies that $\mathbf{b'w} \to \infty$ for all feasible solution $\mathbf{w}$ of the dual, which indicates that there is no feasible $\mathbf{w}$ whose components are all finite. Hence the only conclusion is that the dual has no feasible solution.

Note: But the converse statement is not necessarily true. Because if the primal problem has no feasible solution, then the dual will have either no feasible solution or unbounded solution. We cannot say more than that from the duality theory. But if we have previous information that the dual has a feasible solution set then it will have only unbounded solution.

▶ **Example 12.7.1** *Primal problem,*

$$\text{maximize,} \quad z = 2x_1 + 7x_2$$

subject to

$$\begin{aligned} -x_1 + x_2 &\leq 0 \\ x_1 - x_2 &\leq -2\,, \quad x_1, x_2 \geq 0. \end{aligned}$$

Using graphical method, or by simplex method it can be easily verified that the problem has no feasible solution. Now consider its dual problem. The dual of which is,

$$\text{minimize,} \quad z_w = -2w_2$$

subject to

$$\begin{aligned} -w_1 + w_2 &\geq 2 \\ w_1 - w_2 &\geq 7\,, \quad w_1, w_2 \geq 0. \end{aligned}$$

From the graph or by simplex method it can also be shown that the dual has also no feasible solution.

▶ **Example 12.7.2** *Solve the primal problem,*

$$\text{minimize,} \quad z = -2x_1 - 2x_2$$

subject to

$$\begin{aligned} x_1 - 5x_2 &\geq 3 \\ x_1 - 2x_2 &\leq 1\,, \quad x_1, x_2 \geq 0 \end{aligned}$$

by solving its dual problem.

The dual of which is given by,

$$\text{maximize,} \quad z_w = 3w_1 - w_2$$

subject to

$$\begin{aligned} w_1 - w_2 &\le -2 \\ -5w_1 + 2w_2 &\le -2 \\ \text{or,} \quad -w_1 + w_2 &\ge 2 \\ 5w_1 - 2w_2 &\ge 2, \quad w_1, w_2 \ge 0. \end{aligned}$$

The dual problem is said to have an unbounded solution [Solved in Example 9.2.5]. Then the primal problem will have no feasible solution, which can also be verified by the graphical method.

▶ **Example 12.7.3** *Solve the primal problem,*

$$\textit{minimize,} \quad z = 8x_1 - 100x_2$$

subject to

$$\begin{aligned} 2x_1 - 10x_2 &\ge 2 \\ x_1 - 11x_2 &\ge -3, \quad x_1, x_2 \ge 0 \end{aligned}$$

by solving its dual problem.

The dual of it is given by,

$$\text{maximize,} \quad z_w = 2w_1 - 3w_2$$

subject to

$$\begin{aligned} 2w_1 + w_2 &\le 8 \\ -10w_1 - 11w_2 &\le -100, \quad w_1, w_2 \ge 0. \\ \text{or,} \quad 2w_1 + w_2 &\le 8 \\ 10w_1 + 11w_2 &\ge 100, \quad w_1, w_2 \ge 0. \end{aligned}$$

The dual problem is solved in Example 9.2.6. The dual problem has no feasible solution. Thus the primal will have either no F.S. or unbounded solution. We cannot say more than that by using duality theory. Now from the graphical method, it can be shown that the primal problem has a feasible solution set. Hence we can now say definitely that the primal has an unbounded solution.

Theorem 12.7.3 *If the primal problem has feasible solutions and the dual has no feasible solution then the primal problem is said to have unbounded solution and vice versa.*

Proof. Since the primal problem has feasible solutions then the primal objective function must have some value or values corresponding to the feasible solutions and if the values be finite then the primal problem has definite optimal value from which we can conclude that the dual has also finite optimal solution. But dual has no feasible solution and therefore no question of finite optimal value arises. Thus the only conclusion is that the primal has no finite optimal solution, i.e., the primal problem is said to have an unbounded solution.

Below given the results of the observations in a tabular form.

Primal	Dual	Conclusion
Unbounded solution of the primal (dual)	—	No feasible solution of the dual (primal).
No F.S. of the primal (dual).	—	Either unbounded solution or no feasible of the dual (primal).
Bounded optimal solution	—	Dual has also bounded optimal solution and vice versa.
Feasible solution	Feasible solution	Finite optimal values of both exist.
Feasible solution	No feasible solution	Unbounded solution of the primal.
No feasible solution	Feasible solution	Unbounded solution of the dual.
No feasible solution	No feasible solution	No optimal solution of the either problem.

Rules for obtaining the dual optimal solution from the final simplex table of the primal problem and vice versa [for symmetrical primal problem.]

Rule 1. The optimal value of the dual objective function is equal to the optimal value of the primal objective function, i.e., $\min z_w = \max z_x$ assuming that the primal problem is a problem of maximization. The result has been proved in theorem (12.7.1).

Rule 2. The values of $z_j - c_j$ for the columns corresponding to the slack (surplus) vectors in the final simplex table of the primal problem, are the values of the corresponding dual optimal variables and vice versa provided the problem is solved as a maximization problem. The result is very important to find out the optimal dual variables and vice versa.

Proof. $z_j - c_j = z_j \geq 0$ [as the prices or costs of slack (surplus) variables are zero]

$$= \mathbf{c}_B\mathbf{y}_j = \mathbf{c}_B B^{-1}\mathbf{a}_j = \mathbf{c}_B B^{-1}\mathbf{e}_i \; [\mathbf{a}_j = \mathbf{e}_i \text{ for slack vectors}]$$

$z_j - c_j$ for the columns corresponding to slack vectors are

$$= \mathbf{c}_B B^{-1}(\mathbf{e}_1, \mathbf{e}_2, \cdots, \mathbf{e}_m)$$
$$= \mathbf{c}_B B^{-1} I_m = \mathbf{c}_B B^{-1}$$
$$= \mathbf{w}_0' \text{ [by condition (12.7.10) of the theorem (12.7.1)]}$$

General discussions about the method of finding the primal (dual) optimal solution from the final simplex table of the dual (primal) problem.

In the case of symmetric primal dual problem, we have mentioned in Rule 2, that if the primal (dual) problem be solved as a maximization problem by using usual simplex method, i.e., by making $\mathbf{b} \geq 0$ and taking a unit basis as the initial basis (whether the actual problem is a maximization or minimization problem), the values of $z_j - c_j = z_j$ [as $c_j = 0$ for slack and surplus variables] for the columns of the slack and surplus vectors, in the final simplex table, give the corresponding value of the components of the dual (primal) optimal solution. We have nothing to say if all m vectors be slack vectors, which is evident from theorem (12.7.1) and Rule 2. But why are we getting such a magnificent result if the vectors be surplus vectors? Let us try to explain very carefully the theoretical idea behind this important conclusion. Let the constraints of the primal (maximization) problem be

$$A\mathbf{x} \leq \mathbf{b}, \quad \mathbf{x} \geq \mathbf{0},$$
$$\mathbf{b} = [b_1, b_2, \cdots, b_m],$$

$\mathbf{b}$ is unrestricted and let the constraints, after converting all of them into equations, be

$$A\mathbf{x} + I_m\mathbf{x}_s = \mathbf{b}, \quad \mathbf{x} \geq \mathbf{0}, \quad \mathbf{x}_s \geq \mathbf{0},$$

where $\mathbf{x}_s$ is a set of m slack variables and $\mathbf{b}$ remains unrestricted. For simplicity, we assume that only the last component of $\mathbf{b}$, b_m is negative, i.e., — b_m is positive. Now let us try to solve the problem in two ways:

(i) keeping the requirement vector unrestricted in sign as in the original problem and selecting an arbitrary admissible basis B as an initial basis such that

$$B^{-1}\mathbf{b} \geq \mathbf{0},$$

i.e., initial solution is a basic feasible solution. [It is possible and for that see the definition of admissible basis].

(ii) usual method of making requirement vector non-negative by suitable adjustment and selecting a unit basis as the initial basis.

To solve the problem in the second method, the last constraint

$$a_{m1}x_1 + a_{m2}x_2 + \cdots + a_{mn}x_n + \cdots + x_{n+m} = b_m$$

is to be converted in the following way

$$-a_{m1}x_1 - a_{m2}x_2 - \cdots - a_{mn}x_n - x_{n+m} + x_\alpha = -b_m.$$

Here x_α is an artificial variable and the mth slack vector

$$\mathbf{e}_m = [0, 0, \cdots, 1]$$

in the first case changes to a surplus vector $[0, 0, \cdots, -1]$ in the second case and the requirement vector is then

$$[b_1, b_2 \cdots, -b_m] \geq \mathbf{0}.$$

Of course, we may not require to add the artificial variable x_α at all to get the mth unit vector $\mathbf{e}_m$ of the initial unit basis.

Let B_0 and B^* be the optimal basis for the first and second method respectively. As the optimal solutions remain same [assuming that alternative optimal solution is not present] in each case then the optimal solution

$$\mathbf{x}_0^* = B_0^{-1}[b_1, b_2, \cdots . b_m] = B^{*-1}[b_1, b_2, \cdots, -b_m]$$

from which we can write by assuming $b_1 = b_2 = \cdots = b_{m-1} = 0$ and $b_m = 1$, $B_0^{-1}[0, 0, \cdots, 1] = B^{*-1}[0, 0, \cdots, -1]$. The associated cost vector $\mathbf{c}_B$ remains same in the case of optimal basic feasible solution. Thus $\mathbf{c}_B B^{*-1}[0, 0, \cdots, -1] = \mathbf{c}_B B_0^{-1}[0, 0, \cdots, 1]$, which is the mth dual optimal variable and which can be obtained from the $z_j - c_j = z_j = \mathbf{c}_B B^{*-1}[0, 0, \cdots, -1]$ in the final simplex table corresponding to the surplus vector $[0, 0, \cdots, -1]$ in the second method, i.e., in our usual method. This is the mechanism behind this important conclusion and the artificial vector has nothing contribution in finding the mth dual variable. The same rule is also valid, if any minimization problem is solved as a maximization problem.

Rule 3. If either problem has unbounded solution, then the other will have no feasible solution [which we have established in theorem (12.7.1)].

12.8 Importance of the Duality Theory

When the number of constraints are greater than the number of variables, duality theory is very helpful in solving the problem by simplex method. For example, let the number of constraints be five and the number of variables be two in the primal problem. If we try to solve the primal problem by simplex method, then the basis will be a 5×5 square matrix and it requires enough time to compute in each table. But if we convert the problem into its dual we get only two constraints instead of five and the dual problem can be solved easily. Now solving the dual problem we get the optimal value of the objective function of the primal as well as the primal optimal variables. Hence by using duality theory we can solve the problems (sometimes) easily and more quickly. Duality theory has also much economical importance.

▶ **Example 12.8.1** *Solve the problem, minimize, $z = 3x_1 + x_2$ subject to*

$$\begin{aligned} 2x_1 + x_2 &\geq 14 \\ x_1 - x_2 &\geq 4, \quad x_1, x_2 \geq 0. \end{aligned}$$

by solving its dual problem with the help of simplex method. **[C.U.(P)'88]**

The dual of the problem is,

$$\text{maximize,} \quad z_w = 14w_1 + 4w_2$$

subject to

$$\begin{aligned} 2w_1 + w_2 &\leq 3 \\ w_1 - w_2 &\leq 1 . \quad w_1, w_2 \geq 0. \end{aligned}$$

Solving the dual problem by usual simplex method, we have the simplex tables given below are,

		c	14	4	0	0	
Basis	$\mathbf{c}_B$	b	$\mathbf{a}_1$	$\mathbf{a}_2$	$\mathbf{a}_3(\mathbf{e}_1)$	$\mathbf{a}_4(\mathbf{e}_2)$	Min. ratio
$\mathbf{a}_3$	0	3	2	1	1	0	$\frac{3}{2}$
$\mathbf{a}_4^*$	0	1	1*	−1	0	1	$\frac{1}{1}=1^*$
z_j-c_j		0	−14*	−4	0	0	
$\mathbf{a}_3^*$	0	1	0	3*	1	−2	$\frac{1}{3}$
$\mathbf{a}_1$	14	1	1	−1	0	1	...
z_j-c_j		14	0	−18*	0	14	
$\mathbf{a}_2$	4	$\frac{1}{3}$	0	1	$\frac{1}{3}$	$-\frac{2}{3}$	
$\mathbf{a}_1$	14	$\frac{4}{3}$	1	0	$\frac{1}{3}$	$\frac{1}{3}$	
z_j-c_j		20	0	0	6	2	

All $z_j - c_j \geq 0$. Then the solution is optimal.

$$\max z_w = 20 \text{ at } w_1 = \tfrac{4}{3} \text{ and } w_2 = \tfrac{1}{3}.$$

Now

$$z_3 - c_3 = 6$$

$$\text{and } z_4 - c_4 = 2$$

corresponding to the unit slack vectors $\mathbf{a}_3(\mathbf{e}_1)$ and $\mathbf{a}_4(\mathbf{e}_2)$ at the optimal stage.

Hence the primal optimal solution is $x_1 = 6$ and $x_2 = 2$, i.e., $\min z = \max z_w = 20$ at $x_1 = 6$, $x_2 = 2$.

Again, we have the final basis

$$B = (\mathbf{a}_2, \mathbf{a}_1) = \begin{bmatrix} 1 & 2 \\ -1 & 1 \end{bmatrix} \text{ and } B^{-1} = \begin{bmatrix} 1/3 & -2/3 \\ 1/3 & 1/3 \end{bmatrix}$$

which are nothing but $\mathbf{y}_3, \mathbf{y}_4$ vectors under the unit vectors $\mathbf{a}_3, \mathbf{a}_4$, in the final simplex table. Hence the primal optimal variables are

$$\mathbf{x}'_B = \mathbf{c}_B B^{-1} = (4, 14) \begin{bmatrix} 1/3 & -2/3 \\ 1/3 & 1/3 \end{bmatrix} = [6 \quad 2].$$

Hence the primal optimal solution is $x_1 = 6$ and $x_2 = 2$, i.e., $\min z = \max z_w = 20$ at $x_1 = 6, x_2 = 2$.

Verification of the correctness of the computation in each table, starting from the second table and onwards.

We can verify the correctness of the computation of each table with almost perfectly [but not exactly] by using duality theory. In the second table $z_3 - c_3 = 0$,

$z_4 - c_4 = 14$. The components of the requirement vector [' $\leq$' type constraints] of the dual problem is $[3, 1]$. Now $3 \times 0 + 1 \times 14 = 14$ which is the value of the dual objective function of the second table which is shown under the column **b** and in $z_j - c_j$ row. In the third table $z_3 - c_3 = 6$, $z_4 - c_4 = 4$. Now $3 \times 6 + 1 \times 2 = 20$, which is same as shown in the corresponding position in the final table. Thus we may expect a correct calculation [but not with 100% certainty]. But the verification of the result when one or more constraints are equations is not so easy.

▶ **Example 12.8.2** *Solve the L.P.P.,*

$$\textit{maximize,} \quad z_x = 5x_1 + 4x_2$$

subject to

$$\begin{aligned} 3x_1 + 4x_2 &\leq 24 \\ 3x_1 + 2x_2 &\leq 18 \\ x_2 &\leq 5, \quad x_1 \geq 0, x_2 \geq 0. \end{aligned}$$

by simplex method.

Hence find out the minimum value of the objective function of the dual problem and the corresponding dual optimal solution. Verify the result in solving the dual problem by simplex method.

Primal problem is,

$$\text{maximize,} \quad z_x = 5x_1 + 4x_2$$

subject to

$$\begin{aligned} 3x_1 + 4x_2 &\leq 24 \\ 3x_1 + 2x_2 &\leq 18 \\ x_2 &\leq 5. \end{aligned}$$

Introducing three slack variables x_3, x_4 and x_5 we get the following equations:

$$\begin{aligned} 3x_1 + 4x_2 + x_3 \qquad &= 24 \\ 3x_1 + 2x_2 \qquad + x_4 \qquad &= 18 \\ x_2 \qquad\qquad + x_5 &= 5. \end{aligned}$$

$$z_x = 5x_1 + 4x_2 + 0 \cdot x_3 + 0 \cdot x_4 + 0 \cdot x_5.$$

Initial B.F.S.

$$\begin{aligned} \mathbf{x}_B &= [x_3, x_4, x_5] = [24, 18, 5] \\ \mathbf{c}_B &= (c_3, c_4, c_5) = (0, 0, 0), \\ \mathbf{y}_j &= B^{-1}\mathbf{a}_j = \mathbf{a}_j \\ z_B &= \mathbf{c}_B\mathbf{x}_B = 0 \end{aligned}$$

The optimal simplex table is

		c	5	4	0	0	0
Basis	$\mathbf{c}_B$	b	$\mathbf{a}_1$	$\mathbf{a}_2$	$\mathbf{a}_3(\mathbf{e}_1)$	$\mathbf{a}_4(\mathbf{e}_2)$	$\mathbf{a}_5(\mathbf{e}_3)$
$\mathbf{a}_2$	4	3	0	1	$\frac{1}{2}$	$-\frac{1}{2}$	0
$\mathbf{a}_1$	5	4	1	0	$-\frac{1}{3}$	$\frac{2}{3}$	0
$\mathbf{a}_5$	0	2	0	0	$-\frac{1}{2}$	$\frac{1}{2}$	1
$z_j - c_j$		32	0	0	$\frac{1}{3}$	$\frac{4}{3}$	0

$\max z_x = 32$ for $x_1 = 4$ and $x_2 = 3$.

The dual problem is,

$$\text{minimize,} \quad z_w = 24w_1 + 18w_2 + 5w_3$$

subject to

$$\begin{aligned} 3w_1 + 3w_2 + 0 \cdot w_3 &\geq 5 \\ 4w_1 + 2w_2 + \quad w_3 &\geq 4 , \quad w_1 \geq 0, w_2 \geq 0 \text{ and } w_3 \geq 0. \end{aligned}$$

As $\max z_x = \min z_w$ then $\min z_w = 32$.

12.8.1 Determination of the Dual Optimal Variables

There are three dual variables w_1, w_2 and w_3; x_3, x_4 and x_5 are slack variables in the original primal problem; $\mathbf{a}_3, \mathbf{a}_4$ and $\mathbf{a}_5$ are the unit slack vectors $\mathbf{e}_1, \mathbf{e}_2$ and $\mathbf{e}_3$ respectively.

Hence the dual optimal solution is $w_1 = \frac{1}{3}$, $w_2 = \frac{4}{3}$, $w_3 = 0$ corresponding to $z_j - c_j$ $[j = 3, 4, 5]$ in the final simplex table.

Dual problem: Minimize, $z_w = 24w_1 + 18w_2 + 5w_3$ subject to

$$\begin{aligned} 3w_1 + 3w_2 + 0 \cdot w_3 &\geq 5 \\ 4w_1 + 2w_2 + \quad w_3 &\geq 4 , \quad w_j \geq 0[j = 1, 2, 3.] \end{aligned}$$

Introducing two surplus variables w_4 and w_5, we get the following equations

$$\begin{aligned} 3w_1 + 3w_2 + 0 \cdot w_3 - w_4 \qquad &= 5 \\ 4w_1 + 2w_2 + \quad w_3 \qquad - w_5 &= 4. \end{aligned}$$

Now to get a unit matrix, only one artificial variable w_6 is to be added in the first equation and then the converted equations are

$$\begin{aligned} 3w_1 + 3w_2 + 0 \cdot w_3 - w_4 \qquad + w_6 &= 5 \\ 4w_1 + 2w_2 + \quad w_3 \qquad - w_5 \qquad &= 4. \end{aligned}$$

This is a problem of minimization.

Let

$$\begin{aligned} z'_w &= -z_w = -24w_1 - 18w_2 - 5w_3 \\ \min(z_w) &= -\max(z'_w). \end{aligned}$$

Now the problem is a problem of maximization,

$$z'_w = -24w_1 - 18w_2 - 5w_3 + 0 \cdot w_4 + 0 \cdot w_5 - Mw_6$$

[assigning very large negative price to the artificial variable w_6].

Initial (artificial) solution $= [w_6, w_3] = [5, 4]$

Simplex tables

Basis	$\mathbf{c}_B$	c	-24	-18	-5	0	0	$-M$	
		b	$\mathbf{a}_1$	$\mathbf{a}_2$	$\mathbf{a}_3(\mathbf{e}_2)$	$\mathbf{a}_4$	$\mathbf{a}_5$	$\mathbf{a}_6(\mathbf{e}_1)$	Min. ratio
$\mathbf{a}_6$	$-M$	5	3	3	0	-1	0	1	$\frac{5}{3}$
$\mathbf{a}_3^*$	-5	4	4^*	2	1	0	-1	0	$\frac{4}{4} = 1^*$
$z_j - c_j$			$-3M + 4^*$	$-3M + 8$	0	M	5	0	
$\mathbf{a}_6^*$	$-M$	2	0	$\frac{3}{2}^*$	$-\frac{3}{4}$	-1	$\frac{3}{4}$	1	$2/\frac{3}{2} = \frac{4}{3}^*$
$\mathbf{a}_1$	-24	1	1	$\frac{1}{2}$	$\frac{1}{2}$	0	$-\frac{1}{4}$	0	$1/\frac{1}{2} = 2$
$z_j - c_j$			0	$-\frac{3}{2}M + 6^*$	$\frac{3}{4}M + 1$	M	$-\frac{3}{4}M + 6$	0	
$\mathbf{a}_2$	-18	$\frac{4}{3}$	0	1	$-\frac{1}{2}$	$-\frac{2}{3}$	$\frac{1}{2}$		
$\mathbf{a}_1$	-24	$\frac{1}{3}$	1	0	$\frac{1}{2}$	$\frac{1}{3}$	$-\frac{1}{2}$		
$z_j - c_j$		-32	0	0	2	4	3		

As none of $z_j - c_j < 0$, then the solution is optimal. The artificial vector $\mathbf{a}_6$ is driven out from the basis. Hence the solution is a B.F.S. $\max(z'_w) = -32$. Then $\min(z_w) = -\max(z'_w) = 32$ for $w_1 = \frac{1}{3}$, $w_2 = \frac{4}{3}$ and $w_3 = 0$ [non-basic].

Now $\max z_x = \min z_w = 32$. The dual problem is solved as a maximization problem and the values of $z_j - c_j$ corresponding to the surplus vectors $\mathbf{a}_4(-\mathbf{e}_1)$ and $\mathbf{a}_5(-\mathbf{e}_2)$ in the final (third) simplex table are 4 and 3 respectively.

Therefore, $x_1 = 4$, $x_2 = 3$.

Hence the optimal value of primal objective function is $\max z_x = 32$ for $x_1 = 4$, $x_2 = 3$, which is obtained earlier.

▶ **Example 12.8.3** *Write down the dual of the following problem,*

$$\textit{minimize,} \quad z = 30x_1 + 36x_2$$

subject to

$$\begin{aligned} x_1 + x_2 &\geq 5 \\ 2x_1 + 3x_2 &\geq 2 \\ -2x_1 + x_2 &\geq 2, \quad x_1, x_2 \geq 0 \end{aligned}$$

and solving the dual problem find out the optimal solution and the optimal value of the objective function.

The dual problem is, maximize $5w_1 + 2w_2 + 2w_3$ subject to

$$w_1 + 2w_2 - 2w_3 \le 30$$
$$w_1 + 3w_2 + w_3 \le 36, \quad w_1, w_2, w_3 \ge 0.$$

The problem has been solved in Example (8.9.1) and from the final table we get $\max(5w_1 + 2w_2 + 2w_3) = 174$ at $w_1 = 34$, $w_3 = 2$, $w_2 = 0$ and $z_4 - c_4 = 1$, $z_5 - c_5 = 4$. This $\min z = 174$ at $x_1 = 1$, $x_2 = 4$.

► **Example 12.8.4** *Write down the dual of the following problem and solving the dual problem by simplex method find the optimal solutions and optimal values of the primal and dual as well.*

$$\textit{Maximize}, \quad z = 3x_1 + 4x_2$$

subject to

$$x_1 + x_2 \le 10$$
$$2x_1 + 3x_2 \le 18$$
$$x_1 \le 8$$
$$x_2 \le 6, \quad x_1, x_2 \ge 0.$$

[C.U. (Hons.) '87, '94, '96]

The dual problem is

$$\textit{minimize}, \quad z_w = 10w_1 + 18w_2 + 8w_3 + 6w_4$$

subject to

$$w_1 + 2w_2 + w_3 \ge 3$$
$$w_1 + 3w_2 + w_4 \ge 4, \quad w_1, w_2, \cdots, w_4 \ge 0.$$

To solve the problem by simplex method we have noticed that unit matrix I_2 is already present there and two surplus variables are subtracted from each constraint one to each constraint and the problem reduces to

$$\text{maximize}, \quad (-z_w) = -10w_1 - 18w_2 - 8w_3 - 6w_4 + 0w_5 + 0w_6$$

subject to

$$w_1 + 2w_2 + w_3 - w_5 = 3$$
$$w_1 + 3w_2 + + w_4 - w_6 = 4,$$
$$w_1, w_2, \cdots, w_4 \ge 0, w_5, w_6 \text{ (surplus variables)} \ge 0,$$

where Min $(z_w) = -$ Max $(-w) = -$ Max $w^* [w^* = -w]$ with the same solution set.

		c	-10	-18	-8	-6	0	0	
Basis vector	$\mathbf{c}_B$	**b**	$\mathbf{a}_1$	$\mathbf{a}_2$	$\mathbf{a}_3(\mathbf{e}_1)$	$\mathbf{a}_4(\mathbf{e}_2)$	$\mathbf{a}_5(-\mathbf{e}_1)$	$\mathbf{a}_6(-\mathbf{e}_2)$	Min ratio
$\mathbf{a}_3$	-8	3	1	2	1	0	-1	0	$\frac{3}{2}$
$\mathbf{a}_4$	-6	4	1	3*	0	1	0	-1	$\frac{4}{3}$*
$z_j - c_j$		-48	-4	-16*	0	0	8	6	
$\mathbf{a}_3$	-8	$\frac{1}{3}$							
$\mathbf{a}_2$	-18	$\frac{4}{3}$							
		$-\frac{80}{3}$	$\frac{4}{3}$	0	0	$\frac{16}{3}$	8	$\frac{2}{3}$	

In the second table all $z_j - c_j \geq 0$, $j = 1, 2, \cdots, 6$. Thus we reach at the optimal stage Min $z_w = -$ Max $(w^*) = -(-80/3) = 80/3$ at $w_1 = 0$, $w_2 = 4/3$, $w_3 = 1/3$ and $w_4 = 0$ [w_1, w_4 are non-basic variables]. Thus using the property of the duality theory, Max $z =$ Min $z_w = 80/3$ at $x_1 = 8$ and $x_2 = 2/3$ which are the $z_j - c_j$ corresponding to the surplus vectors $\mathbf{a}_5(-\mathbf{e}_1)$ and $\mathbf{a}_6(-\mathbf{e}_2)$ respectively.

Verification of the correctness of the solution

Max $z = 3 \times 8 + 4 \times \dfrac{2}{3} = \dfrac{80}{3}$, which has been obtained earlier.

Remark: Since all $z_j - c_j \geq 0$, $j = 1, 2, \cdots, 6$, we reach at the optimal stage and we need not complete the body of second table which has been shown in the next table.

▶ **Example 12.8.5** *Solve the primal problem,*

$$\text{minimize,} \quad z = 10x_1 + 2x_2$$

subject to

$$\begin{aligned} x_1 + 2x_2 + 2x_3 &\geq 1 \\ x_1 \qquad - 2x_3 &\geq -1 \\ x_1 - \ x_2 + 3x_3 &\geq \ 3, \quad x_1, x_2, x_3 \geq 0 \end{aligned}$$

by solving its dual problem.

The dual of the following problem is

$$\text{maximize,} \quad z_w = w_1 - w_2 + 3w_3$$

subject to

$$\begin{aligned} w_1 + \ w_2 + \ w_3 &\leq 10 \\ 2w_1 \qquad - \ w_3 &\leq \ 2 \\ 2w_1 - 2w_2 + 3w_3 &\leq \ 0, \quad w_1, w_2, w_3 \geq 0. \end{aligned}$$

The final simplex table of the dual is [See solved example 8.11.3]

		c	1	−1	3	0	0	0
Basis	$\mathbf{c}_B$	b	$\mathbf{a}_1$	$\mathbf{a}_2$	$\mathbf{a}_3$	$\mathbf{a}_4(\mathbf{e}_1)$	$\mathbf{a}_5(\mathbf{e}_2)$	$\mathbf{a}_6(\mathbf{e}_3)$
$\mathbf{a}_2$	−1	6	$\frac{1}{5}$	1	0	$\frac{3}{5}$	0	$-\frac{1}{5}$
$\mathbf{a}_5$	0	6	$\frac{14}{5}$	0	0	$\frac{2}{5}$	1	$\frac{1}{5}$
$\mathbf{a}_3$	3	4	$\frac{4}{5}$	0	1	$\frac{2}{5}$	0	$\frac{1}{5}$
$z_j - c_j$		6	$\frac{6}{5}$	0	0	$\frac{3}{5}$	0	$\frac{4}{5}$

From the optimal table max $z = 6$ at $w_1 = 0$, $w_2 = 6$ and $w_3 = 4$. Therefore min $z_w = 6$. Now $z_3 - c_3 = \frac{3}{5}$, $z_4 - c_4 = 0$ and $z_5 - c_5 = \frac{4}{5}$, thus min $z_w = 6$, at $x_1 = \frac{3}{5}, x_2 = 0$ and $x_3 = \frac{4}{5}$.

12.8.2 Unsymmetric primal-dual problem and mixed primal-dual problem

A primal problem expressed in the following form, maximize, $z_x = \mathbf{cx}$ subject to

$$A\mathbf{x} = \mathbf{b}, \quad \mathbf{x} \geq \mathbf{0} \tag{12.8.1}$$

is called an unsymmetric primal problem.

The dual of the problem (12.8.1) is given by, minimize, $z_w = \mathbf{b}'\mathbf{w}$ subject to

$$A'\mathbf{w} \geq \mathbf{c}' \tag{12.8.2}$$

The dual variables $w_1, w_2, \cdots, w_m$ are unrestricted in sign, i.e., the components of $\mathbf{w}$ may be positive or negative or zero. $A', \mathbf{b}'$ and $\mathbf{c}'$ are the transposes of $A, \mathbf{b}$ and $\mathbf{c}$ respectively. We shall now establish the fact that the dual variables will be unrestricted in sign.

Theorem 12.8.1 *Complementary slackness theorem. (Particular case).*

(i) *If a slack or surplus variable x_{n+i} which has been added to the ith primal constraint appears in an optimal basic solution with the optimal basis B, then for the corresponding optimal solution to the dual, ith dual variable vanishes, i.e., $w_i = 0$.*

(ii) *If the legitimate variable x_j appears in an optimal basic solution to the primal, then in the corresponding optimal solution to the dual, the jth dual constraint reduces an equality, that is, the dual slack or surplus variable $w_{m+j} = 0$.*

Proof. (i) x_{n+i} is in the optimal basis. Its activity vector is $\pm\mathbf{e}_i$ and its cost or price is zero. Then

$$0 = z_{n+i} - c_{n+i} = z_{n+i} = \mathbf{c}_B B^{-1}(\pm\mathbf{e}_i) = \pm(\mathbf{c}_B B_i^{-1})_i = \pm w_i. \tag{12.8.3}$$

(ii) Since x_j is the optimal basis, the $z_j - c_j = 0$ at the optimal stage. Now

$$z_j - c_j = \mathbf{c}_B B^{-1}\mathbf{a}_j - c_j = \mathbf{w}'\mathbf{a}_j - c_j = 0,$$

which implies that $\mathbf{w}$ satisfies the jth dual constraint as a strict equality from which we can conclude that the corresponding dual slack or surplus variable $w_{m+j} = 0$.

Remark: These proofs used hold for any type of primal and its corresponding dual problem.

Theorem 12.8.2 *Complementary slackness theorem. (Modified form).*

Statement of the theorem : For any pair of optimal solutions to the primal, Maximize, $z = \mathbf{cx}$ subject to

$$A\mathbf{x} \leq \mathbf{b}, \quad \mathbf{x} \geq \mathbf{0} \quad [A \text{ being an } m \times n \text{ matrix}] \tag{12.8.4}$$

and its dual, Minimize, $z_w = \mathbf{b}'\mathbf{w}$ subject to

$$A'\mathbf{w} \geq \mathbf{c}', \quad \mathbf{w} \geq \mathbf{0} \tag{12.8.5}$$

$w_i x_{n+i} = 0$, $i = 1, 2, \cdots, m$; $x_j\ w_{m+j} = 0$, $j = (1, 2, \cdots, n)$, i.e., or (i) The product of ith variable (legitimate) of the dual and ith slack variable of the primal vanishes, $i = 1, 2, \cdots, m$.

(ii) The product of the jth legitimate variable of the primal and jth surplus variable of the dual vanishes, $j = 1, 2, \cdots, n$.

Proof. After adding a set of m slack variables $\mathbf{x}_s$ to the constraints of (12.8.4) we have

$$A\mathbf{x} + I_m\mathbf{x}_s = \mathbf{b}. \tag{12.8.6}$$

Pre-multiplying (12.8.6) by $\mathbf{w}' \geq \mathbf{0}$, we get

$$\mathbf{w}'(A\mathbf{x}) + \mathbf{w}'\mathbf{x}_s = \mathbf{w}'\mathbf{b}. \tag{12.8.7}$$

Now if $[\mathbf{x}_0, \mathbf{x}_s]$ is an optimal solution to the primal and $\mathbf{w}_0$ and optimal solution to the dual we have from theorem (12.6.1) and (12.6.2).

Max $z = \mathbf{w}_0'(A\mathbf{x}_0) = $ Min $z_w = \mathbf{w}_0'\mathbf{b}$ from which we get

$$\mathbf{w}'\mathbf{x}_s = 0. \tag{12.8.8}$$

Thus $\sum_{i=1}^{m} w_i \cdot x_{n+i} = 0$; since $w_i \geq 0$ and $x_{n+i} \geq 0$ then

$$w_i x_{n+i} = 0, \quad i = 1, 2, \cdots, m. \tag{12.8.9}$$

(ii) Again after subtracting a set of n surplus variables $\mathbf{w}_s$ from the constraints of (12.8.5) we get

$$A'\mathbf{w} - I_n\mathbf{w}_s = \mathbf{c}'. \tag{12.8.10}$$

Pre-multiplying (12.8.10) by $\mathbf{x}', \mathbf{x}' \geq \mathbf{0}$, we get

$$\mathbf{x}'(A'\mathbf{w}) - I_n\mathbf{x}'\mathbf{w}_s = \mathbf{x}'\mathbf{c}'. \tag{12.8.11}$$

Since all terms of (12.8.1) are scalars, we have

$$\mathbf{w}'(A\mathbf{x}) - \mathbf{x}'\mathbf{w}_s = \mathbf{c}\mathbf{x} \tag{12.8.12}$$

$$[\because \mathbf{x}'(A'\mathbf{w}) = (A\mathbf{x})'\mathbf{w} = \mathbf{w}'(A\mathbf{x})]$$

As $\mathbf{x}_0$ and $[\mathbf{w}_0, \mathbf{w}_s]$ are the optimal solution of the primal and dual, then again, we have

$$\mathbf{w}_0'(A\mathbf{x}_0) = \mathbf{c}\mathbf{x}_0 \Rightarrow \mathbf{x}_0'\mathbf{w}_s = 0. \tag{12.8.13}$$

Since $x_j \geq 0$, $w_{m+j} \geq 0$, then

$$x_j \cdot w_{m+j} = 0, \quad j = 1, 2, \cdots, n, \cdots. \tag{12.8.14}$$

Deductions: From (12.8.11) if $x_{n+i} \neq 0$, then $w_i = 0$ and conversely if $w_i \neq 0$, then $x_{n+i} = 0$. From the deduction, we can state the theorem in the manner.

(i) If a slack variable x_{n+i} has been added to the ith constraint of (12.8.4) is different from zero in any optimal solution of (12.8.4) then the ith dual variable w_i of (12.8.5) will be zero in every optimal solution. Conversely,

(ii) If w_i is different from zero in any optimal solution in (12.8.5), then $x_{n+i} = 0$ in every optimal solution of (12.8.4).

Remark: This statement holds good for any optimal solutions to (12.8.4) and (12.8.5) not merely for basic optimal solutions.

(iii) From (12.8.12) $x_j > 0 \Rightarrow w_{m+j} = 0$, i.e., if any legitimate variable be strictly positive at the optimal stage, then the corresponding dual surplus variable be zero which implies that the corresponding dual constraint is strictly an equality and vice-versa.

(iv) For an optimal solution of the primal if ith constraint be a strict inequality, then the ith slack variable $x_{n+i} > 0$ and hence from (12.8.10) we can conclude that $w_i = 0$ and vice-versa.

12.9 Mixed type problem

When some constraints are equations and some of the variables unrestricted in sign the problem is called as mixed type problem.

12.9.1 Variable unrestricted in sign

The difference of two non-negative variables is a variable which is unrestricted in sign. Let x_1 and x_2 be two non-negative variables. The difference of these two variables is a variable x_3 which is unrestricted in sign. If $x_1 > x_2$, then $x_3 > 0$; if $x_1 < x_2$, then $x_3 < 0$ and if $x_1 = x_2$ then $x_3 = 0$.

12.9.2 Expression of an equation into a set of two inequations

Let

$$\sum_{j=1}^{n} a_{ij}x_j = b_i$$

be a linear equation. It can be expressed as the combination of the two different inequations.

$$\sum_{j=1}^{n} a_{ij}x_j = b_i$$

is equivalent to the intersection of the sets

$$\left.\begin{aligned} &\sum_{j=1}^{n} a_{ij}x_j \le b_i \\ \text{and } &\sum_{j=1}^{n} a_{ij}x_j \ge b_i \end{aligned}\right\} \tag{12.9.1}$$

$$\left.\begin{array}{l} \text{or } \sum_{j=1}^{n} a_{ij}x_j \leq b_i \\ \text{and } \sum_{j=1}^{n} (-a_{ij}x_j) \leq -b_i \end{array}\right\} . \qquad (12.9.2)$$

Theorem 12.9.1 *If the k-th constraint of a primal be an equation, the k-th dual variable will be unrestricted in sign.*

Before establishing the theorem in general let us first consider a numerical example to give a clear idea about the theorem.

Let the primal problem be, maximize, $z_x = 2x_1 + 3x_2 - x_3$ subject to

$$\begin{aligned} x_1 + 4x_2 + 2x_3 &\leq 7 \\ x_1 - 3x_2 + x_3 &= 4\,, \quad x_1, x_2, x_3 \geq 0. \end{aligned}$$

Here the second constraint is an equation. We are to establish the fact that the second dual variable w_2 will be unrestricted in sign. The second constraint (equation) can be written by using (12.9.1) and (12.9.2) as follows:

$$\begin{aligned} x_1 - 3x_2 + x_3 &\leq 4 \\ x_1 - 3x_2 + x_3 &\geq 4 \\ \text{or} \quad x_1 - 3x_2 + x_3 &\leq 4 \\ -x_1 + 3x_2 - x_3 &\leq -4 \end{aligned}$$

Hence ultimately there are three inequations expressed in the standard form

$$\begin{aligned} x_1 + 4x_2 + 2x_3 &\leq 7 \\ x_1 - 3x_2 + x_3 &\leq 4 \\ -x_1 + 3x_2 - x_3 &\leq -4 \end{aligned}$$

If three dual variables are $w_1 \geq 0$, $w_2' \geq 0$ and $w_2'' \geq 0$ then by using the rule given in (12.4.2) the dual of the problem is given by,

$$\text{minimize,} \quad z_w = 7w_1 + 4w_2' - 4w_2''$$

subject to

$$\begin{aligned} w_1 + w_2' - w_2'' &\geq 2 \\ 4w_1 - 3w_2' + 3w_2'' &\geq 3 \\ 2w_1 + w_2' - w_2'' &\geq -1\,, \quad w_1 \geq 0, w_2' \geq 0 \text{ and } w_2'' \geq 0. \end{aligned}$$

Putting $w_2' - w_2'' = w_2$, the problem can be written as, minimize, $z_w = 7w_1 + 4w_2$

$$\begin{aligned} w_1 + w_2 &\geq 2 \\ 4w_1 - 3w_2' &\geq 3 \\ 2w_1 + w_2 &\geq -1 \end{aligned}$$

and $w_1 \geq 0$, and w_2 is unrestricted in sign as w_2 is the difference of two non-negative variables.

We can also write at once the dual of the primal (unsymmetrical) following the rule of transformation given in (12.8.2).

Note: Similarly, if all m constraints of a primal be equations, all m dual variables will be unrestricted in sign.

General proof: As the k-th constraint of the primal is an equation, therefore the primal in the standard form can be written as,

$$\text{maximize,} \quad z_x = c_1x_1 + c_2x_2 + \cdots + c_nx_n$$

subject to

$$\begin{array}{l}
a_{11}x_1 + a_{12}x_2 + \cdots + a_{1n}x_n \le b_1 \\
a_{21}x_1 + a_{22}x_2 + \cdots + a_{2n}x_n \le b_2 \\
\cdots\cdots\cdots\cdots\cdots\cdots\cdots\cdots\cdots\cdots \\
a_{k1}x_1 + a_{k2}x_2 + \cdots + a_{kn}x_n \le b_k \\
-a_{k1}x_1 - a_{k2}x_2 - \cdots - a_{kn}x_n \le -b_k \\
a_{k+1,1}x_1 + a_{k+1,2}x_2 + \cdots + a_{k+1,n}x_n \le b_{k+1} \\
\cdots\cdots\cdots\cdots\cdots\cdots\cdots\cdots\cdots\cdots \\
a_{m1}x_1 + a_{m2}x_2 + \cdots + a_{mn}x_n \le b_m
\end{array}$$

$x_1, x_2, \cdots, x_n \ge 0.$

The dual of the above primal can be written as,

$$\text{minimize,} \ z_w = b_1w_1 + b_2w_2 + \cdots + b_k(w_k' - w_k'') + b_{k+1}w_{k+1} + \cdots + b_mw_m$$

subject to

$$\begin{array}{l}
a_{11}w_1 + a_{21}w_2 + \cdots + a_{k1}(w_k' - w_k'') + a_{k+1,1}w_{k+1} + \cdots + a_{m1}w_m \ge c_1 \\
a_{12}w_1 + a_{22}w_2 + \cdots + a_{k2}(w_k' - w_k'') + a_{k+1,2}w_{k+1} + \cdots + a_{m2}w_m \ge c_2 \\
\cdots\cdots\cdots\cdots\cdots\cdots\cdots\cdots\cdots\cdots\cdots\cdots\cdots\cdots \\
a_{1n}w_1 + a_{2n}w_2 + \cdots + a_{kn}(w_k' - w_k'') + a_{k+1,n}w_{k+1} + \cdots + a_{mn}w_m \ge c_n.
\end{array}$$

Putting $w_k' - w_k'' = w_k$, the dual problem can be written as,

$$\text{minimize,} \ z_w = b_1w_1 + b_2w_2 + \cdots + b_kw_k + \cdots + b_mw_m$$

subject to

$$\begin{array}{l}
a_{11}w_1 + a_{21}w_2 + \cdots + a_{k1}w_k + \cdots + a_{m1}w_m \ge c_1 \\
a_{12}w_1 + a_{22}w_2 + \cdots + a_{k2}w_k + \cdots + a_{m2}w_m \ge c_2 \\
\cdots\cdots\cdots\cdots\cdots\cdots\cdots\cdots\cdots\cdots \\
a_{1n}w_1 + a_{2n}w_2 + \cdots + a_{kn}w_k + \cdots + a_{mn}w_m \ge c_n
\end{array}$$

and w_1, w_2, w_{k-1}, w_{k+1}, $\cdots$, $w_m \ge 0$, but w_k is unrestricted in sign as w_k is expressed as the difference of two non-negative variables, w_k' and w_k''. Hence the theorem is proved.

Theorem 12.9.2 *If any variable of the primal problem be unrestricted in sign, the corresponding dual constraint will be strictly an equation.*

Proof. Let the k-th primal variable x_k be unrestricted in sign. Then putting $x_k = x_k' - x_k''$, where x_k' and x_k'' are non-negative variables, the primal problem can be written as,

$$\text{maximize,} \quad z_x = c_1x_1 + c_2x_2 + \cdots + c_k(x_k' - x_k'') + \cdots + c_nx_n$$

subject to

$$\begin{array}{l} a_{11}x_1 + a_{12}x_2 + \cdots + a_{1k}(x'_k - x''_k) + \cdots + a_{1n}x_n \le b_1 \\ a_{21}x_1 + a_{22}x_2 + \cdots + a_{2k}(x'_k - x''_k) + \cdots + a_{2n}x_n \le b_2 \\ \cdots\cdots\cdots\cdots\cdots\cdots\cdots\cdots\cdots\cdots\cdots\cdots \\ a_{m1}x_1 + a_{m2}x_2 + \cdots + a_{mk}(x'_k - x''_k) + \cdots + a_{mn}x_n \le b_m \end{array}$$

and $x_1, x_2, \cdots, x_{k-1}, x'_k, x''_k, \cdots, x_n \ge 0$.

The dual of the above problem is given by,

$$\text{minimize,} \quad z_w = b_1w_1 + b_2w_2 + \cdots + b_mw_m$$

subject to

$$\begin{array}{l} a_{11}w_1 + a_{21}w_2 + \cdots + a_{m1}w_m \ge \quad c_1 \\ a_{12}w_1 + a_{22}w_2 + \cdots + a_{m2}w_m \ge \quad c_2 \\ \cdots\cdots\cdots\cdots\cdots\cdots\cdots\cdots\cdots \\ \left\{ \begin{array}{l} a_{1k}w_1 + a_{2k}w_2 + \cdots + a_{mk}w_m \ge \quad c_k \\ -a_{1k}w_1 - a_{2k}w_2 - \cdots - a_{mk}w_m \ge -c_k \end{array} \right\} \\ \cdots\cdots\cdots\cdots\cdots\cdots\cdots\cdots\cdots \\ a_{1n}w_1 + a_{2n}w_2 + \cdots + a_{mn}w_m \ge \quad c_n \end{array}$$

and $w_1, w_2, \cdots, w_m \ge 0$.

The two constraints under the bracket { } are equivalent to an equation

$$a_{1k}w_1 + a_{2k}w_2 + \cdots + a_{mk}w_m = c_k.$$

Hence the k-th dual constraint is an equation which proves the theorem.

► **Example 12.9.1** *Find out the dual of the problem, maximize,* $z_x = 2x_1 + 3x_2 - 4x_3$ *subject to*

$$\begin{array}{rl} 3x_1 + x_2 + x_3 &\le 2 \\ -4x_1 \qquad + 3x_3 &\ge 4 \\ x_1 - 5x_2 + x_3 &= 5 \end{array}$$

and $x_1 \ge 0$, $x_2 \ge 0$ *and* x_3 *is unrestricted in sign.* **[C.U., M.Com.'90] [C.U.(P)'93]**

Putting $x_3 = x'_3 - x''_3$ where both x'_3 and x''_3 are non-negative variables, the primal problem can be written in the standard form as,

$$\text{maximize,} \quad z_x = 2x_1 + 3x_2 - 4(x'_3 - x''_3)$$

subject to

$$\begin{array}{rl} 3x_1 + x_2 + (x'_3 - x''_3) &\le 2 \\ 4x_1 \qquad - 3(x'_3 - x''_3) &\le -4 \\ x_1 - 5x_2 + (x'_3 - x''_3) &\le 5 \\ -x_1 + 5x_2 - (x'_3 - x''_3) &\le -5 \qquad x_1 \ge 0, x_2 \ge 0, x'_3 \ge 0, x''_3 \ge 0. \end{array}$$

Remark. Note that the problem is now reduced to a symmetric primal problem.

The dual of above primal problem can be written as,

$$\text{minimize,} \quad z_w = 2w_1 - 4w_2 + 5w_3' - 5w_3''$$

subject to

$$\begin{aligned} 3w_1 + \ 4w_2 + (w_3' - w_3'') &\geq 2 \\ w_1 - 0 \cdot w_2 - 5(w_3' - w_3'') &\geq 3 \end{aligned}$$

$$\begin{Bmatrix} w_1 - 3w_2 + (w_3' - w_3'') \geq -4 \\ -w_1 + 3w_2 - (w_3' - w_3'') \geq \ \ 4 \end{Bmatrix}$$

and $w_1, w_2, w_3', w_3'' \geq 0$.

Putting $w_3' - w_3'' = w_3$ and expressing the last two constraints in the form of an equation, the dual problem is,

$$\text{minimize,} \quad z_w = 2w_1 - 4w_2 + 5w_3$$

subject to

$$\begin{aligned} 3w_1 + 4w_2 + \ w_3 &\geq \ \ 2 \\ w_1 \qquad\quad - 5w_3 &\geq \ \ 3 \\ w_1 - 3w_2 + \ w_3 &= -4 \end{aligned}$$

and $w_1 \geq 0, w_2 \geq 0$ and w_3 is unrestricted is sign.

The dual of the problem can be written at once if we follow the rules established in theorem (12.9.1) and (12.9.2).

(i) As the primal variable x_3 is unrestricted in sign, then the third dual constraint will be an equation {by Theorem (12.9.2)}.

(ii) As the third constraint of the primal is an equation, then the third dual variable will be unrestricted in sign {by Theorem (12.9.1)}.

Now transforming the problem in the form,

$$\text{maximize,} \quad z_w = 2x_1 + 3x_2 - 4x_3$$

subject to

$$\begin{aligned} 3x_1 + \ x_2 + \ x_3 &\leq \ \ 2 \\ 4x_1 \qquad\quad - 3x_3 &\leq -4 \\ x_1 - 5x_2 + \ x_3 &= \ \ 5 \end{aligned} \qquad \text{[all inequations will be of "}\leq\text{" type]}$$

and $x_1 \geq 0, x_2 \geq 0$ and x_3 is unrestricted is sign.

The dual of it can be written as,

$$\text{minimize,} \quad z_w = 2w_1 - 4w_2 + 5w_3$$

subject to

$$\begin{aligned} 3w_1 + 4w_2 + \ w_3 &\geq \ \ 2 \\ w_1 \qquad\quad - 5w_3 &\geq \ \ 3 \\ w_1 - 3w_2 + \ w_3 &= -4 \end{aligned}$$

$w_1 \geq 0, w_2 \geq 0$ and w_3 is unrestricted in sign.

▶ **Example 12.9.2** *Find the dual of the following primal problem,*

$$\textit{maximize,} \quad z = x_1 + 4x_2 + 3x_3$$

subject to

$$\begin{aligned} 2x_1 + 3x_2 - 5x_3 &\le 2 \\ 3x_1 - x_2 + 6x_3 &\ge 1 \\ x_1 + x_2 + x_3 &= 4. \end{aligned}$$

$x_1, x_2 \ge 0$, x_3 *is unrestricted.* **[C.U.(P)'92]**

Solution. Since x_3 is unrestricted then $x_3 = x_3' - x_3''$, x_3', $x_3'' \ge 0$ and third constraint is equation which can be written in the manner

$$\begin{aligned} x_1 + x_2 + x_3 &\le 4, \\ x_1 + x_2 + x_3 &\ge 4. \end{aligned}$$

The primal problem after putting the value of x_3, is

$$\text{maximize,} \quad z = x_1 + 4x_2 + 3x_3' - 3x_3''$$

subject to

$$\begin{aligned} 2x_1 + 3x_2 - 5x_3' + 5x_3'' &\le 2 \\ -3x_1 + x_2 - 6x_3' + 6x_3'' &\le -1 \\ x_1 + x_2 + x_3' - x_3'' &\le 4 \\ -x_1 - x_2 - x_3' + x_3'' &\le -4 \end{aligned}$$

$x_1, x_2, x_3', x_3'' \ge 0.$

The dual problem is,

$$\text{minimize,} \quad z_w = 2w_1 - w_2 + 4(w_3 - w_4)$$

subject to

$$\begin{aligned} 2w_1 - 3w_2 + (w_3 - w_4) &\ge 1 \\ 3w_1 + w_2 + (w_3 - w_4) &\ge 4 \\ \left.\begin{aligned} -5w_1 - 6w_2 + (w_3 - w_4) &\ge 3 \\ 5w_1 + 6w_2 - (w_3 - w_4) &\ge -3 \end{aligned}\right\} \end{aligned}$$

$w_1, w_2, w_3, w_4 \ge 0.$

Putting $w_3 - w_4 = w_3^0$ (which is unrestricted) the dual problem is

$$\text{maximize,} \quad z_w = 2w_1 - w_2 + 4w_3^0$$

subject to

$$\begin{aligned} 2w_1 - 3w_2 + w_3^0 &\ge 1 \\ 3w_1 + w_2 + w_3^0 &\ge 4 \\ -5w_1 - 6w_2 + w_3^0 &= 3\,, \quad w_1 \ge 0, w_2 \ge 0, w_3^0 \text{ is unrestricted.} \end{aligned}$$

Remark: As the 3rd variable x_3 is unrestricted then the 3rd constraint is an equation and as the 3rd constraint of the primal is unrestricted then the 3rd dual variable is unrestricted in sign.

▶ **Example 12.9.3** *Set up the dual of the primal problem,*

$$\text{minimize,}\quad z = x_1 - 2x_2 + x_3$$

subject to

$$\begin{aligned} 3x_1 - x_2 + 5x_3 &\le 17 \\ x_1 + 2x_2 - x_3 &= 25 \\ 2x_1 - x_2 + 5x_3 &\ge 57\,, \quad x_1 \ge 0, x_2 \ge 0 \end{aligned}$$

x_3 is unrestricted in sign. **[C.U.(P)'94]**

Solution. Since x_3 is unrestricted then $x_3 = x_3' - x_3''$ and $x_3', x_3'' \ge 0$. Now as the 2nd constraint is equation then we can write the 2nd constraint in the manner

$$\begin{aligned} x_1 + 2x_2 - x_3 &\le 25, \\ x_1 + 2x_2 - x_3 &\ge 25. \end{aligned}$$

Now putting the value of x_3 and since it is a minimization problem, primal can be written

$$\text{minimize,}\quad z = x_1 - 2x_2 + x_3' - x_3''$$

subject to

$$\begin{aligned} -3x_1 + x_2 - 5x_3' + 5x_3'' &\ge -17 \\ -x_1 - 2x_2 + x_3' - x_3'' &\ge -25 \\ x_1 + 2x_2 - x_3' + x_3'' &\ge 25 \\ 2x_1 - x_2 + 5x_3' - 5x_3'' &\ge 57\,, \quad x_1, x_2, x_3', x_3'' \ge 0. \end{aligned}$$

The dual of it is

$$\text{maximize,}\quad z_w = -17w_1 - 25w_2 + 25w_3 + 57w_4$$

subject to

$$\begin{aligned} -3w_1 - w_2 + w_3 + 2w_4 &\le 1 \\ w_1 - 2w_2 + 2w_3 - w_4 &\le -2 \\ \left.\begin{aligned} -5w_1 + w_2 - w_3 + 5w_4 &\le 1 \\ 5w_1 - w_2 + w_3 - 5w_4 &\le -1 \end{aligned}\right\} \end{aligned}$$

w_1, w_2, w_3, $w_4 \ge 0$, which can be written as

$$\text{maximize,}\quad z_w = -17w_1 - 25(w_2 - w_3) + 57w_4$$

subject to

$$\begin{aligned} -3w_1 - (w_2 - w_3) + 2w_4 &\le 1 \\ w_1 - 2(w_2 - w_3) - w_4 &\le -2 \\ -5w_1 + (w_2 - w_3) + 5w_4 &= 1\,, \quad w_1, w_2, w_3 w_4 \ge 0. \end{aligned}$$

Putting $w_2 - w_3 = w_2^0$ the problem is

$$\text{maximize,}\quad z_w = -17w_1 - 25w_2^0 + 57w_4$$

subject to

$$\begin{aligned} -3w_1 - w_2^0 + 2w_4 &\leq 1 \\ w_1 - 2w_2^0 - w_4 &\leq -2 \\ -5w_1 + w_2^0 - 5w_4 &= 1, \quad w_1 \geq 0, w_4 \geq 0 \end{aligned}$$

and w_2^0 is unrestricted in sign.

▶ **Example 12.9.4** *Find out the dual of the problem,*

$$\textit{minimize,} \quad z = 2x_1 - x_2 + 3x_3$$

subject to

$$\begin{aligned} -x_1 + 2x_2 - 4x_3 &\leq 2 \\ 2x_1 - 3x_2 + x_3 &= 6 \end{aligned}$$

x_1 *is unrestricted* $x_2 \geq 0, x_3 \geq 0.$

Solution. (1) In this problem first constraint is to be converted in the manner

$$x_1 - 2x_2 + 4x_3 \geq -2$$

[as in the minimization problem, all inequations must be "$\geq$"]. (2) The primal has two constraints with three variables. Thus the dual will contain three constraints with two variables. (3) In the primal problem x_1 is unrestricted. Therefore first dual constraint will be an equation. (4) In the primal second constraint is an equation. Therefore, the second dual variable will be unrestricted in sign. The constraints are

$$\begin{aligned} x_1 - 2x_2 + 4x_3 &\geq -2 \\ 2x_1 - 3x_2 + x_3 &= 6. \end{aligned}$$

The dual problem is thus,

$$\text{maximize,} \quad z_w = -2w_1 + 6w_2$$

subject to

$$\begin{aligned} w_1 + 2w_2 &= 2 \\ -2w_1 - 3w_2 &\leq -1 \\ 4w_1 + w_2 &\leq 3, \quad w_1 \geq 0, w_2 \text{ unrestricted.} \end{aligned}$$

▶ **Example 12.9.5** *Find the dual of the problem,*

$$\textit{maximize,} \quad z = 2x_1 + x_3$$

subject to

$$\begin{aligned} 4x_1 - 5x_2 + x_3 &= 0 \\ x_1 + 2x_2 + 3x_3 &\leq 7, \quad x_1 \geq 0, x_2 \geq 0, \end{aligned}$$

x_3 *is unrestricted.*

Using the property of the theorems (12.9.1) and (12.9.2) the dual problem can be written at once as given by,

$$\text{minimize,} \quad z_w = 0 \cdot w_1 + 7w_2$$

subject to

$$\begin{aligned} 4w_1 + w_2 &\geq 2 \\ -5w_1 + 2w_2 &\geq 0 \\ w_1 + 3w_2 &= 1 \end{aligned}$$

w_1 is unrestricted, $w_2 \geq 0$.

▶ **Example 12.9.6** *Solve the L.P.P.*

$$\textit{Maximize,} \quad z_x = 4x_1 + 3x_2$$

subject to

$$\begin{aligned} x_1 \qquad &\leq 6 \\ x_2 &\leq 8 \\ x_1 + x_2 &\leq 7 \\ 3x_1 + x_2 &\leq 15 \\ -x_2 &\leq 1, \quad x_1, x_2 \geq 0 \end{aligned}$$

by using duality theory.

We shall first solve the dual problem of it.

The dual of the given problem is,

$$\text{minimize,} \quad z_w = 6w_1 + 8w_2 + 7w_3 + 15w_4 + w_5$$

subject to

$$\begin{aligned} w_1 + 0 \cdot w_2 + w_3 + 3w_4 + 0 \cdot w_5 &\geq 4 \\ 0 \cdot w_1 + w_2 + w_3 + w_4 - w_5 &\geq 3, \quad w_1, w_2, w_3, w_4, w_5 \geq 0. \end{aligned}$$

Let

$$-z_w = z_w^* = -6w_1 - 8w_2 - 7w_3 - 15w_4 - w_5.$$

Introducing the surplus variables w_6 and w_7 one to each of the constraints, the converted equations are

$$\begin{aligned} w_1 + 0 \cdot w_2 + w_3 + 3w_4 + 0 \cdot w_5 - w_6 \qquad &= 4 \\ 0 \cdot w_1 + w_2 + w_3 + w_4 - w_5 \qquad - w_7 &= 3. \end{aligned}$$

The problem is a problem of maximization of z_w^* and $\min(z_w) = -\max(z_w^*)$.

No artificial vector is required to get a unit basis. $\mathbf{a}_1(\mathbf{e}_1)$ and $\mathbf{a}_2(\mathbf{e}_2)$ constitute the basis. Initial B.F.S. $\mathbf{w}_B = [w_1, w_2] = [4, 3]$

$$\begin{aligned} \mathbf{c}_B &= (c_1, c_2) = (-6, -8) \\ z^* &= \mathbf{c}_B \mathbf{w}_B = -48. \end{aligned}$$

Simplex Tables

		c	−6	−8	−7	−15	−1	0	0	
Basis	c_B	b	$\mathbf{a}_1$	$\mathbf{a}_2(\mathbf{e}_2)$	$\mathbf{a}_3(\mathbf{e}_1)$	$\mathbf{a}_4$	$\mathbf{a}_5$	$\mathbf{a}_6(-\mathbf{e}_1)$	$\mathbf{a}_7(-\mathbf{e}_2)$	Min. ratio
$\mathbf{a}_1^*$	−6	4	1	0	1	3*	0	−1	0	$\frac{4}{3}^*$
$\mathbf{a}_2$	−8	3	0	1	1	1	−1	0	−1	$\frac{3}{1}$
$z_j - c_j$		−48	0	0	−7	−11*	9	6	8	
$\mathbf{a}_4$	−15	$\frac{4}{3}$	$\frac{1}{3}$	0	$\frac{1}{3}$	1	0	$-\frac{1}{3}$	0	$\frac{4}{3}/\frac{1}{3}$
$\mathbf{a}_2^*$	−8	$\frac{5}{3}$	$-\frac{1}{3}$	1	$\frac{2}{3}^*$	0	−1	$\frac{1}{3}$	−1	$\frac{5}{3}/\frac{2}{3}^*$
$z_j - c_j$		$-\frac{100}{3}$	$\frac{11}{3}$	0	$-\frac{10}{3}^*$	0	9	$\frac{7}{3}$	8	
$\mathbf{a}_4$	−15	$\frac{1}{2}$	$\frac{1}{2}$	$-\frac{1}{2}$	0	1	$\frac{1}{2}$	$-\frac{1}{2}$	$\frac{1}{2}$	
$\mathbf{a}_3$	−7	$\frac{5}{2}$	$-\frac{1}{2}$	$\frac{3}{2}$	1	0	$-\frac{3}{2}$	$\frac{1}{2}$	$-\frac{3}{2}$	
$z_j - c_j$		−25	2	5	0	0	4	4	3	

As none of $z_j - c_j < 0$ therefore the solution is optimal.

Maximum value of $z_w^* = -25$ for $w_3 = \frac{5}{2}$, $w_4 = \frac{1}{2}$ [other variables are non-basic].

Hence min $z_w = -\max z_w^* = -(-25)$ for $w_3 = \frac{5}{2}$ and $w_4 = \frac{1}{2}$.

The maximum value of the primal objective function $\max z_x = \min z_w = 25$. The dual problem is solved as a maximization problem.

12.9.3 Determination of primal optimal solution

$\mathbf{a}_6(-\mathbf{e}_1)$ and $\mathbf{a}_7(-\mathbf{e}_2)$ are the surplus vectors.

$z_6 - c_6 = 4$ and $z_7 - c_7 = 3$ in the final dual simplex table.

Therefore $x_1 = 4$ and $x_2 = 3$ corresponding to $-\mathbf{e}_1$ and $-\mathbf{e}_2$ respectively.

Hence $\max z_x = 25$ for $x_1 = 4$ and $x_2 = 3$.

Note: No artificial variable is introduced to get the initial unit basis. But this does not make any hurdle to determine the primal optimal solution.

► **Example 12.9.7** *Solve the following L.P.P. by usual simplex method without using any artificial variable.*

$$\textit{Minimize,} \quad z = x_1 + x_2$$

subject to

$$x_1 + 2x_2 \geq 12$$
$$5x_1 + 6x_2 \geq 48\,, \quad x_1 \geq 0, x_2 \geq 0.$$

Solution. In the usual simplex method, generally initial basis is a unit basis. But in this problem, there will be no unit basis in the absence of artificial variables. Let us now write down the dual of the problem and dual is given by,

$$\text{maximize,} \quad z_w = 12w_1 + 48w_2$$

subject to

$$w_1 + 5w_2 \le 1$$
$$2w_1 + 6w_2 \le 1\,, \quad w_1 \ge 0, w_2 \ge 0.$$

In this problem we shall get a unit basis after the introduction of two slack variables only and the constraints are thus

$$w_1 + 5w_2 + w_3 \qquad = 1$$
$$2w_1 + 6w_2 \qquad + w_4 = 1\,, \quad w_j \ge 0, [j = 1, 2, \cdots, 4.]$$

Now solving as usual, the simplex tables are

		c	12	48	0	0
Basis	$\mathbf{c}_B$	b	$\mathbf{a}_1$	$\mathbf{a}_2$	$\mathbf{a}_3(\mathbf{e}_1)$	$\mathbf{a}_4(\mathbf{e}_2)$
$\mathbf{a}_3$	0	1	1	5	1	0
$\mathbf{a}_4^*$	0	1	2	6*	0	1
$z_j - c_j$		0	−12	−48*	0	0
$\mathbf{a}_3$	0	$\frac{1}{6}$	$-\frac{2}{3}$	0	1	$-\frac{5}{6}$
$\mathbf{a}_2$	48	$\frac{1}{6}$	$\frac{1}{3}$	1	0	$\frac{1}{6}$
$z_j - c_j$		8	4	0	0	8

From the above table we have min $z = \max z_w = 8$ at $w_1 = 0, w_2 = 1/6$ and $x_1 = 0, x_2 = 8$. Thus the first problem has an optimal solution at $x_1 = 0, x_2 = 8$.

► **Example 12.9.8** *Below given the optimal table of the L.P.P.*

$$\textit{Maximize,} \quad Z = 5x_1 + 2x_2 + 3x_3$$

subject to

$$x_1 + 5x_2 + 2x_3 \le b_1$$
$$x_1 - 5x_2 - 6x_3 \le b_2\,, \quad x_1, x_2, x_3 \ge 0.$$

[C.U.(H) '83]

Basis	b	$\mathbf{a}_1$	$\mathbf{a}_2$	$\mathbf{a}_3$	$\mathbf{a}_4$	$\mathbf{a}_5$
$\mathbf{a}_1$	30	1				0
$\mathbf{a}_2$	10	0				1
$z_j - c_j$	150	0				0

Complete the table and write down the optimal solution of the dual L.P.P. from the table and find the values of b_1 and b_2.

We do not know the values of b_1 and b_2. Then in general adding two slack variables we get the equations and therefore $\mathbf{a}_5 = \begin{bmatrix} 0 \\ 1 \end{bmatrix}$ and hence final basis is $(\mathbf{a}_1, \mathbf{a}_5) = \begin{bmatrix} 1 & 0 \\ 1 & 1 \end{bmatrix} = B.$

$$\mathbf{x}_B = \begin{bmatrix} 30 \\ 10 \end{bmatrix}, \quad \mathbf{b} = B\mathbf{x}_B = \begin{bmatrix} 1 & 0 \\ 1 & 1 \end{bmatrix}\begin{bmatrix} 30 \\ 10 \end{bmatrix} = \begin{bmatrix} 30 \\ 40 \end{bmatrix}.$$

Thus $b_1 = 30$ and $b_2 = 40$.

$$\begin{bmatrix} 1 & 5 & 2 & 1 & 0 \\ 1 & -5 & -6 & 0 & 1 \end{bmatrix} \quad \text{and} \quad B^{-1} = \begin{bmatrix} 1 & 0 \\ -1 & 1 \end{bmatrix}.$$

Thus $[\mathbf{y}_1, \mathbf{y}_2, \mathbf{y}_3, \mathbf{y}_4, \mathbf{y}_5] = B^{-1}A = \begin{bmatrix} 1 & 5 & 2 & 1 & 0 \\ 0 & -10 & -8 & -1 & 1 \end{bmatrix}$.

Hence the complete final table is

		c	5	2	3	0	0
Basis	$\mathbf{c}_B$	b	$\mathbf{a}_1$	$\mathbf{a}_2$	$\mathbf{a}_3$	$\mathbf{a}_4$	$\mathbf{a}_5$
$\mathbf{a}_1$	5	30	1	5	2	1	0
$\mathbf{a}_5$	0	10	0	−10	−8	−1	1
$z_j - c_j$		150	0	23	7	5	0

Thus the dual optimal value is 150 corresponding to $w_1 = 5$ and $w_2 = 0$ as $\mathbf{a}_4$ and $\mathbf{a}_5$ are slack vectors.

12.9.4 Solution of mixed type primal dual problem

► **Example 12.9.9** *Find the optimal solution of the dual of the problem, maximize z where $z = 5x_1 + 12x_2 + 4x_3$, subject to*

$$\begin{aligned} x_1 + 2x_2 + x_3 &\le 5 \\ 2x_1 - x_2 + 3x_3 &= 2, \quad x_1, x_2, x_3 \ge 0. \end{aligned}$$

from the simplex table of the primal. **[C.U.(P)'90]**

Solving the primal problem using the simplex method, the optimal simplex table is

		c	5	12	4	0	$-M$
Basis	$\mathbf{c}_B$	b	$\mathbf{a}_1$	$\mathbf{a}_2$	$\mathbf{a}_3$	$\mathbf{a}_4(\mathbf{e}_1)$	$\mathbf{a}_5(\mathbf{e}_2)$
$\mathbf{a}_2$	12	$\frac{8}{5}$	0	1	$-\frac{1}{5}$	$\frac{2}{5}$	$-\frac{1}{5}$
$\mathbf{a}_1$	5	$\frac{9}{5}$	1	0	$\frac{7}{5}$	$\frac{1}{5}$	$\frac{2}{5}$
$z_j - c_j$		$28\frac{1}{5}$	0	0	$\frac{3}{5}$	$\frac{29}{5}$	$M - \frac{2}{5}$

From the optimal table, we have $\max z = 28\frac{2}{5}$ at $x_1 = \frac{9}{5}, x_2 = \frac{8}{5}, x_3 = 0$.

There are two methods of finding the dual optimal solution.

(i) Final basis

$$B = (\mathbf{a}_2, \mathbf{a}_1), B^{-1} = \begin{bmatrix} 2/5 & -1/5 \\ 1/5 & 2/5 \end{bmatrix}, \quad \mathbf{c}_B = (12, 5).$$

The dual optimal solution,

$$\mathbf{c}_B B^{-1} = (12, 5)\begin{bmatrix} 2/5 & -1/5 \\ 1/5 & 2/5 \end{bmatrix} = \begin{bmatrix} 29/5 \\ -2/5 \end{bmatrix}.$$

Thus the dual optimal solution $w_1^* = \frac{29}{5}, w_2^* = -\frac{2}{5}$ and $\min z_w = \max z = 28\frac{1}{5}$.

(ii) Here the column corresponding to the artificial vector $\mathbf{a}_5(\mathbf{e}_2)$ is essential and we shall not remove that column. $\mathbf{a}_4(\mathbf{e}_1)$ and $\mathbf{a}_5(\mathbf{e}_2)$ are the initial unit basis. Then the dual optimal solution is $w_1^* = \frac{29}{5}$ and $w_2^* = -\frac{2}{5}$ [putting $M = 0$ which are $z_4 - c_4$ and $z_5 - c_5$ in the final simplex table]. Here w_2^* is unrestricted (which is $-$ve here).

Note. The dual objective function is

$$z_w = 5w_1 + 2w_2 \text{ and } 5 \times \tfrac{29}{5} + 2\left(-\tfrac{2}{5}\right) = \tfrac{141}{5} = 28\tfrac{1}{5}$$

and thus the final result is being checked or verified.

Objective and short answer type questions with answers Duality theory

1. Write down the dual of the given problem, maximize, $z = \mathbf{cx}$ subject to $A\mathbf{x} \leq \mathbf{b}, \mathbf{x} \geq \mathbf{0}$.

2. Write down the dual of the problems given below:

   (a) Minimize, $z = \mathbf{cx}$ subject to $D\mathbf{x} \geq \mathbf{d}, \mathbf{x} \geq \mathbf{0}$.

   (b) Minimize, $z = \mathbf{cx}$ subject to $D\mathbf{x} \leq \mathbf{b}, \mathbf{x} \geq \mathbf{0}$.

3. Write down the dual of the following problems:

   (a) Maximize, $z = 4x_1 + 2x_2$

   $$\text{subject to } 3x_1 + 4x_2 \leq 7$$
   $$7x_1 - 2x_2 \leq 13, \quad x_1 \geq 0, x_2 \geq 0.$$

   (b) Maximize, $z = 3x_1 - x_2$

   $$\text{subject to } 3x_1 + x_2 \leq 14$$
   $$2x_1 - 4x_2 \geq 4, \quad x_1 \geq 0, x_2 \geq 0.$$

   (c) Minimize, $z_x = 7x_1 - x_2 + x_3$

   $$\text{subject to } 2x_1 + x_2 + x_3 \geq 6$$
   $$x_1 + x_2 + 3x_3 \geq 16, \quad x_1 \geq 0, x_2 \geq 0, x_3 \geq 0.$$

   (d) Minimize, $z_x = 7x_1 + x_2$

   $$\text{subject to } x_1 + 3x_2 \leq 17$$
   $$x_1 - 4x_2 \leq 6, \quad x_1 \geq 0, x_2 \geq 0.$$

4. Write down the dual of the given problem, minimize, $2x_1 + 3x_2$

$$\text{subject to } 4x_1 + 5x_2 \geq 7$$
$$3x_1 - 2x_2 \geq 2, \quad x_1 \geq 0, x_2 \geq 0$$

and hence establish the fact that no artificial variable is needed to solve the dual problem in usual simplex method.

[*Ans.* The dual is: maximize, $z_w = 7w_1 + 2w_2$

$$\text{subject to } 4w_1 + 3w_2 \leq 2$$
$$5w_1 - 2w_2 \leq 3, \quad w_1 \geq 0, w_2 \geq 0$$

and hence two slack vectors will constitute the initial unit basis of the dual problem, $\mathbf{b} \geq \mathbf{0}$ and no artificial variable is needed to get the initial unit basis.]

5. Justify or correct the following statements:

(a) The dual of the maximization problem is a minimization problem.

(b) The dual of a dual problem is the primal problem.

(c) If the kth constraint of a primal be an equation, the kth dual variable will be non-negative.

(d) If the kth variable of a primal problem is unrestricted in sign, the kth dual constraint will be strictly an equation.

(e) If the primal problem is unbounded, the dual may have finite value of the objective function.

(f) If the primal problem has m constraints with n variables the dual will have the same number of m constraints with n variables.

(g) If either of the primal or dual has finite optimal solution then the other has also finite optimal solution.

(h) If the primal problem has no feasible solution, then the dual will also have no feasible solution.

[*Ans.* (a) Correct (b) Correct (c) If the kth constraint of a primal be an equation, the kth dual variable will be unrestricted in sign (d) Correct (e) If the primal problem is unbounded, the dual will have no feasible solution. (f) If the primal problem has m constraints with n variables then dual will have n constraints with m variables. (g) Correct (h) If the primal problem has no feasible solution, then the dual will have either no F.S. or unbounded solution.]

6. Define (i) symmetric primal and dual, (ii) unsymmetrical primal and dual (iii) mixed type primal and dual.

[*Ans.* See 10.4 and 10.9.]

7. If $\mathbf{x}^*$ and $\mathbf{w}^*$ be any two feasible solutions of symmetric primal (maximization problem) and dual problem with objective functions $\mathbf{cx}$ and $\mathbf{b'w}$ respectively, then which one of the statements is correct?

(a) $\mathbf{cx}^* \leq \mathbf{b'w}^*$, (b) $\mathbf{cx}^* = \mathbf{b'w}^*$, (c) $\mathbf{cx}^* \geq \mathbf{b'w}^*$

[*Ans.* $\mathbf{cx}^* \leq \mathbf{b'w}^*$.]

8. What is the relation between the optimal values of primal and dual problems [assume that both exist]?

   [*Ans.* **If the primal be (maximization) with objective function z_x and obviously dual (minimization) with objective function z_w then $\max z_x = \min z_w$.**]

9. You are given a problem (L.P.) with five constraints with two variables. Which one of the problems, primal or dual will you select to solve? Justify your answer.

   [*Ans.* **Dual problem will be selected to solve as the dual will have only two constraints with five variables which can be solved easily.**]

10. In a primal problem, the 4th constraint is an equation and the 3rd variable is unrestricted in sign. What will be nature of the 4th dual variable and the 3rd dual constraint?

    [*Ans.* **4th dual variable will be unrestricted in sign and the 3rd dual constraint will be strictly an equation.**]

11. How can you determine the primal optimal solution from the dual simplex table in the case of symmetric problem?

    [*Ans.* **The values of $z_j - c_j$ for the columns corresponding to the slack and surplus vectors in the final simplex table of dual problem are the values of the corresponding primal optimal variables provided the dual problem is solved as a maximization problem (whether the actual dual problem is a maximization or a minimization problem).**]

12. State the necessary and sufficient condition that both the primal and dual will have finite optimal solutions.

    (a) If either the primal or dual has finite optimal solution then both the problems will have finite optimal solutions.

    Or,

    (b) The necessary and sufficient condition that both problems will have finite optimal solutions is the existence of feasible solutions of both problems.

Exercise 12A

Obtain the dual of the given L.P. problems:

1. Maximize, $z = 2x_1 + 3x_2 - 4x_3$

$$\begin{aligned} \text{subject to } 5x_1 - 2x_2 + x_3 &\leq 4 \\ x_1 + x_2 - 4x_3 &\leq 7, \quad x_1, x_2, x_3 \geq 0. \end{aligned}$$

[C.U.(P)'90 (Old)]

2. (a) Minimize, $z = x_1 - x_2 + 2x_3$

$$\begin{aligned} \text{subject to } x_1 + x_2 + 4x_3 &\geq 7 \\ x_2 - 2x_3 &\geq 10 \\ 3x_1 + x_2 + x_3 &\leq 3, \quad x_1, x_2, x_3 \geq 0. \end{aligned}$$

(b) Maximize, $z = x_1 - x_2 + 3x_3$

$$\begin{aligned} \text{subject to } & x_1 + x_2 + x_3 \le 10 \\ & 2x_1 - x_3 \le 2 \\ & 2x_1 - 2x_2 + 3x_3 \le 6, \quad x_1, x_2, x_3 \ge 0. \end{aligned}$$

[C.U.(P)'83]

3. Maximize, $z = -x_1 + 2x_2 + 7x_3$

$$\begin{aligned} \text{subject to } & 4x_1 + x_2 + x_3 \le 2 \\ & 7x_1 - x_2 + 5x_3 \ge 7 \\ & 2x_1 + x_2 - x_3 \le 4, \quad x_1 \ge 0, x_2 \ge 0 \end{aligned}$$

and x_3 is unrestricted in sign.

4. Maximize, $z = x_1 + x_2 + x_3$

$$\begin{aligned} \text{subject to } & x_1 - 3x_2 + 4x_3 = 5 \\ & x_1 - 2x_2 \ge 3 \\ & 2x_2 - x_3 \le 4, \quad x_1, x_2 \ge 0 \end{aligned}$$

and x_3 is unrestricted.

5. Maximize, $z = 3x_1 + x_2 + x_3 - x_4$

$$\begin{aligned} \text{subject to } & x_1 + 5x_2 + 3x_3 + 4x_4 \le 5 \\ & x_1 + x_2 = -1 \\ & x_3 - x_4 \le -5, \quad x_1, x_2, x_3, x_4 \ge 0. \end{aligned}$$

6. Maximize, $z = 3x_1 - 2x_2$

$$\begin{aligned} \text{subject to } & x_1 \le 4 \\ & x_2 \le 6 \\ & x_1 + x_2 \le 5 \\ & -x_2 \le -1, \quad x_1, x_2 \ge 0. \end{aligned}$$

7. Maximize, $z = 2x_1 - 3x_3$

$$\begin{aligned} \text{subject to } & x_1 - 2x_2 + x_3 \ge 0 \\ & 2x_1 + 4x_2 + x_3 \le 27, \quad x_1 \ge 0, x_2 \ge 0, x_3 \ge 0. \end{aligned}$$

8. Minimize, $z = x_1 + x_2 + x_3$

$$\begin{aligned} \text{subject to } & 3x_1 - x_2 + x_3 = 4 \\ & 2x_1 + x_2 - x_3 \le 8 \end{aligned}$$

x_1 is unrestricted in sign, $x_2 \ge 0, x_3 \ge 0$.

9. Maximize, $z = 2x_1 + x_2$

$$\begin{aligned} \text{subject to } & 2x_1 + x_2 \ge 9 \\ & 3x_1 \ge 6 \\ & 2x_1 - 3x_2 = 1 \end{aligned}$$

$x_1 \ge 0$ and x_2 is unrestricted in sign.

10. Minimize, $z = 2x_2 + 3x_3$

$$\text{subject to } \begin{aligned} 2x_1 + x_2 - x_3 &\le 4 \\ -x_1 - x_2 + 4x_3 &\le 16 \end{aligned}$$

x_1 is unrestricted in sign, $x_2 \ge 0, x_3 \ge 0$.

11. Maximize, $z = x_1 - x_2 - x_4$

$$\text{subject to } \begin{aligned} 2x_1 + x_2 - x_3 \quad &= 10 \\ x_2 + x_3 - x_4 &\le 0 \\ x_1 \quad + x_3 + 2x_4 &\ge 6, \quad x_1 \ge 0, x_2 \ge 0, \end{aligned}$$

x_3 is unrestricted in sign $x_4 \ge 0$.

12. (a) Minimize, $z = 2x_1 - x_3 + x_4$

$$\text{subject to } \begin{aligned} x_1 - x_2 - x_3 + x_4 &= 8 \\ 3x_1 \quad + x_3 - 2x_4 &\ge 9, \quad x_1 \ge 0, x_2 \ge 0, x_3 \ge 0, \end{aligned}$$

x_4 is unrestricted in sign.

(b) Maximize, $z = 2x_1 + 3x_2 + x_3$

$$\text{subject to } \begin{aligned} 4x_1 + 3x_2 + x_3 &= 6 \\ x_1 + x_2 + 5x_3 &\le 4, \quad x_j \ge 0 [j = 1, 2, 3.] \end{aligned}$$

[C.U.(H)'80]

13. (a) Construct the dual of the following problem,

minimize, $z = x_1 + 3x_2$

$$\text{subject to } \begin{aligned} x_1 + 2x_2 &\le -6 \\ x_1 - x_2 &\le 2 \\ 2x_1 - x_2 &= 4 \\ -x_1 + x_2 &\ge 7, \quad x_1 \ge 0, x_2 \ge 0. \end{aligned}$$

[C.U., M.Com,'89]

(b) Write down the dual of the following primal,

minimize, $z = x_1 - 2x_2 - x_3 + 2x_4 + x_5$

$$\text{subject to } \begin{aligned} 3x_1 - x_2 + 5x_3 - 7x_4 + x_5 &\le 17 \\ x_1 + 2x_2 - x_3 + 5x_4 + 2x_5 &= 25 \\ 2x_1 - x_2 + 5x_3 - x_4 + 3x_5 &\ge 57, \quad x_j \ge 0 [j = 1, 2, \cdots, 5.] \end{aligned}$$

[C.U.(P)'87]

14. Find the dual problem for the following problem,

maximize, $z = 7x_1 + 5x_2 - 2x_3$

$$\text{subject to } \begin{aligned} x_1 + x_2 + x_3 &= 10 \\ 2x_1 - x_2 + 3x_3 &\le 16 \\ 3x_1 + x_2 - 2x_3 &\ge -5, \quad x_1 \ge 0, x_2 \ge 0, \end{aligned}$$

x_3 unrestricted. **[C.U.,M.Com'89]**

15. (a) Obtain the dual of the L.P.P., minimize, $z = x_1 + x_2 + x_3$ subject to the constraints

$$\begin{aligned} x_1 - 3x_2 + 4x_3 &= 5 \\ x_1 - 2x_2 \qquad &\leq 3 \\ 2x_2 - x_3 &\geq 4, \quad x_1, x_2 \geq 0 \end{aligned}$$

and x_3 is unrestricted in sign. [C.U.(H)'88]

(b) Find the dual of the following programme,

maximize, $z = 2x_1 + 3x_2 + x_3$

subject to $4x_1 + 3x_2 + x_3 = 6$
$x_1 + 2x_2 + 5x_3 = 4, \quad x_1, x_2, x_3 \geq 0.$

[C.U.,M.Com'86(P),'90,'99]

Answers

1. Minimize $z_w = 4w_1 + 7w_2$

subject to
$$\begin{aligned} 5w_1 + w_2 &\geq 2 \\ -2w_1 + w_2 &\geq 3 \\ w_1 - 4w_2 &\geq -4, \quad w_1 \geq 0, w_2 \geq 0. \end{aligned}$$

2. (a) Maximize, $z_w = 7w_1 + 10w_2 - 3w_3$

subject to
$$\begin{aligned} w_1 \qquad - 3w_3 &\leq 1 \\ w_1 + w_2 - w_3 &\leq -1 \\ 4w_1 - 2w_2 - w_3 &\leq 2, \quad w_1, w_2, w_3 \geq 0. \end{aligned}$$

(b) Minimize, $z_w = 10w_1 + 2w_2 + 6w_3$

subject to
$$\begin{aligned} w_1 + 2w_2 + 2w_3 &\geq 1 \\ w_1 \qquad - 2w_3 &\geq -1 \\ w_1 - w_2 + 3w_3 &\geq 3, \quad w_1 \geq 0, w_2 \geq 0, w_3 \geq 0. \end{aligned}$$

3. Minimize, $z_w = 2w_1 - 7w_2 + 4w_3$

subject to
$$\begin{aligned} 4w_1 - 7w_2 + 2w_3 &\geq -1 \\ w_1 + w_2 + w_3 &\geq 2 \\ w_1 - 5w_2 - w_3 &= 7, \quad w_1 \geq 0, w_2 \geq 0, w_3 \geq 0. \end{aligned}$$

4. Minimize $z_w = 5w_1 - 3w_2 + 4w_3$

subject to
$$\begin{aligned} w_1 - w_2 \qquad &\geq 1 \\ -3w_1 + 2w_2 + 2w_3 &\geq 1 \\ 4w_1 \qquad - w_3 &= 1. \end{aligned}$$

w_1 is unrestricted in sign, $w_2 \geq 0$, $w_3 \geq 0$.

5. Minimize, $z_w = 5w_1 - w_2 - 5w_3$

subject to
$$\begin{aligned} w_1 + w_2 \qquad &\geq 3 \\ 5w_1 + w_2 \qquad &\geq 1 \\ 3w_1 \qquad + w_3 &\geq 1 \\ 4w_1 \qquad - w_3 &\geq -1, \quad w_1 \geq 0, w_3 \geq 0, \end{aligned}$$

w_2 is unrestricted in sign.

6. Minimize, $z_w = 4w_1 + 6w_2 + 5w_3 - w_4$

subject to
$$\begin{aligned} w_1 \qquad + w_3 \qquad &\geq 3 \\ w_2 + w_3 - w_4 &\geq -2, \quad w_1 \geq 0, w_2 \geq 0, w_3 \geq 0, w_4 \geq 0. \end{aligned}$$

7. Minimize, $z_w = 27w_2$

$$\text{subject to } \begin{aligned} -w_1 + 2w_2 &\geq 2 \\ 2w_1 + 4w_2 &\geq 0 \\ -w_1 + w_2 &\geq -3, \quad w_1 \geq 0, w_2 \geq 0. \end{aligned}$$

8. Maximize, $z_w = 4w_1 - 8w_2$

$$\text{subject to } \begin{aligned} 3w_1 - 2w_2 &= 1 \\ -w_1 - w_2 &\leq 1 \\ w_1 + w_2 &\leq 1 \end{aligned}$$

w_1 is unrestricted in sign, $w_2 \geq 0$.

9. Minimize, $z_w = -9w_1 - 6w_2 + w_3$

$$\text{subject to } \begin{aligned} -2w_1 - 3w_2 + 2w_3 &\geq 2 \\ -w_1 \qquad - 3w_3 &= 1, \quad w_1 \geq 0, w_2 \geq 0, \end{aligned}$$

w_3 is unrestricted in sign.

10. Maximize, $z_w = -4w_1 - 16w_2$

$$\text{subject to } \begin{aligned} -2w_1 + w_2 &= 0 \\ -w_1 + w_2 &\leq 2 \\ w_1 - 4w_2 &\leq 3, \quad w_1 \geq 0, w_2 \geq 0. \end{aligned}$$

11. Minimize, $z_w = 10w_1 - 6w_3$

$$\text{subject to } \begin{aligned} 2w_1 \qquad - w_3 &\geq 1 \\ w_1 + w_2 \qquad &\geq -1 \\ -w_1 + w_2 - w_3 &= 0 \\ - w_2 - 2w_3 &\geq -1, \end{aligned}$$

w_1 is unrestricted in sign $w_2 \geq 0, w_3 \geq 0$.

12. (a) Maximize, $z_w = 8w_1 + 9w_2$

$$\text{subject to } \begin{aligned} w_1 + 3w_2 &\leq 2 \\ -w_1 \qquad &\leq 0 \\ -w_1 + w_2 &\leq -1 \\ w_1 - 2w_2 &= 1, \end{aligned}$$

w_1 is unrestricted in sign, $w_2 \geq 0$.

(b) Minimize, $z_w = 6w_1 + 4w_2$

$$\text{subject to } \begin{aligned} 4w_1 + w_2 &\geq 2 \\ 3w_1 + w_2 &\geq 3 \\ w_1 + 5w_2 &\geq 1, \end{aligned}$$

w_1 is unrestricted, $w_2 \geq 0$.

13. (a) Maximize, $z_w = 6w_1 - 2w_2 + 4w_3 + 7w_4$

$$\text{subject to } \begin{aligned} -w_1 - w_2 + 2w_3 - w_4 &\leq 1 \\ -2w_1 + w_2 - w_3 + w_4 &\leq 3, \quad w_1, w_2, w_4 \geq 0, \end{aligned}$$

w_3 is unrestricted.

(b) Maximize, $z_w = -17w_1 + 25w_2 + 57w_3$

$$\text{subject to } \begin{aligned} -3w_1 + w_2 + 2w_3 &\leq 1 \\ w_1 + 2w_2 - w_3 &\leq -2 \\ -5w_1 - w_2 + 5w_3 &\leq -1 \\ 7w_1 + 5w_2 - w_3 &\leq 2 \\ -w_1 + 2w_2 + 3w_3 &\leq 1, \quad w_1, w_2 \geq 0, \end{aligned}$$

w_2 is unrestricted.

14. Minimize, $z_w = 10w_1 + 16w_2 + 5w_3$

$$\begin{aligned} \text{subject to } w_1 + 2w_2 - 3w_3 &\geq 7 \\ w_1 - w_2 - w_3 &\geq 5 \\ w_1 + 3w_2 + 2w_3 &= -2, \end{aligned}$$

w_1 unrestricted, $w_2, w_3 \geq 0$.

15. (a) Maximize, $z_w = 5w_1 - 3w_2 + 4w_3$

$$\begin{aligned} \text{subject to } w_1 - w_2 &\leq 1 \\ -3w_1 + 2w_2 + 2w_3 &\leq 1 \\ 4w_1 - w_3 &= 1, \end{aligned}$$

w_1 unrestricted, $w_2, w_3 \geq 0$.

(b) Minimize, $z_w = 6w_1 + 4w_2$

$$\begin{aligned} \text{subject to } 4w_1 + w_2 &\geq 2 \\ 3w_1 + 2w_2 &\geq 3 \\ w_1 + 5w_2 &\geq 1 \text{ , } w_1, w_2 \text{ unrestricted in sign.} \end{aligned}$$

Exercise 12B

Solving the dual problem obtain the optimum solution, if any to each of the following L.P.P.

1. Minimize, $z = 4x_1 + 3x_2$

$$\begin{aligned} \text{subject to } x_1 + 2x_2 &\geq 8 \\ 3x_1 + 2x_2 &\geq 12 \text{ , } \quad x_1, x_2 \geq 0. \end{aligned}$$

2. Minimize, $z = 15x_1 + 10x_2$

$$\begin{aligned} \text{subject to } 3x_1 + 5x_2 &\geq 5 \\ 5x_1 + 2x_2 &\geq 3 \text{ , } \quad x_1, x_2 \geq 0. \end{aligned}$$

[Cal. Applied Math.'79]

3. Minimize, $z = 3x_1 + x_2$

$$\begin{aligned} \text{subject to } x_1 + x_2 &\geq 1 \\ 2x_1 + 3x_2 &\geq 2 \text{ , } \quad x_1, x_2 \geq 0. \end{aligned}$$

4. Minimize, $z = 4x_1 - 3x_2$

$$\begin{aligned} \text{subject to } 2x_1 - x_2 &\geq 4 \\ 4x_1 + 3x_2 &\leq 28 \text{ , } \quad x_1, x_2 \geq 0. \end{aligned}$$

5. Maximize, $z = 2x_1 + 2x_2$

$$\begin{aligned} \text{subject to } 2x_1 + 4x_2 &\geq 1 \\ x_1 + 2x_2 &\geq 1 \\ 2x_1 + x_2 &\geq 1 \text{ , } \quad x_1, x_2 \geq 0. \end{aligned}$$

6. **Minimize,** $z = 3x_1 + 2x_2$

$$\begin{aligned} \text{subject to } x_1 + \ x_2 &\geq 1 \\ x_1 + \ x_2 &\leq 7 \\ x_1 + 2x_2 &\geq 10 \\ x_2 &\leq 3, \quad x_1, x_2 \geq 0. \end{aligned}$$

7. **Maximize,** $z = 3x_1 - 4x_2$

$$\begin{aligned} \text{subject to } x_1 + x_2 &\leq 7 \\ x_1 - x_2 &\leq 13 \\ x_1 \quad &\geq 3 \\ x_2 &\leq 8, \quad x_1, x_2 \geq 0. \end{aligned}$$

8. **Maximize,** $z = 3x_1 - 2x_2$

$$\begin{aligned} \text{subject to } x_1 + x_2 &\leq 5 \\ x_1 \quad &\leq 4 \\ x_2 &\leq 6 \\ -x_2 &\leq -1, \quad x_1, x_2 \geq 0. \end{aligned}$$

9. **Minimize,** $z = 2x_1 + 3x_2$

$$\begin{aligned} \text{subject to } -x_1 + 2x_2 &\leq 4 \\ x_1 + \ x_2 &\leq 6 \\ x_1 + 3x_2 &\geq 9, \quad x_1, x_2 \geq 0. \end{aligned}$$

10. **Maximize,** $z = 3x_1 + 4x_2$

$$\begin{aligned} \text{subject to } x_1 - \ x_2 &\leq 1 \\ x_1 + \ x_2 &\geq 4 \\ x_1 - 3x_2 &\leq 3, \quad x_1, x_2 \geq 0. \end{aligned}$$

From the dual solution, discuss the nature of the primal problem. Determine the exact nature of the problem by using graphical method to the primal problem. **[C.U. (Hons.) '87]**

11. Solve the primal problem, minimize, $z = -2x_1 + 3x_2$

$$\begin{aligned} \text{subject to } 2x_1 - 5x_2 &\leq 7 \\ 4x_1 + \ x_2 &\geq 8 \\ 7x_1 + 2x_2 &\leq 16, \quad x_j \geq 0, \ j = 1, 2. \end{aligned}$$

by simplex method and find the dual optimal variables from the final simplex table of the primal problem.

12. (a) Solve the following primal problem by using simplex method and discuss the nature of the dual problem from the nature of the primal problem, maximize, $2x_2 + x_3$

$$\begin{aligned} \text{subject to } \quad x_1 + x_2 - 2x_3 &\leq 7 \\ -3x_1 + x_2 + 2x_3 &\leq 3, \quad x_1, x_2, x_3 \geq 0. \end{aligned}$$

Verify the nature of the dual problem by using graphical method.

[C.U. (Hons.) '96]

(b) Use simplex method to show that the linear programming problem, maximize, $x_1 - 2x_2$

$$\text{subject to} \quad \begin{aligned} x_1 + 2x_2 &\geq 1 \\ -x_1 + \ x_2 &\leq 1. \end{aligned}$$

has an unbounded solution. What can you conclude about the solution to the dual of the above L.P.P.? **[C.U. (Hons.) '86, '88]**

(c) Write down the dual of the following problem,
minimize, $z = -12x_1 + 6x_2 - 4x_3$

$$\text{subject to} \quad \begin{aligned} -3x_1 + x_2 + x_3 &\geq 3 \\ 4x_1 + x_2 + x_3 &\leq 2, \quad x_1, x_2, x_3 \geq 0. \end{aligned}$$

Solving the dual problem, discuss the nature of the primal problem.

Answers

1. min $z = 17$ at $x_1 = 2$, $x_2 = 3$. 2. min $z = \frac{235}{19}$, for $x_1 = \frac{5}{19}$, $x_2 = \frac{16}{19}$. 3. Min $z = 1$ for $x_1 = 0$, $x_2 = 1$. 4. Min $z = 4$ for $x_1 = 4$, $x_2 = 4$. 5. Min $z = \frac{4}{3}$ for $x_1 = \frac{1}{3}$, $x_2 = \frac{1}{3}$. 6. Max $z = 18$ for $x_1 = 4$, $x_2 = 3$. 7. Max $z = 21$ for $x_1 = 7$, $x_2 = 0$. 8. Max $z = 10$ for $x_1 = 4$, $x_2 = 1$. 9. Max $z = \frac{27}{2}$ for $x = \frac{9}{2}$, $x_2 = \frac{3}{2}$.

10. Dual has no F.S. Then the primal problem will have either no F.S. on unbounded solution. Using duality theory we cannot say more than that. But using graphical method to the primal [See Ex. 7.1.1(b)] it can be shown that primal has a set of feasible solutions. Hence we can now definitely come to the conclusion that primal has an unbounded solution.

1. Min $z = -4$ for $x_1 = 2$, $x_2 = 0$, and the dual optimal variables are $w_1 = 0$, $w_2 = \frac{1}{2}$, $w_3 = 0$.

12. (a) The primal has an unbounded solution. Thus the dual of it will have no F.S. (b) Dual has no feasible solution. (c) The dual is said to have an unbounded solution. Hence the primal has no F.S.

12.9.5 Economical importance of duality theory

Shadow prices or Accounting prices: We have already discussed the computational advantage of the duality theory. The duality has also much economical importance also and the determination of the shadow prices etc. (which will be defined just now) of the available raw-materials etc., a business man can arrange his plan so that he can make on overall maximum profit etc.

Shadow price: Let the primal problem be, Maximize $z = \mathbf{cx}$ subject to $A\mathbf{x} \leq \mathbf{b}, \mathbf{x} \geq 0$, $\mathbf{b}$ is unrestricted] and assume that the problem has a finite optimal solution which can be obtained by usual simplex method. Let Max $z = z_B$ corresponding to the final basis B.

Then

$$z_B = \mathbf{c}_B B^{-1}\mathbf{b}.$$

Now

$$\frac{\partial z_B}{\partial b_i} = \frac{\partial(\mathbf{c}_B B^{-1}\mathbf{b})}{\partial b_i} = \mathbf{c}_B B^{-1}\mathbf{e}_i = w_i \quad [i = 1, \cdots, m] \qquad (12.9.3)$$

which is the ith dual optimal variable, which will be obtained from the $z_j - c_j$ corresponding to the ith slack vector. From (12.9.3) it is enough clear that increase of one unit of b_i, will make the increase of optimal value of z to w_i which is known as the shadow price etc. $[i = 1, 2, \cdots, m]$

But to study the nature of increase of profit etc. from the shadow prices we must be very careful about the following two points:

(1) The result is valid so long as the optimal basis remains same for any change of b_i, if the optimal basis changes then our analysis fails.

(2) Each and every constraint must be converted in such a way that the coefficient of variables in each constraint are in their lowest term, i.e., all coefficients must be integers and their L.C.M. will be unity, b_i etc., are to be taken as the components of requirement vector after such proper reduction. Otherwise it leads to erroneous result.

From the problem solved below, we can get a clear idea of the shadow price etc. and how the concept of shadow price will help in determining the strategy of a business etc.

► **Example 12.9.10** ***A unit uses scarce material of 60 units to produce two products of P_1 and P_2. Each unit of P_1 and P_2 requires 2 units and 1 unit of scarce material and yield Rs. 60 and Rs. 40 as profit respectively. Market condition suggests that P_1 should not be less than 25 units and P_2 should be less than 35 units. (i) Find a suitable production plan to maximize profit by expressing the problem in linear programming problem and use simplex method for solution. (ii) Find the shadow price for the scarce material and how it is read directly from the simplex table (Final). (iii) Explain the concept of shadow price in planning and in determination for scarce resources.*** **[C.U. M. Com. '86]**

(i) If x_1 and x_2 units of P_1 and P_2 be produced to maximize to profit, then the L.P.P. is

$$\text{maximize,} \quad 60x_1 + 40x_2$$

$$\begin{aligned} \text{subject to } 2x_1 + x_2 &\le 60 \\ x_1 \quad &\ge 25 \\ x_2 &\le 35, \quad x_1, x_2 \ge 0. \end{aligned}$$

Using the usual method of solving the problem by Big-M method the final simplex table is given below. [x_5 artificial variable]

	c		60	40	0	0	$-M$	0
Basis	$\mathbf{c}_B$	$\mathbf{b}$	$\mathbf{a}_1$	$\mathbf{a}_2$	$\mathbf{a}_3(\mathbf{e}_1)$	$\mathbf{a}_4(-\mathbf{e}_2)$	$\mathbf{a}_5$	$\mathbf{a}_6(\mathbf{e}_3)$
$\mathbf{a}_2$	40	10	0	1	1	2		0
$\mathbf{a}_1$	60	25	1	0	0	-1		0
$\mathbf{a}_6$	0	25	0	0	-1	-2		1
$z_j - c_j$		1900	0	0	40	20		0

Thus max z = Rs. 1,900 at $x_2 = 10, x_1 = 25$ and $x_6 = 25$, i.e., for $x_1 = 25$ and $x_2 = 10$ the original problem attains its maximum. Hence 25 and 10 units of P_1 and P_2 be produced to get maximum profit.

(ii) The shadow prices are the components of dual optimal solution which are $w_1 = 40, w_2 = 20$ and $w_3 = 0$ which can be obtained from the $z_j - c_j$ of the final table corresponding to the slack and surplus vectors provided the optimal basis remains unchanged due to the increase of b_i etc.

12.9.6 Concept of shadow price and planning

Here the coefficients of the variables in the first constraint are in their lowest terms. Now if it is possible to increase one unit of b_1, i.e., for $b_1 = 61$, the increase in profit will be Rs. 40 (provided the optimal basis remains unchanged). Hence if the price of one unit of scarce material be less than Rs. 40 per unit, the business man will try to collect more units of scarce material to increase his over all profit. But there is some limit in the process. If the value of b_1 be greater than 85, the optimal basis will change and the optimal schedule will also alter and the shadow prices etc. also change. [The range of b_i for which the optimal basis remains unchanged can be determined by using Sensitivity Analysis.]

Exercise 12C

1. Define the term 'shadow price' in duality theory. Explain how by using shadow price a planning in production can be made to make an overall maximum profit.

2. A dye manufacturer is considering the production of two need products. Each product requires manufacturing time on two machines (see table below). The profit per unit products 1 and 2 are respectively, Rs. 50 and Rs. 60. If restrictions (such as coordinating production on two machines) are neglected, find the units of each product that should be manufactured every week so that the profit be maximum. Also find the optimal values of dual variables.

Table : Date for the manufacture of two products

Product	Hour needed on Machine I (Per unit of product)	Hours needed on Machine II (Per unit of product)
1	10	5
2	15	5
	120	50

Total time available (hours per week).

Find also the increase of the value of the objective function of the total available hours (per week) for the first machine be increased to 125 hours (provided the optimal basis remains same).

Ans. Max z = Rs. 540 for $x_1 = 6$ and $x_2 = 4$

Dual optimal variables $w_1 = 10, x_2 = 30$.

First constraint is $10x_1 + 15x_2 \le 120$. After proper scaling the constraint is $2x_1 + 3x_2 \le 24$. Hence for unit increase from 24 to 25, increase in value of z will be 10 units. Hence for 5 units from 120 to 125, increase in value of z will be,

$$\frac{125-120}{120} \times \frac{24}{25-24} \times 10 = 10 \text{ units.}$$

Dual optimal solution will be $w_1 = 2, w_2 = 6$ if the scale be not changed.

3. A company produces two types of product say type A and type B. Product B is superior quality and the product A is of lower quality. Profits on the two types of product are Rs. 30 and Rs. 40 respectively. The data on resource required and the availability of resources are given below:

	Requirement		Capacity available
	Product A	Product B	per month
Raw-materials (Kgs)	60	120	12,000
Machining (hours per piece)	8	5	610
Assembly (Man hour)	3	4	500

How shall the company manufacture the two types of product in order to produce a maximum overall profit? [C.U. M. Com. '87]

Find also the shadow prices and the increase in value of the objective function if the available machining hour he increased to 632 hours. Also prove that the maximum profit remains stationary even if the assembly hour be reduced up to 420 hours. [provided the optimal basis remains same]

Ans. Max profit Rs. 4,200 for manufacturing 20 and 90 units of A and B respectively. Shadow prices are $w_1 = \frac{170}{11}$, $w_2 = \frac{20}{11}$, $w_3 = 0$. Increase in value of z is 40 units.

Since $w_3 = 0$, then the third constraint $3x_1 + 4x_2 \le 500$ may be an inequality and this will be a strictly equality for $x_1 = 20$ and $x_2 = 90$ and for that, assembly hour will be $3 \times 20 + 4 \times 90 = 420$ and basis remains unchanged so long and hence the assembly hour can be reduced up to 420 hours without changing the value of the maximum profit.

Chapter 13

Transportation, Assignment and Travelling Salesman Problems

13.1 Definition

A transportation problem (simply T.P.) is a particular type linear programming problem. Here, a particular commodity which is stored at different warehouses (origins) is to be transported to different distribution centers (destinations) in such a way that the transportation cost is minimum. Consider a particular example. Let there be m origins $O_1, O_2, \cdots, O_i, \cdots, O_m$ and the quantity available at origin O_i be $a_i[i = 1, 2, \cdots, m]$ and let there be n destinations $D_1, D_2, \cdots, D_j, \cdots, D_n$ and the quantity required, i.e., the demand at D_j be $b_j[j = 1, 2, \cdots, n]$. Let us make an assumption that

$$\sum_{i=1}^{m} a_i = M = \sum_{j=1}^{n} b_j. \tag{13.1.1}$$

This assumption is not restrictive. In a particular problem when

$$\sum_{i=1}^{m} a_i = \sum_{j=1}^{n} b_j,$$

i.e., the total available quantity is equal to total demand, it is called as *balanced transportation problem* and when

$$\sum_{i=1}^{m} a_i \neq \sum_{j=1}^{n} b_j$$

it is called as *unbalanced transportation problem.* We shall initially discuss first type problems.

$$
\begin{array}{cc|ccccccc|cc}
 & & & & & \text{Destinations} & & & & & \\
 & & D_1 & \cdots & D_2 & \cdots & D_j & \cdots & D_n & & \\
\hline
 & O_1 & x_{11} & \cdots & x_{12} & \cdots & x_{1j} & \cdots & x_{1n} & a_1 & \\
 & O_2 & x_{21} & \cdots & x_{22} & \cdots & x_{2j} & \cdots & x_{2n} & a_2 & \\
 & \vdots & \vdots & & \vdots & & \vdots & & \vdots & \vdots & \\
\text{Origins} & O_i & x_{i1} & \cdots & x_{i2} & \cdots & x_{ij} & \cdots & x_{in} & a_i & \text{Capacities} \\
 & \vdots & \vdots & & \vdots & & \vdots & & \vdots & \vdots & \\
 & O_m & x_{m1} & \cdots & x_{m2} & \cdots & x_{mj} & \cdots & x_{mn} & a_m & \\
\hline
 & & b_1 & \cdots & b_2 & \cdots & b_j & \cdots & b_n & & \\
 & & & & & \text{Demands} & & & & &
\end{array}
$$

$c_{ij}[i = 1, 2, \cdots, m;\ j = 1, 2, \cdots, n]$, the cost of transporting per unit of commodity from the ith origin to jth destination is a known quantity. It is assumed in general that $c_{ij} \geq 0$ for all i and j. But it may be negative under some special conditions. The problem before us is to determine the quantity $x_{ij}[i = 1, 2, \cdots, m;$ $j = 1, 2, \cdots, n]$ which is to be transported from ith origin to jth destination such that the transportation cost is minimum provided the condition (13.1.1) is satisfied. Mathematically, the problem can be written as,

$$\text{minimize,} \quad z = \sum_{i=1}^{m} \sum_{j=1}^{n} c_{ij} x_{ij} \tag{13.1.2}$$

subject to the constraints

$$\sum_{j=1}^{n} x_{ij} = a_i, \ i = 1, 2, \cdots, m \tag{13.1.3}$$

$$\sum_{i=1}^{m} x_{ij} = b_j, \ j = 1, 2, \cdots, n \tag{13.1.4}$$

$$\text{and} \quad \sum_{i=1}^{m} a_i = \sum_{j=1}^{n} b_j.$$

From the above diagram, the constraints (13.1.3) and (13.1.4) can be written easily. The sum of the variables of the ith row is equal to a_i and the sum of the variables of the jth column is equal to b_j.

It is evident that $x_{ij} \geq 0$ for all i and j.

The problem is a minimization problem and

$$z = \sum_{i=1}^{m} \sum_{j=1}^{n} c_{ij} x_{ij}$$

is the *objective function* which is to be minimized.

In this problem, there are $(m + n)$ constraints of which all are equations of mn variables

$$x_{ij}[i = 1, 2, \cdots, m; j = 1, 2, \cdots, n].$$

Since, in general, in a L.P.P. the number of variables are greater than the number of constraints, therefore m and n both must be ≥ 2. Since

$$\sum_{i=1}^{m} a_i = \sum_{j=1}^{n} b_j,$$

all constraint equations are not linearly independent. There are only

$$(m+n)-1=(m+n-1)$$

linearly independent equations. This can be easily verified by considering the particular case given in (13.1.5) where the $R(A)=R(Ab)$ of A given in (13.1.7) is 4. Hence the problem ultimately is a L.P.P. with $(m+n-1)$ independent constraints and mn variables

$$x_{ij}[i=1,2,\cdots,m;\ j=1,2,\cdots,n].$$

Consider a particular T.P. involving 2 origins and 3 destinations.

Here the problem can be written as

$$\text{minimize,}\quad z=c_{11}x_{11}+c_{12}x_{12}+c_{13}x_{13}+c_{21}x_{21} \\ +c_{22}x_{22}+c_{23}x_{23}$$

subject to the constraints

$$\left.\begin{array}{llllllll}
(1) & x_{11}+x_{12}+x_{13} & & & & & = a_1 \\
(2) & & & x_{21}+x_{22}+x_{23} & & & = a_2 \\
(3) & x_{11} & & +x_{21} & & & = b_1 \\
(4) & & x_{12} & & +x_{22} & & = b_2 \\
(2) & & & x_{13} & & +x_{23} & = b_3
\end{array}\right\} \tag{13.1.5}$$

and $x_{ij}\geq 0$ for all i and j and

$$a_1+a_2=b_1+b_2+b_3. \tag{13.1.6}$$

In this problem there are $2\times 3=6$ variables and though apparently there are $2+3=5$ constraints but actually there are only $2+3-1=4$ independent constraints. For example, the constraint (1) can be expressed as,

$$(1)=(3)+(4)+(5)-(2)\quad [\text{as } a_1+a_2=b_1+b_2+b_3].$$

Hence the first equation is redundant. Similarly it can be shown that any one of the constraints can be reduced to a redundant constraint.

But still if we keep the first constraint intact in its due position, the co-efficient matrix A can be written symmetrically as

$$A=\begin{bmatrix} 1 & 1 & 1 & 0 & 0 & 0 \\ 0 & 0 & 0 & 1 & 1 & 1 \\ 1 & 0 & 0 & 1 & 0 & 0 \\ 0 & 1 & 0 & 0 & 1 & 0 \\ 0 & 0 & 1 & 0 & 0 & 1 \end{bmatrix} \tag{13.1.7}$$

There are six column vectors in the co-efficient matrix A.

A can be expressed as

$$A=(\mathbf{a}_{11},\mathbf{a}_{12},\mathbf{a}_{13},\mathbf{a}_{21},\mathbf{a}_{22},\mathbf{a}_{23}) \tag{13.1.8}$$

where $\mathbf{a}_{11}$ is the column vector associated with the variable x_{11} etc. Again each column vector can be expressed as the sum of two unit vectors. For example,

$$\mathbf{a}_{11} = \mathbf{e}_1 + \mathbf{e}_{2+1}; \quad \mathbf{a}_{12} = \mathbf{e}_1 + \mathbf{e}_{2+2}; \quad \mathbf{a}_{23} = \mathbf{e}_2 + \mathbf{e}_{2+3} \text{ etc.}$$

where all unit vectors are $(2+3)$ component unit column vectors.

In general, it can be written as

$$\mathbf{a}_{ij} = \mathbf{e}_i + \mathbf{e}_{m+j} \quad [i = 1, 2; j = 1, 2, 3]. \tag{13.1.9}$$

The maximum number of vectors which are linearly independent in this set of six vectors is $(2+3-1) = 4$.

Ultimately this T.P. problem can be written in the matrix form given below:

$$\text{minimize,} \quad z = \mathbf{cx} \text{ subject to}$$

$$A\mathbf{x} = \mathbf{b}, \quad \mathbf{x} \geq \mathbf{0} \tag{13.1.10}$$

where A is the co-efficient matrix given in (13.1.7) or (13.1.8),

$\mathbf{c}$ is an $2 \times 3 = 6$ component row vector,

$\mathbf{x}$ is an 6 component column vector, and

$\mathbf{b} = [a_1, a_2, b_1, b_2, b_3]$ is an $2 + 3 = 5$ component column vector.

The cost vector $\mathbf{c}$ can be written in matrix form

$$\mathbf{c} = \begin{bmatrix} c_{11} & c_{12} & c_{13} \\ c_{21} & c_{22} & c_{23} \end{bmatrix} \tag{13.1.11}$$

which is called as a *cost matrix*.

Similarly a general T.P. involving m origins and n destinations can be written in matrix form given below:

$$\text{minimize,} \quad z = \mathbf{cx}$$

$$\text{subject to} \quad A\mathbf{x} = \mathbf{b}, \mathbf{x} \geq \mathbf{0}$$

where $A = (\mathbf{a}_{11}, \mathbf{a}_{12}, \cdots, \mathbf{a}_{ij}, \cdots, \mathbf{a}_{mn})$ is the co-efficient matrix,

$\mathbf{c} = (c_{11}, c_{12}, \cdots, c_{ij}, \cdots, c_{mn})$ is an mn component row vector (cost vector),

$\mathbf{x} = (x_{11}, x_{12}, \cdots, x_{ij}, \cdots, x_{mn})$ is an mn component column vector (variable vector), and

$\mathbf{b} = [a_1, a_2, \cdots, a_m, b_1, b_2, \cdots, b_n]$ is an $(n+m)$ component column vector.

Each column vector $\mathbf{a}_{ij}$ can be expressed as

$$\mathbf{a}_{ij} = \mathbf{e}_i + \mathbf{e}_{m+j} \tag{13.1.12}$$

where both $\mathbf{e}_i$ and $\mathbf{e}_{m+j}$ are $(m+n)$ component unit column vectors.

As there are only $(m+n-1)$ linearly independent equations, the maximum number of vectors which will be linearly independent is $m+n-1$ in the set of mn vectors $\mathbf{a}_{ij} [i = 1, 2, \cdots, m; j = 1, 2, \cdots, n]$.

Any set of $(m+n)$ of more vectors $\mathbf{a}_{ij}$ are always linearly dependent.

Theorem 13.1.1 *There exists a feasible solution in each T.P. which is given by,*

$$x_{ij} = \frac{a_i b_j}{M} \quad [i = 1, 2, \cdots, m;\ j = 1, 2, \cdots, n]$$

where $M = \sum_{i=1}^{m} a_i = \sum_{j=1}^{n} b_j$.

Proof. Since all a_i and b_j are non-negative quantities therefore $x_{ij} \geq 0$ for all i and j.

$$\sum_{j=1}^{n} x_{ij} = \frac{a_i \sum_{j=1}^{n} b_j}{M} = \frac{a_i M}{M} = a_i$$

$$\text{and} \quad \sum_{i=1}^{m} x_{ij} = \frac{b_j \sum_{i=1}^{m} a_i}{M} = \frac{b_j M}{M} = b_j$$

which satisfy the conditions given in (13.1.3) and (13.1.4).

Hence in each T.P. there exists a feasible solution and

$$x_{ij} = \frac{a_i b_j}{M} \quad [i = 1, 2, \cdots, m; j = 1, 2, \cdots, n] \tag{13.1.13}$$

Theorem 13.1.2 *In each T.P. there exists at least one B.F.S. which makes the objective function a minimum.*

Proof. Since there exists a feasible solution, it has at least one B.F.S. (by theorem 6.1.1). As all c_{ij} and x_{ij} are finite quantities, therefore z is a function which is bounded from below and above. Hence the objective function must have finite minimum value. As the objective function has an optimal value, there exists at least one B.F.S. which will be an optimal solution to the problem.

Theorem 13.1.3 *In a balanced T.P. having m origins and n destinations $(m, n \geq 2)$ the exact number of basic variables is $m + n - 1$.*

The balanced T.P. is

$$\textit{minimize,} \quad z = \sum_{j=1}^{n} \sum_{i=1}^{m} c_{ij} x_{ij}$$

$$\textit{subject to} \quad \sum_{j=1}^{n} x_{ij} = a_i \quad [i = 1, 2, \cdots, m] \tag{13.1.14}$$

$$\sum_{i=1}^{m} x_{ij} = b_j \quad [j = 1, 2, \cdots, n] \tag{13.1.15}$$

$$\textit{and} \quad \sum_{i=1}^{m} a_i = \sum_{j=1}^{n} b_j$$

There are $m+n$ linear constraints with mn variables and $mn > m+n-1$ ($\because\ m, n \geq 2$).

From (13.1.14)

$$\sum_{i=1}^{m}\sum_{j=1}^{n} x_{ij} = \sum_{i=1}^{m} a_i = \sum_{j=1}^{n} b_j. \tag{13.1.16}$$

Now summing the first $(n-1)$ constraints of (13.1.15) we get,

$$\sum_{j=1}^{n-1}\sum_{i=1}^{m} x_{ij} = \sum_{j=1}^{n-1} b_j. \tag{13.1.17}$$

Now subtracting (13.1.17) from (13.1.16) we get,

$$\sum_{j=1}^{n}\sum_{i=1}^{m} x_{ij} - \sum_{j=1}^{n-1}\sum_{i=1}^{m} x_{ij} = \sum_{j=1}^{n} b_j - \sum_{j=1}^{n-1} b_j = b_n. \tag{13.1.18}$$

Thus we get,

$$\sum_{i=1}^{m}\left(\sum_{j=1}^{n} x_{ij} - \sum_{j=1}^{n-1} x_{ij}\right) = b_n, \quad \text{or,} \quad \sum_{i=1}^{m} x_{in} = b_n, \tag{13.1.19}$$

which is the last or nth constraint of (13.1.4). Therefore, there are only $(m+n-1)$ linearly independent equations with mn variables ($mn > m+n-1$). Thus from the definition of the basic solution, we can say that the number of basic variables is exactly $(m+n-1)$.

Remark. All basic variables may not be positive, some of them may be zero. When all basic variables are positive, the solution is called a non-degenerate B.F.S. When at least one basic variable is zero, the solution is called a degenerate B.F.S.

The number of basic cells will be exactly $(m+n-1)$ all of which contain $(m+n-1)$ basic variables which are either all positive basic variables or some variables may be 0 (zero) which has been shown later on.

13.2 Transportation Table

T.P. is a special case of a L.P.P. Therefore a T.P. can be solved by using simplex method. But the method is not suitable in solving a T.P. Here a specially designed table is constructed to solve the problem systematically which is called as a *transportation table*. A specimen of a transportation table with m origins and n destinations is given below.

In this table there are mn squares or rectangles arranged in m rows and n columns. Each square or rectangle is called as a *cell*. The cell which is in the ith row and jth column is called as (i, j)th cell or cell (i, j). Each cost component c_{ij} is displayed at the *south-east* corner of the corresponding cell. A component of a feasible solution x_{ij} (if any) is to be displayed inside a small square situated at the *north-west* corner of the cell (i, j). The different origin capacities and destination demands (requirements) are listed in the outer column and outer row respectively as given in the table (13.1). These quantities are called as *rim requirements*.

Table 13.1 : Transportation table

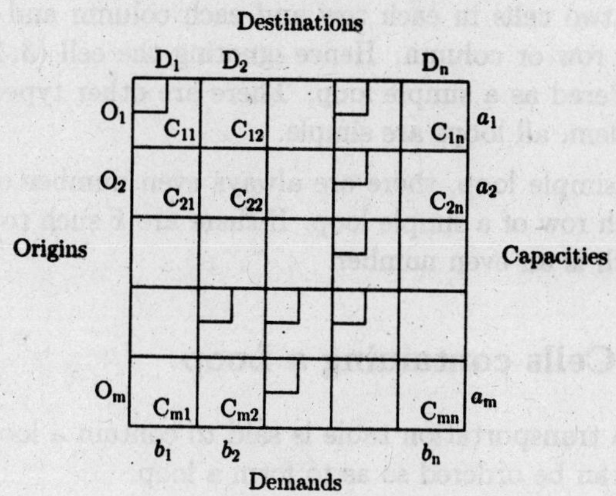

13.2.1 Loops in a Transportation Table and their properties

In a transportation table, an ordered set of four or more cells are said to form a loop (i) if and only if two consecutive cells in the ordered set lie either in the same row or in the same column and if (ii) the first and the last cell of the set also lie either in the same row or in the same column.

In the following figure one closed circuit is formed in each of the three transportation tables.

Table : 13.2

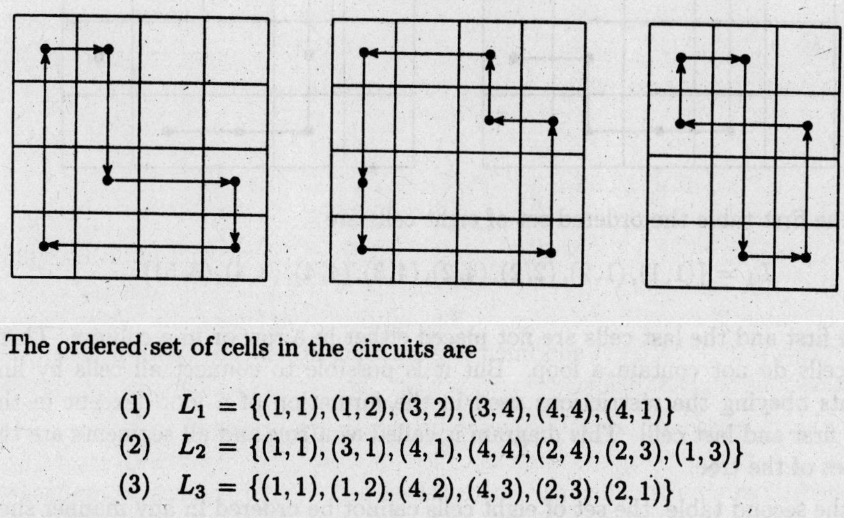

The ordered set of cells in the circuits are

$$\text{(1)} \quad L_1 = \{(1,1),(1,2),(3,2),(3,4),(4,4),(4,1)\}$$
$$\text{(2)} \quad L_2 = \{(1,1),(3,1),(4,1),(4,4),(2,4),(2,3),(1,3)\}$$
$$\text{(3)} \quad L_3 = \{(1,1),(1,2),(4,2),(4,3),(2,3),(2,1)\}$$

[Observe the ordered sets very carefully]

In the first and third table there are only two cells in each row and each column and the first and last cell are in the same row or same column. These loops are called simple loops. In the second table three cells are in the first column. But if we ignore or omit the cell (3, 1) and ordered the set of cells in the manner

$$L_2 = \{(1,1),(4,1),(4,4),(2,4),(2,3),(1,3)\}$$

then there are just two cells in each row and each column and the first and last cell are in the same row or column. Hence ignoring the cell (3, 1), the 2nd closed circuit is also considered as a simple loop. There are other types of loops. But in transportation problem, all loops are simple.

Note. (1) In a simple loop, there are always even number of cells. There are only two cells in each row of a simple loop. If there are k such rows, the number of cells will be $2k$ which is an even number.

13.2.2 Set of Cells containing a Loop

A set C of cells of a transportation table is said to contain a loop if the cells of C or any subset of C can be ordered so as to form a loop.

13.2.3 Tree and other form of Ordered Set of Cells

In the following two transportation tables with four origins and five destinations, $(4+5-1) = 8$ cells are selected. The cells of the first table are ordered in the following manner:

Table : 13.3

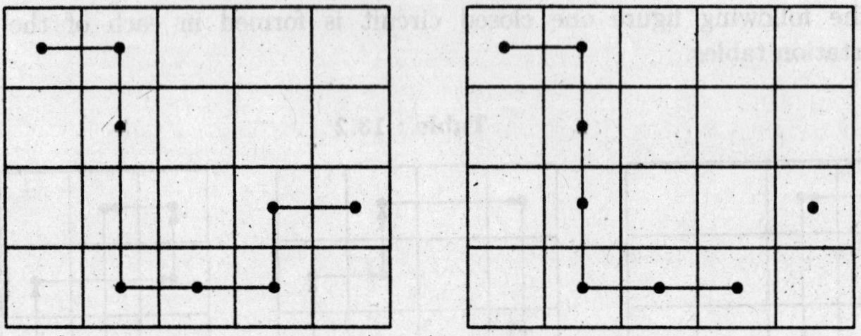

In the first table the ordered set of eight cells are

$$L_1 = \{(1,1),(1,2),(2,2),(4,2),(4,3),(4,4),(3,4),(3,5)\}.$$

The first and the last cells are not placed either in a row or in a column. These set of cells do not contain a loop. But it is possible to connect all cells by line segments obeying the restrictions used in the formation of a loop [except in the case of first and last cell]. This diagram is called as a *tree* and all segments are the branches of the tree.

In the second table, the set of eight cells cannot be ordered in any manner such that even a tree diagram cannot be constructed. But *it is interesting to note that if we select only one more cell out of remaining 12 cells arbitrarily, it is always possible to form a loop connecting either by all cells or by a subset of the cells in both tables* [verify the statement by selecting any cell]. From this geometrical idea we can establish a very important property regarding the linear dependence of a set of column vectors of the co-efficient matrix A of a transportation problem.

We know that the maximum number of column vectors $\mathbf{a}_{ij}$ which will be linearly independent in a T.P. having m origins, n destinations is $(m + n - 1)$. Hence for $m = 4, n = 5$ the maximum number of vectors which will be linearly independent is $(4 + 5 - 1) = 8$. Any set of nine or more vectors are always linearly dependent for $m = 4, n = 5$. We have established in the above figures that any set of nine or more cells can be ordered so as to form a loop and there is at least one set of eight cells which cannot be ordered so as to form a loop. From this, we conclude that if a set of four or more cells can be ordered so as to form a loop, the set of corresponding vectors are linearly dependent and there is at least one set of $(m+n-1)$ cells which cannot be ordered so as to form a loop and the set of corresponding vectors are linearly independent.

Theorem 13.2.1 *A set X of column vectors of the co-efficient matrix of a T.P. will be linearly dependent if their corresponding cells in the transportation table contain a loop.*

Theorem 13.2.2 *A set C of cells of a transportation table contains a loop, if their corresponding column vectors of the co-efficient matrix are linearly dependent (Proof is beyond the scope of the book).*

13.3 Determination of an Initial B.F.S.

So long we have discussed some fundamental properties of a transportation problem which will help in solving a problem. Next problem before us is to determine the initial B.F.S. of the problem and from this we proceed to find another B.F.S. which will improve the value of the objective function. There are various methods of finding an initial B.F.S. *It is interesting to note that in all cases the solution is a B.F.S.*. Of course, the solution may be non-degenerate or degenerate. If the solution is a B.F.S. the *cells to which some allocation is made are called as basic cells.* Obviously the allocated values are the components of the B.F.S. The methods of determining an initial B.F.S. are (1) North-west corner rule, (2) Row minima method, (3) Column minima method, (4) Matrix minima method and (5) Vogel's approximation method (VAM).

13.3.1 North-West Corner Rule:

Step 1. Compute $\min(a_1, b_1)$. If $a_1 < b_1 \cdot \min(a_1, b_1) = a_1$ and if $b_1 < a_1 \cdot \min(a_1, b_1) = b_1$. Select $x_{11} = \min(a_1, b_1)$ and allocate the value of x_{11} in the cell $(1, 1)$, i.e., in the cell situated in the *north-west corner* of the transportation table.

Step 2. If $a_1 < b_1$, the capacity of the origin O_1 will be exhausted completely which indicates that all other cells of the first row will remain vacant. But there remain some demand in the destination D_1. Compute $\min(a_2, b_1 - a_1)$. Select $x_{21} = \min(a_2, b_1 - a_1)$ and allocate the value of x_{21} in the cell $(2,1)$. Let us now make an assumption that $b_1 - a_1 < a_2$ which indicates that the demand of D_1 is satisfied completely. Of course, this assumption is not essential. With this assumption, the next cell for which some allocation is to be made, is the cell $(2,2)$ etc.

If $b_1 < a_1$, the demand of the destination D_1 will be satisfied exactly which indicates that all other cells of the first column will remain vacant. But the capacity of origin O_1 will not be exhausted. Compute, $\min(a_1 - b_1, b_2)$. Select $x_{12} = \min(a_1 - b_1, b_2)$ and allocate the value of x_{12} in cell $(1,2)$. Let us now make an assumption that $a_1 - b_1 < b_2$ which indicates that the capacity of O_1 is exhausted completely. With this assumption the next cell for which some allocation is to be made, is the cell $(2,2)$ etc. If $a_1 = b_1$, the capacity of the origin O_1 will be exhausted as well as the demand of D_1 will be satisfied simultaneously. In that case, the solution will be *degenerate*. Select either x_{12} or $x_{21} = \min(a_1 - b_1, b_2) = \min(a_2, a_1 - b_1) = 0$ and allocate the value 0 only in one of the two cells $(1,2)$ or $(2,1)$. The next cell for which some allocation is to be made is cell $(2,2)$. In this way proceed step by step until all the rim requirements are satisfied. In general, if an allocation is made in the cell (i,j) in the current step, the next allocation will be made either in cell $(i+1,j)$ or $(i,j+1)$. The feasible solution obtained by this method is always a B.F.S. In North-West Corner Rule, a tree diagram can be constructed by connecting all basic cells but no loop will be formed.

▶ **Example 13.3.1** *Determine an initial B.F.S. of the following problems by the method of North-West Corner Rule.*

(i)

	D_1	D_2	D_3	D_4	a_i
O_1	4	6	9	5	16
O_2	2	6	4	1	12
O_3	5	7	2	9	15
b_j	12	14	9	8	43

(ii)

	D_1	D_2	D_3	D_4	a_i
O_1	2	5	4	7	4
O_2	6	1	2	5	6
O_3	4	5	2	4	8
b_j	3	7	6	2	18

Note. Inside the rectangle, the cost matrix is given and outside the rectangle, the rim requirements are given.

Solution: The initial B.F.S. are displayed in the following two tables:

Table 13.4(A)

[12] 4	[4] 6	9	5
2	[10] 6	[2] 4	1
5	7	[7] 2	[8] 9

Table 13.4(B)

[3] 2	[1] 5	4	7
6	[6] 1	[0] 2	5
4	5	[6] 2	[2] 4

Explanation: (i) B.F.S. is displayed in the Table 13.4A : $\min(16, 12) = 12$. Therefore $x_{11} = 12$ and allocate it in the cell $(1,1)$. The demand of D_1 is satisfied and hence all other cells in first column remain vacant.

As $b_1 = 12 < 16 = a_1$, therefore next allocation will be in cell (1, 2) and $x_{12} = \min(16 - 12, 14) = 4$. Now the capacity of O_1 is exhausted. Next allocation will be in cell (2, 2) and $x_{22} = \min(12, 14 - 4) = 10$. Proceeding similarly we get $x_{23} = 2, x_{33} = 7$ and $x_{34} = 8$ and all the rim requirements are satisfied. The solution obtained is a B.F.S. because the set of cells (which contain components of F.S.) do not contain a loop (number of variables $= 4 + 3 - 1 = 6$) and the cost due to this assignment is $4 \times 12 + 6 \times 4 + 6 \times 10 + 4 \times 2 + 2 \times 7 + 9 \times 8 = 226$ units.

(ii) B.F.S. is displayed in Table 13.4B : $\min(4, 3) = 3$. Therefore $x_{11} = 3$ and allocate it in the cell (1, 1). The demand of D_1 is satisfied and hence all other cells of first column remain vacant. As $b_1 = 3 < 4 = a_1$, therefore next allocation will be in the cell (1, 2) and $x_{12} = \min(4 - 3, 7) = 1$. Now the capacity of O_1 is exhausted. Next allocation will be in cell (2, 2) $x_{22} = \min(6, 7 - 1) = 6$. Now the capacity of O_3 is exhausted and the demand of D_2 is satisfied simultaneously. Therefore either x_{23} or x_{32} will be zero. Let us take $x_{23} = 0$ and proceed similarly until all rim requirements are satisfied. The solution obtained is a degenerate B.F.S. as the number of variables is $4 + 3 - 1 = 6$ (one of them being zero) and the set of corresponding cells do not contain a loop.

13.3.2 Row Minima Method

Step 1. Select the smallest cost in the first row. Let it be c_{1j}; compute $\min(a_1, b_j)$. Set $x_{1j} = \min(a_1, b_j)$ and allocate it in the cell $(1, j)$. This is the maximum feasible amount which can be allocated in the cell $(1, j)$. If the smallest cost is not unique, select any one of the minimum cost arbitrarily.

Step 2. If $a_1 < b_j$, the capacity of the origin O_1 will be exhausted. But the demand of destination D_j remains unsatisfied. Cross off[1] the first row and diminish b_j by a_1. Proceeding similarly, allocate the maximum feasible amounts in the cells of the remaining rows starting from the second until all rim requirements are satisfied. If $b_j < a_1$, the total demand of the destination D_j is satisfied but the capacity of the origin O_1 is not exhausted completely. Cross off the jth column and diminish a_1 by b_j. Reconsider the first row and select the next smallest cost of this row. Let it be c_{1k}. Compute, $\min(a_1 - b_j, b_k)$. Set $x_{1k} = \min(a_1 - b_j, b_k)$ and allocate it in the cell $(1, k)$. Let us now make an assumption that $a_1 - b_j < b_k$. Therefore the capacity of O_1 will be exhausted completely [assumption is not restrictive]. Cross off the first row and repeat the above procedure for the second row and so on as in the above method until all rim requirements are satisfied.

Step 3. If $a_1 = b_j; \min(a_1, b_j) = a_1 = b_j$. Set $x_{1j} = a_1 = b_j$ and allocate it in the cell $(1, j)$. Due to this allocation, the capacity of origin O_1 will be exhausted as well the demand of D_j will be satisfied completely. In that case solution will be degenerate. Set $x_{1k} = 0$ and display it in the cell $(1, k)$ with the assumption, that the cost of $(1, k)$ cell is the next minimum cost. Now cross off both the first row and jth column and proceed similarly until all rim requirements are satisfied.

[1] By "cross off" a row or column we mean that we shall ignore that row or column for further consideration.

▶ **Example 13.3.2** *Find out an initial B.F.S. of the following balanced T.P. using row minima method.*

	D_1	D_2	D_3	D_4	a_i	
O_1	4	2	5	3	6	
O_2	5	4	3	2	13	Capacity
O_3	1	4	6	5	9	
b_j	7	8	5	8	28	
	Demand					

Solution: It is displayed in the transportation table given below:

Table 13.5

	D_1	D_2	D_3	D_4	a_i			
O_1	4	[6] 2	5	3	6			
O_2	5	[0] 4	[5] 3	[8] 2	13	13	5	
O_3	[7] 1	[2] 4	6	5	9	9	9	9
b_j	7	8	5	8	28 / 28			
	7	2	5	8				
	7	2	5					
	7	2						

Explanation. B.F.S. is given in the Table 13.5.

Lowest cost in first row is $c_{12} = 2$. $\min(a_1, b_2) = \min(6, 8) = 6$. Set $x_{12} = 6$ and allocate it in cell $(1, 2)$, $a_1 < b_2$; therefore the capacity of O_1 is exhausted completely and hence cross off the first row and diminish b_2 by a_1 which is shown in Table 13.5. Give a shade on the first row and ignore it for future computation.

Lowest cost in the second row is $c_{24} = 2$. $\min(a_2, b_4) = \min(13, 8) = 8$. Set $x_{24} = 8$ and allocate it in the cell $(2, 4)$; $b_4 < a_2$. Therefore the capacity of O_2 will not be exhausted; give a shade on the fourth column and ignore the column for future computation since the demand of D_4 is satisfied completely. Cross off the fourth column and diminish a_2 by b_4. Reconsider the second row. The next lowest cost in the row is $c_{23} = 3$. $\min(a_2 - b_4, b_2) = \min(13 - 8, 5) = 5$. Set $x_{23} = 5$ and allocate it in cell the $(2, 3)$. As $a_2 - b_4 = 5 = b_3$, the capacity of O_2 is exhausted as well as the demand of D_3 is satisfied completely. Therefore the solution will be degenerate. The next lowest cost of the second row is $c_{22} = 4$. Set $x_{22} = 0$ and allocate it in the cell $(2, 2)$. Cross off the second row and third column simultaneously. Now complete the Table 13.5 and all the rim requirements are satisfied now. Solution obtained is a B.F.S., because the number of variables are $3 + 4 - 1 = 6$ and the set of corresponding cells do not contain a loop.

Column Minima Method: The technique used in the column minima method is same as that in the row minima method.

13.3.3 Matrix Minima Method or (Cost Minima Method)

Step 1. Select the smallest cost in the cost matrix. Let it be c_{ij}. Set $x_{ij} = \min(a_i, b_j)$ and allocate it in the cell (i, j). This is the maximum feasible amount that can be allocated in the cell (i, j).

Step 2. If $a_i < b_j$, the capacity of the origin O_i will be exhausted completely. Cross off the ith row and diminish b_j by a_i.

If $b_j < a_i$, the demand of the destination D_j will be satisfied completely. Cross off the jth column and diminish a_i by b_j.

Step 3. If $a_i = b_j$, the capacity of the origin O_i will be exhausted and the demand of D_j will be satisfied simultaneously. Set $x_{ij} = a_i = b_j$; allocate it in the cell (i, j). Cross off either the ith row or jth column but not the both. Of course, we may drop both the ith row and jth column by inserting a basic variable 0 (zero) at a cell corresponding to the lowest cost of those row and column.

Step 4. Apply the same technique in the reduced transportation table until all rim requirements are satisfied. At any stage, if the minimum cost is not unique, make any arbitrary choice among the minima.

► **Example 13.3.3** *Determine an initial B.F.S. to the following balanced T.P. using matrix minima method:*

	D_1	D_2	D_3	D_4	a_i	
O_1	5	3	6	2	19	
O_2	4	7	9	1	37	Capacity
O_3	3	4	7	5	34	
b_j	16	18	31	25	90	
	Demand					

where O_i and D_j denote the ith origin and jth destination respectively.

Explanation: B.F.S. is given in the Table 13.6.

Table 13.6

	D_1	D_2	D_3	D_4	a_i			
O_1	5	[18] 3	[1] 6	2	19	19	1	1
O_2	4	7	[12] 9	[25] 1	37	12	12	12
O_3	[16] 3	4	[18] 7	5	34	34	34	18
b_j	16	18	31	25	90 / 90			
	16	18	31					
	16		31					

The smallest cost is $c_{24} = 1$. Set $x_{24} = \min(a_2, b_4) = \min(37, 25) = 25$ and allocate it in the cell $(2, 4)$. $b_4 < a_2$, therefore cross off the fourth column and diminish a_2 by b_4 and give a shade on the fourth column and ignore it for future computation.

The smallest cost in the reduced table is c_{12} or c_{31}. Let us select $c_{12} = 3$ as the smallest cost and allocate $x_{12} = \min(a_1, b_2) = \min(19, 18) = 18$ in the cell $(1, 2), b_2 < a_1$, therefore cross off the second column; give shade on the 2nd column and diminish a_1 by b_2. Proceed similarly until all rim requirements are satisfied and Table 13.6 gives the B.F.S. It is a B.F.S. because the number of variables are $4 + 3 - 1 = 6$ and the cells corresponding to the feasible solution do not contain a loop. Here the solution is not unique.

13.3.4 Vogel's Approximation Method: (Unit Penalty Method)

Step 1. Select the lowest and next to lowest cost for each row and determine the difference between them for each row and display them within the first bracket against the respective rows. If there are two or more with same lowest costs, difference may be taken to be zero (in some cases it may produce good result). Compute, similarly the difference for each column and display them within the bracket against the respective columns.

Step 2. Find the largest value of the differences and find out the row or column for which the difference is maximum. Let the maximum difference corresponds to ith row. Select the lowest cost in the ith row. Let it be c_{ij}. Allocate $x_{ij} = \min(a_i, b_j)$ in the cell (i, j) which is the maximum feasible amount that can be allocated in the cell (i, j). If the maximum difference is not unique, select any one of them.

Step 3. If $a_i < b_j$, cross off the ith row and diminish b_j by a_i. If $b_j < a_i$, cross off the jth column and diminish a_i by b_j. If $a_i = b_j$ allocate $x_{ij} = a_i = b_j$ in cell (i, j) and cross off either ith row or jth column but not the both. Of course, we can omit both the ith row and jth column simultaneously by inserting a basic variable 0 (zero) to one of the cells of the corresponding row or column possessing the next minimum cost and the solution will be degenerate then.

Step 4. Recompute the row and column differences for the reduced transportation table. Repeat the procedure discussed above until all rim requirements are satisfied.

► **Example 13.3.4** *Obtain an initial B.F.S. to the balanced T.P. given below using Vogel's approximation method.*

Warehouses

		D_1	D_2	D_3	D_4	a_i	
	O_1	19	30	50	10	7	
Factory	O_2	70	30	40	60	9	Factory capacity
	O_3	40	8	70	20	18	
	b_j	5	8	7	14	34	

Demands

Initial B.F.S.: Here the initial basic feasible solution using by VAM is being shown in a single table in a very compact manner which will save some time and labour.

The initial B.F.S. obtained by using the method is non-degenerate and unique and the initial solution is $x_{11} = 5$, $x_{14} = 2$, $x_{23} = 7$, $x_{24} = 2$, $x_{32} = 8$ and $x_{34} = 10$.

Explanation: Step 1: Select the lowest and next to lowest cost for each row and each column and determine the difference between them for each row and column and display them within the first bracket against the respective rows and columns. Here all the differences have been shown within the first compartment. Maximum difference is 22 which occurs at the second column and the minimum cost of that column is $c_{32} = 8$. Allocate $\min(18, 8) = 8$ in the cell $(3, 2)$. The demand of D_2 has been satisfied and shade the second column as shown in the Table 13.7. The resulting cost matrix will be obtained after deleting the cost components of the second column.

Table 13.7

	D_1	D_2	D_3	D_4	a_i			
O_1	5 / 19	30	50	2 / 10	7(9)	7(9)	2(40)	2(40)
O_2	70	30	7 / 40	2 / 60	9(10)	9(20)	9(20)	9(20)
O_3	40	8 / 8	70	10 / 20	18(12)	10(20)	10(50)	
b_j	5(21)	8(22)	7(10)	14(10)				
	5(21)		7(10)	14(10)				
			7(10)	14(10)				
			7(10)	4(50)				

Step 2: Applying the same technique in the resulting matrix, the capacities, demands and the differences of the cost components have been shown in the second compartment. Maximum difference is 21 which occurs in the first column and the lowest cost of that column is $c_{11} = 19$. Allocate $\min(7, 5) = 5$ in the cell $(1, 1)$. The demand of D_1 has been satisfied and shade the first column as shown in the table. The resulting cost matrix will be obtained deleting the cost components of the first column.

Step 3: Proceeding in the same way, we get the maximum difference 50 which occurs in the third row and minimum cost is $c_{34} = 20$. Allocate $\min(10, 14) = 10$ in the cell $(3, 4)$ and the capacity of O_3 will be exhausted and the resulting matrix will be obtained deleting the cost components of the third row; shade the third row as shown in the table. Using the same technique, ultimately we obtain the initial B.F.S. where all capacities have been exhausted and all the demands will be met [as the problem is a balanced transportation problem].

▶ **Example 13.3.5** *Find the initial basic feasible solution of the following balanced T.P. by Vogel's approximation method and calculate the cost for that solution.*

	A	B	C	a_i
F_1	10	9	8	8
F_2	10	7	10	7
F_3	11	9	7	9
F_4	12	14	10	4
b_j	10	10	8	

[Kalyani (H)'82]

Solution: I.B.F.S. by VAM.

Table 13.8

	A	B	C	a_i				
F_1	6 (10)	2 (9)	(8)	8(1)	8(1)	8(1)	8(1)	6
F_2	(10)	7 (7)	(10)	7(3)				
F_3	(11)	1 (9)	8 (7)	9(2)	9(2)	1(2)		
F_4	4 (12)	(14)	(10)	4(2)	4(2)	4(2)	4(2)	4
b_j	10(0)	10(2)	8(1)	28 / 28				
	10(1)	3(0)	8(1)					
	10(1)	3(0)						
	10(2)	2(5)						
	10							

Here the solution is not unique since in the second compartment maximum difference occurs in two rows. I.B.F.S. is $x_{11} = 6$, $x_{12} = 2$, $x_{22} = 7$, $x_{32} = 1$, $x_{33} = 8$, $x_{41} = 4$. There are six components of B.F.S. and $6 = 4+3-1$ and all the components are positive. Hence the solution is a non-degenerate B.F.S. and the cost $= 6\times10+2\times9+7\times7+1\times9+8\times7+4\times12 = 60+18+49+9+56+48 = 240$.

▶ **Example 13.3.6** *Find the initial basic feasible solution of the following balanced T.P. with the help of matrix minima method and VAM method and compare their corresponding costs.*

	D_1	D_2	D_3	D_4	a_i
O_1	15	28	13	21	18
O_2	22	15	19	14	14
O_3	16	12	14	31	13
O_4	24	23	15	30	20
b_j	16	15	10	24	

Solution: (i) Matrix Minima Method:

Table 13.9

	D_1	D_2	D_3	D_4	a_i				
O_1	8 (15)	(28)	10 (13)	(21)	18	18	8	8	
O_2	(22)	(15)	(19)	14 (14)	14	14	14		
O_3	(16)	13 (12)	(14)	(31)	13				
O_4	8 (24)	2 (23)	(15)	10 (30)	20	20	20	20	20
b_j	16	15	10	24	65 / 65				
	16	2	10	24					
	16	2		24					
	16	2		10					
	8	2		10					

Here $m+n = 4+4 = 8$, and the number of basic cells $7 = 4+4-1$, and the set of cells containing positive components do not form a loop. Hence the solution is a non-degenerate basic feasible solution. I.B.F.S. by matrix minima method which is unique and the solution is $x_{11} = 8$, $x_{13} = 10$, $x_{24} = 14$, $x_{32} = 13$, $x_{41} = 8$, $x_{42} = 2$, $x_{44} = 10$ and the cost

$$= 8 \times 15 + 10 \times 13 + 14 \times 14 + 13 \times 12 + 8 \times 24 + 2 \times 23 + 10 \times 30$$
$$= 120 + 130 + 196 + 156 + 192 + 46 + 300 = 1140.$$

(ii) I.B.F.S. by VAM

Table 13.10

(Allocations shown in brackets, followed by the cost.)

	D_1	D_2	D_3	D_4	a_i				
O_1	[16] 15	28	13	[2] 21	18(2)	18(6)	18(6)	18(6)	2(7)
O_2	22	15	19	[14] 14	14(1)	14(1)			
O_3	16	[13] 12	14	31	13(2)	13(4)	13(4)		
O_4	24	[2] 23	[10] 15	[8] 30	20(8)	10(1)	10(1)	10(1)	10(7)
b_j	16(1)	15(3)	10(1)	24(7)	65 / 65				
	16(1)	15(3)		24(7)					
	16(1)	15(11)		10(9)					
	16(9)	2(5)		10(9)					
		2(5)		10(9)					

Here also the solution is a non-degenerate basic feasible solution. The solution is $x_{11} = 16$, $x_{14} = 2$, $x_{24} = 14$, $x_{32} = 13$, $x_{42} = 2$, $x_{43} = 10$, $x_{44} = 8$ and here the solution is not unique. Cost is

$$= 16 \times 15 + 2 \times 21 + 14 \times 14 + 13 \times 12 + 2 \times 23 + 10 \times 15 + 8 \times 30$$
$$= 240 + 42 + 196 + 156 + 46 + 150 + 240 = 1070 \text{ units}.$$

VAM gives a better solution than that of the matrix minima method.

► **Example 13.3.7** *Find the initial basic feasible solution of the following balanced T.P. by row minima method and VAM and compare their corresponding costs.*

	D_1	D_2	D_3	D_4	D_5	a_i
O_1	4	7	0	3	6	14
O_2	1	2	−3	3	8	9 Supply
O_3	3	−1	4	0	5	17
b_j	8	3	8	13	8	

Demand

Solution: (i) Row minima method:

Table 13.11

	D_1	D_2	D_3	D_4	D_5	a_i				
O_1	4	7	[8] 0	[6] 3	6	14	6			
O_2	[8] 1	[1] 2	−3	3	8	9	9	9	1	
O_3	3	[2] −1	4	[7] 0	[8] 5	17	17	17	17	17
b_j	8	3	8	13	8	40 / 40				
	8	3		13	8					
	8	3		7	8					
		3		7	8					
		2		7	8					

The number of rows and columns are five and three; $5 + 3 - 1 = 7$. Here the number of basic cells are seven and the set of cells do not form a loop. And all components of the basic variables are positive and hence the solution is a non-degenerate B.F.S. and unique. I.B.F.S. $= x_{13} = 8$, $x_{14} = 6$, $x_{21} = 8$, $x_{22} = 1$, $x_{32} = 2$, $x_{34} = 7$, $x_{35} = 8$ and the cost is

$$= 8 \times 0 + 6 \times 3 + 8 \times 1 + 1 \times 2 + 2 \times (-1) + 7 \times 0 + 8 \times 5$$
$$= 0 + 18 + 8 + 2 - 2 + 0 + 40 = 66 \text{ units.}$$

(ii) By VAM:

Table 13.12

	D_1	D_2	D_3	D_4	D_5	a_i				
O_1	[7] 4	7	0	3	[7] 6	14(3)	14(1)	14(1)	14(2)	14(2)
O_2	[1] 1	2	[8] −3	3	8	9(4)	1(1)	1(2)	1(7)	
O_3	3	[3] −1	4	[13] 0	[1] 5	17(1)	17(1)	14(3)	1(2)	1(2)
b_j	8(2)	3(3)	8(3)	13(3)	8(1)	40 / 40				
	8(2)	3(3)		13(3)	8(1)					
	8(2)			13(3)	8(1)					
	8(2)				8(1)					
	7(1)				8(1)					

Here the solution is not unique but here also the solution is non-degenerate and the B.F.S. $x_{11} = 7$, $x_{15} = 7$, $x_{21} = 1$, $x_{23} = 8$, $x_{32} = 3$, $x_{34} = 13$, $x_{35} = 1$ and the cost

$$= 7 \times 4 + 7 \times 6 + 1 \times 1 + 8 \times (-3) + 3 \times (-1) + 13 \times 0 + 1 \times 5$$
$$= 28 + 42 + 1 - 24 - 3 + 0 + 5 = 49 \text{ units.}$$

VAM gives better result than that of the row minima method.

▶ **Example 13.3.8** ***Obtain the initial B.F.S. to the following transportation problem by Vogel's approximation method and prove that the solution is degenerate.***

	D_1	D_2	D_3	D_4	a_i
O_1	4	6	5	2	6
O_2	6	4	1	4	10
O_3	5	2	3	1	12
O_4	4	6	7	8	14
b_j	9	16	10	7	42

Table 13.13

	D_1	D_2	D_3	D_4	a_i			
O_1	4	6	5	[6] 2	6(2)	6(2)	6(2)	
O_2	6	4	[10] 1	4	10(3)			
O_3	5	[12] 2	3	1	12(1)	12(1)		
O_4	[9] 4	[4] 6	[0] 7	[1] 8	14(2)	14(2)	14(2)	14(2)
b_j	9(0)	16(2)	10(2)	7(1)	42 / 42			
	9(0)	16(4)	0(2)	7(1)				
	9(0)	4(0)	0(2)	7(6)				
	9	4	0	1				

Here the solution is a degenerate B.F.S. The solution is not unique because the basic variables can be put in other cells so that the set of cells do not contain a loop. I.B.F.S. $x_{14} = 6$, $x_{23} = 10$, $x_{32} = 12$, $x_{41} = 9$, $x_{42} = 4$, $x_{43} = 0$, $x_{44} = 1$ and the cost

$$= 6 \times 2 + 10 \times 1 + 12 \times 2 + 9 \times 4 + 4 \times 6 + 0 \times 7 + 1 \times 8$$
$$= 12 + 10 + 24 + 36 + 24 + 0 + 8 = 114 \text{ units.}$$

Explanation. Taking the difference of two lowest costs of the first column as 0 (zero) all the differences are placed in the first compartment. Maximum difference is 3 which occurs in the second row and the lowest cost of the second row is $c_{23} = 1$. Allocate $\min(10, 10) = 10$ in the cell $(2, 3)$ and the supply of O_2 will be exhausted and the demand of D_3 will be met simultaneously. Then we can cross off either the second row and put '0' in the third column or cross off the third column and put '0' in the second row. Proceeding similarly as in the above problem ultimately we get a degenerate B.F.S. which is not unique one; $m + n = 4 + 4 = 8$; therefore $m + n - 1 = 7$, there is one of which is zero and the set of cells do not form a loop. Hence the B.F.S. is degenerate one.

Note. In some cases, if the difference of two lowest costs be taken as zero, we may get a better reult which we can notice in some problems. But Prof. Hadley had taken the difference between the lowest and next to lowest cost which had been shown in the problem solved in the table (9 – 14) of page 309 of 'Linear Programming' written by him.

Remarks. We have discussed five methods of finding an initial B.F.S. to a problem. From the discussions it is clear that the solutions are not unique. The basic feasible solution obtained by using north-west corner rule may be far from optimal as the costs are totally ignored during computation. The solutions obtained by matrix minima method and Vogel's approximation method are good enough because generally these require lesser number of iterations to reach at the optimal stage. Therefore in order to solve a problems it is better to find an initial B.F.S. by using either matrix minima method or VAM method. Out of these, generally VAM requires lesser number of iterations. Hence if there is no instruction in finding the initial B.F.S. of a problem by using a particular method, we should always find the initial B.F.S. by VAM.

Exercise 13A

1. Obtain an initial B.F.S. to the T.P. using north-west corner rule.

(a)

	D_1	D_2	D_3	D_4	a_i
O_1	4	3	2	5	6
O_2	6	1	4	3	9
O_3	7	2	4	6	7
b_j	4	6	6	6	

(b)

	D_1	D_2	D_3	D_4	a_i
O_1	9	8	5	7	12
O_2	4	6	8	7	14
O_3	5	8	9	5	16
b_j	8	18	13	3	

Prove that initial B.F.S. to the T.P. (b) is degenerate. Find an initial B.F.S. to the problem (b) using matrix minima method and prove that the B.F.S. will not be degenerate then.

2. Determine an initial B.F.S. to the following problem using (a) row minima method, (b) matrix minima method and (c) Vogel's approximation method.

	D_1	D_2	D_3	D_4	a_i	
O_1	21	16	25	13	11	
O_2	17	18	14	23	13	Capacity
O_3	32	27	18	41	19	
b_j	6	10	12	15	43	

Demand **[C.U.M.Com'86;C.U.(P)'89]**

3. Obtain an initial B.F.S. to the transportation problem given in Ex. (13.3.8) by north-west corner rule and matrix minima method and prove that the solution is degenerate in the second case.

4. For the following problem obtain the different starting solutions by adopting the north-west corner rule and Vogel's approximation method respectively and find out which solution is better?

				a_i
	5	1	8	12
	2	4	0	14
	3	6	7	4
b_j	9	10	11	

[C.U.(P)'81]

[*Hints.* In north-west corner rule solution is $x_{11} = 9$, $x_{12} = 3$, $x_{22} = 7$, $x_{23} = 7$, $x_{33} = 4$ and the corresponding cost is $5\times9+1\times3+7\times4+0\times7+7\times4 = 104$ units and the solution due to VAM is $x_{11} = 2$, $x_{12} = 10$, $x_{21} = 3$, $x_{23} = 11$, $x_{31} = 4$ and the corresponding cost is $5\times2+1\times10+2\times3+0\times11+3\times4 = 38$ units.]

5. Find the initial basic feasible solution to the following transportation problem by adopting (a) row minima method, (b) matrix minima method, (c) VAM and prove that row minima method and VAM will give better result.

	D_1	D_2	D_3	D_4	a_i
O_1	9	5	4	7	12
O_2	2	5	8	3	15
O_3	6	2	3	4	13
b_j	6	11	13	10	40

6. For a problem in which $a_1 = 3$, $a_2 = 4$, $a_3 = 7$ and $b_1 = 1$, $b_2 = 3$, $b_3 = 3$, $b_4 = 2$, $b_5 = 5$, find the initial B.F.S. and show that it is a degenerate B.F.S.

[C.U.M.Com.'88,'91]

Answers

1. (a) $x_{11} = 4$, $x_{12} = 2$, $x_{22} = 4$, $x_{23} = 5$, $x_{33} = 1$, $x_{34} = 6$.
   (b) $x_{11} = 8$, $x_{12} = 4$, $x_{22} = 14$, $x_{23} = 0$, $x_{33} = 13$, $x_{34} = 3$.
   (c) $x_{13} = 12$, $x_{21} = 8$, $x_{22} = 5$, $x_{23} = 1$, $x_{32} = 13$, $x_{34} = 3$.
2. (a) $x_{14} = 11$, $x_{21} = 1$, $x_{23} = 12$, $x_{31} = 5$, $x_{32} = 10$, $x_{34} = 4$.
   (b) Same as (a).
   (c) $x_{14} = 11$, $x_{21} = 6$, $x_{22} = 3$, $x_{24} = 4$, $x_{32} = 7$, $x_{33} = 12$.

3. (a) $x_{11} = 6, x_{21} = 3, x_{22} = 7, x_{32} = 9, x_{33} = 3, x_{43} = 7, x_{44} = 7$.

   (b) $x_{11} = 6, x_{22} = 0, x_{23} = 10, x_{32} = 5, x_{34} = 7, x_{41} = 3, x_{42} = 11$, solution is not unique.

5. Solution is $x_{13} = 12, x_{21} = 6, x_{24} = 9, x_{32} = 11, x_{33} = 1, x_{34} = 1$ in row minima method and VAM. Cost = 116 units. $x_{13} = 11, x_{14} = 1, x_{21} = 6, x_{24} = 9, x_{32} = 11, x_{33} = 2$, cost 118 units.

6. As the cost components are not given, we shall have to find the initial B.F.S. using only North-West Corner Rule. I.B.F.S. is $x_{11} = 1, x_{12} = 2, x_{22} = 1, x_{23} = 3$, either x_{24} or $x_{33} = 0, x_{34} = 2, x_{35} = 3$. Solution is degenerate.

13.4 Optimality Conditions

So long we have discussed the methods of determining the initial B.F.S. The next problem before us is to find whether the solution obtained, is optimal or not. A transportation problem is a maximization problem. Hence at the optimal stage, $c_{ij} - z_{ij} \geq 0$ for all cells corresponding to non-basic variables, i.e., $z_{ij} - c_{ij} \leq 0$ for all cells corresponding to non-basic variables. Therefore to test the optimality of the problem, we shall have to determine the values of $z_{ij} - c_{ij}$ for all cells corresponding to non-basic variables. If $z_{ij} - c_{ij} \leq 0$ for all cells corresponding to non-basic variables, the solution is optimal. We know from theorem (A.2.14) that for all basic cells $z_{ij} - c_{ij} = 0$. This property is very useful in determining the values of $z_{ij} - c_{ij}$ for all cells corresponding to non-basic variables. If the conditions $z_{ij} - c_{ij} \leq 0$ are not satisfied for all non-basic cells, we shall have to proceed further to get an optimal solution which we shall discuss in future.

13.4.1 Determination of net evaluation ($\mathbf{z_{ij} - c_{ij}}$) (u v method)

In determining the net evaluation we shall make use of the property of the duality theory.

The original T.P. is

$$\text{minimize,} \quad z = \sum_{i=1}^{m}\sum_{j=1}^{n} c_{ij}x_{ij}$$

subject to

$$\sum_{j=1}^{n} x_{ij} = a_i, i = 1, 2, \cdots, m$$

$$\sum_{i=1}^{m} x_{ij} = b_j, j = 1, 2, \cdots, n,$$

where there are $(m+n)$ constraints all of which are equations and out of them only $(m+n-1)$ equations are independent. Hence there are $(m+n)$ dual variables to the primal (original) problem of which *one variable can be selected arbitrarily and all variables are unrestricted* in sign [as all the primal constraints are equations]. If the dual variables are

$$\mathbf{w} = (u_1, u_2, \cdots, u_m, v_1, \cdots, v_n) = (\mathbf{u}, \mathbf{v}) \tag{13.4.1}$$

the dual constraints are given by

$$u_i + v_j \leq c_{ij} \tag{13.4.2}$$

$u_i[i = 1, 2, \cdots, m]$ and $v_j[j = 1, 2, \cdots, n]$ are unrestricted in sign.

Now if B^{-1} is the basis inverse of the primal problem at the optimal stage and $\mathbf{c}_B$ is the associated cost vector, the dual optimal solution is given by

$$\mathbf{w} = \mathbf{c}_B B^{-1} \tag{13.4.3}$$

[from (theorem 10.7.1)] $\mathbf{w}$ is expressed here as a row vector with $(m+n)$ components. If $\mathbf{a}_j$ is a column vector corresponding to a non-basic variable then

$$z_j - c_j = \mathbf{c}_B B^{-1} \mathbf{a}_j - c_j. \tag{13.4.4}$$

Therefore the net evaluation $z_{ij} - c_{ij}$, corresponding to non-basic cells in a transportation problem are given by

$$\begin{aligned} z_{ij} - c_{ij} &= \mathbf{c}_B B^{-1} \mathbf{a}_{ij} - c_{ij} \\ &= \mathbf{w}\mathbf{a}_{ij} - c_{ij} \\ &= (\mathbf{u}, \mathbf{v})\mathbf{a}_{ij} - c_{ij}, \end{aligned} \tag{13.4.5}$$

where $\mathbf{a}_{ij}$ is the column vector of the co-efficient matrix associated with x_{ij}.

$\mathbf{a}_{ij} = \mathbf{e}_i + \mathbf{e}_{m+j}$. Hence from (13.4.5) we get

$$\begin{aligned} z_{ij} - c_{ij} &= (u_1, u_2, \cdots, u_m, v_1, v_2, \cdots, v_n)(\mathbf{e}_i + \mathbf{e}_{m+j}) - c_{ij} \\ &= u_i + v_j - c_{ij}[i = 1, 2, \cdots, m; j = 1, 2, \cdots, n.] \end{aligned} \tag{13.4.6}$$

For the basic cells, the net evaluations $z_{ij} - c_{ij} = 0$, i.e., if x_{rs} be a basic variable then

$$z_{rs} - c_{rs} = 0 \tag{13.4.7}$$

$$\begin{aligned} \text{or,} \quad & u_r + v_s - c_{rs} = 0 \\ \text{or,} \quad & u_r + v_s = c_{rs} \text{ for all basic cells.} \\ \text{or,} \quad & u_r = c_{rs} - v_s \text{ if } v_s \text{ is known} \\ \text{or,} \quad & v_s = c_{rs} - u_r \text{ if } u_r \text{ is known} \end{aligned} \tag{13.4.8}$$

If we select arbitrarily one of the values of the dual variables $(u_1, u_2, \cdots, u_m, v_1, v_2, \cdots, v_n)$ as zero, all other values can be determined using the relation (13.4.8). And once all quantities are known, the net evaluations $z_{ij} - c_{ij}$, given by the formula in (13.4.6) can be calculated. All these calculations can be done easily from the transportation table.

13.4.2 Numerical Calculation of the Net Evaluations corresponding to the Non-basic Cells

Below given is a transportation table involving 4 origins and 4 destinations in which the cost components are displayed in their proper places and all basic cells are marked by circular black spots. The solution (not given in the table) is a *basic feasible solution.* The problem before us is to calculate numerically all $z_{ij} - c_{ij}$ corresponding to the non-basic cells.

Table 13.14

				u_i
• 6	• 9	(−1) 8	(3) 7	3
(−1) 4	• 6	• 4	• 7	0
(−6) 9	(1) 5	• 4	(5) 2	0
(1) 3	(6) 1	(−1) 6	• 8	1
v_j: 3	6	4	7	

From (13.4.8) we get for a basic cell (r, s)

$$u_r + v_s = c_{rs}$$

and for non-basic cells (i, j)

$$z_{ij} - c_{ij} = u_i + v_j - c_{ij} \quad \text{[From (13.4.6)].}$$

We know that we can select arbitrarily one of the values of $u_i [i = 1, 2, \cdots, m]$ and $v_j [j = 1, 2, \cdots, n]$.

For simple computation we may take the value of one variable equal to zero.

In the above table, cells (1,1), (1,2), (2,2), (2,3), (2,4), (3,3) and (4,4) are marked with circular black spots. The cells are basic cells and the solution will be a B.F.S. because the set of cells do not contain a loop and the number of cells $(4 + 4 - 1) = 7$.

In the second row there are three basic cells. Let us take $u_2 = 0$ [in general a variable u_i or v_j is taken to be zero, the corresponding row or column of which contains the maximum basic cells].

For basic cell (2, 2)

$$u_2 + v_2 = c_{22}$$

$$\text{or,} \quad 0 + v_2 = 6 \text{ which implies that } v_2 = 6.$$

Similarly for basic cell (2, 3)

$$u_2 + v_3 = c_{23}$$

$$\text{or,} \quad 0 + v_3 = 4 \text{ which implies that } v_3 = 4.$$

and for basic cell (2, 4)

$$u_2 + v_4 = c_{24}$$

$$\text{or,} \quad 0 + v_4 = 7 \text{ which implies that } v_4 = 7.$$

Now for basic cell (3, 3)

$$\begin{aligned} u_3 + v_3 &= c_{33} \\ \text{or,} \quad u_3 + 4 &= 4 \text{ which implies that } u_3 = 0. \end{aligned}$$

For basic cell (4, 4)

$$\begin{aligned} u_4 + v_4 &= c_{44} \\ \text{or,} \quad u_4 - 7 &= 8 \text{ which implies that } u_4 = 1. \end{aligned}$$

Again for basic cell (1, 2),

$$\begin{aligned} u_1 + v_2 &= c_{12} \\ \text{or,} \quad u_1 - 6 &= 9 \text{ which implies that } u_1 = 3. \end{aligned}$$

and last of all for basic cell (1, 1),

$$\begin{aligned} u_1 + v_1 &= c_{11} \\ \text{or,} \quad 3 + v_1 &= 6 \text{ which implies that } v_1 = 3. \end{aligned}$$

Thus we have calculated all values of variables $u_1, u_2, u_3, u_4; v_1, v_2, v_3, v_4$. Then using the relation

$$z_{ij} - c_{ij} = u_i + v_j - c_{ij}$$

we can calculate all net evaluations corresponding to the non-basic cells.

For example,

$$\begin{aligned} z_{13} - c_{13} &= u_1 + v_3 - c_{13} = 3 + 4 - 8 = -1 \\ z_{14} - c_{14} &= u_1 + v_4 - c_{14} = 3 + 7 - 7 = 3 \end{aligned}$$

and thus $c_{14} - z_{14} = -3$ etc. and so on.

In the transportation table, the values of all u_i and v_j are listed outside of the block as given in the table and all net evaluations corresponding to non-basic cells are displayed on the north-east corner of the respective cells. Following this rule, all net evaluations corresponding to the non-basic cells can be calculated.

Note. There is another method of computing $z_{ij} - c_{ij}$ which is known as *stepping stone algorithm*. But in digital computer *u v method* is widely used.

13.4.3 Determination of the Entering Cell and the Entering Vector

To test the optimality of a B.F.S. at any stage, we require to calculate the values of all net evaluations corresponding to the non-basic cells. If all net evaluations are non-positive quantities, the solution is optimal. But if at least one of them is positive, the solution is not optimal. As in the simplex method, now we shall have to proceed further to get an optimal solution. The first problem before us is to select a vector which will enter in the basis and will move the solution towards optimality. The vector which will enter in the basis is the entering vector and the corresponding cell in the transportation table is the new *basic cell.* If $\mathbf{a}_{pk}$ be the entering vector, the cell (p, k) will be the new basic cell which will enter in the set of basic cells. And due to this, a cell will leave the set of basic cells. This cell is known as the *leaving cell or departing cell* and the vector corresponding to the departing cell is known as *departing vector.* The entering vector is selected in the following manner.

If $\max_{ij}(z_{ij} - c_{ij}, z_{ij} - c_{ij} > 0) = z_{pk} - c_{pk} > 0$, i.e., if the positive maximum of the net evaluations occurs at cell $(p, k), \mathbf{a}_{pk}$ is the entering vector and cell (p, k) is the entering cell. If the maximum is not unique we may select any one cell corresponding to the maximum value of the net evaluations.

In the table (13.14) $\max_{ij}(z_{ij} - c_{ij}, z_{ij} - c_{ij} > 0) = 6 > 0$ which occurs at the cell $(4, 2)$. Therefore, in the next iteration, the cell $(4, 2)$ will be the new basic cell, and$\mathbf{a}_{42}$ will be the vector to enter in the next basis.

13.4.4 Determination of the Departing Cell and the value of the Basic Variable in the Entering Cell

After the selection of entering cell, we can identify the cell which will leave the set of basic cells, geometrically from the transportation table. Let the cell (p, k) be the entering cell. The number of (basic) cells including the cell (p, k) is $m + n - 1 + 1 = m + n$. Evidently the set of column vectors corresponding to these $(m + n)$ cells are linearly dependent. Therefore it is always possible to construct a loop connecting the cell (p, k) and the set or any subset of the basic cells. Construct the loop by trial and error method and the loop is unique.

Now allocate a value $\theta > 0$ [a variable] in the cell (p, k) and readjust the basic variables in the ordered set of cells containing the simple loop by adding and subtracting the value θ alternately from the corresponding quantities such that all rim requirements are satisfied. If the ordered set of cells containing the simple loop are

$$(p, k), (p, t), (t, r), (r, s) \text{ etc.}$$

the new variables corresponding to the cells are θ, $x_{pt} - \theta$, $x_{tr} + \theta$, $x_{rs} - \theta$ etc. where x_{pt} etc. are the basic variables in the current iteration.

Now select the maximum value of θ in such a way that readjusted values of the variables vanish at least in one cell {excluding the cell (p, k)} of the ordered set *and all other variables remain non-negative.* Let us assume that the variable vanishes in the cell (r, s) satisfying all conditions stated above, i.e., $x_{rs} - \theta = 0$ which gives the value of $\theta = x_{rs}$. This is the value of new basic variable to be allocated in the new basic cell (p, k). The cell (r, s) is the departing cell and it will be a non-basic cell during next iteration. The vector $\mathbf{a}_{rs}$ will leave the basis and the new basic variables will be $x_{rs}, x_{pt} - x_{rs}, x_{tr} + x_{rs}$, i.e., $\theta, x_{pt} - \theta, x_{tr} + \theta$ etc. corresponding to the cells $(p, k), (p, t), (t, r)$ respectively. All basic variables in the cells not in the loop remain unchanged. It may so happen that for maximum value of θ, the readjusted variables may vanish for more than one cell. In that case it is not possible to select uniquely the cell which will leave the set of basic cells. Select arbitrarily one of the cells as a departing cell and write down the value 0 (zero) as new basic variable in all other such cells and the solution in the next iteration will be degenerate. The method of solving a degenerate problem will be discussed in future.

13.4.5 Computational Procedure

To solve a transportation problem, proceed step by step as mentioned below:

Step 1. Determine an initial B.F.S. of the given problem using any one of methods discussed previously. But to get the optimal solution quickly, in general use either matrix minima method or Vogel's approximation method.

Step 2. Calculate all net evaluations corresponding to non-basic cells and display them on the north-east corner of the corresponding non-basic cells. If all net evaluations are non-positive quantities at any iteration, the solution is optimal. Then calculate the corresponding minimum cost using the relation $\min z = \hat{z} = \Sigma \hat{c}_{ij}\hat{x}_{ij}$ where $\hat{x}_{ij}$ are the components of optimal solution and $\hat{c}_{ij}$ are the corresponding transportation cost per unit commodity. If at least one net evaluation is positive, the solution is not optimal.

Step 3. If the solution is not optimal, determine the entering cell and the value of θ which will be allocated in the entering cell and the cell which will leave the basis. Construct a new transportation table with readjusted basic variables. Calculate again, all net evaluations corresponding to the non-basic cells and display them on the north-east corner of the corresponding non-basic cells. Test the optimality of the solution. If the solution is optimal, calculate the minimum cost of transportation. If the solution is not optimal, proceed similarly until the optimality conditions are satisfied.

This method of solving a transportation problem is known as MODI METHOD.

Solved Problems

▶ **Example 13.4.1** *Obtain the optimum basic feasible solution to the transportation problem given in* Ex. 13.3 3 *and find out the corresponding cost of transportation.* [C.U.,App.Math.'77; C.U.(P)'83,'92,'93]

Solution: In table (13.6) an initial B.F.S. is given which is obtained using matrix minima method.

We have to test first whether the solution is optimal or not. For this we shall have to calculate the net evaluations for all non-basic cells. The calculated values are displayed on north-east corner of respective non-basic cells in the table 13.15(A).

Table 13.15(A)

				u_i
−3 / 5	[18] / 3	[1] / 6	−4 / 2	0
θ, 1 / 4	−1 / 7	[12] − θ / 9	[25] / 1	3
[16] − θ / 3	0 / 4	[18] + θ / 7	−6 / 5	1
v_j: 2	3	6	−2	

Table 13.15(B)

				u_i
−3 / 5	[18] / 3	[1] / 6	−3 / 2	0
[12] / 4	−2 / 7	−1 / 9	[25] / 1	2
[4] / 3	0 / 4	[30] / 7	−5 / 5	1
v_j: 2	3	6	−1	

During calculation of the net evaluations we have taken arbitrarily $u_1 = 0$ and have calculated all net evaluations. All net evaluations corresponding to the non-basic cells are not less than or equal to zero. Net evaluation corresponding to non-basic cell $(2, 1)$ is 1. Therefore the solution is not optimal.

13.4.6 Selection of Entering Vector and Entering Cell

Since there is only one cell (non-basic) $(2, 1)$ for which net evaluation is positive therefore the vector $\mathbf{a}_{21}$ corresponding to the cell $(2, 1)$ is the entering vector and cell $(2, 1)$ will be the entering cell (new basic cell in the next iteration).

13.4.7 Value of the Variable which is to be allocated in the new basic cell and the Vector which will leave the basis

Construct a loop connecting the cell $(2, 1)$ and the set of basic cells or any subset of the basic cells of the origin solution. In this problem the ordered set of cells $(2, 1), (2, 4), (3, 3), (3, 1)$ are said to form a loop. Now allocate a value $\theta > 0$ [a variable] in the cell $(2, 1)$ and readjust the basic variables in the ordered set of cells forming the loop, by adding and subtracting θ alternately as given in the table 13.15(A), such that all rim requirements are satisfied properly. Now select the maximum value of θ in such a way that the values of the readjusted variables vanish at least in one cell containing the loop $\{$except the cell $(2, 1)\}$ and *variables remain non-negative in other cells.*

From the table, it is clear that value of $\theta = 12$ and for that the value of variable in the cell $(2, 3)$ is zero. Therefore the cell $(2, 3)$ will leave the set of basic cells and the vector $\mathbf{a}_{23}$ will leave the basis. With this known value of $\theta = 12$ construct the new transportation table 13.15(B). All basic variables in the cells not in the loop remain unchanged and again calculate the net evaluations corresponding to the non-basic cells. Calculated values are displayed in table 13.15(B). All calculations are performed with the assumption that $u_1 = 0$. Here all net evaluations are non-positive quantities. Therefore the solution obtained is optimal. The optimal solution (non-degenerate) is given by $x_{12} = 18$, $x_{13} = 1$, $x_{21} = 12$, $x_{24} = 25$, $x_{31} = 4$ and $x_{33} = 30$ and the corresponding cost of transportation is given by

$$\hat{z} = 3 \times 18 + 6 \times 1 + 4 \times 12 + 1 \times 25 + 3 \times 4 + 7 \times 30 = 355 \text{ units.}$$

Note 1. As at the optimal stage, net evaluation, corresponding to the non-basic cell $(3, 2)$ is zero, then alternative optimal solution exists.

2. If we calculate $c_{ij} - z_{ij}$, the vector to enter in the next basis is $\mathbf{a}_{pk}$, where $\min_{ij}(c_{ij} - z_{ij}, c_{ij} - z_{ij} < 0) = c_{pk} - z_{pk}$ and at the optimal stage, all $c_{ij} - z_{ij} \geq 0$ for all non-basic cells.

► **Example 13.4.2** *Determine the optimal solution to the problem given in* Ex. 13.3.4 *and find the minimum cost of transportation.* **[C.U.(P)'99]**

Table 13.16(A)

				u_i
[5] / 19	−32 / 30	−60 / 50	[2] / 10	0
−1 / 70	θ 18 / 30	[7] / 40	[2] − θ / 60	50
−11 / 40	[8] − θ / 8	−70 / 70	[10] + θ / 20	10
v_j 19	−2	−10	10	

Table 13.16(B)

				u_i
[5] / 19	−32 / 30	−42 / 50	[2] / 10	0
−19 / 70	[2] / 30	[7] / 40	−18 / 60	32
−11 / 40	[6] / 8	−52 / 70	[12] / 20	10
v_j 19	−2	8	10	

First we have to test whether the initial B.F.S. is optimal or not. For this we shall have to calculate the net evaluations corresponding to all non-basic cells. The calculated values are displayed in the table 13.16(A). During computation of net evaluations we have taken arbitrarily $u_1 = 0$. All net evaluations corresponding to non-basic cells are not less than or equal to zero. Net evaluation corresponding to the non-basic cell $(2, 2)$ is 18. Therefore the solution is not optimal. We shall have to construct the next table to get an optimal solution.

13.4.8 Selection of Entering Vector and Entering Cell

Since there is only one cell $(2, 2)$ for which net evaluation is positive, therefore the vector $\mathbf{a}_{22}$, corresponding to the cell $(2, 2)$ is the entering cell. This cell $(2, 2)$ will be a basic cell in the next iteration.

13.4.9 Value of the Variable which is to be allocated in the new basic cell and the vector which will leave the basis

Construct a loop connecting the cell $(2, 2)$ and the set of basic cells or any subset of basic cells. In this problem the ordered set of cells $(2, 2)$, $(2, 4)$, $(3, 4)$ and $(3, 2)$ are said to form a simple loop [Ignore the intermediate basic cell (2, 3)]. Now allocate a value $\theta > 0$, a variable in the cell $(2, 2)$ and readjust the basic variables in the ordered set of cells forming a simple loop by adding and subtracting θ alternately as given in the table 13.16(A) such that all rim requirements are satisfied properly. Now select the maximum value of θ in such a way that the values of the readjusted variables vanish at least in one cell containing the loop [except in the cell $(2, 2)$] and *variables remain non-negative in other cells.* From the table it is clear that $\theta = 2$ and for that the value of the variable in the cell $(2, 4)$ is zero. Therefore the cell $(2, 4)$ will leave the set of basic cells and the vector $\mathbf{a}_{24}$ will leave the basis in the next iteration. With the known value $\theta = 2$ construct the new transportation table and again calculate the net evaluations corresponding to non-basic cells. Calculated values are displayed in the table 13.16(B) and all calculations are non-positive quantities. Hence the solution obtained is optimal. The optimal solution is given by $x_{11} = 5$, $x_{14} = 2$, $x_{22} = 2$, $x_{23} = 7$, $x_{32} = 6$ and $x_{34} = 12$ and the corresponding cost of transportation is given by

$$\hat{z} = 19 \times 5 + 10 \times 2 + 30 \times 2 + 40 \times 7 + 8 \times 6 + 20 \times 2 = 743 \text{ units.}$$

► **Example 13.4.3** *Obtain the initial B.F.S. to the following transportation problem by matrix minima method and then find out an optimal solution and the corresponding cost of transportation.*

	D_1	D_2	D_3	D_4	a_i
O_1	5	4	6	14	15
O_2	2	9	8	6	4
O_3	6	11	7	13	8
b_j	9	7	5	6	27

Solution: Initial B.F.S. calculated with the help of matrix minima method is given in table 13.17(A). Now calculate all net evaluations corresponding to the non-basic cells with the assumption $u_1 = 0$. All net evaluations are not non-positive. Hence the solution is not optimal.

Cell $(2, 4)$ has the positive net evaluation 3. Thus in the next iteration, cell $(2, 4)$ will be the new basic cell. Construct the loop as shown in the table 13.17(A). Loop is simple and unique. Insert the value $\theta > 0$ in the cell $(2, 4)$ and readjust the basic variables in the cells containing the loop accordingly as given in the table 13.17(A).

Table 13.17(A)

					u_i
	5 +θ 5	7 4	3 −θ 6	−2 14	0
	4 −θ 2	−8 9	−5 8	θ 3 6	−3
	0 6	−6 11	2 +θ 7	6 −θ 13	1
v_j	5	4	6	12	

Table 13.17(B)

					u_i
	8 5	7 4	−3 6	−5 14	0
	1 −θ 2	−8 9	−8 8	3 +θ 6	−3
	θ 3 6	−3 11	5 7	3 −θ 13	4
v_j	5	4	3	9	

Now the maximum value of θ will be 3 and cell $(1, 3)$ will leave the basic cell and all other variables remain non-negative. Construct the table 13.17(B) with the value of $\theta = 3$. Calculate all net evaluations corresponding to the non-basic cells with the assumption $u_1 = 0$. Cell $(3, 1)$ will be the new basic cell. Construct a simple loop with cell $(3, 1)$ as shown in the table 13.17(B). Insert the value $\theta > 0$ in the cell $(3, 1)$ and readjust the basic variables as shown in the table 13.17(B). Maximum value of $\theta = 1$, cell $(2, 1)$ will leave the set of basic cells and all other variables remain non-negative. With $\theta = 1$ construct the table 13.17(C).

Table 13.17(C)

					u_i
	8 5	7 4	0 6	−2 14	−1
	−3 2	−11 9	−8 8	4 6	−7
	1 6	−6 11	5 7	2 13	0
v_j	6	5	7	13	

Calculate all net evaluations corresponding to the non-basic cells in the table 13.17(C) with the assumption $u_2 = 0$. All net evaluations are non-positive. Hence the solution is optimal and the optimal solution is $x_{11} = 8$, $x_{12} = 7$, $x_{24} = 4$, $x_{31} = 1$, $x_{33} = 5$, $x_{34} = 2$ and the minimum cost of transportation is $\hat{z} = 5 \times 8 + 4 \times 7 + 6 \times 4 + 6 \times 1 + 7 \times 5 + 13 \times 2 = 159$ units.

Note. As the net evaluation corresponding to the non-basic cell (1, 3) is zero, an alternative optimal solution exists.

► **Example 13.4.4** *Solve the following balanced T.P. by using VAM to determine the initial B.F.S.*

	D_1	D_2	D_3	D_4	a_i
O_1	4	2	7	–1	27
O_2	3	0	2	4	33
O_3	5	3	4	5	23
O_4	3	5	4	–2	17
b_j	31	24	25	20	100 / 100

Solution: Determination of I.B.F.S. by VAM.

Table 13.18

	D_1	D_2	D_3	D_4	a_i				
O_1	[24] 4	2	7	[3] –1	27(3)	27(3)	24(2)	24(3)	
O_2	[7] 3	[24] 0	[2] 2	4	33(2)	33(2)	33(2)	9(1)	9(1)
O_3	5	3	[23] 4	5	23(1)	23(1)	23(1)	23(1)	23(1)
O_4	3	5	4	[17] –2	17(5)				
b_j	31(0)	24(2)	25(2)	20(1)	100 / 100				
	31(1)	24(2)	25(2)	3(5)					
	31(1)	24(2)	25(2)						
	31(1)		25(2)						
	7(2)		25(2)						

Here the solution is non-degenerate but not unique.

Optimality Test:

Table 13.19

	D_1	D_2	D_3	D_4	u_i
O_1	[24] 4	−1 2	−4 7	[3] −1	1
O_2	[7] 3	[24] 0	[2] 2	−6 4	0
O_3	0 5	−1 3	[23] 4	−5 5	2
O_4	0 3	−5 5	−2 4	[17] −2	0
v_j	3	0	2	−2	

Since there are three basic variables in the second row then let us take $u_2 = 0$; with u_2 zero we calculate all u_i and v_j which are shown in the table 13.19. Now we calculate all $z_{ij} - c_{ij}$ for all non-basic cells. All $z_{ij} - c_{ij} \leq 0$; and for non-basic cell $(4,1)$ $z_{41} - c_{41} = 0$. Then the alternative optimal solution exists. One optimal solution is $x_{11} = 24$, $x_{14} = 3$, $x_{21} = 7$, $x_{22} = 24$, $x_{23} = 2$, $x_{33} = 23$, $x_{44} = 17$ and the min cost $= 24 \times 4 + 3 \times (-1) + 7 \times 3 + 24 \times 0 + 2 \times 2 + 23 \times 4 + 17 \times -2 = 96 - 3 + 21 + 0 + 4 + 92 - 34 = 196$ units.

► **Example 13.4.5** *Solve the following balanced T.P. by using VAM to determine the initial B.F.S.*

	D_1	D_2	D_3	D_4	a_i
O_1	9	8	5	7	12
O_2	4	6	8	7	14
O_3	5	8	9	5	16
b_j	8	18	13	3	42 / 42

Solution: Determination of I.B.F.S. by VAM.

Table 13.20

	D_1	D_2	D_3	D_4	a_i		
O_1	9	8	[12] 5	7	12(2)		
O_2	4	[14] 6	8	7	14(2)	14(2)	
O_3	[8] 5	[4] 8	[1] 9	[3] 5	16(0)	16(0)	16
b_j	8(1)	18(2)	13(3)	3(2)	42 / 42		
	8(1)	18(2)	1(1)	3(2)			
	8	4	1	3			

I.B.F.S. by VAM is not unique but it is a non-degenerate B.F.S.

Optimality Test: In the third row there are four basic variables then we take $u_3 = 0$; with u_3 zero we calculate all u_i and v_j which are shown in the table (13.21). Now we calculate all $z_{ij} - C_{ij}$ for non-basic cells which are shown in the table (13.21).

Table 13.21

	D_1	D_2	D_3	D_4	u_i
O_1	−8 / 9	−4 / 8	[12] 5	−6 / 7	−4
O_2	−1 / 4	[14] 6	−1 / 8	−4 / 7	−2
O_3	[8] 5	[4] 8	[1] 9	[3] 5	0
v_j	5	8	9	5	

All $z_{ij} - c_{ij} < 0$ at all non-basic cells. Then we reach at the optimal stage. Optimal solution is $x_{13} = 12$, $x_{22} = 14$, $x_{31} = 8$, $x_{32} = 4$, $x_{33} = 1$, $x_{33} = 3$ and the optimal solution is unique. The min cost $= 12\times5+14\times6+8\times5+4\times8+1\times9+3\times5 = 60 + 84 + 40 + 32 + 9 + 15 = 240$ units.

▶ **Example 13.4.6** *Find the I.B.F.S. to the following transportation problem using north-west corner rule and prove that the optimal solution is non-degenerate though the initial solution is degenerate.*

	D_1	D_2	D_3	D_4	a_i
O_1	9	8	5	7	12
O_2	4	6	8	7	14
O_3	5	8	9	5	16
b_j	8	18	13	3	42

Solution: Initial B.F.S. is given in the table 13.21(A). Solution is degenerate with $x_{23} = 0$. Calculate the net evaluations corresponding to the non-basic cells with the assumption $u_1 = 0$. Net evaluations are 5, 3 and 3 corresponding to the non-basic cells (1, 3), (2, 1) and (3, 1) respectively. Hence the solution given is not optimal. Put $\varepsilon > 0$ [very small +ve quantity] in the cell (2, 3) instead of zero. As the net evaluation 5 in the cell (1, 3) is the positive maximum, then the cell (1, 3) will be the new basic cell. Construct a loop as given in the table 13.21(A). Put $\theta > 0$ in the cell (1, 3) and readjust the basic variables in the cells containing the loop. Now the maximum value of θ will be ϵ such that basic variable in the cell (2, 3) will be zero and all other variables remain positive. Thus the cell (2, 3) will

Table 13.21(A)

				u_i
[8] / 9	[4]−θ / 8	[θ] 5 / 5	−1 / 7	0
3 / 4	[14]+θ / 6	[ε]−θ / 8	−3 / 7	−2
3 / 5	−1 / 8	[13] / 9	[3] / 5	−1
v_j 9	8	10	6	

Table 13.21(B)

				u_i
[8]−θ / 9	[4]−ε / 8	[ε]+θ / 5	−6 / 7	0
3 / 4	[14]+ε / 6	−5 / 8	−8 / 7	−2
θ 8 / 5	4 / 8	[13]−θ / 9	[3] / 5	4
v_j 9	8	5	1	

leave the set of basic cells. With the value of $\theta = \epsilon$, construct the table 13.21(B) and again calculate the net evaluations corresponding to the non-basic cells. All net evaluations are not non-positive. Thus the solution is not optimal. Net evaluation corresponding to the cell (3, 1) is 8 which is a positive maximum. Thus the cell (3, 1) will be the new basic cell. Proceed accordingly as given in the table 13.21(B). Maximum value of θ will be 8 and cell (1, 1) will leave the set of basic cells. With the value of $\theta = 8$ construct the table 13.21(C). Now it is interesting to note that the solution will remain non-degenerate if we put $\epsilon = 0$ at this stage. Thus we put $\epsilon = 0$ and get the basic variables in the table 13.21(C).

Table 13.21(C)

				u_i
−8 / 9	[4]−θ / 8	[8]+θ / 5	−6 / 7	−4
−5 / 4	[14] / 6	−5 / 8	−8 / 7	−6
[8] / 5	θ 4 / 8	[5]−θ / 9	[3] / 5	0
v_j 5	12	9	5	

Table 13.21(D)

				u_i
−8 / 9	−4 / 8	[12] / 5	−6 / 7	−4
−1 / 4	[14] / 6	−1 / 8	−4 / 7	−2
[8] / 5	[4] / 8	[1] / 9	[3] / 5	0
v_j 5	8	9	5	

Now proceed step by step and in the table 13.21(D) we get the optimal solution because all net evaluations corresponding to the non-basic cells are non-positive. The optimal solution is

$x_{13} = 12$, $x_{22} = 14$, $x_{31} = 8$, $x_{32} = 4$, $x_{33} = 1$, $x_{34} = 3$ and $\min z = 5 \times 12 + 6 \times 14 + 5 \times 8 + 8 \times 4 + 9 \times 1 + 5 \times 3 = 240$.

The solution is non-degenerate as the number of positive basic variables is $(6 = 3 + 4 - 1)$ and the set of cells do not contain a loop.

Remark. Treat this problem as degenerate and see sec. 13.4.12.

13.4.10 Unbalanced Transportation Problems

There are two types of unbalanced T.P.:

1. **When $\sum_{i=1}^{m} a_i > \sum_{j=1}^{n} b_j$, i.e., the total available capacities of m-origins are greater than the total demands of n-destinations. But this problem can be converted into a balanced transportation problem using the following device:**

   (a) Imagine a fictitious or fake $(n+1)$th destination D_{n+1}.

   (b) Assume that the demand of the destination D_{n+1} is

$$b_{n+1} = \sum_{i=1}^{m} a_i - \sum_{j=1}^{n} b_j.$$

   (c) Assume that the cost components $c_{i,n+1} = 0$ for all i $[i = 1, 2, \cdots, m]$ i.e., $c_{1,n+1} = c_{2,n+1} = \cdots = c_{m,n+1} = 0$.

With this assumption, the T.P. will be a balanced problem having m-origins and $(n+1)$ destinations. Due to the assumption $c_{i,n+i} = 0$, $[i = 1, 2, \cdots, m]$ the minimum cost of transportation remains unaffected and the total demands of the destinations will be satisfied completely though the capacities of the origins will not be exhausted completely. Below an unbalanced T.P. is given:

	D_1	D_2	D_3	a_i
O_1	2	3	4	6
O_2	4	3	1	8
O_3	2	2	5	6
b_j	5	4	7	

In the problem,

$$\sum_{i=1}^{3} a_i = a_1 + a_2 + a_3 = 6 + 8 + 6 = 20$$

and $$\sum_{j=1}^{3} b_j = b_1 + b_2 + b_3 = 5 + 4 + 7 = 16.$$

$$\sum_{i=1}^{3} a_i = 20 > 16 = \sum_{j=1}^{3} b_j.$$

This problem can be converted into a balanced T.P. in the following manner:

	D_1	D_2	D_3	D_4	a_i
O_1	2	3	4	0	6
O_2	4	3	1	0	8
O_3	2	2	5	0	6
b_j	5	4	7	4	20

$$b_4 = \sum_{i=1}^{3} a_i - \sum_{j=1}^{3} b_j = 20 - 16 = 4 \text{ and } c_{14} = c_{24} = c_{34} = 0.$$

Now this transportation problem can be solved as in the previous methods.

In this problem, an initial B.F.S. obtained, using the matrix minima method, is $x_{11} = 2$, $x_{14} = 4$, $x_{22} = 1$, $x_{23} = 7$, $x_{31} = 3$ and $x_{32} = 3$, and the minimum cost of transportation such that the total demands of the destinations will be satisfied is 25. Multiple optimal solutions exist in the problem and one of the optimal solutions is $x_{11} = 3$, $x_{14} = 3$, $x_{23} = 7$, $x_{24} = 1$, $x_{31} = 2$ and $x_{32} = 4$, i.e., $x_{11} = 3$, $x_{23} = 7$, $x_{31} = 2$ and $x_{32} = 4$ as actually no quantity is transported to the destination D_4.

2. When $\sum_{j=1}^{n} b_j > \sum_{i=1}^{m} a_i$, i.e., the total demands of n destinations are greater than the total available capacities of m-origins. Here also the problem can be converted into an ordinary balanced transportation problem but the total demands of n-destinations will not be satisfied completely. Though we cannot satisfy all demands, we can still allocate the materials available at the origins to the destinations in such a way that minimizes the cost of transportation.

   To convert it into a balanced transportation problem:

   (a) Imagine a fake $(m+1)$th origin O_{m+1}.

   (b) Assume that the capacity of the origin O_{m+1} is

$$a_{m+1} = \sum_{j=1}^{n} b_j - \sum_{i=1}^{m} a_i.$$

   (c) Assume that the components $c_{m+1,j} = 0$ for all j.

The problem will be ultimately a transportation problem having $(m+1)$ origins and n-destinations. Now the problem can be solved as in the previous case.

13.4.11 Solution of Unbalanced Transportation Problems

► **Example 13.4.7** *Determination of the initial B.F.S of the following unbalanced transportation problem by VAM.*

	D_1	D_2	D_3	D_4	a_i	
O_1	14	19	11	20	10	
O_2	19	12	14	17	15	Supply
O_1	14	16	11	18	12	
b_j	8	12	16	14		
		Demand				

Here $\Sigma b_j = 50 > \Sigma a_i = 37$ then the problem is an unbalanced problem. $(50 - 37) = 13$; we assume a fictitious or fake origin O_4 having supply capacity 13 and the cost components of transporting these 13 units is 0, and we can rewrite the problem in the manner.

	D_1	D_2	D_3	D_4	a_i
O_1	14	19	11	20	10
O_2	19	12	14	17	15
O_3	14	16	11	18	12
O_4	0	0	0	0	13
b_j	8	12	16	14	50 / 50

Initial basic feasible by VAM.

Table 13.22

	D_1	D_2	D_3	D_4	a_i				
O_1	14	19	[10] 11	20	10(3)	10(3)	10(3)		
O_2	19	[12] 12	[2] 14	[1] 17	15(2)	15(2)	3(3)	3(3)	3(3)
O_3	[8] 14	16	[4] 11	18	12(3)	12(3)	12(3)	12(2)	4(7)
O_4	0	0	0	[13] 0	13(0)				
b_j	8(14)	12(12)	16(11)	14(17)	50 / 50				
	8(0)	12(4)	16(0)	1(1)					
	8(0)		16(0)	1(1)					
	8(5)		6(3)	1(1)					
			6(3)	1(1)					

Here the solution is a non-degenerate basic feasible solution.

Solve the above problem by taking the initial basic feasible solution obtain by VAM. Find the amount which is not supplied and find the destination to which the fake amount has been supplied.

Optimality test: This is a minimization problem.

Table 13.23

	D_1	D_2	D_3	D_4	u_i
O_1	0 / 14	−10 / 19	[10] / 11	−6 / 20	−3
O_2	−2 / 19	[12] / 12	[2] / 14	[1] / 17	0
O_3	[5] / 14	−7 / 16	[4] / 11	−4 / 18	−3
O_4	0 / 0	−5 / 0	−3 / 0	[13] / 0	−17
v_j	17	12	14	17	

In the second row there are three basic variables then we shall take $u_2 = 0$, with $u_2 = 0$ we have calculated all $u_i(i = 1, 2, 3, 4)$ and $v_j(j = 1, \cdots, 4)$ as shown in the table. Now we calculate all $z_{ij} - c_{ij}$ for non-basic cells.

All $z_{ij} - c_{ij}$ are ≤ 0 and $z_{11} - c_{11} = 0$. Then we reach at the optimal stage and alternative optimal solution exist.

One optimal solution is $x_{13} = 10$, $x_{22} = 12$, $x_{23} = 2$, $x_{24} = 1$, $x_{31} = 8$, $x_{33} = 4$ and the fake amount 13 units is not supplied to the destination D_4, and

$$\begin{aligned}\text{Min cost } &= 10 \times 11 + 12 \times 12 + 2 \times 14 + 1 \times 17 + 8 \times 14 + 4 \times 11 \\ &= 110 + 144 + 28 + 17 + 112 + 44 = 455 \text{ units.}\end{aligned}$$

Remark: Destination D_4 will be want of 13 units.

► **Example 13.4.8** *Identical products are produced in three factories and sent to four warehouses for delivery to the customers. The costs of transportation and capacities are given by the cost matrix as*

Factories	W_1	W_2	W_3	W_4	a_i (Capacities)
F_1	3	8	7	4	30
F_2	5	2	9	5	50
F_3	4	3	6	2	80
b_j (Demands)	20	60	55	40	

(a) *Find an optimal schedule of delivery for minimization of cost of transportation.*

(b) *Find the idle capacity of the warehouses.* **[C.U.(Pass)'95]**

(c) *Do you anticipate any alternative optimum solution for the problem? How can the same be identified?* **[C.U.M.Com.'89]**

Solution: Total supply $= 30 + 50 + 80 = 160 \neq 175 = 20 + 60 + 55 + 40 =$ Total demands. Thus, it is an unbalanced transportation problem with

$$\sum_{i=1} a_i < \sum_{j=1} b_j.$$

Now $175 - 160 = 15$. Now assuming a fictitious factory F_4 with production capacity 15 with the cost of transportation per unit zero we can adjust it as a balanced T.P. with the cost matrix with supply and demand as given below:

	W_1	W_2	W_3	W_4	a_i
F_1	3	8	7	4	30
F_2	5	2	9	5	50
F_3	4	3	6	2	80
F_4	0	0	0	0	15
b_j	20	60	55	40	175 / 175

Initial B.F.S. with VAM: Initial B.F.S. using VAM is given in the following table in a compact form which will be able to save both labour and time with B.F.S. $x_{11} = 20$, $x_{13} = 10$, $x_{22} = 50$, $x_{32} = 10$, $x_{33} = 30$, $x_{34} = 40$ and $x_{43} = 15$ which is non-degenerate one.

Table 13.24

	W_1	W_2	W_3	W_4	a_i				
F_1	(20) 3	8	(10) 7	4	30(1)	30(1)	30(1)	30(1)	30(4)
F_2	5	(50) 2	9	5	50(3)	50(3)			
F_3	4	(10) 3	(30) 6	(40) 2	80(1)	80(1)	80(1)	70(2)	30(2)
F_4	0	0	(15) 0	0	15(0)				
b_j	20(3)	60(2)	55(6)	40(2)					
	20(1)	60(1)	40(1)	40(2)					
	20(1)	10(5)	40(1)	40(2)					
	20(1)		40(1)	40(2)					
	20(1)		40(1)						

Optimality Test: Initial B.F.S. is the optimal solution which has been shown in the following table where all $z_{ij} - c_{ij} \leq 0$ for all non-basic cells which indicates that the solution is optimal.

Table 13.25

				u_i
[20] 3	−4 8	[10] 7	−1 4	1
−4 5	[50] 2	−4 9	−4 5	−1
−2 4	[10] 3	[30] 6	[40] 2	0
−4 0	−3 0	[15] 0	−4 0	−6
v_j: 2	3	6	2	

Then (a) the optimal schedule is $x_{11} = 20$, $x_{13} = 10$, $x_{22} = 50$, $x_{32} = 10$, $x_{33} = 30$, $x_{34} = 40$ and $x_{43} = 15$ and Min cost $= 3 \times 20 + 7 \times 10 + 2 \times 50 + 3 \times 10 + 6 \times 30 + 2 \times 40 + 0 \times 15 = 520$ units.

(b) Actually there is no factory F_4 and there is 15 units demand short and the optimal variable $x_{43} = 15$ is nothing but the idle capacity of the warehouse W_3, i.e., W_3 will be 15 units short of demand to make it full.

(c) The optimal solution is unique, i.e., there is no alternative optimum solution which we can anticipate from the fact that there exists no $z_{ij} - c_{ij} = 0$ corresponding to a non-basic cell at the optimal stage.

► **Example 13.4.9** *Solve the following balanced T.P. by finding I.B.F.S by matrix minima method and by VAM.*

Problem is shown in the table which the readers can easily realize.

Solution: Determination of I.B.F.S. by matrix minima method.

Table 13.26

	D_1	D_2	D_3	D_4	a_i				
O_1	2	5	0	[12] −3	12				
O_2	[7] 4	6	8	[8] 1	15	15	15	7	7
O_3	[3] 4	[8] 0	[3] 4	5	14	14	6	6	6
O_4	2	6	[9] 1	4	9	9	9	9	
b_j	10	8	12	20	50/50				
	10	8	12	8					
	10		12	8					
	10		12						
	10		3						

Optimality Test: Initial basic feasible solution is not unique.

Table 13.27

	D_1	D_2	D_3	D_4	u_i
O_1	−2 / 2	−9 / 5	[θ] 0 / 0	[12] −θ / −3	−4
O_2	[7] −θ / 4	−6 / 6	−4 / 8	[8] +θ / 1	0
O_3	[3] +θ / 4	[8] / 0	[3] −θ / 4	−6 / 5	0
O_4	−1 / 2	−9 / 6	[9] / 1	−8 / 4	−3
v_j	4	0	4	1	

All $z_{ij} - c_{ij} \leq 0$ in the table 13.27; $z_{13} - c_{13} = 0$. Thus the solution $x_{14} = 12$, $x_{21} = 7$, $x_{24} = 8$, $x_{31} = 3$, $x_{32} = 8$, $x_{33} = 3$, $x_{43} = 9$ is optimal and min cost = 33 units and alternative optimal solution exists. We now find an alternative optimal solution. We put θ in this cell (1, 3) and draw a simple loop as shown in the table 13.27 and readjusted the basic variables as shown in the table 13.27. Now θ to be taken in such a way that all basic variables remain positive and vanishes in a cell. If we take $\theta = 3$ then the conditions will be satisfied, variable in the cell (3, 3) vanishes, with the value of $\theta = 3$, the next solution is

Table 13.28

	D_1	D_2	D_3	D_4	u_i
O_1	−2 / 2	−9 / 5	[3] / 0	[9] / −3	0
O_2	[4] / 4	−6 / 6	−4 / 8	[11] / 1	4
O_3	[6] / 4	[8] / 0	0 / 4	−4 / 5	4
O_4	−3 / 2	−11 / 6	[9] / 1	−8 / 4	−1
u_j	0	−4	0	−3	

Here in the table 13.28 all $z_{ij} - c_{ij} \leq 0$ for the non-basic cells and $z_{33} - c_{33} = 0$. Thus alternative optimal solution exists. One alternative optimal solution is $x_{13} = 3$, $x_{14} = 9$, $x_{21} = 4$, $x_{24} = 11$, $x_{31} = 6$, $x_{32} = 8$, $x_{43} = 9$ and

$$\begin{aligned} \text{min cost} &= 3 \times 0 + 9 \times (-3) + 4 \times 4 + 11 \times 1 + 6 \times 4 + 8 \times 0 + 9 \times 1 \\ &= 0 - 27 + 16 + 11 + 24 + 0 + 9 = 33 \text{ units.} \end{aligned}$$

I.B.F.S. by VAM.

Table 13.29

	D_1	D_2	D_3	D_4	a_i				
O_1	2	5	0	[12] −3	12(3)	12(3)			
O_2	[7] 4	6	8	[8] 1	15(3)	15(3)	15(3)	7(4)	
O_3	[3] 4	[8] 0	[3] 4	5	14(4)	6(0)	6(0)	6(0)	6(0)
O_4	2	6	[9] 1	4	9(1)	9(1)	9(1)	9(1)	9(1)
b_j	10(0)	8(5)	12(1)	20(4)	50 / 50				
	10(0)		12(1)	20(4)					
	10(2)		12(3)	8(3)					
	10(2)		12(3)						
	3(2)		12(3)						

The initial basic feasible solution is not unique.

Optimality Test: Same as in the case of matrix minima method.

13.4.12 Solution to a degenerate problem

Degeneracy may occur at the initial stage or at any subsequent iteration. Here we shall discuss a problem where the initial B.F.S. is degenerate and only one basic variable is zero. The problem can be solved similarly if more than one basic variables are zero. Allocate a quantity $\varepsilon > 0$ (very small) instead of the basic variable 0 [zero] in the cell and readjust all basic variables in the cells such that

$$\sum_{i=1}^{m} a_i = \sum_{i=1}^{n} b_j = M$$

is satisfied and it may be assumed that $x_{ij} + \varepsilon = x_{ij}$. Now solve the problem treating it as a non-degenerate problem. You may drop ε at any subsequent iteration if it has been found that the solution at the stage will be non-degenerate on the assumption that $\varepsilon = 0$. Otherwise put $\varepsilon = 0$ after finding the optimal solution.

► **Example 13.4.10** *Obtain the optimal solution to the following transportation problem by using the North-West-Corner rule to find I.B.F.S.* **[C.U. (Hons.) '85]**

Table 13.30

	D_1	D_2	D_3	D_4	a_i
O_1	2	5	4	7	4
O_2	6	1	2	5	6
O_3	4	5	2	4	8
b_j	3	7	6	2	

The I.B.F.S. by North-West-Corner rule.

Table 13.31

(Boxed allocations in brackets, followed by cost.)

	D_1	D_2	D_3	D_4
O_1	[3] 2	[1] 5	4	7
O_2	6	[6] 1	[0] 2	5
O_3	4	5	[6] 2	[2] 4

Remark. Either (2, 3) or (3, 2) contains the basic variable 0, which indicates that the I.B.F.S. is degenerate. 0, cannot be inserted in any other cells and in that case solution will not be by N.W.C.R.

Optimality test. Initially keep '0' in its own position as '0' and not by $\varepsilon > 0$ such that $x_{ij} \pm \varepsilon = x_{ij}$ etc.

Table 13.32

(Boxed allocations in brackets; other values: $c_{ij} - z_{ij}$ above, cost below.)

					u_i
	[3] 2	[1] 5	−2 / 4	−1 / 7	0
	8 / 6	[6] 1	[0] 2	1 / 5	−4
	6 / 4	1 / 5	[6] 2	[2] 4	−4
v_j	2	5	6	8	

Minimum value $c_{ij} - z_{ij}$, $c_{ij} - z_{ij} < 0$ is -2 which is in the cell (1, 3). In the first table put $\varepsilon > 0$ instead of zero such that $x_{ij} \pm \varepsilon = x_{ij}$ [you may not put ϵ instead of 0 in the first table] and the first table becomes

Table 13.33

	D_1	D_2	D_3	D_4
O_1	[3] 2	[1] $-\theta$ 5	[θ] -2 4	-1 7
O_2	8 6	[6] $+\theta$ θ 1	[ε] $-\theta$ 2	1 5
O_3	6 4	1 5	[6] 2	[2] 4

Put $\theta > 0$ in (1, 3) where $c_{ij} - z_{ij} = -2$, Min $(c_{ij} - z_{ij},\ c_{ij} - z_{ij} < 0)$. Now construct the loop as shown in the table. Now if we take $\theta = \varepsilon$, then $6 + \theta$ and $1 - \theta > 0$ and the readjusted B.F.S. is

Table 13.34

	D_1	D_2	D_3	D_4	u_i
O_1	[3] 2	[1] $-\varepsilon$ 5	[ε] 4	1 7	0
O_2	8 6	[6]+ ε 1	2 2	3 5	-4
O_3	4 4	2 5	[6] 2	[2] 4	-2
v_j	2	5	4	6	

All $c_{ij} - z_{ij} \geq 0$. Thus we reach at the optimal stage. Now put $\varepsilon = 0$ and the optimal solution is $x_{11} = 3, x_{12} = 1, x_{13} = 0, x_{22} = 6, x_{33} = 6, x_{34} = 2$ and the min cost $= 3 \times 2 + 1 \times 5 + 0 \times 4 + 6 \times 1 + 6 \times 1 + 6 \times 2 + 2 \times 4 = 37$ units. The optimal solution is degenerate as at least one optimal basic variables $x_{13} = 0$.

► **Example 13.4.11** *Solve the following transportation problem by using VAM to determination the I.B.F.S. and show that the optimal solution is a degenerate solution.*

Table 13.35

	D_1	D_2	D_3	D_4	a_i
O_1	10	20	5	7	15
O_2	18	9	12	8	25
O_3	15	14	16	18	5
b_j	5	15	15	10	

[C.U. (Hons.) '86]

Solution: Initial solution by VAM.

Table 13.36

	D_1	D_2	D_3	D_4	a_i			
O_1	10	20	[15] 5	7	15(2)			
O_2	18	[15] 9	[0] 12	[10] 8	25(1)	25(1)	15(3)	6(0)
O_3	[5] 15	14	[0] 16	18	5(1)	5(1)	5(1)	5(1)
b_j	5(5)	15(5)	15(7)	10(1)				
	5(3)	15(5)	0(4)	10(10)				
	5(3)	15(5)	0(4)					
	5(3)		0(4)					

The above table is self explanatory. Initial solution is not unique.

Optimality test

Table 13.37

	D_1	D_2	D_3	D_4	u_i
O_1	6 / 10	18 / 20	[15] 5	6 / 7	5
O_2	7 / 18	[15] 9	[0] 12	[10] 8	12
O_3	[5] 15	1 / 14	[0] 16	6 / 18	16
v_j	−1	−3	0	−4	

For optimality test, we calculate $c_{ij} - z_{ij} = c_{ij} - u_i - v_j$ to get more positive $c_{ij} - z_{ij}$ gradually.

The initial basic feasible solution is the optimal solution. The optimal solution is $x_{13} = 15$, $x_{22} = 15$, $x_{23} = 0$, $x_{24} = 10$, $x_{31} = 5$, $x_{33} = 0$, the solution is degenerate.

Min cost $= 15 \times 5 + 15 \times 9 \times 0 \times 12 + 10 \times 8 + 5 \times 15 + 0 \times 16 = 365$ units.

► **Example 13.4.12** *Solve the following transportation problem using north-west corner rule to find the starting solution.* **[C.U. (Hons.) '81, '83, '93]**

	D_1	D_2	D_3	a_i
O_1	0	2	1	5
O_1	2	1	5	10
O_1	2	4	3	5
b_j	5	5	10	

Initial solution obtained by north-west corner rule is degenerate. Either x_{12} or x_{21} is zero. Assume $x_{12} = 0$ and put $\varepsilon > 0$ in the cell (1, 2). Calculate all net evaluations corresponding to the non-basic cells which are shown on the north-east corner of the corresponding cells in the table. Net cell evaluation in the cell (1, 3) is 5. Hence the solution is not optimal. Put θ in the cell (1, 3) and construct the loop as given in the table and readjust the basic variables in the cell (1, 3), (2, 3), (2, 2) and (1, 2). Maximum value of θ satisfying all conditions is ε. With this value of θ

Table 13.38.1 **Table 13.38.2** **Table 13.38.3**

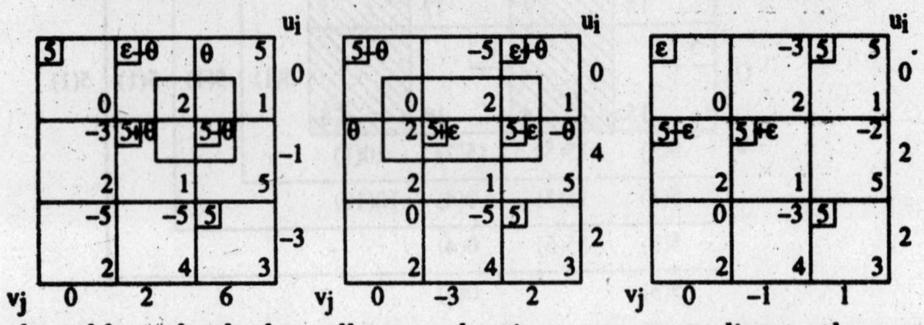

construct the table and calculate all net evaluations, corresponding to the non-basic cells. Net evaluation corresponding to the cell (2, 1) is 2. Hence the solution is not optimal. Put θ in the cell (2, 1) and readjust accordingly as given in the table. Maximum value of θ is $5 - \varepsilon$. With this value of θ construct the table. Calculate the net evaluations corresponding to the non-basic cells. All values are non-positive and the solution is optimal. Put now $\varepsilon = 0$ and then the optimal solution is also degenerate which is $x_{11} = 0$ $x_{13} = 5$, $x_{21} = 5$, $x_{22} = 5$, $x_{33} = 5$ and Min cost $=$ $5 \times 1 + 5 \times 2 + 5 \times 1 + 5 \times 3 = 35$ units.

13.4.13 Solution of a Maximum problem as a transportation problem

So long we have solved all minimization problems using the theories of transportation problem. But it is important to note the maximization problems can also be solved by using the above technique which has been shown in the solved problem given below.

► Example 13.4.13 ***Four products are produced in three machines and their profit margins are given by the table below.***

	P_1	P_2	P_3	P_4	Capacity
M_1	6	4	1	5	14
M_2	8	9	2	7	18
M_3	4	3	6	2	7
Requirements	6	10	15	8	39/39

(a) You are required to find a suitable production plan of products in machines so that the capacities and requirements are met and the profit is maximized.

(b) How can you identify an alternative optimal solution? Can you find a similar situation in the problem?

(c) While evaluating alternative solution, you may meet a case of degeneracy for testing the optimality criteria. Explain the step to avoid degeneracy in the problem.

[C.U. M. Com. '87]

Solution: (a) We know that a T.P. is in general a minimization problem. But here it is a maximization problem. This can be solved by using the relation, Max $z = -$ Min $(-z)$. Thus we try to find the value of Min $(-z)$ and hence to solve the problem, change the sign of each element of the profit matrix and taking that matrix, calculate the I.B.F.S. by matrix minima method. The I.B.F.S. is $x_{13} = 8$, $x_{14} = 6$, $x_{21} = 6$, $x_{22} = 10$, $x_{24} = 2$ and $x_{33} = 7$, which is unique.

Optimality test:

Table 13.39

					u_i
	0 / −6	−3 / −4	[8] / −1	[6] / −5	2
	[6] / −8	[10] / −9	−1 / −2	[2] / −7	0
	−7 / −4	−9 / −3	[7] / −6	−8 / −2	−3
v_j	−8	−9	−3	−7	

Now using the optimality test, we can find that the initial B.F.S. is optimal as all $z_{ij} - c_{ij} \leq 0$ for all non-basic cells. Hence the optimal allocation is $x_{13} = 8$, $x_{14} = 6$, $x_{21} = 6$, $x_{22} = 10$, $x_{24} = 2$ and $x_{33} = 7$.

Min $(-z) = -1 \times 8 + (-5) \times 6 + (-8) \times 6 + (-9) \times 10 + (-7) \times 2 + (-6) \times 7 = -232$.

Hence Max $z = -$ Min $(-z) = -(-232) = 232$ units (Max profit).

(b) If $z_{ij} - c_{ij} = 0$ for a non-basic cell at the optimal stage then we are able to say definitely that the problem has an alternative optimal solution. In this problem $z_{ij} - c_{ij} = 0$ for a non-basic cell (1, 1). Hence alternative optimal solutions exist.

Table 13.40

					u_i
	[6] / −6	−3 / −4	[8] / −1	0 / −5	0
	[ε] / −8	[10] / −9	−1 / −2	[8] / −7	−2
	−7 / −4	−9 / −3	[7] / −6	−8 / −2	−5
v_j	−6	−7	−1	−5	

(c) To find an alternative optimal solution, we shall select the cell (1, 1) as the basic cell [as the value of $z_{11} - c_{11} = 0$] and putting a basic variable $\theta > 0$ we are able to form a loop connecting the cell (1, 1) and the other basic cells (1, 4), (2, 4) and (2, 1) and taking $\theta = 6$, the basic variables of the cells (2, 1) ad (1, 4) vanish simultaneously and hence degeneracy occurs at that stage. Now to test the optimality, we shall put $\varepsilon > 0$ instead to zero in the cell (2, 1) with the cost component minimum among the cells (2, 1) and (1, 4) [considering cell (1, 4) as a non-basic cell] and ε be such that $x_{ij} \pm \varepsilon = x_{ij}$ and performing an optimality test we get that is also an optimal table and now putting $\varepsilon = 0$, the optimal degenerate solution is $x_{11} = 6$, $x_{13} = 8$, $x_{21} = 0$, $x_{22} = 10$, $x_{24} = 8$ and $x_{33} = 7$ and the value of the maximum profit remains unaltered which is Max $z = -\{-6 \times 6(-1) \times 8 + (-8) \times 0 + (-9) \times 10 + (-7) \times 8 + (-6) \times 7\} = 232$ units.

Objective and Short Answer Type Questions with Answers Transportation Problems

1. (a) Formulate mathematically a transportation problem (balanced) as a L.P.P. having m origins and n destinations $(m, n \geq 2)$.

   (b) What are the number of constraints and variables in a balanced T.P.?

   (c) What is the number of independent constraints in a balanced T.P.?

   (d) Is a constraint of a T.P. (balanced) an equation or an inequation?

   [All problems have m origins and n destinations] **[C.U.(H)'94]**

   [*Ans.* (b) $m + n$, mn; (c) $m + n - 1$; (d) Equations.]

2. What is an unbalanced T.P.? How can you convert it into a balanced T.P.?

3. Which one of the statements is true?

   (a) A T.P. is strictly a maximization problem.

   (b) A T.P. is strictly a minimization problem.

   (c) A T.P. may be a maximization or a minimization problem.

   [*Ans.* (c) is true. T.P. is generally a minimization problem. But a maximization problem can also be solved using T.P. technique]

4. Prove that there exists at least one feasible solution in a balanced T.P.

   [*Ans.* Theorem (13.1.1).]

5. Prove that there exists a finite optimal solution in each balanced T.P.

   [*Ans.* Theorem (13.1.2).]

6. Define a loop in a transportation table. What is the nature of a loop in a transportation table?

   [*Ans.* All loops are simple.]

7. (a) What is the number of basic variables in a balanced T.P. with m origins and n destinations? **[C.U.(H)'96]**

   (b) What is the maximum number of positive components in a B.F.S. of a T.P. (balanced) with m origins and n destinations?

   [*Ans.* (a) $m + n - 1$; (b) $m + n - 1$.]

8. How can you establish geometrically that a set of vectors $\mathbf{a}_{ij}$ associated with the variables x_{ij} in a T.P. are linearly independent?

[*Ans.* If the set of cells c_{ij} or any subset of the cells corresponding to the vectors $\mathbf{a}_{ij}$ of the transportation table cannot be ordered so as to form a loop, the set of vectors are said to be linearly independent.]

9. How can you detect that in a T.P. the solution is optimal? What is the criterion for the existence of multiple optimal solutions?

[*Ans.* If at any table all $z_{ij} - c_{ij} \leq 0$ corresponding to non-basic cells, we can say that the solution is optimal.

If at the optimal stage, at least one $z_{ij} - c_{ij} = 0$ corresponding to a non-basic variable, the problem is said to have multiple optimal solutions.]

10. In solving a T.P. what is the utility of constructing loops in a transportation table?

[*Ans.* First, it is possible to determine whether a given set of solution is a B.F.S. Secondly, it is possible to determine which vector will leave the basis and the maximum possible amount that can be allocated in the new basic cell.]

11. Which one of the statements is true?

(a) In a T.P. with m origins and n destinations (balanced) the number of independent constraints is (a) $m+n$, (b) mn, (c) $m+n-1$.

(b) In a transportation table $(m, n \geq 2)$ (a) it is always possible to construct a loop containing $(m+n)$ cells, (b) it is not possible.

[*Ans.* (a) $m+n-1$, (b) It is always possible.]

12. In a transportation problem with 3 origins and 4 destinations, may the variables $x_{11}, x_{22}, x_{33}, x_{34}, x_{24}, x_{23}$ be considered as basic variables? Give reasons.

[*Ans.* No; The set of corresponding cell $(1,1)$, $(2,2)$, $(3,3)$, $(3,4)$, $(2,4)$ and $(2,3)$ are said to form a loop. Thus the vectors of the variables are not linearly independent and thus the set of variables cannot be considered as basic variables.]

13. In a transportation problem with 4 origins and 3 destinations, may the variables $x_{11}, x_{21}, x_{22}, x_{23}, x_{32}$ and x_{43} be considered as a set of basic variables? Give reasons.

[*Ans.* Yes; The number of variables is $6 = 4+3-1$ and the set of corresponding cells $(1,1)$, $(2,1)$, $(2,2)$, $(2,3)$, $(3,2)$ and $(4,3)$ cannot be ordered so as to form a loop. Hence the variables may be considered as a set of basic variables.]

14. In a T.P. with 4 origins and 4 destinations, may the variables x_{11}, x_{22}, x_{23}, x_{24}, x_{31}, x_{34}, x_{42}, x_{44} be considered as basic variables? Give reasons.

[*Ans.* No; The number of variables is greater than $(4+4-1=7)$. Hence the variables cannot be considered as a set of basic variables.]

15. Find the initial B.F.S. of the following T.P. using North-West corner rule and prove that (ii) is degenerate.

(a)

	D_1	D_2	D_3	D_4	a_i
O_1	4	6	9	5	16
O_2	4	2	7	1	14
O_3	2	5	2	8	10
b_j	12	7	6	15	

(b)

	D_1	D_2	D_3	D_4	a_i
O_1	9	7	4	2	20
O_2	2	9	8	6	15
O_3	6	7	5	8	15
b_j	14	21	6	9	

[*Ans.* (a) $x_{11} = 12, x_{12} = 4, x_{22} = 3, x_{23} = 6, x_{24} = 5, x_{34} = 10.$

(b) $x_{11} = 14, x_{12} = 6, x_{22} = 15, x_{33} = 6, x_{34} = 9$ and either $x_{23} = 0$ or $x_{32} = 0.$]

16. Find the initial B.F.S. of the following T.P. using (a) Row minima method (b) Matrix-minima method and (c) VAM.

(a)

			a_i
30	20	10	50
5	15	25	50
30	30	40	b_j

(b)

			a_i
6	7	8	28
9	3	5	12
13	17	10	b_j

(a) $x_{13} = 40, x_{12} = 10, x_{21} = 30, x_{22} = 20$ [solution is same in matrix minima method and VAM].

(b) (i) $x_{11} = 13, x_{12} = 15, x_{22} = 2, x_{23} = 10.$

(ii) $x_{11} = 13, x_{12} = 5, x_{13} = 10, x_{22} = 12.$

(iii) Same as (ii).

Exercise 13B

1. (a) Determine the initial B.F.S. of the problem using row minima method and hence obtain the optimal feasible solution and the corresponding minimum cost of transportation.

(b) Determine the initial B.F.S. of the problem using matrix minima method and then find out the optimal feasible solution and the corresponding minimum cost of transportation.

(a)

	D_1	D_2	D_3	a_i
O_1	5	4	3	5
O_2	2	1	5	4
O_3	6	2	1	5
b_j	2	6	6	14

(b)

	D_1	D_2	D_3	a_i
O_1	7	5	4	12
O_2	5	4	3	10
O_3	4	1	5	8
b_j	10	9	11	30

2. Find the minimum cost of transportation in the following problem:

[C.U.(P)'84,'88,2000] [C.U.M.Com.'88 costs are multiplied by 4]

	D_1	D_2	D_3	D_4	a_i
O_1	10	7	3	6	3
O_2	1	6	8	3	5
O_3	7	4	5	3	7
b_j	3	2	6	4	15

Use row minima method in determining an initial B.F.S. Solve the problem also using north-west corner rule of determination of an initial B.F.S.

3. A company has three plants A, B, C and three warehouses X, Y, Z. The number of units available at the plants are 60, 70, 80 respectively. The demands at X, Y, Z are 50, 80, 80 respectively. The unit cost of transportation are as follows: **[C.U.'80]**

	X	Y	Z
A	8	7	3
B	3	8	9
C	11	3.	5

Find the allocation so that the total transportation cost is minimum.

4. Find the optimal solution and the corresponding cost of transportation in the following T.P.

(a)

	D_1	D_2	D_3	a_i
O_1	2	4	3	2
O_2	1	5	4	4
O_3	2	6	3	6
O_4	4	2	5	8
b_j	7	6	7	20

(b)

	1	2	3	4	
1	23	27	16	18	30
2	12	17	20	51	40
3	22	28	12	32	53
b_j	22	35	25	41	

[J.U.'84; C.U.(P)'87]

(c)

	1	2	3	4	
1	5	3	6	4	30
2	3	4	7	8	15
3	9	6	5	8	15
b_j	10	25	18	7	

[C.U.(P)'84,'86,'90; J.U.M.Sc.'83]

5. Solve the following balanced transportation problem:

	D_1	D_2	D_3	D_4	a_i	
O_1	6	4	2	7	8	
O_2	5	1	4	6	14	
O_3	6	5	2	5	9	Supply
O_4	4	3	2	1	11	
b_j	7	13	12	10	42	
		Demand				

[C.U.(H) 2000]

6. Solve the following unbalanced transportation problem:

	D_1	D_2	D_3	a_i
O_1	4	5	6	12
O_2	3	1	5	11
O_3	2	4	4	7
b_j	6	5	8	

7. A firm manufacturing a single product has three plants I, II, III. The three plants have produced 60, 35 and 40 units respectively during this month. The firm had made a commitment to sell 22 units to customer A, 45 units to customer B, 20 units to customer C, 18 units to customer D and 30 units to customer E. Find the minimum cost of shifting the manufactured product to the five customers. The cost matrix is given below:

		Customer				
		A	B	C	D	E
	I	4	1	3	4	4
Plant	II	2	3	2	2	3
	III	3	5	2	4	4

8. Solve the transportation problem and prove that optimal solutions degenerate and multiple optimal solutions exist.

	D_1	D_2	D_3	D_4	a_i
O_1	5	2	4	6	9
O_2	2	4	1	5	11
O_3	4	2	3	1	8
b_j	6	7	7	8	28

9. A company has factories at A, B and C which supply warehouses at D, E, F and G. Monthly factory capacities are 160, 150 and 190 units respectively. Monthly warehouse requirements are 80, 90, 110 and 160 units respectively. Unit transportation cost (in rupees) are as follows:

	D	E	F	G
A	42	48	38	37
B	40	49	52	51
C	39	38	40	43

Determine the optimum distribution for this company to minimize the transportation cost and determine the idle factory capacity or capacities.

10. Origins O_1, O_2, O_3 and O_4 have surpluses of 30, 50, 75 and 20 empty freight cars respectively and the destinations D_1, D_2, D_3, D_4, D_5 and D_6 are in need of 20, 40, 30, 10, 50 and 25 empties respectively. The cost per unit of moving an empty from the origins to the destinations is given below:

		Destinations					
		D_1	D_2	D_3	D_4	D_5	D_6
	O_1	1	2	1	4	5	2
	O_2	2	3	2	1	4	3
Origins	O_3	4	2	5	9	6	2
	O_4	3	1	7	3	4	6

Determine the minimum possible cost of transporting empty cars from excess origins to "deficit" destinations.

11. Solve the following two balanced T.P. problems by using VAM determine the initial B.F.S.

(a)

	D_1	D_2	D_3	D_4	a_i
O_1	1	2	3	4	6
O_2	4	3	2	0	8
O_3	0	2	2	1	10
b_j	4	6	8	6	24

(b)

	D_1	D_2	D_3	D_4	a_i
O_1	5	8	3	6	30
O_2	4	5	7	4	50
O_3	6	2	4	5	40
b_j	30	20	40	30	120

[C.U.(H)'80]

12. (a) An oil corporation has got three refineries P, Q and R and it has to send petrol to four different depots A, B, C and D. The cost of shipping 1 gal. of petrol from each refinery to each depot is given below. The requirement of the depots and the available petrol at the refineries are also given. Find the minimum cost of shipping after obtaining an initial solution by VAM.

		Depot				
		A	B	C	D	a_i
	P	10	12	15	8	130
Refinery	Q	14	11	9	10	150
	R	20	5	7	18	170
	b_j	90	100	140	120	

(b) In a transportation problem the cost matrix is given:

	D_1	D_2	D_3
O_1	7	3	4
O_2	2	2	3
O_3	3	4	6

A basic F.S. is given, $x_{13} = 2$, $x_{22} = 1$, $x_{23} = 2$, $x_{31} = 4$, $x_{33} = 1$. Is the solution optimal? Find the minimum cost of the problem.

13. There are three sources which store a given product. The sources supply these products to four dealers. The capacities of the sources, and the demands of the dealers are given. Capacities $S_1 = 150$, $S_2 = 40$, $S_3 = 80$. Demands $D_1 = 90$, $D_2 = 70$, $D_3 = 50$, $D_4 = 60$. The cost matrix is given. Find the minimum cost of transportation.

	D_1	D_2	D_3	D_4
S_1	27	23	31	69
S_2	10	45	40	32
S_3	30	54	35	57

[C.U.(H)'84,'87]

14. Solve the following balanced T.P.

(a)

	D_1	D_2	D_3	D_4	a_i
F_1	15	20	13	21	15
F_2	22	15	19	14	12
F_3	16	12	14	31	13
F_4	24	23	15	30	20
b_j	16	10	10	24	

(b)

	D_1	D_2	D_3	D_4	a_i
O_1	21	13	17	24	11
O_2	19	15	13	17	15
O_3	23	11	22	20	18
O_4	16	18	21	14	11
b_j	13	17	11	14	

(c)

	1	2	3	4	a_i
1	2	3	11	7	6
2	1	0	6	1	1
3	5	8	15	9	10
b_j	7	5	3	2	

[Kalyani(H)'87]

(d)

	D_1	D_2	D_3	D_4	D_5	a_i
O_1	3	5	6	9	7	9
O_2	5	2	4	3	6	6
O_3	4	1	3	2	5	10
b_j	2	5	6	7	5	

[*Ans.* (a) Optimal solution $x_{11} = 3$, $x_{14} = 12$, $x_{24} = 12$, $x_{31} = 3$, $x_{32} = 10$, $x_{41} = 10$, $x_{43} = 10$. Alternative optimal solutions exist. Min cost = 1023 units.

(b) Optimal solution $x_{11} = 11$, $x_{23} = 11$, $x_{24} = 4$, $x_{32} = 17$, $x_{34} = 1$, $x_{41} = 2$, $x_{44} = 9$. Min cost = 807 units. Alternative optimal solution exists.

(c) Optimal solution $x_{12} = 5$, $x_{13} = 1$, $x_{23} = 1$, $x_{31} = 7$, $x_{33} = 1$, $x_{34} = 2$. Min cost = 100.

(d) Optimal solution $x_{11} = 2$, $x_{13} = 2$, $x_{15} = 5$, $x_{24} = 6$, $x_{32} = 5$, $x_{33} = 4$, $x_{34} = 1$. Min cost = 90 units. Alternative optimal solutions exist.]

15. A company has three factories A, B, C which supply five wholesale dealers with small car fans. The production capacities of the factories, the demands of the customers assumed constant and the distribution costs are given below. The object is to supply the wholesalers with their demands in the cheapest way. Find the optimal solution.

	a	b	c	d	e	a_i
A_1	5	7	10	5	3	5
A_2	8	6	9	12	14	10
A_3	10	9	8	10	15	10
b_j	3	3	10	5	4	25

16. Solve the following unbalanced T.P. problem:

	D_1	D_2	D_3	D_4	a_i
O_1	4	6	2	3	12
O_2	8	2	3	4	10
O_3	5	7	6	5	8
b_j	9	8	6	5	

17. Solve the following balanced transportation problem:

(a)

	D_1	D_2	D_3	D_4	a_i
O_1	18	22	14	19	15
O_2	23	16	15	20	22
O_3	18	24	15	24	13
b_j	10	10	18	12	50 / 50

(b) Prove that the initial solution obtained by N.W.C rule is the optimal solution but initial solution will not be an optimal one in VAM or Matrix minima method.

	D_1	D_2	D_3	
O_1	6.1	5	6.1	40
O_2	3	1	2	50
O_3	2	1	1	10
	20	50	30	

(c) Obtain the optimal solution of the following transportation problem:

	D_1	D_2	D_3	D_4	D_5	a_i
O_1	2	1	7	4	3	5
O_2	2	1	4	5	3	9
O_3	5	2	9	9	1	9
b_j	5	2	6	7	3	23

[C.U.(H)'89,'96]

[*Hints.* Initial solution by VAM (taking the difference between two lowest costs zero) $x_{14} = 5$, $x_{21} = 1$, $x_{23} = 6$, $x_{24} = 2$, $x_{31} = 4$, $x_{32} = 2$, $x_{35} = 3$ is the optimal solution and min cost = 83 units.]

18. A manufacturer has distribution centers located at Agra, Allahabad and Calcutta. These centers have available 40, 20 and 40 units of his product. His retail outlets require the following number of units; $A = 25$, $B = 10$, $C = 20$, $D = 30$ and $E = 25$. The cost per unit in rupees between each center and outlet is given in the following table:

Distribution centers	Retail outlets A	B	C	D	E
Agra	55	30	40	50	40
Allahabad	35	30	100	45	60
Calcutta	40	60	95	35	30

(a) Determine the optimal shipping cost.

(b) Do you anticipate any alternative solution? Give reason for your answer.

(c) Find the idle capacity of the retail outlets, if any. [C.U.,M.Com.'89]

19. Initial B.F.S. are given using VAM in the problems. Find the optimal solution and minimum cost in each case. See Solved Ex. 13.4.2.

20. Messrs Hindustan Construction Company Limited require 3, 3, 4 and 5 million cubic feet of fill at four earthen dam sites D_1, D_2, D_3 and D_4 in the district of Birbhum, West Bengal. The company can transfer the fill from three mounds M_1, M_2 and M_3 where 2, 6 and 7 million cubic feet of fill are available respectively. The cost of transporting one million cubic feet of fill from the mounds to the dam sites (expressed in Rs. Lakhs) are shown in the cost matrix given below:

		Dam sites			
		D_1	D_2	D_3	D_4
	M_1	15	10	17	18
Mounds	M_2	16	13	12	13
	M_3	12	17	20	11

(a) Formulate the above problem as a linear programming problem.

(b) Find the optimal solution of the problem. [C.U.,M.Com.'85]

21. The cost matrix of a transportation problem is given below. There is a pre-condition that if all demands of a particular destination be not met one unit of penalty is to be charged for each unit of goods not supplied. Find the minimum cost of transportation and mention in which destination or destinations, demand will not be met.

	D_1	D_2	D_3	D_4		
O_1	12	9	13	7	14	
O_2	9	7	11	7	10	Capacity
O_3	12	10	8	9	11	
	9	12	13	6		
	Demands					

[*Hints.* This is actually an unbalanced problem which can be made balanced in the following manner:

	D_1	D_2	D_3	D_4	
O_1	12	9	13	7	14
O_2	9	7	11	7	10
O_3	12	10	8	9	11
O_4	1	1	1	1	5
	9	12	13	6	

where O_4 is a fake origin but the cost components are 1 (one) each instead of 0 (zero).]

Answers

1. (a) $\min z = 30$ for $x_{13} = 5$, $x_{21} = 2$, $x_{22} = 2$, $x_{32} = 4$, $x_{33} = 1$ (multiple solutions exist).

   (b) $\min z = 107$ for $x_{12} = 1$, $x_{13} = 11$, $x_{21} = 10$, $x_{23} = 0$, $x_{32} = 8$.

2. min $z = 47$ for $x_{13} = 3, x_{21} = 3, x_{24} = 2, x_{32} = 2, x_{33} = 3, x_{34} = 2$.

3. min $z = 750$ for $x_{13} = 60, x_{21} = 50, x_{22} = 0, x_{23} = 20, x_{32} = 80$.

4. (a) min $z = 47$ for $x_{11} = 2, x_{21} = 4, x_{31} = 1, x_{33} = 5, x_{42} = 6, x_{43} = 2$ (multiple solutions exist).
   (b) Min cost = 2221 at $x_{14} = 30, x_{21} = 5, x_{22} = 35, x_{31} = 17, x_{33} = 25, x_{34} = 11$.
   (c) Min cost = 231 at $x_{12} = 23, x_{14} = 7, x_{21} = 10, x_{22} = 2, x_{23} = 3, x_{33} = 15$ (Multiple solution exist).

5. min $z = 86$ for $x_{11} = 5, x_{13} = 3, x_{21} = 1, x_{22} = 13, x_{33} = 9, x_{41} = 1, x_{44} = 10$ (alternative solutions exist).

6. min $z = 57$ for $x_{13} = 1, x_{22} = 5, x_{23} = 6, x_{31} = 6, x_{33} = 1$.

7. min $z = 290$ for $x_{12} = 45, x_{15} = 15, x_{21} = 2, x_{24} = 18, x_{25} = 15, x_{31} = 20, x_{33} = 20$ (alternative solution exists).

8. min $z = 47$ for $x_{11} = 2, x_{12} = 7, x_{14} = 0, x_{21} = 4, x_{23} = 7, x_{34} = 8$.

9. min $z = 17050$ for $x_{14} = 160, x_{21} = 80, x_{22} = 10, x_{32} = 80$ and $x_{33} = 110$; 60 units remain idle in the factory B.

10. min $z =$ Rs. 430 for $x_{11} = 20, x_{13} = 10, x_{23} = 20, x_{24} = 10, x_{25} = 20, x_{32} = 40, x_{35} = 10, x_{36} = 25, x_{45} = 20$ (alternative solutions exists).

11. (a) min $z = 28$ for $x_{12} = 6, x_{23} = 2, x_{24} = 6, x_{31} = 4, x_{32} = 0, x_{33} = 6$.
    (b) min $z = 420$ for $x_{13} = 30, x_{21} = 30, x_{24} = 20, x_{32} = 20, x_{33} = 10, x_{34} = 10$.

12. (a) min cost = 3640 for $x_{11} = 90, x_{14} = 40, x_{23} = 70, x_{24} = 80, x_{32} = 100, x_{33} = 70$.
    (b) Solution is not optimal. Min cost = 33 for $x_{13} = 2, x_{23} = 3, x_{31} = 4, x_{32} = 1, x_{33} = 0$.

13. min $z = 8190$ for $x_{11} = 30, x_{12} = 70, x_{13} = 50, x_{24} = 40, x_{31} = 60, x_{34} = 20$.

15. min cost = 183 units for $x_{14} = 1, x_{15} = 4, x_{21} = 3, x_{22} = 3, x_{23} = 4, x_{33} = 6, x_{34} = 4$.

16. min cost = 87 for $x_{11} = 1, x_{13} = 6, x_{14} = 5, x_{22} = 8, x_{31} = 8$ (alternative solution exists).

17. (a) min cost = 835 units at $x_{13} = 3, x_{14} = 12, x_{22} = 10, x_{23} = 12, x_{31} = 10, x_{33} = 3$ (alternative solution exists).
    (b) $x_{11} = 20, x_{12} = 20, x_{22} = 30, x_{23} = 20, x_{33} = 10$. Min cost = 302.

18. (a) Optimal shipping cost = 3525 for $x_{12} = 10, x_{13} = 20, x_{15} = 10, x_{21} = 20, x_{34} = 25$ and $x_{35} = 15$.
    (b) No alternative optimal solution as no $z_{ij} - c_{ij} = 0$ for non-basic cells.
    (c) 5 units at A and 5 units at D.

20. Minimize, $z = 15x_{11} + 10x_{12} + \cdots + 18x_{14} + 16x_{21} + \cdots + 13x_{24} + 12x_{31} + \cdots + 11x_{34}$ subject to

$$\sum_{j=1}^{4} x_{ij} = a_i, a_1 = 2, a_2 = 6, a_3 = 7[i = 1, 2, 3]$$

$$\sum_{i=1}^{3} x_{ij} = b_j, b_1 = 3, b_2 = 3, b_3 = 4, b_4 = 5[j = 1, 2, 3, 4]$$

and $\sum a_i = \sum b_j = 15$.
Min cost = 174 for $x_{12} = 2, x_{22} = 1, x_{23} = 4, x_{24} = 1, x_{31} = 3, x_{34} = 4$.

21. Optimal solution $x_{12} = 8, x_{14} = 6, x_{21} = 6, x_{22} = 4, x_{33} = 11$; 3 and 2 units of demands will not be met for destinations D_1 and D_3 respectively. Min cost = 289 units.

13.5 Assignment Problems

Assignment problem is a particular type of a transportation problem where n origins are to be assigned to an equal number of destinations *in one to one basis* such that the assignment cost (or profit) is minimum (or maximum). Consider the following examples. In a factory there are n jobs and n workers. $c_{ij}, [i, j = 1, 2, \cdots, n]$ is the assignment cost if the ith job is assigned to jth man. The problem is to find the minimum assignment cost such that all n jobs are assigned to n workers in one to one basis, i.e., only one job is assigned to a particular worker. In a medical firm there are n medical representatives and they are to be sent in n different localities for business purpose. c_{ij}, is the profit component if the ith person is sent to the jth destination. Determine the assignment such that the total profit is maximum.

13.5.1 Mathematical Formulation of an Assignment Problem

Let x_{ij} be a variable defined by:

$x_{ij} = 1$ if ith origin is assigned to jth destination.

$x_{ij} = 0$ if ith origin is not assigned to jth destination.

Now the assignment problem is

Optimize (minimize or maximize)

$$z = \sum_{i=1}^{n} \sum_{j=1}^{n} c_{ij} x_{ij}$$

subject to

$$\sum_{j=1}^{n} x_{ij} = a_i = 1, \quad i = 1, 2, \cdots, n$$

$$\sum_{i=1}^{n} x_{ij} = b_j = 1, \quad j = 1, 2, \cdots, n$$

$$\text{and} \quad \sum_{i=1}^{n} a_i = \sum_{j=1}^{n} b_j = n.$$

From the above discussions it is clear that the problem before us is to select n cells in a $(n \times n)$ transportation table, only one cell in each row and each column, such that the sum of the corresponding costs (or profits) be minimum (or maximum). Obviously, the solution obtained is a degenerate solution. To get the solution, we shall state and prove an important theorem.

Thus an assignment problem is a particular type of L.P.P. with B.F.S. always degenerate.

Theorem 13.5.1 *Given a cost or profit matrix $C = [c_{ij}]_{n \times n}$, if we form another matrix $C^* = [c^*_{ij}] = c_{ij} - u_i - v_j, u_i$ and v_j are arbitrary chosen constants, the solution of C will be identical with that of C^*.*

Let

$$z = \sum_{i=1}^{n}\sum_{j=1}^{n} c_{ij}x_{ij} \quad \text{and} \quad z^* = \sum_{i=1}^{n}\sum_{j=1}^{n} c_{ij}^* x_{ij}$$

$$\begin{aligned} z^* &= \sum_{i=1}^{n}\sum_{j=1}^{n}(c_{ij} - u_i - v_j)x_{ij} \\ &= \sum_{i=1}^{n}\sum_{j=1}^{n} c_{ij}x_{ij} - \sum_{i=1}^{n}\sum_{j=1}^{n} u_i x_{ij} - \sum_{i=1}^{n}\sum_{j=1}^{n} v_j x_{ij} \\ &= z - \sum_{i=1}^{n} u_i - \sum_{j=1}^{n} v_j \quad \left[\because \sum_{j=1}^{n} x_{ij} = 1 = \sum_{i=1}^{n} x_{ij}\right]. \end{aligned}$$

As $\sum_{i=1}^{n} u_i$ and $\sum_{j=1}^{n} v_j$ are constants then z^* and z differ only by a constant. From the above theorem we conclude that if in an assignment problem we subtract a number from every element of a row or column of the cost (or profit) matrix, then an assignment which is optimum for the transformed matrix will be optimum for the original matrix also.

13.5.2 Computational Procedure

Without going details into the theoretical developments of the problems, we shall chalk out the computational procedure for a minimization problem, step by step based mainly on the property discussed above.

(1) If the problem is a maximization problem, then convert it into a minimization problem by changing the cost matrix C to $-C$. Subtract the minimum cost element of each row from all other elements of the respective row. Then subtract the minimum element of each column from all other elements of the respective column of the resulting cost matrix. [These two operations can be interchanged]. A set of k zeros ($k \geq n$) will be obtained in the new cost matrix. Connect all the zeros by a *minimum* number of horizontal and vertical lines. Let the number of lines be N. There are two cases to discuss.

(i) If $N = n$, then the optimality conditions are satisfied, i.e., it is possible to select a single zero in each row and each column of the transportation table and hence $c_{ij}^* x_{ij} = 0$ which is the optimal value of z^* (the minimization problem). Now select only n zeros in such a way that there is one and only one zero in each row and each column. To do this, initially cross off all zeros which lie on the points of intersection of the lines. Next select a row or a column containing only one zero and cross off all zeros of the corresponding column or row. Proceed in the way to get n zeros. The sum of the cost components of the original cost matrix corresponding to n zeros of the final matrix gives the minimum cost. If the selection of n zeros is unique, the solution is unique. Otherwise the solution is not unique.

(ii) If $N < n$, then the optimality conditions will not be satisfied. Imagine a new cost matrix, *whose elements are the elements of the remaining rows and columns of the penultimate table which are not covered by the lines.* Subtract the smallest element of the new matrix from all the elements of this matrix and add that smallest element of the matrix to all the elements which lie on the points of intersection of the lines. Now connect all the zeros of the new matrix by a *minimum* number of horizontal and vertical lines. If the number of lines be $N = n$, then the optimality conditions are satisfied and then proceed as in the above case (i) to get the minimum cost. If $N < n$, then proceed similarly as in the case (ii) until the number of lines is n. Then find out the minimum cost etc. and max $(z) = -\min(-z)$.

N.B. This method of solving an assignment problem is called Hungarian Method.

Worked out Examples

▶ **Example 13.5.1** *Find out the optimal (minimum) assignment cost from the following cost matrix.*

	I	*II*	*III*	*IV*
A	9	6	6	5
B	8	7	5	6
C	8	6	5	7
D	9	9	8	8

This is a problem of minimization.

Step 1. Subtract the lowest element 5 of the first row from all the elements of the first row. Similarly subtract the lowest element of each row from all the elements of we corresponding row to get the matrix in table 13.41(A).

Step 2. Subtract the lowest element of each column from all elements of the corresponding column of matrix in table 13.41(A) to get the matrix in table 13.41(B)

Table 13.41(A) **Table 13.41(B)**

$$\begin{bmatrix} 4 & 1 & 1 & 0 \\ 3 & 2 & 0 & 1 \\ 3 & 1 & 0 & 2 \\ 1 & 1 & 0 & 0 \end{bmatrix} \quad \begin{bmatrix} 3 & 0 & 1 & 0 \\ 2 & 1 & 0 & 1 \\ 2 & 0 & 0 & 2 \\ 0 & 0 & 0 & 0 \end{bmatrix}$$

Step 3. Connect all the zeros of matrix in table 13.41(B) by a minimum number of horizontal and vertical lines. Here the number of lines are 4. Hence the optimality conditions are satisfied.

Step 4. Write all zeros in a 4×4 transportation table 13.42(B) in their respective positions and cross off (omit) the zeros at the points of intersection of the lines.

Table 13.42(A)

3	0	1	0
2	1	0	1
2	0	0	2
0	0	0	0

Table 13.42(B)

			√0
		√0	
	√0	$\not{0}$	
√0			$\not{0}$

Step 5. Select 4 zeros in such a way that there is only one zero in each row and each column. Select the first column containing only a single zero at cell $(4, 1)$ and then cross off all the zeros of the fourth row. Now select the single zero at cell $(3, 2)$ in the 2nd column and cross off all zeros of the third row. Then select the zeros at $(2, 3)$ and cross off all zeros of the third row. Then select the zeros at $(2, 3)$ and $(1, 4)$. Each of the zeros is the single zero in each row and each column.

Step 6. Hence the optimal assignment is unique and assignment is $D \to I$, $C \to II$, $B \to III$ and $A \to IV$ and the optimal assignment cost = The sum of the cost components of the original cost matrix corresponding to 4 zeros in the final table $= 9 + 6 + 5 + 5 = 25$ units. Here the solution is unique.

▶ **Example 13.5.2** *Find the optimal assignment and the corresponding assignment cost from the following cost matrix.* **[C.U.(P)'94]**

	A	B	C	D	E
1	9	8	7	6	4
2	5	7	5	6	8
3	8	7	6	3	5
4	8	5	4	9	3
5	6	7	6	8	5

This is a minimization problem.

Step 1. Subtract the lowest element 4 from all the elements of the first row. Similarly subtract the lowest element of each row from all the elements of the corresponding row to get the matrix in table 13.43.

Table 13.43

5	4	3	2	0
0	2	0	1	3
5	4	3	0	2
5	2	1	6	0
1	2	1	3	0

Step 2. Now subtract the lowest element of each column from all the elements of the corresponding column in table 13.43 to get the matrix in table 13.44.

Table 13.44(A)

5	2	3	2	0
0	0	0	1	3
5	2	3	0	2
5	0	1	6	0
1	0	1	3	0

Table 13.44(B)

4	2	2	1	0
0	1	0	1	4
5	3	3	0	3
4	0	0	5	0
0	0	0	2	0

Table 13.44(C)

				✓0
⊗		✓0		
			✓0	
	✓0	⊗		
✓0	⊗	⊗		

Step 3. Connect all the zeros of the matrix of the table 13.44(A) by a minimum number of horizontal and vertical lines. Here the number of lines are 4 which is less than the number of rows or columns of the matrix. Hence the optimality conditions are not satisfied. Now subtract the smallest element 1 (one) from the elements of the remaining rows and columns of the table 13.44(A) which are *not covered by the lines* and add that smallest element to all the elements which lie on the points of intersection of the lines to get the matrix in table 13.44(B).

Step 4. Now connect again all zeros of the matrix in table 13.44(B) by a *minimum* number of horizontal and vertical lines. The number of lines is five now. Hence the optimality conditions are satisfied.

Step 5. Write down all the zeros in a 5 × 5 transportation table in their respective positions after removing all zeros at the points of intersection of the lines from the table 13.44(C).

Step 6. Now select five zeros in such a way that there is only one zero in each row and each column. Select the first row containing only a single zero at cell $(1, 5)$. Now select the zero at cell $(3, 4)$. There are no other zeros in the third row and fourth column. Now further selection of zeros is not unique. Now select arbitrarily the zero (0) at cell $(2, 3)$ and cross off the zeros of the second row and third column. Then select the zero at cell $(5, 1)$ and cross off the remaining zero in the fifth row. Last of all select the zero at cell $(4, 2)$ and the selection is complete. The optimal assignment is $1 \to E$, $2 \to C$, $3 \to D$, $4 \to B$, $5 \to A$ and the optimal assignment cost is $4 + 5 + 3 + 5 + 6 = 23$ units.

Note. Here the solution is not unique. Other optimal assignments are

$$1 \to E, 2 \to A, 3 \to D, 4 \to C, 5 \to B$$
$$\text{and} \quad 1 \to E, 2 \to A, 3 \to D, 4 \to B, 5 \to C$$

and the minimum cost remains same in each case.

► **Example 13.5.3** *Find the optimal assignment and the optimal assignment cost from the following cost matrix.*

Table 13.45

	A	*B*	*C*	*D*	*E*
1	−14	−6	−22	−11	−6
2	−18	−22	−14	−15	−9
3	−18	−12	−9	−12	−12
4	−10	−22	−16	−22	−8
5	−16	−16	−14	−10	−10

Step 1. Select the lowest cost (−22) from the first row and subtract this element from all other elements of the first row. Now the same procedure will be followed for all other rows and the resulting matrix is.

Table 13.46

	A	*B*	*C*	*D*	*E*
1	8	16	0	11	16
2	4	0	8	7	13
3	0	6	9	6	6
4	12	0	6	0	14
5	0	0	2	6	6

Step 2. Now the same procedure will be followed in each column and the resulting matrix is

Table 13.47

	A	B	C	D	E
1	8	16	0	11	10
2	4	0	8	7	7
3	0	6	9	6	0
4	12	0	6	0	8
5	0	0	2	6	0

Step 3. Now try to connect all zeros by minimum numbers of horizontal and vertical lines. Here the minimum number of lines are five. Thus we reach at the optimal stage.

Table 13.48

	A	B	C	D	E
1			√ 0		
2		√ 0			
3	0				√ 0
4				√ 0	
5	√ 0				0

Now select all the zeros (except the zeros at the points of intersection) and put them at their respective positions.

Step 4. There is one zero in cell $(1, 3)$, there are no other zeros in the first row and the third column. Similarly select the zero in cell $(2, 2)$, now we select the zero in $(4, 4)$. Henceforth the selection of zeros are not unique; one selection of zeros is $(5, 1)$ and $(3, 5)$. Therefore one optimal assignment is $1 \to C$, $2 \to B$, $3 \to E$, $4 \to D$, and $5 \to A$ and the optimal assignment cost is $= -[22 + 22 + 12 + 22 + 16] = -94$ units.

Optimal assignment is not unique. Another optimal assignment is $1 \to C$, $2 \to B$, $3 \to A$, $4 \to D$ and $5 \to E$ and the optimal assignment cost is $= -[22 + 22 + 18 + 22 + 10] = -94$ units.

► **Example 13.5.4** *Find the optimal assignment profit from the following profit matrix:*

	D_1	D_2	D_3	D_4	D_5
O_1	2	4	3	5	4
O_2	7	4	6	8	4
O_3	2	9	8	10	4
O_4	8	6	12	7	4
O_5	2	8	5	8	8

This is a maximization problem, We know that $\max z = -\min(-z) = -\min z^*$ where $z^* = -z$. Hence first of all, change the sign of each component of the profit matrix and the new matrix is

Table 13.49

	D_1	D_2	D_3	D_4	D_5
O_1	−2	−4	−3	−5	−4
O_2	−7	−4	−6	−8	−4
O_3	−2	−9	−8	−10	−4
O_4	−8	−6	−12	−7	−4
O_5	−2	−8	−5	−8	−8

Step 1. Subtract the lowest element (-5) from all other elements of the first row. Similarly subtract the lowest element of each row from all other elements of that row to get the matrix in table 13.50.

Table 13.50

	D_1	D_2	D_3	D_4	D_5
O_1	3	1	2	0	1
O_2	1	4	2	0	4
O_3	8	1	2	0	6
O_4	4	6	0	5	8
O_5	6	0	3	0	0

Step 2. Now subtract the lowest elements of each column from all other elements of that column and the resulting matrix is

Table 13.51

	D_1	D_2	D_3	D_4	D_5
O_1	2	1	2	0	1
O_2	0	4	2	0	4
O_3	7	1	2	0	6
O_4	3	6	0	5	8
O_5	5	0	3	0	0

Step 3. Connect all the zeros of the matrix of the table 13.51 by a *minimum* number of horizontal and vertical lines. Here the number of lines are 4 which is less than the number of columns or rows of the matrix. Hence the optimality conditions are not satisfied. Now subtract the smallest element 1 (one) from the elements of the remaining rows and columns of penultimate table 13.52(A) which are not covered by the lines and add that smallest element (one) to all the elements which lie on the points of intersection of the lines to get the new matrix in table 13.52(B).

Table 13.52(A)

2	1	2	0	1
0	4	2	0	4
7	1	2	0	6
3	6	0	5	8
5	0	3	0	0

Table 13.52(B)

2	0	2	0	0
0	3	2	0	3
7	0	2	0	5
3	5	0	5	7
6	0	4	1	0

Step 4. Now connect all the zeros of the matrix in table 13.52(B) by a minimum number of horizontal and vertical lines. The number of lines are 5. Hence the optimality conditions are satisfied.

Table 13.53

	~~0~~		~~0~~	✓0
✓0			~~0~~	
	~~0~~		✓0	
		✓0		
	✓0			~~0~~

Step 5. Write down all the zeros in a 5 × 5 transportation table in their respective positions in table 13.53.

Step 6. Select 5 zeros in such a way that there is only one zero in each row and each column. Select first column containing only a single zero at cell (2, 1) and cross off all zeros remaining in the second row. Now select the zero at cell (4, 3). There are no other zeros in the fourth row and 3rd column. Now select arbitrarily the zero at cell (3, 4) and cross off all the remaining zeros in 3rd row and fourth column. In the same way select the zeros at cells (1, 5) and (5, 2). Here the selection of zeros is not unique. Hence multiple optimal solutions exist.

Step 7. One of the optimal assignments is

$$O_1 \to D_5, O_2 \to D_1, O_3 \to D_4, O_4 \to D_3 \text{ and } O_5 \to D_2$$

and the maximum profit $\max(z) = -\min z^* = -\min(-z)$

$$= -(-4 - 7 - 10 - 12 - 8) = 41 \text{ units.}$$

Other optimal assignments are

$$O_1 \to D_2, O_2 \to D_1, O_3 \to D_4, O_4 \to D_3, O_5 \to D_5$$

$$\text{and } O_1 \to D_4, O_2 \to D_1, O_3 \to D_2, O_4 \to D_3, O_5 \to D_5.$$

► **Example 13.5.5** *There are five pumps available for developing five wells. The efficiency of each pump in producing the maximum yield at each well is shown in the following table. In what way the pumps be assigned to the wells so as to maximize the over all efficiency?* [C.U.(H)'96]

	I	*II*	*III*	*IV*	*V*
A	45	40	65	25	55
B	50	30	25	60	30
C	25	20	10	20	40
D	35	25	30	25	20
E	80	60	50	70	50

This is a profit maximization problem. We know that $\max z = -\min(-z)$. Therefore, initially we shall have to change the sign of the all components of the profit matrix and the resulting matrix is

Table 13.54

	I	*II*	*III*	*IV*	*V*
A	−45	−40	−65	−25	−55
B	−50	−30	−25	−60	−30
C	−25	−20	−10	−20	−40
D	−35	−25	−30	−25	−20
E	−80	−60	−50	−70	−50

Step 1. Now subtract the lowest element of each row from all other elements of that row and the resulting matrix is:

Table 13.55

	I	*II*	*III*	*IV*	*V*
A	20	25	0	40	10
B	10	30	35	0	30
C	15	20	30	20	0
D	0	10	5	10	15
E	0	20	30	10	30

Step 2. Now subtract the lowest element of each column from all other elements of that column and the resulting matrix is

Table 13.56

	I	II	III	IV	V
A	20	15	0	40	10
B	10	20	35	0	30
C	15	10	30	20	0
D	0	0	5	10	15
E	0	10	30	10	30

Step 3. Now try to connect all the zeros by minimum number horizontal and vertical lines. Here the minimum number of lines is five, thus we reach at the optimal stage.

Table 13.57

	I	II	III	IV	V
A			✓0		
B				✓0	
C					✓0
D	⊗	✓0			
E	✓0				

Step 4. Now put all the zeros at their corresponding positions after omitting the zeros at the points of intersection. Now select the zero at the cell $(1,3)$ and cross off all the zeros of the first row and third column. Here there is no such zero. In the same way we shall select the zeros in cell $(2,4)$, $(3,5)$ and $(5,1)$ and $(4,2)$ such that there is only one zero in each row and each column and the selection is unique. The optimal assignment is $A \to III$, $B \to IV$, $C \to V$, $D \to II$ and $E \to I$.
$\min(-z) = -65 - 60 - 40 - 25 - 80 = -270$, $\max z = -\min(-z) = -(-270) = 270$ units with the optimal assignment given above.

13.6 Restricted Assignment

During assignment, some restrictions are being adopted considering from the economical, commercial or physical points of view. Such types of assignment problems are called restricted assignment problems. The following solved problem will give some idea about it.

▶ **Example 13.6.1** ***In a factory there are five operators A, B, C, D, E and the five machines I, II, III, IV, V the operating cost is given if ith operator operates the jth machine $[i, j = 1, 2, \cdots, 5]$. But there is a restriction that C cannot be allowed to operate the third machine and similarly B cannot be allowed to operate the fifth machine. The cost matrix is given below. Find the optimal assignment and the optimal assignment cost.***

Machines

		I	II	III	IV	V
	A	24	29	18	32	19
	B	17	26	34	22	21
Operators	C	27	16	28	17	25
	D	22	18	28	30	24
	E	28	16	31	24	27

Solution: Step 1. Since C cannot operate the third machine and B cannot operate the fifth machine we simply ignore the the cells $(3, 3)$ and $(2, 5)$ and ignoring the cells we simply solve in the usual manner and the matrix is there fore

Table 13.58

	I	II	III	IV	V
A	24	29	18	32	19
B	17	26	34	22	—
C	27	16	—	17	25
D	22	18	28	30	24
E	28	16	31	24	27

Step 2. After the row and column operations the resulting matrix is:

Table 13.59

	I	II	III	IV	V
A	6	11	0	13	0
B	0	9	17	4	
C	11	0		0	8
D	4	0	10	11	5
E	12	0	15	7	10

Now try to connect all zeros by minimum number of horizontal and vertical lines. Here the minimum number of lines is four, less then five. Thus we do not reach at the optimal stage. Now select the lowest element from the uncovered matrix and subtract it from all other uncovered elements and add it with the elements of the points of intersection of the lines. And the resulting matrix is:

Table 13.60

	I	II	III	IV	V
A	6	15	0	13	0
B	0	13	17	4	
C	11	4		0	8
D	4	0	6	7	1
E	8	0	11	3	6

Here the minimum number of lines are four; thus we do not reach at the optimal stage. The same operation will be followed and the resulting matrix is:

Table 13.61

	I	II	III	IV	V
A	7	16	0	13	0
B	0	13	16	3	
C	12	5		0	8
D	4	0	5	6	0
E	8	0	10	2	5

Here the number of lines is five; thus we reach at the optimal stage. Now we shall put the zeros at their corresponding position after omitting the zeros at the points of intersection.

Table 13.62

	I	II	III	IV	V
A			√ 0		8
B	√ 0				
C				√ 0	
D					√ 0
E		√ 0			

The selection of zeros is given in the table such that there is one zero in each row and each column and the selection is unique. The optimal assignment is $A \to III$, $B \to I$, $C \to IV$, $D \to V$ and $E \to II$ and the optimal assignment cost $= 18 + 17 + 17 + 24 + 16 = 92$.

13.7 Unbalanced Assignment Problem

An assignment problem is said to be unbalanced if the number of *Assignee* and the number of *Assignment* be not same and the cost or profit matrix be not a square matrix. Here either the number of assignee is greater than the number of assignment and vice-versa. But the problem can also be solved by converting the matrix (cost or profit) into a square matrix by introducing either fake assignee or assignment whichever is necessary and assuming all costs components zero corresponding to that assignee or assignment. The following solved problem will give a clear idea how to solve such type of problems.

▶ **Example 13.7.1** *In a factory, there are six machines (of same type) and five workers. The handling costs for the ith worker* ($i = 1, \cdots, 5$) *to handle the jth machine* ($j = 1, 2, \cdots, 6$) *are given below in the form of a matrix. Find the optimal assignment and the minimum cost of handling the machines and find which machine will remain unused.*

	M_1	M_2	M_3	M_4	M_5	M_6
W_1	12	8	10	14	11	18
W_2	14	14	8	15	17	12
W_3	9	11	13	15	6	12
W_4	11	9	9	11	8	14
W_5	10	12	15	13	10	12

The problem can be made a balanced problem by introducing a fake worker W_6 with all cost component $c_{6j} = 0[j = 1, 2, \cdots, 6]$ and the square matrix is given below.

Table 13.63

	M_1	M_2	M_3	M_4	M_5	M_6
W_1	12	8	10	14	11	18
W_2	14	14	8	15	17	12
W_3	9	11	13	15	6	12
W_4	11	9	9	11	8	14
W_5	10	12	15	13	10	12
W_6	0	0	0	0	0	0

Proceeding in the similar manner as in the case of the previous solved problems, we obtain the optimal assignment table (as the minimum number of lines required to connect all zeros is six) and the optimal assignments which are given in the following tables:

Table 13.64(A)

	M_1	M_2	M_3	M_4	M_5	M_6
W_1	4	0	2	4	4	8
W_2	6	6	0	3	10	2
W_3	2	4	6	6	0	3
W_4	2	0	0	0	0	3
W_5	0	2	5	1	1	0
W_6	2	2	2	0	3	0

Table 13.64(B)

	M_1	M_2	M_3	M_4	M_5	M_6
W_1		$\checkmark 0$				
W_2			$\checkmark 0$			
W_3					$\checkmark 0$	
W_4		$\otimes$		$\checkmark 0$		
W_5	$\checkmark 0$					$\otimes$
W_6				$\otimes$		$\checkmark 0$

The optimal unique assignment is $W_1 \to M_2$, $W_2 \to M_3$, $W_3 \to M_5$, $W_4 \to M_4$, $W_5 \to M_1$, M_6 remains unused.

The min cost $= 8 + 8 + 6 + 11 + 10 = 43$ units.

Exercises 13C

1. Find the optimal assignment cost from the following cost matrices:

(a)

	A	B	C	D
I	4	5	4	3
II	3	2	2	6
III	4	5	3	5
IV	2	4	2	6

(b)

	A	B	C	D
I	9	12	10	10
II	8	12	7	9
III	4	6	7	8
IV	8	4	5	5

2. Find the optimal assignment profit from the following (a), (b) profit matrices and optimal cost for (c), (d) cost matrices.

(a)

	I	II	III	IV
A	6	4	3	12
B	5	7	6	2
C	4	5	8	7
D	6	7	8	12

(b)

	I	II	III	IV
A	7	5	4	3
B	8	2	6	4
C	5	3	2	1
D	5	4	1	8

(c)

	I	II	III	IV
A	5	3	1	8
B	7	9	2	6
C	6	4	5	7
D	5	7	7	6

[C.U.(P)'87,'89]

(d)

	I	II	III	IV
A	10	12	19	11
B	5	10	7	8
C	12	14	13	11
D	8	15	11	9

[V.U.(P)'91] [V.U.(P)'90]

3. A department has four sub-ordinates and four tasks are to be performed. The sub-ordinates differ in efficiency and the tasks differ in their intrinsic difficulties. The estimate of time (in man hours) each man would take to perform each task is given by:

		Task			
		I	*II*	*III*	*IV*
	1	18	26	17	11
	2	13	28	14	26
Sub-ordinate	3	38	19	18	15
	4	19	26	24	10

[C.U.(H)'87;C.U.(P)'90]

How should the tasks be allotted to men so as to optimize the total man hours?

4. A marketing manager has five salesmen and 5 sales districts. Considering the capabilities of the salesman and the nature of the districts, the marketing manager estimates that sales per month (in hundred rupees) for each salesman in each district would be as follows:

		District				
		A	*B*	*C*	*D*	*E*
	1	32	38	40	28	40
	2	40	24	28	21	36
Salesman	3	41	27	33	30	37
	4	22	38	41	36	36
	5	29	33	40	35	39

Find the assignment of the salesmen to districts that will produce maximum sales.

5. A car hire company has one car in each of the five depots a, b, c, d and e. A customer in each of the five towns A, B, C, D and E requires a car. The distance (in km) between the depots (origins) and the towns (destinations) where the customers are, given by the following distance matrix:

	a	*b*	*c*	*d*	*e*
A	20	40	30	50	40
B	70	40	60	80	40
C	20	90	80	100	40
D	80	60	120	70	40
E	20	80	50	80	80

How should the cars be assigned to the customers so as to minimize the distance travelled?

[Cal. M.Sc. (App. Math.)'77]

6. A company has five jobs to be done on five machines; any job can be done on any machine. The time in hours taken by the machines for the different jobs are as given below. Assign the machine to a job so as to minimize the total machine hours.

		Job 1	2	3	4	5
Machine	A	11	6	14	16	17
	B	7	13	22	7	10
	C	10	7	2	2	2
	D	4	10	8	6	11
	E	13	15	16	10	18

7. A company has 5 jobs to be done. The following matrix shows the return in Rs. of assigning ith machine ($i = 1, 2, \cdots, 5$) to the jth job ($j = 1, 2, \cdots, 5$). Assign the five jobs to five machines so as to maximize the expected profit.

		Job 1	2	3	4	5
Machine	A	5	11	10	12	4
	B	2	4	6	3	5
	C	3	1	5	14	6
	D	6	14	4	11	7
	E	7	9	8	12	5

8. Find the optimal assignment for the assignment problem with the given cost matrix. [C.U.(P)'80]

	M_1	M_2	M_3	M_4	M_5
J_1	8	4	2	6	1
J_2	0	9	5	5	4
J_3	3	8	9	2	6
J_4	4	3	1	0	3
J_5	9	5	8	9	5

9. Find the optimal assignment for the assignment problem with the given cost matrix. [C.U.(H)'80]

	J_1	J_2	J_3	J_4
M_1	10	9	7	8
M_2	5	8	7	7
M_3	5	4	6	5
M_4	2	3	4	3

10. The heads of the department has five jobs V, W, X, Y and Z. The number of hours each man would take to perform each job is given in the matrix. How would the jobs be allocated to minimize the total time? **[C.U.(P) 2000]**

	V	W	X	Y	Z
A	3	5	10	15	8
B	4	7	15	18	8
C	8	12	20	20	12
D	5	5	8	10	6
E	10	10	15	25	10

11. Find the optimal assignments for the assignment problem with the following cost matrix: **[C.U.(P)'81,'99]**

	M_1	M_2	M_3	M_4	M_5
J_1	3	8	2	10	3
J_2	8	7	2	9	7
J_3	6	4	2	7	5
J_4	8	4	2	3	5
J_5	9	10	6	9	10

12. Solve the following unbalanced assignment problem for the assignment with the following cost matrix:

	J_1	J_2	J_3	J_4
A	10	12	8	6
B	6	9	12	14
C	3	8	7	12

	J_1	J_2	J_3	J_4
A	10	12	8	6
B	6	9	12	14
C	3	8	7	12
D	0	0	0	0

[*Hints.* Convert the cost matrix as given in the second matrix as a balanced (4×4) assignment problem with a fictitious row with all costs zero and then solve it in the usual method.]

13. Consider the problem of assigning four operators to four machines. The assignment costs in rupees are given below. Operator 1 cannot be assigned to machine 3 and operator 3 cannot be assigned to machine 4. Find the optimal cost of assignment. **[C.U.(H)'83,'93,'95]**

Operator \ Machine	1	2	3	4
1	5	5	–	2
2	7	4	2	3
3	9	3	5	–
4	7	2	6	7

[*Hints.* As operator 1 cannot be assigned to machine 3 and operator 3 cannot be assigned to machine 4 then during computation no zero will be considered in the cells (1, 3) and (3, 4) and thus solve in the usual method.]

14. Find the optimal assignment for the assignment problem with the following cost matrix: **[C.U.(P)'83,'90]**

	1	2	3	4
1	1	4	6	3
2	9	7	10	9
3	4	5	11	7
4	8	7	8	5

15. In a factory there are four workers and five machines and the operating cost for each worker to handle each machine are given below. Find the optimal assignment and the assignment cost and find which machine will remain idle:

	I	*II*	*III*	*IV*	*V*
A	12	15	14	11	16
B	15	12	14	13	15
C	16	18	12	15	15
D	12	14	13	13	14

16. Five workers perform five jobs and the operating cost are given below, but there is restriction that the worker *A* cannot perform the third job and the worker *C* cannot perform the second job. Find the optimal assignment and the optimal assignment cost.

	I	*II*	*III*	*IV*	*V*
A	14	18	–	16	16
B	12	17	15	15	16
C	14	–	13	16	20
D	12	15	16	14	19
E	15	22	13	14	16

17. Solve the following unbalanced assignment problems where the cost matrix is given below:

	I	*II*	*III*	*IV*
A	13	16	14	21
B	19	16	17	23
C	15	24	13	17
D	11	14	21	23
E	15	17	19	17

18. Solve the two following assignment problems where in the first problem matrix is a cost matrix and the second problem the matrix is a profit matrix:

(a)

	1	2	3	4	5
A	13	15	19	17	20
B	23	26	21	19	15
C	13	17	21	24	15
D	17	13	19	26	17
E	26	19	24	29	11

(b)

	1	2	3	4	5
A	10	2	−7	8	6
B	0	8	−5	5	4
C	8	3	2	1	5
D	1	1	4	2	3
E	8	2	1	−1	4

19. Solve the following restricted assignment problems where the assignment cost matrix is given below:

	I	II	III	IV	V
A	24	29	18	32	19
B	17	26	34	22	–
C	27	16	–	17	25
D	22	18	28	30	24
E	28	16	31	24	27

Solve the above problem taking the matrix to be a profit matrix.

Answers

1. (a) $I \to D$, $II \to B$, $III \to C$, $IV \to A$ and Min. cost = 10 units.
   (b) $I \to D$, $II \to C$, $III \to A$, $IV \to B$ and Min. cost = 25 units.
2. (a) $A \to IV$, $B \to II$, $C \to III$, $D \to I$ and Max. profit = 33 units. Multiple solutions exist.
   (b) $A \to II$, $B \to III$, $C \to I$, $D \to IV$ and Max. profit = 24 units. Multiple solutions exist.
   (c) $A \to III$, $B \to IV$, $C \to II$, $D \to I$ and Min. cost = 16 units.
   (d) $A \to II$, $B \to III$, $C \to IV$, $D \to I$ and Min. cost = 38 units.
3. $1 \to III$, $2 \to I$, $3 \to II$, $4 \to IV$ and Min. time 59 hours.
4. $1 \to B$, $2 \to A$, $3 \to E$, $4 \to G$, $5 \to D$ and Max. profit = Rs.19,100. Multiple solutions exist.
5. $a \to E$, $b \to B$, $c \to A$, $d \to D$, $e \to C$ and Min. distance 200 km. Multiple solutions exist.
6. $A \to 2$, $B \to 5$, $C \to 3$, $D \to 1$, $E \to 4$ and Min. time is 32 hours.
7. $A \to 3$, $B \to 5$, $C \to 4$, $D \to 2$, $E \to 1$ and Max. profit is Rs.50.
8. $J_1 \to M_5$, $J_2 \to M_1$, $J_3 \to M_4$, $J_4 \to M_3$, $J_5 \to M_2$ and Min. cost = 9 units.
9. $M_1 \to J_3$, $M_2 \to J_1$, $M_3 \to J_2$, $M_4 \to J_4$ and Min. cost = 19 units.
10. $A \to X$, $B \to W$, $C \to V$, $D \to Y$, $E \to Z$ and Min. time = 45 hours.

11. $J_1 \rightarrow M_5$, $J_2 \rightarrow M_3$, $J_3 \rightarrow M_2$, $J_4 \rightarrow M_4$, $J_5 \rightarrow M_1$ Min cost 21 units.

12. $A \rightarrow J_4$, $B \rightarrow J_2$, $C \rightarrow J_1$ and Min. cost = 18 units.

13. $1 \rightarrow 4$, $2 \rightarrow 3$, $3 \rightarrow 2$, $4 \rightarrow 1$ and Min. cost = Rs.14.

14. $1 \rightarrow 1$, $2 \rightarrow 3$, $3 \rightarrow 2$, $4 \rightarrow 4$ and Min. cost = 21 units.

15. $A \rightarrow IV$, $B \rightarrow II$, $C \rightarrow III$, $D \rightarrow I$ and the fifth machine will remain idle.

16. $A \rightarrow V$, $B \rightarrow I$, $C \rightarrow III$, $D \rightarrow II$ and $E \rightarrow IV$, Min cost = 70 units.

17. $A \rightarrow I$, $C \rightarrow III$, $D \rightarrow II$, $E \rightarrow IV$ and Min. cost = 57 units. (There will be no-assignment for B). Multiple solutions exist.

18. (a) $A \rightarrow 4$, $B \rightarrow 3$, $C \rightarrow 1$, $D \rightarrow 2$, $E \rightarrow 5$ and Min. cost = 75 units.
    (b) $A \rightarrow 4$, $B \rightarrow 2$, $C \rightarrow 5$, $D \rightarrow 3$, $E \rightarrow 1$ and Max. profit = 33 units.

19. $A \rightarrow III$, $B \rightarrow I$, $C \rightarrow IV$, $D \rightarrow V$, $E \rightarrow II$ and Min. cost = 92 units.
    $A \rightarrow II$, $B \rightarrow III$, $C \rightarrow I$, $D \rightarrow IV$, $E \rightarrow V$ and Max. profit = 147 units.

13.8 Travelling Salesman problems

Travelling salesman problem is a more restricted type of assignment problem where a salesman has to come back to his starting station travelling all the stations once only starting from any station by travelling all the stations once only starting from any station by travelling a minimum distance. Obviously there will be no assignment from the ith origin to jth destination for $i = j$ $[i, j = 1, \cdots, n]$ which is one type of restriction adopted to the problem. Moreover, he has to travel in such a way that the circuit be closed (touching each station once only), i.e., he may start from any station and he has to come back that station touching all stations once only and the distance travelled be minimum. But to solve such type of problems, there is no definite optimality test and we shall have to solve such type of problems by using trial and error method and therefore the technique of solving such type of problems cannot be considered as most scientific one. [There is another technique which is known as branch and bound technique which is a difficult one]. From the solved problem given below we shall get some idea about the nature of procedure of solving this particular type of problems. If the matrix be given in term of distance, it should be symmetric [ignoring the principal diagonal elements]. But this is not strictly necessary from the theoretical point of view because the distance from A station to D station may not be same when travelled in reversed direction as he may be compelled to move in two different routes.

13.8.1 Travelling Salesman problems

▶ **Example 13.8.1** *A salesman has to visit five towns A, B, C, D and E. He does not wish to visit any town twice before completing his tour of all towns and wants to return to the starting town by travelling a minimum distance. The distances from one town to another are given below. Find the optimal route and the minimum distance to be travelled by the salesman.* **[C.U. (Hons.) 2000]**

To town / From town	A	B	C	D	E
A	∞	4	7	3	4
B	4	∞	6	3	4
C	7	6	∞	7	5
D	3	3	7	∞	7
E	4	4	5	7	∞

Note: The matrix [symmetric] of the problem is taken from the [C.U., M. Com. '89] Question paper, but the problem is given in a different manner, the solution procedure is same which is given in the next problem.

Initially we shall solve the problem (obeying the restriction given above). Following the general rule of assignment, the optimal assignment table and the optimal assignment is given below. Here the optimal assignment is not unique and one of the optimal assignments is given in the table 13.65. The optimal assignment is $A \to D, D \to B, B \to A, C \to E, E \to C$ and the min distance is 3+3+4+5+5 = 20 units. But this assignment does not satisfy the restriction of the travelling salesman problem as the circuit is being closed in the manner $A \to D, D \to B, B \to A$ etc.

Table 13.65

∞	~~0~~	2	✓0	~~0~~
✓0	∞	1	~~0~~	~~0~~
2	1	∞	3	✓0
~~0~~	✓0	3	∞	4
~~0~~	~~0~~	✓0(4)	4	∞

Now we shall have to assign in the trial and error basis, if possible by including 1 element in (2, 3) or (3, 2) or by including the next minimum element 2, such that the whole circuit be completed and the distance travelled be minimum. There may exist many possible such circuits and we shall have to select such circuit for which the sum of the cost components of the new assignment be minimum. Below given two such circuits. Now to construct a complete circuit we may stick to the assignments $A \to D \to B$ which are already obtained by assigning cells with cost zero (0). The two tables with different complete circuits are given below.

Table 13.66

∞	~~0~~	2	✓0(1)	~~0~~
~~0~~	∞	1	~~0~~	✓0(3)
✓2(5)	1	∞	3	0
~~0~~	✓0(2)	3	∞	4
~~0~~	~~0~~	✓0(4)	4	∞

Table 13.67

∞	~~0~~	2	✓0(1)	~~0~~
~~0~~	∞	1(3)	~~0~~	~~0~~
2	1	∞	3	✓0(4)
~~0~~	✓0(2)	3	∞	4
✓0(5)	~~0~~	~~0~~	4	∞

In the first table, the assignment is $A \to D \to B \to E \to C \to A$ and the assignment cost is 2 units. (obtained from the final table).

In the second table, the assignment is $A \to D \to B \to C \to E \to A$ and the cost is 1 unit.

Obviously, we can say that the 2nd assignment is the optimal assignment for the salesman and the minimum distance = 3 + 3 + 6 + 5 + 4 = 21 units.

Another optimal assignment is $A \to E \to C \to B \to D \to A$, and min distance = 21 units.

Remark: In this method it is not possible to say definitely in which circuit, the cost or distance be minimum. We shall have to use 'trial and error' method to get the optimal assignment. Hence the method cannot be considered as a most scientific one.

► **Example 13.8.2** *A machine operator processes five types of items on his machine each week and must choose a sequence for them. The set up cost per change depends on the item presently on the machine and the set up to be made according to the following table:*

To item / From item	A	B	C	D	E
A	∞	4	7	3	4
B	4	∞	6	3	4
C	7	6	∞	7	5
D	3	3	7	∞	7
E	4	4	5	7	∞

If he processes each type of time once and only once in each week, how should be sequence the items on his machine in order to minimize the total set up cost?

[C.U. M. Com. '89]

The optimal sequence of items so that the cost be minimum is same as in the above solved problem and that is $A \to D \to B \to C \to E \to A$ and Min cost = 21 units.

Remark: If the cost matrix be a symmetric matrix, then it is very easy to find the optimal route and the optimal cost of the travelling salesman problems. I solve the next problem in such a way and this is one of the easiest methods of solving the problem.

► **Example 13.8.3** *Solve the following T.S. problem where the cost matrix is a symmetric matrix.*

	A	B	C	D	E
A	—	8	12	14	10
B	8	—	7	11	14
C	12	7	—	19	20
D	14	11	19	—	15
E	10	14	20	15	—

Here the working principal is very simple.

(1) We initially start from A and go the destination whose cost in lowest. Thus we shall go to B; then from $B \to C$, the lowest cost of that row. Then from C to D since otherwise the loop (Path) will be closed. Then from $D \to E$ and finally $E \to A$ and the path is then $A \to B \to C \to D \to E \to A$ and the cost is $8 + 7 + 19 + 15 + 10 = 59$ units. The cost may or may not be minimum. Now we start from B and the path is; $B \to C \to A \to E \to D \to B$ [the rule followed is the selection of lowest cost after due consideration that the path be not completed before attaining each and every station and the cost = 55 units, which is 4 units less than the previous one]. Now starting from C and following the same rule, the path is $C \to B \to A \to E \to D \to C$ and the cost is 59 units. Similarly starting from D, and if the path be $D \to B \to C \to A \to E \to D$, cost is 55 units; and finally $E \to A \to B \to C \to D \to A$ and the cost is 59 units. Hence the min cost is 55 units and there is one such path.

► **Example 13.8.4** *Solve the following travelling salesman problem. Find the optimal route and the minimum cost needed by the salesman.*

From town \ To town	A	B	C	D	E
A	–	8	12	6	9
B	4	–	10	8	4
C	7	10	–	6	9
D	11	7	8	–	4
E	5	9	6	9	–

Initially we try to follow the rule which has been done in the case of restricted assignment problem and the optimal assignment table is given below.

Optimal assignment table.

Table 13.68

	A	B	C	D	E
A	–	0	5	0	4
B	0	–	4	3	0
C	2	2	–	0	4
D	7	0	2	–	0
E	1	2	0	4	–

The optimal assignment is unique and the optimal assignment cost is 28 for the assignment $A \to B$, $B \to A$, $C \to D$, $D \to E$, $E \to C$. Here the assignment of travelling salesman has not been followed. Now we find the route by 'trial and error' method.

One travelling salesman assignment cost $A \to B \to E \to C \to D \to A$ and the extra cost is 7 units. We try to include 1 or 2 in the optimal assignment. Other optimal assignment following the restriction of the travelling salesman problem is $A \to D \to E \to C \to B \to A$ and in that case there will be increment of the cost 2 units more. Hence the cost is (28 + 2) units = 30 units.

Let us try to include cost 1 unit in the path and such route the cost is more than two units. Hence the optimal cost in the travelling salesman problem is 30 units and the optimal path is $A \to D \to E \to C \to B \to A$.

▶ **Example 13.8.5** *Solve the following travelling salesman problem.*

	A	B	C	D	E
A	–	12	15	10	8
B	8	–	15	12	8
C	9	11	–	15	11
D	7	12	19	–	11
E	9	12	16	10	–

Initially we try to solve this problem as a restricted assignment problem.

Subtracting the lowest element of each row from all other elements of the corresponding row we get the new table.

Table 13.69

	A	B	C	D	E
A	–	4	7	2	0
B	0	–	7	4	0
C	0	2	–	7	2
D	7	5	12	–	4
E	0	3	7	1	–

Now subtract the lowest element of each column from the all other elements of the corresponding column we get the following table.

Table 13.70

	A	B	C	D	E
A	–	2	0	1	0
B	0	–	0	3	0
C	0	0	–	6	1
D	0	3	5	–	3
E	0	1	0	0	–

All zeros can only be covered by minimum five lines which are shown in the table.

Now from the table, construct the table containing only all zeros at their proper position after removing all zeros at the points of intersection of the lines and the new table is

Table 13.71

	A	B	C	D	E
A	–		$\boxed{0}$		$\not{0}$
B		–	$\not{0}$		$\boxed{0}$
C		$\boxed{0}$	–		
D	$\boxed{0}$				
E			$\not{0}$	$\boxed{0}$	

Selection of only one zero in each row and each column is not unique. One such selection is given above. The restricted optimal assignment is $A \to C$, $B \to E$, $C \to B$, $D \to A$, $E \to D$ and the min cost is $15 + 8 + 11 + 7 + 10 = 49$ units.

Now if we arrange to route in the manner $A \to C \to B \to E \to D \to A$ then it follows the route of the travelling salesman problem and hence the optimal cost of the travelling salesman problem is 49 units and one route is $A \to C \to B \to E \to D \to A$.

► **Example 13.8.6** *A salesman has to visit five cities A, B, C, D and E. The distances (in hundred km.) between the five cities are given below in matrix form. If the salesman starts from city A and has to come back to that city (covering all the cities once only) which route should he follow so that the total distance travelled be minimum.*

Note. Here the matrix C (distance) is not symmetric.

		To				
		A	*B*	*C*	*D*	*E*
	A	—	2.5	4	6	1
	B	4	—	3	7	2
From	*C*	8	6	—	4	6
	D	10	4	6	—	6
	E	1	3	2	8	—

[**Hints.** Here the initial optimal assignment is $A \to E, E \to A, B \to C, C \to D, D \to B$, and the min. distance = 13 units. But to satisfy the condition of travelling salesman problem, there exist so many complete circuits the min. cost of one such table (which will be determined by trial and error method) is 2, and that assignment is

$A \to E \to C \to D \to B \to A$ and the min. distance travelled by the salesman is $1 + 2 + 4 + 4 + 4 = 15$ [in hundred km.]. Other optimal route is $A \to C \to D \to B \to E \to A$.]

► **Example 13.8.7** *A medical representative has to visit five stations A, B, C, D and E. He does not want to visit any station twice before completing his tour of all stations and wishes to return to the starting station. Cost of going from one station to another are given below. Determine the optimal route and the minimum expenditure to be done by the representative.*

		To				
		A	B	C	D	E
	A	—	1	4	7	1
	B	3	—	2	7	2
From	C	8	6	—	4	6
	D	9	3	5	—	7
	E	1	2	2	7	—

Optimal route $A \to E \to C \to D \to B \to A$, min. cost = 13 units.

► **Example 13.8.8** *Solve the following travelling salesman problem.*

	A	B	C	D	E
A	—	7	6	8	4
B	7	—	8	5	6
C	6	8	—	9	7
D	8	5	9	—	8
E	4	6	7	8	—

[C.U. (Hons.) '96, '98]

Ans. Optimal route: $A \to E \to B \to D \to C \to A$

or, $A \to C \to D \to B \to E \to A$. Min. cost = 30 units.

► **Example 13.8.9** *A salesman estimates that the following would be the costs (Rs.) on his route by visiting the five cities as shown in the table beside. The salesman can visit each of the cities once. He does not want to visit any city twice before completing his tour of all the cities and wishes, to return to the starting city. Determine the optimum sequence he should follow to minimise the total cost. Determine the total cost.*

		To city				
		A	B	C	D	E
	A	—	2	5	7	1
	B	6	—	3	8	2
From city	C	8	7	—	4	7
	D	12	4	6	—	5
	E	1	3	2	8	—

[V.U. (Hons.) '89]

Ans. $A \to B \to C \to D \to E \to A$. Min. cost = 15 units.

► **Example 13.8.10** *Solve the following travelling salesman problem.*

		To city				
		A	B	C	D	E
	A	—	5	8	4	5
	B	5	—	7	4	5
From city	C	8	7	—	8	6
	D	4	4	8	—	8
	E	5	5	6	8	—

[V.U. (Hons.) '95]

Ans. $A \to E \to C \to B \to D \to A$. Min. cost = 26 units.

▶ **Example 13.8.11** *Solve the following travelling salesman problem.*

To city

	A	*B*	*C*	*D*	*E*
A	—	14	10	24	41
B	6	—	10	12	10
C	7	13	—	8	15
D	11	14	30	—	17
E	6	8	12	16	—

[V.U. (Hons.) '88]

Ans. $A \to C \to D \to B \to E \to A$. Min. cost = 48 units.

▶ **Example 13.8.12** *Solve the following travelling salesman problem.*

(i)

	A	B	C	D	E
A	–	24	17	11	19
B	24	–	18	16	11
C	17	18	–	15	9
D	11	16	15	–	21
E	19	11	9	21	–

(ii)

	A	B	C	D	E
A	–	14	7	17	9
B	8	–	13	20	12
C	10	10	–	15	12
D	13	11	13	–	14
E	9	11	10	12	–

[V.U. (Hons.) '98]

[**Ans.** (i) $A \to D \to B \to E \to C \to A$. (ii) $A \to C \to D \to B \to E \to A$.

Min cost = 64 units. Min cost = 48 units.

Objective and short answer type questions with answers Assignment problems

1. (a) Define an assignment problem. (b) Formulate an assignment problem mathematically. **[Ans. See page 313.]**

2. Can we consider an assignment problem as a transportation problem; if so, state the nature of the optimal solution.

   Ans. Yes, we can consider an assignment problem as a particular case of transportation problem; but here in each case the optimal solution will be degenerate one.

3. Which of the statements is true?

   (a) An assignment problem is a minimization problem.

   (b) An assignment problem may be a maximization or a minimization problem.

   (c) An assignment problem is a miximization problem.

   Ans. (ii) is true.

4. What is an unbalanced assignment problem? How can you convert into a balanced assignment problem?

Ans. An assignment problem is said to be an unbalanced one, if the number of assignee and the number of assignments be not same. For example, if the number of assignee be n and the number of assignments be $n-1$ (one less than the number of assignee), assuming a fictitious assignment B_n with the cost components $c_{in} = 0, i = 1, 2, \cdots, n$ we can make it a balanced assignment problem and in that case one of the assignee will remain idle etc.

5. How can we solve a maximization assignment problem?

Ans. Here, first of all change the sign of all cost components. Then solve it considering it as a minimization problem and finally $\max z = -\min(-Z)$ etc. For example, if the profit matrix be

$$\begin{array}{c} \\ A_1 \\ A_2 \\ A_3 \end{array} \begin{array}{c} \begin{matrix} M_1 & M_2 & M_3 \end{matrix} \\ \begin{pmatrix} 4 & 0 & 3 \\ 2 & 3 & 5 \\ 3 & 4 & 2 \end{pmatrix} \end{array}$$

then convert it in the manner and consider it now as a cost matrix which is

$$\begin{bmatrix} -4 & 0 & -3 \\ -2 & -3 & -5 \\ -3 & -4 & -2 \end{bmatrix}$$

Now find the minimum cost of the second matrix. If it be k (say) then max. profit $= -k$ with the same optimal assignments.

6. What is a restricted assignment problem? How can you solve such type of problems?

Ans. A particular assignment problem is said to be a restricted assignment problem if one or more assignee are restricted not to accept particular assignment or assignments. For example, if the i-th assignee be not allowed to receive the j-th assignment, then during computation to solve the problem by the Hungarian method, we are forced not to consider zero in the cell (i, j) and thus we simply treat that in no case there will be zedro in that cell and assuming that we shall be able to solve any such problem.

7. (a) What is a travelling salesman problem?

   (b) What are the restrictions required to adopt in a travelling salesman problem?

   (c) How can you solve a travelling salesman problem?

Ans. For (i) see sec. 13.8.1.

(ii) There are two restrictions:

(a) i-th assignee will not be assigned to j-th assignment for $i = j$ and

(b) the mode of assignment be such that the circuit be complete, i.e., starting from any station, and touching all stations only once, a salesman will have to travel such that the distance or the cost of travelling be minimum.

(iii) Practically, to solve a travelling salesman problem we are to follow the trial and error method to get the optimal solution.

8. Is the optimal solution of an assignment problem unique?

Ans. Not necessarily. Solution may or may not be unique one.

Chapter 14

Revised Simplex Algorithm

14.1 Introduction

The simplex method of solving a L.P.P. is a straight forward algebraic procedure. But during computation, we require to transform many quantities in each iteration all of which have no current use. It was unnecessary to transform all $\mathbf{y}_j$, $\mathbf{x}_B$, $z_j - c_j$ and z at each iteration. In fact all quantities can be computed directly from their definitions if B^{-1} is known, i.e., if only the basis inverse is transformed. The *revised simplex algorithm* or *revised simplex method* makes use of this observation. In this method, only B^{-1}, $\mathbf{x}_B$, $\mathbf{c}_B B^{-1}$ and z are transformed in each iteration. None of $\mathbf{y}_j$ is transformed. Practically, only $\mathbf{y}_j$ that is determined at each iteration, is that one for the vector which is going to enter in the next basis. This type of transformation scheme has some advantages over that used in the simplex method. In revised simplex method, initially a basic feasible solution is obtained. Next we move from that solution to an optimal feasible solution by changing a single vector in the basis at a time. The methods of selecting vector to enter in the next basis and the vector to leave the current basis are precisely the same as those used in the ordinary simplex method.

14.2 Revised Simplex Method in Standard Form I

In this case, it is assumed that an identity basis matrix is present after introducing slack and surplus variables and there is no need of using artificial variables and $b_i \geq 0$.

$[i = 1, \cdots, m]$

Let the linear programming problem be

maximize, $z = c_1x_1 + c_2x_2 + \cdots + c_nx_n + 0.x_{n+1} + 0.x_{n+2} + \cdots + 0.x_{n+m}$,

subject to

$$\left.\begin{array}{l} a_{11}x_1 + a_{12}x_2 + \cdots + a_{1n}x_n + x_{n+1} = b_1 \\ a_{21}x_1 + a_{22}x_2 + \cdots + a_{2n}x_n \qquad + x_{n+2} = b_2 \\ \cdots \cdots \cdots \cdots \cdots \cdots \cdots \cdots \\ a_{m1}x_1 + a_{m2}x_2 \qquad\qquad + x_{n+m}, = b_m \end{array}\right\} \quad (14.2.1)$$

and $x_1, x_2, \cdots, x_{n+m} \geq 0$,

after the introduction of slack and surplus variables.

The initial basis matrix is an $(m \times m)$ unit or identity matrix and $\mathbf{x}_B = B^{-1}\,\mathbf{b} = \mathbf{b} \geq 0$. In the revised simplex method $z = \mathbf{cx}$ is being treated as an additional constraint for which z is to be made as large as possible. Expressing the objective function as a following constraint

$$\begin{array}{l} z - c_1x_1 - c_2x_2 - \cdots - c_nx_n - 0.x_{n+1} \cdots - 0.x_{n+m} = 0 \\ \text{or, } z - c_1x_1 - c_2x_2 \qquad - c_nx_n - c_{n+1}x_{n+1} \quad - c_{n+m}x_{n+m} = 0 \end{array}$$

where $c_{n+1} = c_{n+2} = \cdots = c_{n+m} = 0$

the set of $(m+1)$ constraints can be written in the manner as (using the constraint of the objective function as the first constraint)

$$\left.\begin{array}{l} z - c_1x_1 - c_2x_2 \cdots - c_nx_n - c_{n+1}x_{n+1} - \cdots - c_{n+m}x_{n+m} = 0 \\ \quad a_{11}x_1 + a_{12}x_2 \quad + a_{1n}x_n + \quad x_{n+1} \qquad = b_1 \\ \quad \cdots \cdots \cdots \cdots \cdots \cdots \cdots \\ \quad a_{m1}x_1 + a_{m2}x_2 \quad + a_{mn}x_n + \qquad + \quad x_{m+n} = b_n \end{array}\right\} \quad (14.2.2)$$

which can be considered as $(m+1)$ linear simultaneous equations with $(n+m+1)$ variables $z, x_1, x_2, \cdots, x_n, \cdots, x_{n+m}$. The variable z is as large as possible and unrestricted in sign and all other variables are non-negative quantities. The problem before us is to solve (14.2.2) such that z is as large as possible and unrestricted in sign.

The set of $(m+1)$ equations in (14.2.2) can be rewritten as follows :

$$\left.\begin{array}{l} 1.x_0 - c_1x_1 - c_2x_2 - \cdots - c_nx_n - c_{n+1}x_{n+1} \qquad - c_{n+m}x_{n+m} = 0 \\ 0.x_0 + a_{11}x_1 + a_{12}x_2 + \cdots + a_{1n}x_n + \quad x_{n+1} \qquad = b_1 \\ 0.x_0 + a_{21}x_1 + a_{22}x_2 + \cdots + a_{2n}x_n \qquad + x_{n+2} \qquad = b_2 \\ \cdots \cdots \cdots \cdots \cdots \cdots \cdots \\ 0.x_0 + a_{m1}x_1 + a_{m2}x_2 + \cdots + a_{mn}x_n \qquad + \qquad x_{n+m} = b_m \end{array}\right\} \quad (14.2.3)$$

where z is taken as x_0.

Using matrix notation, the equations can be written as :

$$\begin{bmatrix} 1 : - & c_1 & - & c_2 & - \cdots - & c_n & - & c_{n+1} & - & c_{n+m} \\ \cdots & \cdots & \cdots & \cdots & \cdots & \cdots & \cdots & \cdots & \cdots & \cdots \\ 0 : & a_{11} & & a_{12} & \cdots & a_{1n} & & 1 & & 0 \\ \cdots & \cdots & \cdots & \cdots & \cdots & \cdots & \cdots & \cdots & \cdots & \cdots \\ 0 : & a_{m1} & & a_{m2} & \cdots & a_{mn} & & 0 & & 1 \end{bmatrix} \begin{bmatrix} x_0 \\ x_1 \\ \cdots \\ \cdots \\ x_{n+m} \end{bmatrix} = \begin{bmatrix} 0 \\ b_1 \\ b_2 \\ \cdots \\ b_m \end{bmatrix} \tag{14.2.4}$$

and hence expressing in partitioned method it can be written as :

$$\begin{bmatrix} 1 & -\mathbf{c} \\ 0 & A \end{bmatrix} \begin{bmatrix} x_0 \\ \mathbf{x} \end{bmatrix} = \begin{bmatrix} 0 \\ \mathbf{b} \end{bmatrix} \quad \text{or,} \quad \begin{bmatrix} 1 & -\mathbf{c} \\ 0 & A \end{bmatrix} \begin{bmatrix} z \\ \mathbf{x} \end{bmatrix} = \begin{bmatrix} 0 \\ \mathbf{b} \end{bmatrix} \tag{14.2.5}$$

where $\mathbf{c} = (c_1, c_2, \cdots, c_n, \cdots, c_{n+m})$ is an $(n+m)$ component row vector and the rest of the symbols have their usual meanings.

Writing $-c_j = a_{0j}[j = 1, 2, \cdots, n+m]$ and $-\mathbf{c} = \mathbf{a}^0$, (14.2.5) can be written as

$$\begin{bmatrix} 1 & \mathbf{a}^0 \\ 0 & A \end{bmatrix} \begin{bmatrix} z \\ \mathbf{x} \end{bmatrix} = \begin{bmatrix} 0 \\ \mathbf{b} \end{bmatrix} \tag{14.2.6}$$

14.3 Change of Usual Notations

We have already seen that all column vectors have $(m+1)$ components instead of m components. We shall use the super-script on all the vectors to indicate that they have $(m+1)$ components. Hence corresponding to each $\mathbf{a}_j$ in A we shall represent a new $(m+1)$ component vector $\mathbf{a}_j^*$ given by

$$\begin{aligned} \mathbf{a}_j^* &= [-c_j, a_{1j}, a_{2j}, \cdots, a_{mj}] \quad j = 1, 2, \cdots, n+m \\ &= [a_{0j}, a_{1j}, a_{2j}, \cdots, a_{mj}] \\ &= [a_{0j}, \mathbf{a}_j] \end{aligned} \tag{14.3.1}$$

where a_{ij} have their usual meanings.

Similarly corresponding to m-component column vector $\mathbf{b}$ in $A\mathbf{x} = \mathbf{b}$ we shall represent a new $(m+1)$ component vector $\mathbf{b}^*$ given by

$$\mathbf{b}^* = [0, b_1, b_2, \cdots, b_m] = [0, \mathbf{b}]. \tag{14.3.2}$$

14.4 Basis Matrix in the Revised Simplex Method

The basis matrix B in the set of equations $A\mathbf{x} = \mathbf{b}$ is given by

$$B = (\beta_1, \beta_2, \cdots, \beta_m) = \begin{bmatrix} \beta_{11} & \beta_{12} & \cdots & \beta_{1m} \\ \beta_{21} & \beta_{22} & \cdots & \beta_{2m} \\ \cdots & \cdots & \cdots & \cdots \\ \beta_{m1} & \beta_{m2} & \cdots & \beta_{mm} \end{bmatrix}$$

The column vector corresponding to z (or x_0) in (14.2.4) is an $(m+1)$ component unit vector $\mathbf{e}_1$. As we always try to maximize z, then $\mathbf{e}_1$ must be present in the basis matrix of the revised simplex method at each stage. Let us assume that $\mathbf{e}_1$ is present in the first column of the basis matrix. The basis matrix B^* of the revised simplex method is a $(m+1) \times (m+1)$ matrix.

The basis matrix B^* is given by

$$B^* = (\mathbf{e}_1, \beta_1^*, \beta_2^*, \cdots, \beta_m^*) = (\beta_0^*, \beta_1^*, \beta_2^*, \cdots, \beta_m^*)$$

$$= \left[\begin{array}{c|cccc} 1 & -c_{\beta 1} & -c_{\beta 2} & \cdots & -c_{\beta m} \\ \hline 0 & \beta_{11} & \beta_{12} & \cdots & \beta_{1m} \\ 0 & \beta_{21} & \beta_{22} & \cdots & \beta_{2m} \\ \cdots & \cdots & \cdots & \cdots & \cdots \\ 0 & \beta_{m1} & \beta_{m2} & \cdots & \beta_{mm} \end{array}\right] \tag{14.4.1}$$

where $-c_{\beta i}$ is the co-efficient of $x_{\beta i}$ (basic variable) in the equation

$$z - \sum_{j=1}^{n+m} c_j x_j = 0, \beta_i^* = [-c_{\beta i}, \beta_{1i}, \beta_{2i}, \cdots, \beta_{mi}]$$

is an $(m+1)$ component column vector $[i = 1, 2, \cdots, m]$ and $\beta_0^* = \mathbf{e}_1$, is an $(m+1)$ component unit column vector. Expressing the basis matrix B^* in partitioned method, it can be written as

$$B^* = \begin{bmatrix} I & -\mathbf{c}_B \\ \mathbf{0} & B \end{bmatrix} \tag{14.4.2}$$

where $\mathbf{c}_B$ is the associated cost vector in the original simplex method and $\mathbf{c}_B = (c_{B1}, c_{B2}, \cdots, c_{Bm})$.

14.5 Computation of $\mathbf{B}^{*-1}$

B^{*-1} is very essential in solving the problem by revised simplex method.

We know that if M is a matrix which can be expressed in partitioned method

$$M = \begin{bmatrix} I & P \\ \mathbf{0} & Q \end{bmatrix}$$

where I is an identity matrix, P is a matrix, $\mathbf{0}$ is a null matrix, Q is a non-singular sub-matrix and Q^{-1} is known then M^{-1} is given by

$$M^{-1} = \begin{bmatrix} I & -PQ^{-1} \\ \mathbf{0} & Q^{-1} \end{bmatrix}$$

and using this formula, we can find B^{*-1}.

$$B^* = \begin{bmatrix} 1 & -\mathbf{c}_B \\ \mathbf{0} & B \end{bmatrix} \tag{14.5.1}$$

and B^{-1} is known to us.

Taking $I = 1$, $P = -\mathbf{c}_B$ and $Q = B$, we have

$$B^{*-1} = \begin{bmatrix} 1 & \mathbf{c}_B B^{-1} \\ \mathbf{0} & B^{-1} \end{bmatrix}. \tag{14.5.2}$$

This result is very helpful in solving the problems.

14.6 Expression of $\mathbf{a}_j^*$ as a Linear Combination of the Vectors in $\mathbf{B}^*$

Any column vector $\mathbf{a}_j^*$ which is not in B^* can be expressed as a linear combination of vectors present in B^*.

Let

$$\mathbf{a}_j^* = \beta_0^* y_{0j} + \beta_1^* y_{1j} + \cdots + \beta_m^* y_{mj} = B^* \mathbf{y}_j^*, \tag{14.6.1}$$

where

$$\mathbf{y}_j^* = [y_{0j}, y_{1j}, \cdots, y_{mj}]. \tag{14.6.2}$$

From (14.6.1) we get,

$$\mathbf{y}_j^* = B^{*-1} \mathbf{a}_j^* \tag{14.6.3}$$

Now substituting the value of B^{*-1} from (14.5.1) in (14.6.3) we get,

$$\mathbf{y}_j^* = \begin{bmatrix} 1 & \mathbf{c}_B B^{-1} \\ 0 & B^{-1} \end{bmatrix} \begin{bmatrix} -c_j \\ \mathbf{a}_j \end{bmatrix} \quad \{\mathbf{a}_j^* = [-c_j, \mathbf{a}_j]\}$$

$$= \begin{bmatrix} - \quad c_j & + & \mathbf{c}_B B^{-1} \mathbf{a}_j \\ 0 & + & B^{-1} \mathbf{a}_j \end{bmatrix}$$

$$= \begin{bmatrix} z_j - c_j \\ \mathbf{y}_j \end{bmatrix} \quad (\text{as } \mathbf{c}_B B^{-1} \mathbf{a}_j = z_j)$$

$$\text{i.e.,} \quad \mathbf{y}_j^* = \begin{bmatrix} z_j - c_j \\ \mathbf{y}_j \end{bmatrix}. \tag{14.6.4}$$

It is important to note that the first component of $\mathbf{y}_j^*$ is $z_j - c_j = y_{0j}$ which is the net evaluation of the original simplex method and which has been always used to test the optimality of the problem.

Hence from (14.6.3) and (14.6.4) we can say that $z_j - c_j$ can be found from the product

$$\text{(First row of } B^{*-1}) \times (\mathbf{a}_j^* \text{ not in the basis}) = z_j - c_j. \tag{14.6.5}$$

14.7 Basic Solution to the Revised Simplex Problem

Basic solution of the revised simplex problem is denoted by $\mathbf{x}_B^*$ and given by

$$\mathbf{x}_B^* = B^{*-1}\mathbf{b}^* \quad \text{or} \quad \mathbf{x}_B^* = \begin{bmatrix} 1 & \mathbf{c}_B B^{-1} \\ 0 & B^{-1} \end{bmatrix} \begin{bmatrix} 0 \\ \mathbf{b} \end{bmatrix} = \begin{bmatrix} \mathbf{c}_B(B^{-1}\mathbf{b}) \\ \mathbf{x}_B \end{bmatrix}$$

$$= \begin{bmatrix} \mathbf{c}_B \mathbf{x}_B \\ \mathbf{x}_B \end{bmatrix} = \begin{bmatrix} z_B \\ \mathbf{x}_B \end{bmatrix} \quad [\text{as } \mathbf{c}_B\mathbf{x}_B = z_B] \tag{14.7.1}$$

Hence the first component of the basic solution $\mathbf{x}_B^*$ is the value of the objective function z in the original problem for the solution set $\mathbf{x}_B$. Therefore the value of the objective function at any stage can be determined from the solution set $\mathbf{x}_B^*$.

The solution set $\mathbf{x}_B^*$ may not be feasible at each stage (as the value of z may be negative).

14.8 Computation of Initial Basis Inverse B^{*-1} [Standard Form I]

We have from (14.5.1)

$$B^{*-1} = \begin{bmatrix} 1 & \mathbf{c}_B B^{-1} \\ 0 & B^{-1} \end{bmatrix}$$

We have assumed that $B = I_m$ is an identity matrix of order m. If we further assume that all $b_i \geq 0$ and only slack variables are needed to get a unit basis I_m then we have initially $c_{B1} = c_{B2} = \cdots = c_{Bm} = 0$, i.e., $\mathbf{c}_B = \mathbf{0}$.

Then

$$B^{*-1} = \begin{bmatrix} 1 & \mathbf{c}_B B^{-1} \\ 0 & B^{-1} \end{bmatrix} = \begin{bmatrix} 1 & 0 \\ 0 & I_m \end{bmatrix} \quad (\text{as } B^{-1} = I_m)$$

$$= I_{m+1} \tag{14.8.1}$$

Therefore the inverse of the initial basis B^* is I_{m+1} and $B^{*-1} = B^* = I_{m+1}$ as the initial basis matrix B^* is given by

$$B^* = \left[\begin{array}{c|ccccc} 1 & 0 & 0 & \cdots & 0 \\ \hline 0 & 1 & 0 & \cdots & 0 \\ 0 & 0 & 1 & \cdots & 0 \\ \cdots & \cdots & \cdots & \cdots & \cdots \\ 0 & 0 & 0 & \cdots & 1 \end{array}\right] \text{ from (14.4.1) [putting } c_{Bi} = 0 \text{ etc.]}$$

The initial basic solution of revised simplex problem is given by

$$\mathbf{x}_B^* = B^{*-1}\mathbf{b}^* = I_{m+1}\mathbf{b}^* = \mathbf{b}^* = \begin{bmatrix} 0 \\ \mathbf{b} \end{bmatrix} \quad (14.8.2)$$

which is feasible also.

14.9 Computational Procedure of the Revised Simplex Method in the Standard Form I

Step 1. If the problem be a problem of minimization, convert it to a maximization problem.

Step 2. After ensuring that all $b_i \geq 0$, initial basis $B = I_m$ and $\mathbf{c}_B = \mathbf{0}$, the cost components corresponding to the basic variables of the ordinary simplex method, we have then $B^* = I_{m+1}$.

Step 3. Determine the initial basic solution $\mathbf{x}_B^*$.

Step 4. Construct the initial table for the revised simplex method.

Initial revised simplex table (Standard Form I)

Variables in the basis	$\beta_0^* = \mathbf{e}_1$	B^{*-1}				$\mathbf{x}_B^*$	$\mathbf{y}_k^* = B^{*-1}\mathbf{a}_k^*$
z	1	0	0	$\cdots$	0	0	$z_k - c_k$
x_{B1}	0	1	0	$\cdots$	0	b_1	y_{1k}
x_{B2}	0	0	1	$\cdots$	0	b_2	y_{2k}
$\cdots$	$\cdots$	$\cdots$	$\cdots$	$\cdots$	$\cdots$	$\cdots$	$\cdots$
x_{Bm}	0	0	0	$\cdots$	1	b_m	y_{mk}

Note. We shall have to read in the manner $x_{B1} = b_1, x_{B2} = b_2$ etc.

The method of finding the value of $\mathbf{y}_k^*$ is given in Step 6. After calculating it, fill up the last column of the table to complete it.

Step 5. *Test of optimality :* Compute all $z_j - c_j$ corresponding to all $\mathbf{a}_j$ which are not in the basis B by using the formula.

$z_j - c_j =$ (First row of B^{*-1}) $\times \mathbf{a}_j^*$: The solution will be optimal if all $z_j - c_j \geq 0$. Otherwise proceed further to get an optimal solution.

Step 6. If $\min_j(z_j - c_j\ z_j - c_j < 0) = z_k - c_k$ then $\mathbf{a}_k$ is the vector to enter in the next basis.

Now, to calculate $\mathbf{y}_k^* = B^{*-1}\mathbf{a}_k^*$, calculate $\mathbf{y}_k = B^{-1}\mathbf{a}_k$ and then

$$\mathbf{y}_k^* = [z_k - c_k, y_{1k}, y_{2k}, \cdots, y_{mk}].$$

If at least one of $y_{ik} > 0$, then proceed further. Hence, calculate

$$\min_i \left(\frac{x_{Bi}}{y_{ik}}, y_{ik} > 0 \right).$$

If the minimum occurs for $i = r$, then y_{rk} is the key element and β_r^* is the vector to leave the basis and $x_k = x_{Br}$ is the new component of basic solution. If the minimum is not unique, select arbitrarily any vector corresponding to the minimum and the next solution is bound to be degenerate. If none of $y_{ik} > 0$, then the conclusion will be of having an unbounded solution.

Step 7. *Determination of the improved solution :* To get the improved value of the objective function, construct the second revised simplex table. The first table is to be converted in such a way that the column vector corresponding to $\mathbf{y}_k^*$, i.e., the last column of the first table be a unit vector $\mathbf{e}_r$. The method of transformation is the same as in the case of ordinary simplex method. In this way we get an improved value (or at least same) of the objective function.

It is important to note that due to the transformation, the column vectors corresponding to B^{*-1} of the initial table constitute the basis inverse $\hat{B}^{*-1}$ of the next table [proof is obvious from the theory of matrix inverse].

Step 8. Now test the above improved solution for optimality. If all $z_j - c_j \geq 0$ then the solution is optimal. Otherwise proceed further and ultimately we get either a finite optimal value of the objective function or the problem is said to have an unbounded solution.

For clear understanding of the procedure, some worked out examples are given below.

► **Example 14.9.1** *Solve the problem by using revised simplex method.*

$$\text{Maximize,} \quad z = x_1 + x_2.$$

subject to

$$3x_1 + 2x_2 \leq 6$$
$$x_1 + 4x_2 \leq 4, \ x_1, x_2 \geq 0.$$

Solution : Step 1. This is a maximization problem.

Step 2. All $b_i \geq 0$ and two slack variables x_3 and x_4 are required to introduce one to each constraint to get equations and thus a unit matrix is present in the co-efficient matrix. The problem in the revised simplex form is

$$\begin{aligned} z - x_1 - x_2 + 0.x_3 + 0.x_4 &= 0 \\ 3x_1 + 2x_2 + x_3 + 0.x_4 &= 6 \\ x_1 + 4x_2 + x_4 &= 4, \; x_1, x_2 \geq 0, \end{aligned}$$

z is unrestricted.

In matrix notation, the system of constraints can be written as

$$\begin{bmatrix} \mathbf{e}_1 & \mathbf{a}_1^* & \mathbf{a}_2^* & \mathbf{a}_3^* & \mathbf{a}_4^* \\ 1 & -1 & -1 & 0 & 0 \\ 0 & 3 & 2 & 1 & 0 \\ 0 & 1 & 4 & 0 & 1 \end{bmatrix} \begin{bmatrix} z \\ x_1 \\ x_2 \\ x_3 \\ x_4 \end{bmatrix} = \begin{matrix} \mathbf{b}^* \\ \begin{bmatrix} 0 \\ 6 \\ 4 \end{bmatrix} \end{matrix}$$

Step 3. $B = (\mathbf{a}_3, \mathbf{a}_4) = \begin{bmatrix} 1 & 0 \\ 0 & 1 \end{bmatrix}$ is the unit basis of the problem in the ordinary simplex method, $\mathbf{c}_B = (c_3, c_4) = (0, 0)$.

As $\mathbf{c}_B = \mathbf{0}$, then $B^* = I_3$ and $B^{*-1} = I_3$, an identity matrix of order 3.

$$\mathbf{x}_B^* = I_3\mathbf{b}^* = \begin{bmatrix} z \\ x_{B1} \\ x_{B2} \end{bmatrix} = \begin{bmatrix} z \\ x_3 \\ x_4 \end{bmatrix} = \begin{bmatrix} 0 \\ 6 \\ 4 \end{bmatrix}$$

and with these data we can construct the initial revised simplex table. Last column of the table remains vacant initially.

Step 4. *Construction of starting simplex table :*

First revised simplex table

Variables in the basis	$\mathbf{e}_1$			$\mathbf{x}_B^*$	$\mathbf{y}_k^* = \mathbf{y}_1^*$
z	1	0	0	0	-1
$x_3(x_{B1})^*$	0	1	0	6	3^*
$x_4(x_{B2})$	0	0	1	4	1

Step 5. *Test of optimality :* Here $\mathbf{a}_1^*, \mathbf{a}_2^*$ are not on the basis. Therefore we require to compute $z_j - c_j [j = 1, 2]$.

Now

$$(z_1 - c_1, z_2 - c_2) = \text{First row of } B^{*-1} \times [\mathbf{a}_1^*, \mathbf{a}_2^*]$$

$$= (1, 0, 0) \begin{bmatrix} -1 & -1 \\ 3 & 2 \\ 2 & 4 \end{bmatrix} = (-1, -1)$$

Hence $z_1 - c_1 = -1$, $z_2 - c_2 = -1$ both of which are negative.

Step 6. *Determination of the vector to enter in the next basis :*

$\min(z_1 - c_1, z_2 - c_2) = \min(-1, -1) = -1$ which occurs both for $j = 1$ and $j = 2$. Then either $\mathbf{a}_1$ or $\mathbf{a}_2$ be the vector to enter in the next basis. Let us take at random $\mathbf{a}_1$, as the vector to enter in the next basis

$$\mathbf{y}_1 = B^{-1}\mathbf{a}_1 = \begin{bmatrix} 1 & 0 \\ 0 & 1 \end{bmatrix} \begin{bmatrix} 3 \\ 1 \end{bmatrix} = \begin{bmatrix} 3 \\ 1 \end{bmatrix}$$

and therefore obviously

$$\mathbf{y}_1^* = \begin{bmatrix} z_1 - c_1 \\ y_{11} \\ y_{21} \end{bmatrix} = \begin{bmatrix} -1 \\ 3 \\ 1 \end{bmatrix}$$

as $z_1 - c_1$ is already obtained. With this value of $\mathbf{y}_1^*$ fill up the last column of the starting table to complete it.

Vector to leave the basis : At least one component $\mathbf{y}_1$ is positive. Then we can improve the value of the objective function further.

Now,

$$\frac{x_{Br}}{y_{r1}} = \min\left(\frac{x_{Bi}}{y_{i1}}, y_{i1} > 0\right) = \min\left(\frac{6}{3}, \frac{4}{1}\right) = 2$$

which occurs for $i = 1$. Hence $y_{11} = 3$ is the key element and $\mathbf{a}_3$ is the vector to leave the basis. Therefore x_3 will leave the basic solution set and x_1 will enter in the solution set.

Step 7. *Determination of the improved solution :*

To get the improved solution, we shall have to construct the second revised simplex table. We shall have to transform the first table similarly as we have done in the simplex method in such a way that the column corresponding to $\mathbf{y}_1^*$, i.e., last column be a unit vector $\mathbf{e}_2$. The key element is $y_{11} = 3$.

Second revised simplex table

Variables in the basis	$\mathbf{e}_1$			$\mathbf{x}_B^*$	$\mathbf{y}_k^* = \mathbf{y}_2^*$
z	1	$\frac{1}{3}$	0	2	$-\frac{1}{3}$
$x_1(x_{B1})$	0	$\frac{1}{3}$	0	2	$\frac{2}{3}$
$x_4(x_{B2})^*$	0	$-\frac{1}{3}$	1	2	$\frac{10}{3}^*$

Note. Initially the last column remains vacant.

Step 8. Computation of the second revised simplex table from the initial table by using the formula given in (8.4.1). Initially the last column of the second table will remain vacant.

Calculation :

Second row. Divide the elements of the 2nd row of the first table by the key element to get the elements of the second of row of the second table.

Here $x_{\beta 1} = x_1$ instead of $x_{\beta 1} = x_3$ in the first table.

First row. First element: $\dfrac{3 \times 1 - (0 \times -1)}{3} = 1,$

Second element: $\dfrac{3 \times 0 - (1 \times -1)}{3} = \dfrac{1}{3},$

Third element: $\dfrac{3 \times 0 - 0 \times (-1)}{3} = 0,$

Fourth element: $\dfrac{3 \times 0 - (6 \times -1)}{3} = 2.$

Third row. First element: $\dfrac{(3 \times 0) - (0 \times 1)}{3} = 0,$

Second element: $\dfrac{3 \times 0 - (1 \times 1)}{3} = -\dfrac{1}{3},$

Third element: $\dfrac{3 \times 1 - 0 \times 1}{3} = 1,$

Fourth element: $\dfrac{3 \times 4 - 6 \times 1}{3} = 2.$

Step 9. *Test of optimality :* The column vector $\mathbf{a}_2^*$ and $\mathbf{a}_3^*$ are not on the basis. Then

$$(z_2 - c_2, z_3 - c_3) = \text{First row of } B^{*-1} \times [\mathbf{a}_2^*, \mathbf{a}_3^*]$$

$$= \left(1, \ \frac{1}{3}, \ 0\right)\begin{bmatrix} -1 & 0 \\ 2 & 1 \\ 4 & 0 \end{bmatrix} = \left(-\frac{1}{3}, \frac{1}{3}\right)$$

The only negative $z_j - c_j$ is $z_2 - c_2$ which is equal to $-\frac{1}{3}$. Then $\mathbf{a}_2$ may enter in the next basis.

Now

$$\mathbf{y}_2 = B^{-1}\mathbf{a}_2 = \begin{bmatrix} 1/3 & 0 \\ -1/3 & 1 \end{bmatrix}\begin{bmatrix} 2 \\ 4 \end{bmatrix} = \begin{bmatrix} 2/3 \\ 10/3 \end{bmatrix}$$

at least one component of which is positive.

We shall have to proceed further.

Therefore

$$\mathbf{y}_2^* = \begin{bmatrix} -1/3 \\ 2/3 \\ 10/3 \end{bmatrix}$$

as $z_2 - c_2 = -1/3$ is already determined.

Fill in the last column of the second table with $\mathbf{y}_2^*$.

Now

$$\frac{x_{Br}}{y_{r2}} = \min_i \left(\frac{x_{Bi}}{y_{i2}}, y_{i2} > 0\right) = \min\left(\frac{2}{2/3}, \frac{2}{10/3}\right) = \frac{3}{5}$$

which occurs for $i = 2$. Therefore, $y_{22} = 10/3$ is the key element and the second vector of the basis $\mathbf{a}_4$ will leave the basis, x_2 will enter in the basic solution set.

Step 10. *Determination of the Improved Value of the Objective Function :*

We shall have to transform the second table in such a way that the column corresponding to $\mathbf{y}_2^*$, i.e., last column of the 2nd table be a unit vector $\mathbf{e}_3$.

Variables in the basis	$\mathbf{e}_1$			$\mathbf{x}_B^*$
z	1	$\frac{3}{10}$	$\frac{1}{10}$	$\frac{11}{5}$
$x_1(x_{B1})$	0	$\frac{2}{5}$	$-\frac{1}{5}$	$\frac{8}{5}$
$x_2(x_{B2})$	0	$-\frac{1}{10}$	$\frac{3}{10}$	$\frac{3}{5}$

Step 11. *Test of Optimality :*

The column vectors $\mathbf{a}_3^*$ and $\mathbf{a}_4^*$ are not in the basis and therefore we require to compute $z_3 - c_3, z_4 - c_4$.

$$(z_3 - c_3, z_4 - c_4) = \text{First row of } B_{cur}^{*-1} \times [\mathbf{a}_3^*, \mathbf{a}_4^*]$$

$$= \left(1, \frac{3}{10}, \frac{1}{10}\right) \begin{bmatrix} 0 & 0 \\ 1 & 0 \\ 0 & 1 \end{bmatrix} = \left(\frac{3}{10}, \frac{1}{10}\right)$$

Here both $z_3 - c_3$ and $z_4 - c_4$ are ≥ 0. Hence we reach at the optimal stage.

The optimal solution is $x_1 = \frac{8}{5}, x_2 = \frac{3}{5}$ and $\max z = \frac{11}{5}$.

▶ **Example 14.9.2** *Solve the L.P.P.*

$$\textit{maximize,} \quad z = 5x_1 + 2x_2 + 2x_3$$

subject to

$$\begin{aligned} x_1 + 2x_2 - 2x_3 &\leq 30 \\ x_1 + 3x_2 + \ \ x_3 &\leq 36, \ x_1, x_2, x_3 \geq 0. \end{aligned}$$

Solution : Step 1. This is a maximization problem.

Step 2. All $b_i \geq 0$ and the two slack variables x_4 and x_5 are required to introduce one to each constraint to get the equations and thus a unit matrix is present in the co-efficient matrix. The problem in the revised simplex from is

$$\begin{aligned} z - 5x_1 - 2x_2 - 2x_3 \qquad\qquad &= 0 \\ x_1 + 2x_2 - 2x_3 + x_4 \qquad &= 30 \\ x_1 + 3x_2 + \ \ x_3 \qquad + x_5 &= 36, \ x_1, x_2, x_3 \geq 0. \end{aligned}$$

z is unrestricted.

In matrix notation, the system of constraints can be written as

$$\begin{bmatrix} \mathbf{e}_1 & \mathbf{a}_1^* & \mathbf{a}_2^* & \mathbf{a}_3^* & \mathbf{a}_4^* & \mathbf{a}_5^* \\ 1 & -5 & -2 & -2 & 0 & 0 \\ 0 & 1 & 2 & -2 & 1 & 0 \\ 0 & 1 & 3 & 1 & 0 & 1 \end{bmatrix} \begin{bmatrix} z \\ x_1 \\ x_2 \\ x_3 \\ x_4 \\ x_5 \end{bmatrix} = \begin{bmatrix} \mathbf{b}^* \\ 0 \\ 30 \\ 36 \end{bmatrix}$$

Step 3. $B = (\mathbf{a}_4, \mathbf{a}_5) = \begin{bmatrix} 1 & 0 \\ 0 & 1 \end{bmatrix}$ is the unit basis of the problem in the ordinary simplex method, $\mathbf{c}_B = (c_4, c_5) = (0, 0)$. As $\mathbf{c}_B = 0$ then $B^* = I_2$ and $B^{*-1} = I_3$, an identity matrix of order 3.

$$\mathbf{x}_B^* = I_3\mathbf{b}^* = \begin{bmatrix} z \\ x_{B1} \\ x_{B2} \end{bmatrix} = \begin{bmatrix} z \\ x_4 \\ x_5 \end{bmatrix} = \begin{bmatrix} 0 \\ 30 \\ 36 \end{bmatrix} \quad \text{by (14.7.1)}$$

and with these data, we can construct the initial revised table. The last column of the table remains vacant initially.

Step 4. ***Construction of the starting simplex table.***

First revised simplex table

Variables in the basis	$\mathbf{e}_1$			$\mathbf{x}_B^*$	$\mathbf{y}_k^* = \mathbf{y}_1^*$
z	1	0	0	0	−5
$x_4(x_{B1})^{\bullet}$	0	1	0	30	1^*
$x_5(x_{B2})$	0	0	1	36	1

Step 5. ***Test of optimality.***

Here $\mathbf{a}_1^*, \mathbf{a}_2^*, \mathbf{a}_3^*$ are not in the basis. Therefore we require to compute $z_j - c_j$ $[j = 1, 2, 3]$.

Now, $(z_1 - c_1, z_2 - c_2, z_3 - c_3) =$ First row of $B^{*-1} \times (\mathbf{a}_1^*, \mathbf{a}_2^*, \mathbf{a}_3^*)$

$$(1, 0, 0) \begin{bmatrix} -5 & -2 & -2 \\ 1 & 2 & -2 \\ 1 & 3 & 1 \end{bmatrix} = (-5, -2, -2).$$

Here all three $z_j - c_j$, $z_1 - c_1 = -5$, $z_2 - c_2 = -2$, $z_3 - c_3 = -2$ are negative.

Step 6. ***Determination of the vector to enter in the next basis***

$$\min(z_1 - c_1, z_2 - c_2, z_3 - c_3, z_j - c_j < 0) = \min(-5, -2, -2) = -5$$

which occurs for $j = 1$. Then $\mathbf{a}_1$ is the vector to enter in the next basis,

$$\mathbf{y}_1 = B^{-1}\mathbf{a}_1 = \begin{bmatrix} 1 & 0 \\ 0 & 1 \end{bmatrix} \begin{bmatrix} 1 \\ 1 \end{bmatrix} = \begin{bmatrix} 1 \\ 1 \end{bmatrix}$$

and then obviously

$$\mathbf{y}_1^* = \begin{bmatrix} z_1 - c_1 \\ y_{11} \\ y_{21} \end{bmatrix} = \begin{bmatrix} -5 \\ 1 \\ 1 \end{bmatrix}$$

as $z_1 - c_1$ is already obtained. With value of $\mathbf{y}_1^*$ fill up the last column of the starting table to complete it.

Step 7. *Vector to leave the current basis.*

At least one component of $\mathbf{y}_j$ is positive. Then we can improve or at least remains same the value of the objective further

$$\frac{x_{Br}}{y_{r1}} = \min\left(\frac{x_{Bi}}{y_{i1}}, y_{i1} > 0\right) = \min\left(\frac{30}{1}, \frac{36}{1} = 30\right)$$

which occurs for $i = 1$. Hence $y_{11} = 1$ is the key element and $\mathbf{a}_4$ is the vector to leave the current basis. Therefore $x_4 = (x_{B1})$ will leave the basic solution set and x_1 will enter in the new solution set.

Step 8. *Determination of the improved solution.* To get the improved solution, we shall have to transform the first table similarly as we have done in the simplex method in such a way that the column corresponding to $\mathbf{y}_1^*$, i.e., last column be a unit vector $\mathbf{e}_3$. The element is $y_{11} = 1$.

Second revised simplex table

Variables in the basis	$\mathbf{e}_1$			$\mathbf{x}_B^*$	$\mathbf{y}_K^* = \mathbf{y}_3^*$
z	1	5	0	150	−12
$x_1(x_{B1})$	0	1	0	30	−2
$x_5(x_{B2})^*$	0	−1	1	6	3*

Note. Initially the last column remains vacant.

Step 9. Computation of the second table.

Second row. Divide the second row of the first table by $1 = y_{11}$ = key element of the first table to get the second row of the second table.

First row : First element, $\dfrac{1 \times 1 - (-5 \times 0)}{1} = 1$

Second element, $\dfrac{1 \times 0 - (-5 \times 1)}{1} = 5$

Third element, $\dfrac{1 \times 0 - (-5 \times 0)}{1} = 0$

Fourth element, $\dfrac{1 \times 0 - (-5 \times 30)}{1} = 150$

Third row : First element, $\dfrac{1 \times 0 - 1 \times 0}{1} = 0$

Second element, $\dfrac{1 \times 0 - (1 \times 1)}{1} = -1$

Third element, $\dfrac{1 \times 1 - (1 \times 0)}{1} = 1$

Fourth element, $\dfrac{1 \times 36 - 1 \times 30}{1} = 6$

Step 10. ***Test of optimality.*** The column vectors $\mathbf{a}_2^*, \mathbf{a}_3^*$ and $\mathbf{a}_4^*$ are not in the basis. Then

$$(z_2 - c_2, z_3 - c_3, z_4 - c_4) = \text{First row of } B^{*-1} \times [\mathbf{a}_2^*, \mathbf{a}_3^*, \mathbf{a}_4^*]$$

$$= (1, 5, 0)\begin{bmatrix} -2 & -2 & 0 \\ 2 & -2 & 1 \\ 3 & 1 & 0 \end{bmatrix} = \begin{bmatrix} 8 \\ -12 \\ 5 \end{bmatrix}$$

The only negative $z_j - c_j$ is $z_3 - c_3$ which is equal to -12. Then $\mathbf{a}_3$ is the vector to enter in the next basis.

Now

$$\mathbf{y}_3, = B^{-1}\mathbf{a}_3 = \begin{bmatrix} 1 & 0 \\ -1 & 1 \end{bmatrix}\begin{bmatrix} -2 \\ 1 \end{bmatrix} = \begin{bmatrix} -2 \\ 3 \end{bmatrix},$$

at least one component of which is positive. We shall have to proceed further. Therefore,

$$\mathbf{y}_3^* = \begin{bmatrix} -12 \\ -2 \\ 3 \end{bmatrix}$$

as $z_3 - c_3 = -12$.

Fill in the last column of the second table with $\mathbf{y}_3^*$.

Now

$$\frac{x_{Br}}{y_{r2}} = \min_i \left(\frac{x_{Bi}}{y_{i3}}, y_{i3} > 0\right) = \min\left(\frac{x_{B2}}{y_{23}}\right) = \frac{6}{3} = 2$$

which occurs for $i = 2$. Therefore $y_{23} = 3$ is the key element and the second vector of the basis, $\mathbf{a}_5$ will leave the basis B and x_3 will enter in the new basic solution set.

Step 11. *Determination of the improved value of the objective function :*

We shall have to transform in such a way that the column corresponding to $\mathbf{y}_3^*$, i.e., the last column of the second table be a unit vector $\mathbf{e}_3$.

Third simplex (revised) table

Variables in the basis	$\mathbf{e}_1$			$\mathbf{x}_B^*$
z	1	1	4	174
$x_1(x_{B1})$	0	1	1	34
$x_3(x_{B2})$	0	$-\frac{1}{3}$	$\frac{1}{3}$	2

Step 12. *Computation of the elements of the third table.*

Third row. Divide the elements of the third row of second table by 3 = (key element) to get the third row of the third table.

First row. First element; $\dfrac{3 \times 1 - (0 \times -12)}{3} = 1$

Second element; $\dfrac{3 \times 5 - (-1 \times -12)}{3} = 1$

Third element; $\dfrac{3 \times 0 - (1 \times -12)}{3} = 4$

Fourth element; $\dfrac{3 \times 150 - (6 \times -12)}{3} = 174$

Second row. First element; $\dfrac{3 \times 1 - 0 \times (-12)}{3} = 1$

Second element; $\dfrac{3 \times 5 - (-1 \times -12)}{3} = 1$

Third element; $\dfrac{3 \times 0 - (1 \times -2)}{3} = \dfrac{2}{3}$

Fourth element; $\dfrac{3 \times 30 - (6 \times -2)}{3} = 34$

Test of optimality. The column vectors $\mathbf{a}_2^*, \mathbf{a}_4^*$ and $\mathbf{a}_5^*$ are not in the basis and therefore we require to compute $z_2 - c_2, z_4 - c_4$ and $z_5 - c_5$

$$(z_2 - c_2, z_4 - c_4, z_5 - c_5) = \text{First row of } B_{\text{cur}}^{*-1} \times [\mathbf{a}_2^*, \mathbf{a}_4^*, \mathbf{a}_5^*]$$

$$= (1, 1, 4) \begin{bmatrix} -2 & 0 & 0 \\ 2 & 1 & 0 \\ 3 & 0 & 1 \end{bmatrix} = (12, 1, 4)$$

$z_1 - c_2, z_4 - c_4, z_5 - c_5$ all > 0.

Thus the solution is an optimal solution and the solution is unique. Hence $\max z = 174$ at $x_1 = 34$, $x_2 = 0$, $x_3 = 2$.

▶ **Example 14.9.3** *Solve the following problem by revised simplex method and prove that alternative optimal solutions exist and the optimal solutions are degenerate*

$$\text{maximize,} \quad Z = x_1 + 3x_2 + 2x_3$$

subject to

$$\begin{aligned} x_1 + 2x_2 \quad\quad &\leq 10 \\ 2x_1 \quad\quad + x_3 &\leq 8 \\ 2x_2 + x_3 &\leq 6, \; x_1, x_2, x_3 \geq 0. \end{aligned}$$

$b_i \geq 0$. Adding three slack variables x_4, x_5 and x_6 one to each constraint and considering $z - x_1 - 3x_2 - 2x_3 + 0x_4 + 0x_5 + 0x_6 = 0$ as the first constraint we have

$$\begin{aligned} z - x_1 - 3x_2 - 2x_3 + 0x_4 + 0x_5 + 0x_6 &= 0 \\ 0.z + x_1 + 2x_2 + 0x_3 + x_4 + 0x_5 + 0x_6 &= 10 \\ 0.z + 2x_1 + 0x_2 + x_3 + 0x_4 + x_5 + 0x_6 &= 8 \\ 0.z + 0x_1 + 2x_2 + x_3 + 0x_4 + 0x_5 + x_6 &= 6. \end{aligned}$$

The resulting matrix is

$\mathbf{e}_1$	$\mathbf{a}_1^*$	$\mathbf{a}_2^*$	$\mathbf{a}_3^*$	$\mathbf{a}_4^*$	$\mathbf{a}_5^*$	$\mathbf{a}_6^*$	$\mathbf{b}^*$
1	−1	−3	−2	0	0	0	0
0	1	2	0	1	0	0	10
0	2	0	1	0	1	0	8
0	0	2	1	0	0	1	6

Step 1. Now taking $B = (\mathbf{a}_4, \mathbf{a}_5, \mathbf{a}_6) = I_3, \mathbf{x}_B = B^{-1}\mathbf{b} = \mathbf{b} \geq \mathbf{0}$, we can start the revised simplex method by taking $B = (\mathbf{a}_4, \mathbf{a}_5, \mathbf{a}_6) = I_3$ as the basis of the original simplex method and as $\mathbf{c}_B = (0, 0, 0) = \mathbf{0}$. Then $B^* = (\mathbf{e}_1, \mathbf{a}_4^*, \mathbf{a}_5^*, \mathbf{a}_6^*) = I_4$ can be taken as the basis of the revised simplex method. $\mathbf{x}_B^* = B^{*-1}\mathbf{b}^* = \mathbf{b}^* = [0, 10, 8, 6]$, with these data we can construct the starting revised revised simplex table,

Variables in the basis	$\mathbf{e}_1$				$\mathbf{x}_B^*$	$\mathbf{y}_2^*$
z	1	0	0	0	0	−3
$x_{B1}(x_4)$	0	1	0	0	10	2
$x_{B2}(x_5)$	0	0	1	0	8	0
$x_{B3}(x_6)^*$	0	0	0	1	6	2*

Step 2. Calculation of $z_j - c_j$ corresponding to the vectors not in the basis.

$$(z_1 - c_1, z_2 - c_2, z_3 - c_3) = \text{First row of } B^{*-1} \times [\mathbf{a}_1^*, \mathbf{a}_2^*, \mathbf{a}_3^*]$$

$$= (1,0,0,0)\begin{bmatrix} -1 & -3 & -2 \\ 1 & 2 & 0 \\ 2 & 0 & 1 \\ 0 & 2 & 1 \end{bmatrix} = (-1,-3,-2)$$

$$\min_j(z_j - c_j, z_j - c_j < 0) = \min(-1,-3,-2) = -3 = z_2 - c_2.$$

Thus the second vector $\mathbf{a}_2$ will enter in the next basis B

$$\mathbf{y}_2 = B^{-1}\mathbf{a}_2 = I_3\mathbf{a}_2 = \begin{bmatrix} 2 \\ 0 \\ 2 \end{bmatrix}.$$

At least one component of $\mathbf{y}_2$ is > 0. Then the value of the objective function can be improved

$$\mathbf{y}_2^* = \begin{bmatrix} z_2 - c_2 \\ \mathbf{y}_2 \end{bmatrix} = \begin{bmatrix} -3 \\ 2 \\ 0 \\ 2 \end{bmatrix}$$

as $z_2 - c_2 = -3$ which is already obtained.

Step 3. *Vector to leave the basis :*

$$\min_i\left(\frac{x_{Bi}}{y_{i2}}, y_{i2} > 0\right) = \min\left(\frac{x_{B1}}{y_{12}}, \cdots, \frac{x_{B3}}{y_{32}}\right) = \min\left(\frac{10}{2}, \cdots, \frac{6}{2}\right) = 3,$$

which occurs for $i = 3$ $\left[\frac{x_{B2}}{y_{22}} \text{ is not considered as } y_{22} = 0\right]$. Thus the $y_{32} = 2$ is the key element and the third vector $\mathbf{a}_6$ of the basis will leave the basis B and will be replaced by $\mathbf{a}_2$ to form a new basis. Put now the column vector $\mathbf{y}_2^*$ as shown in the initial table.

Step 4. *Construction of the second table*

Variables in the basis	$\mathbf{e}_1$				$\mathbf{x}_B^*$	$\mathbf{y}_1^*$
z	1	0	0	$\frac{3}{2}$	9	−1
$x_{B1}(x_4)$	0	1	0	−1	4	1
$x_{B2}(x_5)^*$	0	0	1	0	8	2*
$x_{B3}(x_2)$	0	0	0	$\frac{1}{2}$	3	0

Step 5. Calculation of $z_j - c_j$ corresponding to the vectors not in the basis.

$$(z_1 - c_1, z_3 - c_3, z_6 - c_6) = \left(1,0,0,\frac{3}{2}\right)[\mathbf{a}_1^*, \mathbf{a}_3^*, \mathbf{a}_6^*] = \left(-1, -\frac{1}{2}, \frac{3}{2}\right).$$

At least one of $z_j - c_j < 0$. Thus the solution is not optimal.

$$\min(z_j - c_j, z_j - c_j < 0) = \min\left(-1, -\frac{1}{2}\right) = -1 = z_1 - c_1.$$

Thus the first vector $\mathbf{a}_1$ will enter in the new basis.

$$\mathbf{y}_1 = B^{-1}\mathbf{a}_1 = \begin{bmatrix} 1 & 0 & -1 \\ 0 & 1 & 0 \\ 0 & 0 & 1/2 \end{bmatrix} \begin{bmatrix} 1 \\ 2 \\ 0 \end{bmatrix} = \begin{bmatrix} 1 \\ 2 \\ 0 \end{bmatrix};$$

at least one component of $\mathbf{y}_1$ is > 0.

Thus we can proceed further to get an optimal solution

$$\mathbf{y}_1^* = \begin{bmatrix} -1 \\ 1 \\ 2 \\ 0 \end{bmatrix} \quad \text{as } z_1 - c_1 = -1.$$

Step 6.

$$\min_i \left(\frac{x_{B1}}{y_{i1}}, y_{i1} > 0 \right) = \min \left(\frac{x_{B1}}{y_{11}}, \frac{x_{B2}}{y_{21}}, \cdots \right) = \left(\frac{4}{1}, \frac{8}{2}, \cdots \right) = 4$$

which occur for $i = 1$ or $i = 2$. Thus either the first vector $\mathbf{a}_4$ or second vector $\mathbf{a}_5$ will leave the basis and replaced by $\mathbf{a}_1$, to form a new basis.

Let us select at random $\mathbf{a}_5$ to leave the basis and $y_{21} = 2$ will be the key element. Put the vector $\mathbf{y}_1^*$ in the place shown in the second table.

Construction of the third table

Variables in the basis	$\mathbf{e}_1$				$\mathbf{x}_B^*$
z	1	0	$\frac{1}{2}$	$\frac{3}{2}$	13
$x_{B1}(x_4)$	0	1	$-\frac{1}{2}$	-1	0
$x_{B2}(x_1)$	0	0	$\frac{1}{2}$	0	4
$x_{B3}(x_2)$	0	0	0	$\frac{1}{2}$	3

Step 7. Calculation of $z_j - c_j$ corresponding to the vectors not in the basis

$$(z_3 - c_3, z_5 - c_5, z_6 - c_6) = \left(1, 0, \frac{1}{2}, \frac{3}{2}\right) [\mathbf{a}_3^*, \mathbf{a}_5^*, \mathbf{a}_6^*] = \left(0, \frac{1}{2}, \frac{3}{2}\right).$$

All $z_i - c_j \geq 0$. Thus the optimality condition has been satisfied.

Thus $\max z = 13$ at $x_4 = 0$ (basic), $x_1 = 4$ (basic), $x_2 = 3$ (basic) which is a degenerate basic feasible solution, i.e., $\max z = 13$ at $x_1 = 4$, $x_2 = 3$ and $x_3 = 0$ (non-basic) of the original problem.

$z_3 - c_3 = 0$ corresponding to a non-basic vector at the optimal stage. Thus there exist alternative optimal solutions.

► **Example 14.9.4** *Solve the following L.P.P. by revised simplex method,*

$$\textit{maximize,} \quad z = 2x_1 - 3x_2 - 2x_3 + 6x_4$$

subject to

$$\begin{aligned} 5x_1 - x_2 + 2x_3 + 6x_4 &\le 20 \\ 2x_1 + 3x_2 + 4x_3 - 5x_4 &\le 16 \\ x_1 + 2x_2 - 3x_3 + x_4 &\le 2, \; x_1, x_2, x_3 \text{ and } x_4 \ge 0. \end{aligned}$$

Step 1. This is a maximization problem.

Step 2. All $b_i \ge 0$ and three slack variables x_5, x_6 and x_7 and thus a unit matrix is present in the co-efficient matrix. The problem in the revised simplex form is

$$\begin{aligned} z - 2x_1 + 3x_2 + 2x_3 - 6x_4 &= 0 \\ 5x_1 - x_2 + 2x_3 + 6x_4 + x_5 &= 20 \\ 2x_1 + 3x_2 + 4x_3 - 5x_4 + x_6 &= 16 \\ x_1 + 2x_2 - 3x_3 + x_4 + x_7 &= 2. \end{aligned}$$

In matrix notation, the system of constraints can be written as

$$\begin{bmatrix} \mathbf{e}_1 & \mathbf{a}_1^* & \mathbf{a}_2^* & \mathbf{a}_3^* & \mathbf{a}_4^* & \mathbf{a}_5^* & \mathbf{a}_6^* & \mathbf{a}_7^* \\ 1 & -2 & 3 & 2 & -6 & 0 & 0 & 0 \\ 0 & 5 & -1 & 2 & 6 & 1 & 0 & 0 \\ 0 & 2 & 3 & 4 & -5 & 0 & 1 & 0 \\ 0 & 1 & 2 & -3 & 1 & 0 & 0 & 1 \end{bmatrix} \begin{bmatrix} z \\ x_1 \\ x_2 \\ x_3 \\ \vdots \\ x_7 \end{bmatrix} = \begin{bmatrix} 0 \\ 20 \\ 16 \\ 2 \end{bmatrix}.$$

Step 3. $B = \begin{bmatrix} 1 & 0 & 0 \\ 0 & 1 & 0 \\ 0 & 0 & 1 \end{bmatrix}$ is the initial admissible unit basis to the ordinary simplex method. $\mathbf{c}_B = (c_5, c_6, c_7) = (0, 0, 0) = \mathbf{0}$. As $\mathbf{c}_B = \mathbf{0}$ then $B^* = I_4$, and $B^{*-1} = I_4$, an identity matrix of order four.

$$\mathbf{x}_B^* = I_4\mathbf{b} = \begin{bmatrix} z \\ x_{B1} \\ x_{B2} \\ x_{B3} \end{bmatrix} = \begin{bmatrix} z \\ x_5 \\ x_6 \\ x_7 \end{bmatrix} = \begin{bmatrix} 0 \\ 20 \\ 16 \\ 2 \end{bmatrix}$$

and with these data we can construct the initial revised simplex table. The last column of the table remains vacant initially.

Step 4. *Construction of the initial simplex table :*

First revised simplex table

Variables in the basis	$\mathbf{e}_1$				$\mathbf{x}_B^*$	$\mathbf{y}_k^* = \mathbf{y}_4^*$
z	1	0	0	0	0	−6
$x_5(x_{B1})$	0	1	0	0	20	6
$x_6(x_{B2})$	0	0	1	0	16	−5
$x_7(x_{B3})^*$	0	0	0	1	2	1*

Step 5. *Test of optimality :* $\mathbf{a}_1^*, \mathbf{a}_2^*, \cdots, \mathbf{a}_4^*$ are not in the basis. Hence we require to calculate $(z_1 - c_1), \cdots, (z_4 - c_4)$, i.e., $z_j - c_j [j = 1, \cdots, 4]$

$$(z_1 - c_1, z_2 - c_2, z_3 - c_3, z_4 - c) = \text{First row of } B^{*-1} \times [\mathbf{a}_1^*, \mathbf{a}_2^*, \cdots, \mathbf{a}_4^*]$$

$$= (1,0,0,0)\begin{bmatrix} -2 & 3 & 2 & -6 \\ 5 & -1 & 2 & 6 \\ 2 & 3 & 4 & -5 \\ 1 & 2 & -3 & 1 \end{bmatrix}$$

$$= (-2, 3, 2, -6).$$

Here $z_1 - c_1$ and $z_4 - c_4$ are negative.

Step 6. *Determination of the vector to enter in the next basis.*

$$\min_j (z_j - c_j, z_j - c_j < 0) = \min(z_1 - c_1, z_4 - c_4) = \min(-2, -6) = -6$$

which occurs at $j = 4$. Then $\mathbf{a}_4$ is the vector to enter in the next basis.

$$\mathbf{y}_4 = B^{-1}\mathbf{a}_4 = \begin{bmatrix} 1 & 0 & 0 \\ 0 & 1 & 0 \\ 0 & 0 & 1 \end{bmatrix}\begin{bmatrix} 6 \\ -5 \\ 1 \end{bmatrix} = \begin{bmatrix} 6 \\ -5 \\ 1 \end{bmatrix}$$

at least one component of $\mathbf{y}_4$ is positive and the problem can be improved further and

$$\mathbf{y}_4^* = \begin{bmatrix} z_1 - c_1 \\ y_{11} \\ y_{21} \\ y_{31} \end{bmatrix} = \begin{bmatrix} -6 \\ 6 \\ -5 \\ 1 \end{bmatrix},$$

as $z_1 - c_1$ is already obtained. With this value of $\mathbf{y}_4^*$ fill in the last column of the initial table to complete it.

Step 7. *Vector to leave the basis :* At least one component of $\mathbf{y}_1$ is positive

$$\frac{x_{\beta r}}{y_{r4}} = \min_i \left(\frac{x_{\beta i}}{y_{i4}}, y_{i4} > 0 \right) = \left(\frac{20}{6}, \cdots, \frac{2}{1} \right) = 2$$

which occurs for $i = 3$.

Hence $y_{34} = 1$ is the key element and $\mathbf{a}_7$ is the vector to leave the basis B which will be replaced by $\mathbf{a}_4$.

Step 8. *Determination of the improved solution :*

To get the improved value of the objective function, we shall have to construct the second revised simplex table. We shall have to transform the initial table corresponding to a $\mathbf{y}_4^*$, a unit vector $\mathbf{e}_4$ i.e., the last column be a unit vector $\mathbf{e}_4$.

The key element is $1 = y_{34}$.

Second revised simplex table

Variables in the basis	$\mathbf{e}_1$				$\mathbf{x}_B^*$	$\mathbf{y}_K^* = \mathbf{y}_3^*$
z	1	0	0	6	12	−16
$x_5(x_{B1})^*$	0	1	0	−6	8	20*
$x_6(x_{B2})$	0	0	1	5	26	−11
$x_4(x_{B3})$	0	0	0	1	2	−3

Step 9. *Test of optimality :* The column vectors $\mathbf{a}_1^*$, $\mathbf{a}_2^*$, $\mathbf{a}_3^*$ and $\mathbf{a}_7^*$ are not in the basis

$$(z_1 - c_1, z_2 - c_2, z_3 - c_3, z_7 - c_7) = \text{First row of } B^{*-1} \times [\mathbf{a}_1^*, \mathbf{a}_2^*, \mathbf{a}_3^*, \mathbf{a}_7^*]$$

$$= (1, 0, 0, 6)\begin{bmatrix} -2 & 3 & 2 & 0 \\ 5 & -1 & 2 & 0 \\ 2 & 3 & 4 & 0 \\ 1 & 2 & -3 & 1 \end{bmatrix}$$

$$= (4, 15, -16, 6).$$

The only negative $z_j - c_j$ is $z_3 - c_3 = -16$. Thus the third vector may enter in the next basis provided at least one $y_{i3} > 0[i = 1, 2, 3]$.

Now,

$$\mathbf{y}_3 = B^{-1}\mathbf{a}_3 = \begin{bmatrix} 1 & 0 & -6 \\ 0 & 1 & 5 \\ 0 & 0 & 1 \end{bmatrix}\begin{bmatrix} 2 \\ 4 \\ -3 \end{bmatrix} = \begin{bmatrix} 20 \\ -11 \\ -3 \end{bmatrix}$$

only $y_{13} = 20 > 0$, then $y_{13} = 20$ is the key element. Then $\mathbf{y}_3^* = \begin{bmatrix} -16 \\ 20 \\ -11 \\ -3 \end{bmatrix}$ and fill in the last column of the second table.

Step 10. The vector to leave the basis (current)

$$\frac{x_{Br}}{y_{r3}} = \min_i \left(\frac{x_{Bi}}{y_{i3}}, y_{i3} > 0 \right) = \min \left(\frac{8}{20} \cdots, \cdots \right) = \frac{2}{5}$$

which occurs at $i = 1$. Then the fifth vector $\mathbf{a}_5$ will leave the basis and is replaced by $\mathbf{a}_3$.

Step 11. *Construction of the third revised simplex table.*

Third revised simplex table

Variables in the basis	$\mathbf{e}_1$				
z	1	$\frac{4}{5}$	0	$\frac{6}{5}$	$\frac{92}{5}$
$x_3(x_{\beta 1})$	0	$\frac{1}{20}$	0	$-\frac{3}{10}$	$\frac{2}{5}$
$x_6(x_{\beta 2})$	0	$\frac{11}{20}$	1	$\frac{17}{10}$	$\frac{152}{5}$
$x_4(x_{\beta 3})$	0	$\frac{3}{20}$	0	$\frac{1}{10}$	$\frac{16}{5}$

Step 12. *Test of optimality :* The column vectors $\mathbf{a}_1^*, \mathbf{a}_2^*, \mathbf{a}_5^*, \mathbf{a}_7^*$ are not on the basis

$$\begin{aligned}(z_1-c_1, z_2-c_2, z_3-c_3, z_7-c_7) &= \text{First row of } B^{*-1} \times [\mathbf{a}_1^*, \mathbf{a}_2^*, \mathbf{a}_5^*, \mathbf{a}_7^*] \\ &= \left(1, \frac{4}{5}, 0, \frac{6}{5}\right)\begin{bmatrix} -2 & 3 & 0 & 0 \\ 5 & -1 & 1 & 0 \\ 2 & 3 & 0 & 0 \\ 1 & 2 & 0 & 1 \end{bmatrix} \\ &= \left(\frac{16}{5}, \frac{23}{5}, \frac{1}{5}, \frac{6}{5}\right).\end{aligned}$$

All $z_j - c_j \geq 0$; $j = 1, 2, \cdots, 7$. Hence we reach at the optimal stage.

$$\max z = \frac{92}{5} \text{ at } x_3 = \frac{2}{5},\ x_6 = \frac{152}{5},\ x_4 = \frac{16}{5},$$

$$\text{i.e., } \max z = \frac{92}{5} \text{ at } x_1 = 0,\ x_2 = 0,\ x_3 = \frac{2}{5},\ x_4 = \frac{16}{5}$$

[value of x_6 need not to be mentioned as it is a slack variable]. $x_1 = 0, x_2 = 0$ are not basic variables.

▶ **Example 14.9.5** *Using revised simplex method, prove that the following problem has said to be an unbounded solution*

$$\textit{maximize,} \quad 2x_2 + x_3$$

subject to

$$\begin{aligned} x_1 + x_2 - 2x_3 &\leq 7 \\ -3x_1 + x_2 + 2x_3 &\leq 3, \ x_1, x_2, x_3 \geq 0. \end{aligned}$$

Step 1. This is a maximization problem.

Step 2. All $b_i \geq 0$ and two slack variables x_3 and x_4 are required to introduce, one to each constraint to get equations and thus a unit matrix is present in the co-efficient matrix.

$$\begin{aligned} z - 0x_1 - 2x_2 - x_3 \quad\quad &= 0 \\ x_1 + x_2 - 2x_3 + x_4 \quad &= 7 \\ -3x_1 + x_2 + 2x_3 \quad + x_5 &= 3. \end{aligned}$$

In matrix notation, the system of constraints can be written as

$$\begin{bmatrix} \mathbf{e}_1 & \mathbf{a}_1^* & \mathbf{a}_2^* & \mathbf{a}_3^* & \mathbf{a}_4^* & \mathbf{a}_5^* \\ 1 & 0 & -2 & -1 & 0 & 0 \\ 0 & 1 & 1 & -2 & 1 & 0 \\ 0 & -3 & 1 & 2 & 0 & 1 \end{bmatrix} \begin{bmatrix} z \\ x_1 \\ x_2 \\ x_3 \\ x_4 \\ x_5 \end{bmatrix} = \begin{bmatrix} 0 \\ 7 \\ 3 \end{bmatrix}$$

Step 3. $B = (\mathbf{a}_4, \mathbf{a}_5) = \begin{bmatrix} 1 & 0 \\ 0 & 1 \end{bmatrix}$ is the unit basis of the problem in the ordinary simplex method

$$\mathbf{c}_B = (c_4, c_5) = (0, 0) = \mathbf{0}.$$

As $\mathbf{c}_B = \mathbf{0}$ then $B^* = I_3$ and $B^{*-1} = I_3$ an identity matrix of order three.

$$\mathbf{x}_B^* = I_3\mathbf{b}^* = \begin{bmatrix} z \\ x_{B1} \\ x_{B2} \end{bmatrix} = \begin{bmatrix} z \\ x_4 \\ x_5 \end{bmatrix} = \begin{bmatrix} 0 \\ 7 \\ 3 \end{bmatrix}.$$

and with these data we can construct the initial revised simplex table. Last column of the table remains vacant initially.

Step 4. *Construction of the starting simplex table*

First revised simplex table

Variables in the basis	$\mathbf{e}_1$			$\mathbf{x}_B^*$	$\mathbf{y}_k^* = \mathbf{y}_2^*$
z	1	0	0	0	−2
$x_4(x_{B1})$	0	1	0	7	1
$x_5^*(x_{B2})$	0	0	1	3	1^*

Step 5. *Test of optimality*

Here $\mathbf{a}_1^*, \mathbf{a}_2^*, \mathbf{a}_3^*$ are not in the basis. Therefore we required to calculate $z_j - c_j$ $[j = 1, 2, 3]$.

Now,

$$(z_1 - c_1, z_2 - c_2, z_3 - c_3) = \text{First row of } \mathbf{B}^{*-1} \times [\mathbf{a}_1^*, \mathbf{a}_2^*, \mathbf{a}_3^*]$$

$$= (1, 0, 0) \begin{bmatrix} 0 & -2 & -1 \\ 1 & 1 & -2 \\ -3 & 1 & 2 \end{bmatrix} = (0, -2, -1)$$

Here $z_2 - c_2 = -2, z_3 - c_3 = -1$ are negative.

Step 6. *Determination of the vector to enter in the next basis*

$$\min(z_2 - c_2, z_3 - c_3) = \min(-2, -1) = -2 = z_2 - c_2,$$

which occurs at $j = 2$. Then the second vector $\mathbf{a}_2$ will enter in the next basis

$$\mathbf{y}_2 = B^{-1}\mathbf{a}_2 = I_2 \begin{bmatrix} 1 \\ 1 \end{bmatrix} = \begin{bmatrix} 1 \\ 1 \end{bmatrix}$$

Then

$$\mathbf{y}_2^* = \begin{bmatrix} z_2 - c_2 \\ y_{21} \\ y_{22} \end{bmatrix} = \begin{bmatrix} -2 \\ 1 \\ 1 \end{bmatrix}$$

as $z_2 - c_2$ is already obtained and with this value of $\mathbf{y}_2^*$ fill up the column of the starting table to complete it.

Step 7. ***Vector to leave the current basis and replaced by $\mathbf{a}_2$.***

At least one component of $\mathbf{y}_2$ is positive. Then we can improve or at least remains same the value of the objective function.

Now,

$$\frac{x_{Br}}{y_{r2}} = \min\left(\frac{x_{Bi}}{y_{i2}}, y_{i2} > 0\right) = \min\left(\frac{7}{1}, \frac{3}{1}\right) = 3$$

which occurs for $j = 2$. Then the second vector of the basis $\mathbf{a}_5$ will be replaced by $\mathbf{a}_2$ to form a new basis and $1 = y_{21}$ is the key element.

Step 8. ***Determination of the improved solution.***

To get the improved solution, we shall have to construct the second revised simplex table. We shall have to transform the first table similarly as we have done in the simplex method in such a way that the column corresponding to $\mathbf{y}_2^*$, i.e., the last column be a unit vector $\mathbf{e}_3$.

The key element is $y_{21} = 1$.

Second revised simplex table

Variables in the basis	$\mathbf{e}_1$			$\mathbf{x}_B^*$	$\mathbf{y}_K^* = \mathbf{y}_1^*$
z	1	0	2	6	−6
$x_4^*(x_{B1})$	0	1	−1	4	4*
$x_2(x_{B2})$	0	0	1	3	−3

Step 9. ***Test of optimality***

The column vectors $\mathbf{a}_1^*, \mathbf{a}_3^*$ and $\mathbf{a}_5^*$ are not in the basis.

Now,

$$(z_1 - c_1, z_3 - c_3, z_5 - c_5) = \text{First row of } \mathbf{B}^{*-1} \times [\mathbf{a}_1^*, \mathbf{a}_3^*, \mathbf{a}_5^*]$$

$$= (1, 0, 2)\begin{bmatrix} 0 & -1 & 0 \\ 1 & -2 & 0 \\ -3 & 2 & 1 \end{bmatrix} = (-6, 3, 2).$$

Here $z_1 - c_1$ is the only negative. Then $\mathbf{a}_1$ may enter in the next basis.

Now,

$$\mathbf{y}_1 = B^{-1}\mathbf{a}_1 = \begin{bmatrix} 1 & -1 \\ 0 & 1 \end{bmatrix}\begin{bmatrix} 1 \\ -3 \end{bmatrix} = \begin{bmatrix} 4 \\ -3 \end{bmatrix}$$

at least one component of which is positive. We shall have to proceed further.

Therefore $\mathbf{y}_1^* = \begin{bmatrix} -6 \\ 4 \\ -3 \end{bmatrix}$ as $z_1 - c_1$ is obtained earlier. Fill in the last column of the second table with $\mathbf{y}_1^*$

$$\frac{x_{Br}}{y_{r1}} = \min_i \left(\frac{x_{Bi}}{y_{i1}}, y_{i1} > 0\right) = \min\left(\frac{4}{4}\right) = 1 \text{ at } i = 1.$$

Then the first vector $\mathbf{a}_4$ of the basis will leave the basis and replaced by $\mathbf{a}_1$ and $x_1(x_{B1})$ will enter in the new basic solution set.

Step 10. *Determination of the improved value of the objective function.*

We shall have to transform the second table in such a way that the column corresponding to $\mathbf{y}_1^*$, i.e., the last column of the second table be a unit vector $\mathbf{e}_2$

Variables in the basis	$\mathbf{e}_1$			$\mathbf{x}_B^*$
z	1	$\frac{3}{2}$	$\frac{1}{2}$	12
$x_1(x_{B1})$	0	$\frac{1}{4}$	$-\frac{1}{4}$	1
$x_2(x_{B2})$	0	$\frac{3}{4}$	$\frac{1}{4}$	6

Step 11. *Test of optimality*

The column vectors $\mathbf{a}_3^*, \mathbf{a}_4^*$ and $\mathbf{a}_5^*$ are not in the basis and therefore we require to compute $z_3 - c_3, z_4 - c_4$ and $z_5 - c_5$

$$(z_3 - c_3, z_4 - c_4, z_5 - c_5) = \text{First row of } \mathbf{B}_{\text{cur}}^{*-1} \times [\mathbf{a}_3^*, \mathbf{a}_4^*, \mathbf{a}_5^*]$$

$$= \left(1, \frac{3}{2}, \frac{1}{2}\right)\begin{bmatrix} -1 & 0 & 0 \\ -2 & 1 & 0 \\ 2 & 0 & 1 \end{bmatrix} = \left(-3, \frac{3}{2}, \frac{1}{2}\right).$$

Here $z_3 - c_3$ is negative. There $\mathbf{a}_3$ may enter in the next basis.

Now

$$\mathbf{y}_3 = B^{-1}\mathbf{a}_3 = \begin{bmatrix} \frac{1}{4} & -\frac{1}{4} \\ \frac{3}{4} & \frac{1}{4} \end{bmatrix}\begin{bmatrix} -2 \\ 2 \end{bmatrix} = \begin{bmatrix} -1 \\ -1 \end{bmatrix}$$

Here both y_{13} and y_{23} are negative. We need not proceed further and the only conclusion is that the problem is said to have an unbounded solution.

Now we shall solve a problem where instead of M, a large positive number which is very large in comparison to other given cost components, be used. But this method of solving a problem cannot be treated as 'Big M method'. Following, is an example which has been solved using this technique. This method has been followed by computers.

14.10 Revised simplex method (Standard Form II)

In the case of ordinary simplex method, in many cases, to get an initial unit basis [with all $b_i \geq 0$] we require to introduce some artificial variables [minimum number of artificial variables be introduced] and the slack vectors and the artificial vectors constitute the initial unit basis. The problems of that type can be tackled either by using, Charnes' "Big M-method" or "method of penalties" and by "two phase method". For hand computation, revised simplex method cannot be considered as a good technique in solving a linear programming problem. Many people still prefer the ordinary simplex method instead of revised simplex method in solving a linear programming problem for hand computation. Thus we here discuss only the cases of revised simplex method which are related to the 'Big M-method' of the ordinary simplex method and ignore the discussion related to 'Two-phase method' which is extremely lengthy and there is every possibility in making mistakes at any stage.

In the case of the problems of standard form I, the initial unit basis of the ordinary simplex method contains only slack vectors and that is why $\mathbf{c}_B = \mathbf{0}$ (null vector) and in that case initial basis and its inverse of the revised simplex method is also a unit basis but its order is $m+1$ instead of m. But for problems of the standard form II (using Big M-method) $\mathbf{c}_B$ will not be a null vector.

In that case

$$\mathbf{B}^* = \begin{bmatrix} 1 & -\mathbf{c}_B \\ \mathbf{0} & B \end{bmatrix} \qquad \text{from (14.5.1)}$$

Here B is a unit basis but $\mathbf{c}_B$ is not a null vector and therefore

$$\mathbf{B}^{*-1} = \begin{bmatrix} 1 & \mathbf{c}_B B^{-1} \\ \mathbf{0} & B^{-1} \end{bmatrix} \qquad \text{from (14.5.2)}$$

will not be a unit basis of order $m+1$. This is the only difference between the problems of standard form I and the problems of standard form II. All other theoretical aspects remain same which will be cleared from the example solved below.

► **Example 14.10.1** *Solve the following problem by revised simplex method,*

$$\textit{minimize,} \quad z = 4x_1 + 3x_2$$

subject to

$$x_1 + 2x_2 \geq 8$$
$$3x_1 + 2x_2 \geq 12, \; x_1, x_2 \geq 0.$$

The problem is, maximize $z' = -4x_1 - 3x_2 + 0x_3 + 0x_4 - Mx_5 - Mx_6$ *when*

$$\min z = -\max z', \mathbf{b} \geq \mathbf{0}.$$

Introducing surplus variables and artificial variables as shown below and considering $z' + 4x_1 + 3x_2 + 0x_3 + 0x_4 + Mx_5 + Mx_6 = 0$ as the first constraint, we have

$$\begin{aligned} z' + 4x_1 + 3x_2 + 0x_3 + 0x_4 + Mx_5 + Mx_6 &= 0 \\ x_1 + 2x_2 - x_3 + 0x_4 + x_5 + 0x_6 &= 8 \\ 3x_1 + 2x_2 + 0x_3 - x_4 + 0x_5 + x_6 &= 12, \end{aligned}$$

x_3 and x_4 are surplus variables, x_5, x_6 are artificial variables.

$$B = (\mathbf{a}_5, \mathbf{a}_6) = \begin{bmatrix} 1 & 0 \\ 0 & 1 \end{bmatrix} = I_2$$

is taken as the initial basis as

$$B^{-1}\mathbf{b} = I_2 \begin{bmatrix} 8 \\ 12 \end{bmatrix} = \begin{bmatrix} 8 \\ 12 \end{bmatrix} \geq 0$$

$\mathbf{c}_B = (c_5, c_6) = (-M, -M)$. Then B^{*-1} [basis inverse] of the revised simplex method is

$$B^{*-1} = \begin{bmatrix} 1 & \mathbf{c}_B B^{-1} \\ 0 & B^{-1} \end{bmatrix} = \begin{bmatrix} 1 & -M & -M \\ 0 & 1 & 0 \\ 0 & 0 & 1 \end{bmatrix}$$

which is not an identity matrix.

$$\mathbf{x}_B^* = B^{*-1} = B^{*-1} \begin{bmatrix} 0 \\ 8 \\ 12 \end{bmatrix} = \begin{bmatrix} -20M \\ 8 \\ 12 \end{bmatrix}$$

and with these data, we are able to construct the initial table.

Initial table

Variables in the basis	$\mathbf{e}_1$			$\mathbf{x}_B^*$	$\mathbf{y}_2^*$
z'	1	$-M$	$-M$	$-20M$	$-4M+3$
$x_{B1}(x_5)^*$	0	1	0	8	2^*
$x_{B2}(x_6)$	0	0	1	12	2

Calculation of $z_j - c_j$ corresponding to the vectors not in the basis B.

$$\begin{aligned} z_j - c_j, (j = 1, 2, 3, 4) &= \text{First row of } B^{*-1} \times [\mathbf{a}_1^*, \mathbf{a}_2^*, \mathbf{a}_3^*, \mathbf{a}_4^*] \\ &= (-4M + 4, -4M + 3, M, M) \end{aligned}$$

$\min(z_j - c_j, z_j - c_j < 0) = -4M + 3 = z_2 - c_2$ and thus $\mathbf{a}_2$ is the vector to enter in the next basis.

$$\mathbf{y}_2 = B^{-1}\mathbf{a}_2 = \mathbf{a}_2 = \begin{bmatrix} 2 \\ 2 \end{bmatrix} \text{ and obviously}$$

$$\mathbf{y}_2^* = \begin{bmatrix} -4M+3 \\ 2 \\ 2 \end{bmatrix}$$

as $z_2 - c_2 = -4M + 3$

$$\min\left(\frac{x_{Bi}}{y_{i2}}, y_{i2} > 0\right) = \min\left(\frac{x_{B1}}{y_{12}}, \frac{x_{B_2}}{y_{22}}\right) = \min\left(\frac{8}{2}, \frac{12}{2}\right) = 4$$

which occurs for $i = 1$. Thus $y_{12} = 2$ is the key element and the first vector of B, $\mathbf{a}_5$ (artificial vector) will leave the basis. Put the column vector $\mathbf{y}_2^*$ in the initial table as shown above.

Construction of the second table

Variables in the basis				$\mathbf{x}_B^*$	$\mathbf{y}_1^*$
z'	1	$M - \frac{3}{2}$	$-M$	$-4M - 12$	$-2M + \frac{5}{2}$
$x_{B1}(x_2)$	0	$\frac{1}{2}$	0	4	$\frac{1}{2}$
$x_{B2}(x_6)^*$	0	-1	1	4	2^*

Calculation of $z_j - c_j$ corresponding to non-basic vectors in B [drop $\mathbf{a}_5$]

$$z_j - c_j, (j = 1, 3, 4) = \text{First row of } B^{*-1} \times [\mathbf{a}_1^*, \mathbf{a}_3^*, \mathbf{a}_4^*]$$

$$= \left(-2M + \frac{5}{2}, -M + \frac{3}{2}, M\right)$$

$$\min_j(z_j - c_j, z_j - c_j < 0) = -2M + \frac{5}{2} = z_1 - c_1.$$

Thus the first vector $\mathbf{a}_1$ will enter in the next basis.

$$\mathbf{y}_1 = B^{-1}\mathbf{a}_1 = \begin{bmatrix} \frac{1}{2} & 0 \\ -1 & 1 \end{bmatrix}\begin{bmatrix} 1 \\ 3 \end{bmatrix} = \begin{bmatrix} \frac{1}{2} \\ 2 \end{bmatrix} \text{ and obviously}$$

$$\mathbf{y}_1^* = \begin{bmatrix} -2M + \frac{5}{2} \\ \frac{1}{2} \\ 2 \end{bmatrix}$$

$$\min_i\left(\frac{x_{Bi}}{y_{i1}}, y_{i1} > 0\right) = \min\left(\frac{4}{\frac{1}{2}}, \frac{4}{2}\right) = 2$$

which occurs for $i = 2$. Thus the second vector $\mathbf{a}_6$ of the basis will leave the basis and replaced by $\mathbf{a}_1$ to form a new basis and $y_{21} = 2$ is the key element. Put the column vector $\mathbf{y}_1^*$ in the second table as shown in the second table.

Third table

Variables in the basis				$\mathbf{x}_B$
z'	1	$-\frac{1}{4}$	$-\frac{1}{4}$	3
$x_{B1}(x_2)$	0	$\frac{3}{4}$	$-\frac{1}{4}$	3
$x_{B2}(x_1)^*$	0	$-\frac{1}{2}$	$\frac{1}{2}$	2

Calculation of $z_j - c_j$ not in the basis. [For artificial vectors we need not calculate] $z_j - c_j(j = 3, 4)$ = First row of $B^{*-1} \times [\mathbf{a}_3^*, \mathbf{a}_4^*] = [\frac{1}{4}, \frac{5}{4}]$.

Thus all $z_j - c_j \geq 0$ and all artificial vectors are removed from the basis. Hence we reach at the optimal stage and the optimal value of $z' = -17$ at $x_1 = 2$ and $x_2 = 3$, i.e., $\min z = -(-17) = 17$ at $x_1 = 2, x_2 = 3$.

► **Example 14.10.2** *Solve the L.P.P. by revised simplex method*

$$\text{maximize}, \quad z = 2x_1 - 3x_2$$

subject to

$$\begin{aligned} x_1 - \ x_2 &\leq 2 \\ 5x_1 + 4x_2 &\leq 46 \\ 7x_1 + 2x_2 &\geq 32, \ x_1, x_2 \geq 0. \end{aligned}$$

Step 1. It is a maximization problem.

Step 2. All $b_i \geq 0$; but one artificial variable is required to introduce to get a unit basis matrix. The constraints in the revised simplex method is

$$z = 2x_1 - 3x_2 + 0x_3 + 0x_4 + 0x_5 - 10x_6[M = 10] \text{ (say)}$$

or the constraints are

$$z - 2x + 3x_2 + 0 \cdot x_3 + 0x_4 + 0x_5 + 10x_6 = 0$$

$$\begin{aligned} x_1 - x_2 + x_3 \qquad\qquad &= 2 \\ 5x_1 + 4x_2 \qquad + x_4 \qquad &= 46 \\ 7x_1 + 2x_2 \qquad\qquad - x_5 + x_6 &= 32 \end{aligned}$$

$x_1, x_2 \geq 0$, x_3, x_4 slack and x_5 surplus variable, x_6 artificial variable ≥ 0, z is unrestricted, $M = 10$ (say) large with comparison to 2 and (-3). The problem in matrix notation is

$$\begin{bmatrix} \mathbf{e}_1 & \mathbf{a}_1^* & \mathbf{a}_2^* & \mathbf{a}_3^* & \mathbf{a}_4^* & \mathbf{a}_5^* & \mathbf{a}_6^* \\ 1 & -2 & 3 & 0 & 0 & 0 & 10 \\ 0 & 1 & -1 & 1 & 0 & 0 & 0 \\ 0 & 5 & 4 & 0 & 1 & 0 & 0 \\ 0 & 7 & 2 & 0 & 0 & -1 & 1 \end{bmatrix} \begin{bmatrix} z \\ x_1 \\ x_2 \\ x_3 \\ x_4 \\ x_5 \\ x_6 \end{bmatrix} = \begin{bmatrix} \mathbf{b}^* \\ 0 \\ 2 \\ 46 \\ 32 \end{bmatrix}$$

$B = (\mathbf{a}_3, \mathbf{a}_4, \mathbf{a}_6) = \begin{bmatrix} 1 & 0 & 0 \\ 0 & 1 & 0 \\ 0 & 0 & 1 \end{bmatrix} = I_3$ which is the unit basis of the original simplex method.

$$\mathbf{x}_B = B^{-1}\mathbf{b} = \begin{bmatrix} 2 \\ 46 \\ 32 \end{bmatrix} \geq 0.$$

$\mathbf{c}_B = (0, 0, -10)$ taking $M = 10$ which is large enough in comparison to 2, −3.

$$\mathbf{c}_B = (0, 0, -10).$$

The basis inverse B^{*-1} of the revised simplex method is $\begin{bmatrix} 1 & 0 & 0 & -10 \\ 0 & 1 & 0 & 0 \\ 0 & 0 & 1 & 0 \\ 0 & 0 & 0 & 1 \end{bmatrix}$

$$\mathbf{x}_B^* = B^{*-1}\mathbf{b}^* = \begin{bmatrix} 1 & 0 & 0 & -10 \\ 0 & 1 & 0 & 0 \\ 0 & 0 & 1 & 0 \\ 0 & 0 & 0 & 1 \end{bmatrix} \begin{bmatrix} 0 \\ 2 \\ 46 \\ 32 \end{bmatrix} = \begin{bmatrix} -320 \\ 2 \\ 46 \\ 32 \end{bmatrix}$$

Initial table

Basic variables	$\mathbf{e}_1$				$\mathbf{x}_B^*$	$\mathbf{y}_1^*$
z	1	0	0	−10	−320	−72
$x_{B1}(x_3)^*$	0	1	0	0	2	1*
$x_{B2}(x_4)$	0	0	1	0	46	5
$x_{B3}(x_6)$	0	0	0	1	32	7

Step 3. Calculation of $z_j - c_j$ corresponding to the vectors not in the basis B. $(z_j - c_j)$ $j = 1, 2, 5$

$$(z_1 - c_1, z_2 - c_2, z_5 - c_5) = \text{First row of } B^{*-1} \times [\mathbf{a}_1^*, \mathbf{a}_2^*, \mathbf{a}_5^*]$$

$$= (1\,0\,0\,-10) \begin{bmatrix} -2 & 3 & 0 \\ 1 & -1 & 0 \\ 5 & 4 & 0 \\ 7 & 2 & -1 \end{bmatrix}$$

$$= (-72, -17, 10)$$

$\min(z_j - c_j, z_j - c_j < 0) = -72$ which is $z_1 - c_1$ and $B = (\mathbf{a}_3, \mathbf{a}_4, \mathbf{a}_6) = I_3$.

$$B^{-1} = I_3, \mathbf{y}_1 = B^{-1}\mathbf{a}_1 = I_3\mathbf{a}_1 = \begin{bmatrix} 1 \\ 5 \\ 7 \end{bmatrix}.$$

Then

$$\mathbf{y}_2^* = \begin{bmatrix} -72 \\ 1 \\ 5 \\ 7 \end{bmatrix}.$$

With this value fill in the last column of the previous table.

Step 4. Construction of the second table

Basic variables	$\mathbf{e}_1$				$\mathbf{x}_{*B}$	$\mathbf{y}_{*1}$
z	1	72	0	−10	−176	−89
$x_4(x_{B1})$	0	1	0	0	2	−1
$X_5(x_{B2})$	0	−5	1	0	36	9
$x_6(x_{B3})*$	0	−7	0	1	18	9*

Step 5. Calculation of $z_j - c_j$ corresponding to the vectors not in the basis $z_j - c_j$, $j = 2, 3, 5$

$$(z_2 - c_2, z_3 - c_3, z_5 - c_5) = \text{First row } B^{*-1} \times (\mathbf{a}_{*2}, \mathbf{a}_{*3}, \mathbf{a}_{*5})$$

$$= (1 \quad 72 \quad 0 \quad -10)\begin{vmatrix} 3 & 0 & 0 \\ -1 & 1 & 0 \\ 4 & 0 & 0 \\ 2 & 0 & -1 \end{vmatrix}$$

$$= (-89, 72, 10)$$

$z_2 - c_2$ is the only negative value. The x_2 will be next basic variable and the second vector will enter in the new basis.

$$\mathbf{y}_2 = B^{-1}\mathbf{a}_2 = \begin{bmatrix} 1 & 0 & 0 \\ -5 & 1 & 0 \\ -7 & 0 & 1 \end{bmatrix}\begin{bmatrix} -1 \\ 4 \\ 2 \end{bmatrix} = \begin{bmatrix} -1 \\ 9 \\ 9 \end{bmatrix}$$

Then

$$\mathbf{y}_2^* = \begin{bmatrix} -89 \\ -1 \\ 9 \\ 9 \end{bmatrix}$$

and with this value of $\mathbf{y}_2^*$ fill up the last column of the second table.

$$\frac{x_{Br}}{y_{r2}} = \left(\frac{x_{Bi}}{y_{i2}}, \frac{x_{Bi}}{y_{i2}} > 0\right) = \left(\frac{36}{9}, \frac{18}{9}\right) = 2$$

which occurs for $i = 3$. Then the artificial variable will leave from the basic solution and the new component is x_2. We shall have to construct the new table in the usual manner, $y_{33} = 9$ is the key element.

Third table

Basic variables	$\mathbf{e}_1$				$\mathbf{x}_B^*$
z	1	$\frac{25}{9}$	0	$\frac{1}{9}$	2
$x_{B1}(x_1)$	0	$\frac{2}{9}$	0	$-\frac{1}{9}$	4
$x_{B2}(x_4)$	0	−2	1	1	18
$x_{B3}(x_2)$	0	$-\frac{7}{9}$	0	$\frac{1}{9}$	2

We require to calculate the $z_j - c_j$ corresponding to non basis vector ignoring $\mathbf{a}_6$

$$(z_3 - c_3, z_5 - c_5) = \text{First row of } B^{*-1} \times [\mathbf{a}_3^*, \mathbf{a}_5^*]$$

$$= \left(1, \ \frac{25}{9}, \ 0, \ -\frac{1}{9}\right) \begin{bmatrix} 0 & 0 \\ 1 & 0 \\ 0 & 0 \\ 0 & -1 \end{bmatrix} = \left(\frac{25}{9}, \frac{1}{9}\right)$$

$z_3 - c_3$, $z_5 - c_5 \geq 0$. Hence we reach at the optimal stage. Max $z = 2$ at $x_1 = 4$ and $x_2 = 2$ [x_4 is a slack variable].

14.11 Relative advantages and disadvantages of the revised simplex method over the original simplex method

This method is very helpful in solving the problem by the electronic computers. This is also helpful in solving the problems manually if the number of constraints and variables be large (say ≥ 4). But in solving a problem by revised simplex method, the original column vectors not in the basis are also required. Hence there are more chances of computational mistakes in revised simplex method than in the original simplex method. That is why, still many people prefer the ordinary simplex method rathan than the revised simplex method for hand computation.

Exercise 14

1. Describe the 'revised simplex algorithm' in solving a linear programming problem.

2. Compare the revised simplex method with the ordinary simplex method and find out the main points of difference.

3. Give a brief outline of the revised simplex method (standard form 1).

4. In what type of problems, revised simplex method is to be used?

5. Describe the relative advantages and disadvantages of the revised simplex method over the ordinary simplex method.

   Solving the following problem by revised simplex method:

   (i) Maximize, $z = 4x_1 + 3x_2$ subject to

$$3x_1 + x_2 \leq 15$$
$$3x_1 + 4x_2 \leq 24, \ x_1, x_2 \geq 0.$$

   (ii) Maximize, $z = 2x_1 + 3x_2$ subject to

$$7x_1 + 4x_2 \leq 28$$
$$7x_1 + 12x_2 \leq 52, \ x_1, x_2 \geq 0.$$

(iii) Maximize, $z = 5x_1 + 3x_2$ subject to

$$\begin{aligned} x_1 + x_2 &\le 2 \\ 5x_1 + 2x_2 &\le 15 \\ 3x_1 + 8x_2 &\le 12, \; x_1, x_2 \ge 0. \end{aligned}$$

(iv) Maximize, $z = 3x_1 + 2x_2 + x_3$ subject to

$$\begin{aligned} x_1 + 2x_2 - 3x_3 &\le 8 \\ 3x_2 + x_3 &\le 6, \; x_1, x_2, x_3 \ge 0. \end{aligned}$$

(v) Minimize, $z = -2x_1 + 3x_2$ subject to

$$\begin{aligned} 2x_1 - 5x_2 &\le 7 \\ 4x_1 + x_2 &\le 8 \\ 7x_1 + 2x_2 &\le 16, \; x_1, x_2 \ge 0. \end{aligned}$$

(vi) Maximize, $z = x_1 + x_2 + 3x_3$ subject to

$$\begin{aligned} x_1 + 2x_2 - x_3 &\le 10 \\ 3x_2 + 2x_3 &\le 8 \\ x_2 + 3x_3 &\le 15, \; x_1, x_2, x_3 \ge 0. \end{aligned}$$

(vii) Maximize, $z = 2x_1 + 5x_2$ subject to

$$\begin{aligned} -2x_1 + x_2 &\le 0 \\ x_1 + 3x_2 &\le 14 \\ x_1 + x_2 &\le 8, \; x_1, x_2 \ge 0. \end{aligned}$$

(viii) Maximize, $z = 2x_2 + x_3$ subject to

$$\begin{aligned} x_1 + x_2 - 2x_3 &\le 7 \\ -3x_1 + x_2 + 2x_3 &\le 3, \; x_1, x_2, x_3 \ge 0. \end{aligned}$$

(ix) Minimize, $z = 2x_1 + x_2$ subject to

$$\begin{aligned} 8x_1 + 5x_2 &\le 50 \\ x_1 + 7x_2 &\le 7, \; x_1, x_2 \ge 0. \end{aligned}$$

(x) Maximize, $z = 5x_1 + 8x_2$ subject to

$$\begin{aligned} 3x_1 + 2x_2 &\ge 3 \\ x_1 + 4x_2 &\ge 4 \\ x_1 + x_2 &\le 5, \; x_1, x_2 \ge 0. \end{aligned}$$

(xi) Minimize, $z = 3x_1 + 5x_2$ subject to

$$\begin{aligned} x_1 + 2x_2 &\le 8 \\ 3x_1 + 2x_2 &\le 12 \\ 5x_1 + 6x_2 &\le 60, \; x_1, x_2 \ge 0. \end{aligned}$$

(xii) Maximize, $z = 2x_1 + x_2 + 3x_3$ subject to

$$\begin{aligned} x_1 + \ x_2 + 2x_3 &\le 5 \\ 2x_1 + 3x_2 + 4x_3 &= 12, \ x_1, x_2, x_3 \ge 0. \end{aligned}$$

(xiii) Maximize, $z = x_1 + 2x_2 + 3x_3 - x_4$ subject to

$$\begin{aligned} x_1 + 2x_2 + 3x_3 \qquad &= 15 \\ 2x_1 + \ x_2 + 5x_3 \qquad &= 20 \\ x_1 + 2x_2 + x_3 + x_4 &= 10, \ x_1, x_2, x_3, x_4 \ge 0. \end{aligned}$$

Answers

(i) max $z = 25$ at $x_1 = 4$, $x_2 = 3$; (ii) max $z = 13\frac{4}{7}$ at $x_1 = \frac{16}{7}$, $x_2 = 3$; (iii) max $z = 10$ at $x_1 = 2$, $x_2 = 0$; (iv) max $z = 84$ at $x_1 = 26$, $x_2 = 0$, $x_3 = 0$; (v) min $z = -4$ at $x_1 = 2$, $x_2 = 0$; (vi) max $z = 26$ at $x_1 = 14$, $x_2 = 0$, $x_3 = 4$; (vii) max $z = 25$ at $x_1 = 5$, $x_2 = 3$; (viii) unbounded; (ix) min $z = 10$ at $x_1 = 0$, $x_2 = 5$; (x) max $z = 40$ for $x_1 = 0$, $x_2 = 5$; (xi) min $z = 21$ at $x_1 = 2$, $x_2 = 3$; (xii) max $z = 8$ for $x_1 = 3$, $x_2 = 2$, $x_3 = 0$; (xiii) max $z = 15$ at $x_1 = \frac{5}{2}$, $x_2 = \frac{5}{2}$, $x_3 = \frac{5}{2}$.

Chapter 15

Dual Simplex Method

Dual simplex method of solving a L.P.P. is one of the methods of solving a L.P.P. in which the theory relating to primal problem has been developed with the help of the properties of its dual problem. That is why the method is known as 'Dual simplex method' or 'Dual Simplex Algorithm'. Using this method we are able to solve many problems without using artificial variables. But the only disadvantage of the method is that, all types of problems cannot be solved by using this method. Only a particular type of problems can be solved by using this method.

15.1 General discussions

In the case of ordinary simplex method, we have the intention to move from the feasibility to the finite optimality. In the 'dual simplex method' we try to move from infeasible optimally to the feasible optimality, i.e., we try to find a basic solution (non-feasible) with all $z_j - c_j \geq 0$ [for a maximization problem] and then by using the procedures which will be developed later, we try to reach at a stage where the basic solution is feasible, keeping all $z_j - c_j \geq 0$ at each and every stage. Thus initially if it is not possible to find a basic solution with all $z_j - c_j \geq 0$, we cannot start the dual simplex method to solve the problem and thus we cannot proceed further.

Let us assume that the primal in the standard form be

Maximize, $z = \mathbf{cx}$ subject to

$$A\mathbf{x} = \mathbf{b}, \quad \mathbf{x} \geq \mathbf{0}, \tag{15.1.1}$$

where A is an $m \times n$ matrix.

The corresponding dual problem is

Minimize, $z_w = \mathbf{b}'\mathbf{w}$ subject to

$$A'\mathbf{w} \geq \mathbf{c}', \quad \text{or,} \quad (A'\mathbf{w})' \geq (\mathbf{c}')', \quad \text{or,} \quad \mathbf{w}'A \geq \mathbf{c}, \quad \mathbf{w} \text{ is unrestricted.} \tag{15.1.2}$$

We have the following properties.

For any feasible solution $\mathbf{x}$ to (15.1.1) and for any solution $\mathbf{w}$ (feasible or not) to the dual (15.1.2), we have,

$$\mathbf{w}'A\mathbf{x} = \mathbf{w}'\mathbf{b} \tag{15.1.3}$$

and as $\mathbf{c} \leq \mathbf{w}'A$ and $\mathbf{x} \geq 0$, then

$$\mathbf{cx} \leq \mathbf{w}'A\mathbf{x} = \mathbf{w}'\mathbf{b} \quad [\because\ A\mathbf{x} = \mathbf{b}] \tag{15.1.4}$$

Theorem 15.1.1 *If $\mathbf{w}' = \mathbf{c}_B B^{-1}$, a solution (feasible or not) to the dual* (15.1.2), *where B is a basis matrix of the primal* (15.1.1), *be not an optimal solution of the dual, then at least one x_{Bi} of the basic solution $\mathbf{x}_B = B^{-1}\mathbf{b}$ of the primal be negative.*

Proof. Let the primal basis $B = B(\beta_1, \beta_2, \cdots, \beta_m)$ where $\beta_1, \beta_2, \cdots, \beta_m$ are basis vectors which are some of $\mathbf{a}_j \in A$.

$$\mathbf{w}' = \mathbf{c}_B B^{-1}, \quad \text{or,} \quad \mathbf{w}'B = \mathbf{c}_B,$$
$$\text{or,} \quad \mathbf{w}'(\beta_1, \beta_2, \cdots, \beta_i, \cdots, \beta_m) = (c_{B1}, c_{B2}, \cdots, c_{Bi}, \cdots, c_{Bm}).$$

Then equating we get,

$$\mathbf{w}'\beta_i = c_{Bi}, \quad i = 1, 2, \cdots, m. \tag{15.1.5}$$

As $\mathbf{w}'$ is solution set of the dual, then for $\mathbf{a}_j$ not in the basis

$$\mathbf{w}'\mathbf{a}_j \geq c_j. \tag{15.1.6}$$

Combining (15.1.5) and (15.1.6) we can write

$$\mathbf{w}'\mathbf{a}_j \geq c_j \text{ for all } j,$$
$$\text{or,} \quad \mathbf{c}_B B^{-1}\mathbf{a}_j \geq c_j, \quad \text{or,} \quad z_j - c_j \geq 0 \text{ for all } j,$$

which implies that the optimality condition of the primal has been satisfied.

Now, if $\mathbf{x}_B = B^{-1}\mathbf{b} \geq 0$, then this feasible solution will be optimal solution of the primal and $\mathbf{w}' = \mathbf{c}_B B^{-1}$ be the optimal solution (feasible or not) of the dual since

$$\mathbf{c}_B\mathbf{x}_B = \mathbf{c}_B B^{-1}\mathbf{b} = \mathbf{w}'\mathbf{b}. \tag{15.1.7}$$

Now under such condition if $\mathbf{w}'$ is not an optimal solution of the dual, we are forced to conclude that at least one x_{Bi} of $\mathbf{x}_B$ be negative.

Theorem 15.1.2 *If for a basis B of the primal, at least one x_{Bi} say $x_{Br} < 0$, with all $y_{rj} \geq 0$, and $\mathbf{w}' = \mathbf{c}_B B^{-1}$ be a solution of the dual, then the primal will have no feasible solution.*

Proof. Let $B = (\beta_1, \beta_2, \cdots, \beta_m)$

$$\mathbf{w}' = \mathbf{c}_B B^{-1}$$

is a solution set of the dual implies

$$\mathbf{w}'\beta_i = c_{Bi} \quad [i = 1, \cdots, m]$$

and

$$\mathbf{w}\mathbf{a}_j \geq c_j \text{ for all } \mathbf{a}_j \text{ not in the basis.} \tag{15.1.8}$$

Let us denote the rows of B^{-1} by β^j.

Then

$$\left.\begin{aligned} \beta^j \beta_i &= \delta_{ij} \\ \text{and } x_{Br} &= \beta^r \mathbf{b} \end{aligned}\right\} \quad [i, j = 1, \cdots, m] \tag{15.1.9}$$

[δ_{ij} is known as kronecker δ and $\delta_{ij} = 1$ when $i = j$ and $\delta_{ij} = 0$ for $i \neq j$]. We now define a new vector for the dual

$$\hat{\mathbf{w}}' = \mathbf{w} - \theta\beta^r \tag{15.1.10}$$

and try to investigate the properties of the new vector $\hat{\mathbf{w}}'$.

$$\hat{\mathbf{w}}'\beta_i = \mathbf{w}'\beta_i - \theta\beta^r\beta_i = c_{Bi} - \theta\delta_{ir}. \tag{15.1.11}$$

Hence

$$\left.\begin{aligned} \hat{\mathbf{w}}'\beta_i &= c_{Bi} \quad (i \neq r) \quad [\because \ \delta_{ir} = 0] \\ \text{and } \mathbf{w}'\beta_r &= c_{Br} - \theta \geq c_{Br} \text{ if } \theta \leq 0 \end{aligned}\right\}. \tag{15.1.12}$$

Again for all $\mathbf{a}_j$ not in the primal basis

$$\hat{\mathbf{w}}'\mathbf{a}_j = \mathbf{w}'\mathbf{a}_j - \theta\beta^r\mathbf{a}_j. \tag{15.1.13}$$

Now $\beta^r\mathbf{a}_j = y_{rj}$. If we now assume that $y_{rj} \geq 0$ for every j then for any $\theta \leq 0$, we have

$$\hat{\mathbf{w}}'\mathbf{a}_j \geq \mathbf{w}'\mathbf{a}_j \geq c_j. \tag{15.1.14}$$

Now under the conditions stated above, $\hat{\mathbf{w}}'$ is also a solution set of the dual. The value of the objective function $\hat{z}_w$ corresponding to the solution set $\hat{\mathbf{w}}'$ is given by

$$\hat{z}_w = \hat{\mathbf{w}}'\mathbf{b} = \mathbf{w}'\mathbf{b} - \theta\beta^r\mathbf{b} = z_w - \theta x_{Br}. \tag{15.1.15}$$

Now selecting $\theta < 0$, and as $x_{Br} < 0$, then $\hat{z}_w < z_w$, i.e., $\hat{\mathbf{w}}'$ will produce smaller value of z_w.

Now $\hat{\mathbf{w}}'$ is a solution set of the dual constraints and selecting θ, a large negative number, $\hat{\mathbf{w}}'$ remains a solution of the dual and z_w can be made arbitrarily small which implies that the dual problem has an unbounded solution. Therefore, the only conclusion is that, the primal has no feasible solution and the theorem is proved.

15.1.1 General discussion

Let us now consider the case when $x_{Br} < 0$ and at least one $y_{rj} < 0$, we have,

$$\hat{\mathbf{w}}'\mathbf{a}_j = \mathbf{w}'\mathbf{a}_j - \theta\beta^r\mathbf{a}_j = \mathbf{w}'\mathbf{a}_j - \theta y_{rj}.$$

Now $\hat{\mathbf{w}}'$ will be a solution set of the dual constraints if

$$\begin{aligned} &\mathbf{w}'\mathbf{a}_j - \theta y_{rj} \geq c_j \text{ for all } j \\ \Rightarrow\ &\mathbf{w}'\mathbf{a}_j - c_j \geq \theta y_{rj}. \end{aligned} \tag{15.1.16}$$

To satisfy the condition, we require that

$$\theta \geq \frac{\mathbf{w}'\mathbf{a}_j - c_j}{y_{rj}}, \quad \text{since } y_{rj} < 0. \tag{15.1.17}$$

Now to decrease z_w to the utmost, θ must be (i) as large as possible in its absolute value and (ii) and satisfies the condition (15.1.17).

The two conditions will be satisfied simultaneously if we select θ in the manner

$$\theta = \max_j \frac{\mathbf{w}'\mathbf{a}_j - c_j}{y_{rj}}, \quad y_{rj} < 0. \tag{15.1.18}$$

Now

$$\mathbf{w}'\mathbf{a}_j - c_j = \mathbf{c}_B B^{-1}\mathbf{a}_j - c_j = z_j - c_j \geq 0. \tag{15.1.19}$$

Hence if θ be given by

$$\theta = \frac{z_k - c_k}{y_{rk}} = \max_j \frac{z_j - c_j}{y_{rj}}, \quad y_{rj} < 0 \tag{15.1.20}$$

then we have found a new solution to the dual given by

$$\left.\begin{aligned} &\hat{\mathbf{w}}'\beta_i = c_{Bi}, \quad i \neq r \\ &\hat{\mathbf{w}}'\beta_r \geq c_{Br} \\ &\hat{\mathbf{w}}'\mathbf{a}_j \geq c_j \text{ for all } \mathbf{a}_j \text{ not in the basis } j \neq k \\ &\hat{\mathbf{w}}'\mathbf{a}_k = \mathbf{w}'\mathbf{a}_k - \theta y_{rk} = c_k, \quad \theta \leq 0 \end{aligned}\right\} \tag{15.1.21}$$

and $\hat{z}_w = z_w - \theta x_{Br}$. Thus for $\theta < 0$, $\hat{z}_w < z_w$. Again as $y_{rk} \neq 0$, $\mathbf{a}_k$ together with the $\beta_i (i \neq r)$ constitute a new primal basis $\hat{B}$ in which the column r of the basis has been replaced by $\mathbf{a}_k$.

If $\hat{\mathbf{x}}_B = B^{-1}\mathbf{b}$ still contains one or more negative components, the entire process can be repeated in the same manner until we reach at the optimal feasible stage or we come to the conclusion that the problem has no feasible solution.

It is intersecting to note that the actual computation do not require the dual problem at all. Although the changes to be made were developed from the consideration of the dual, we see that actually they can be expressed entirely in terms of the primal.

15.1.2 Dual Simplex Algorithm

Step 1. We start with a basic solution to the primal with one or more $x_{Bi} < 0$ but with $z_j - c_j \geq 0$ for all j. If it is not possible, we cannot start the Dual Simplex Algorithm.

Step 2. Vector to be removed from the basis.

If $x_{Br} = \min_i x_{Bi}$, $x_{Bi} < 0$, then vector β_r is to be removed from the basis. If the value of r be not unique, select any vectors among the tie vectors.

Step 3. Vector to enter in the next basis.

The vector $\mathbf{a}_k$ to enter in the next basis will be determined from

$$\theta = \frac{z_k - c_k}{y_{rk}} = \max_j \frac{z_j - c_j}{y_{rj}}, \quad y_{rj} < 0. \tag{15.1.22}$$

The value of θ may not be negative in all cases. It may be zero. Thus z_w may not change and we have a case of dual degeneracy and cycling may theoretically occur in the dual simplex algorithm. We avoid such type of problems. y_{rk} is the key element.

Step 4. If a primal has an unbounded solution, its dual will have no feasible solution and therefore we can never find a solution to the dual problem. Hence if the primal has an unbounded solution, we can never find a basic solution to the primal with all $z_j - c_j \geq 0$. But as it is not an easy task to come to the above conclusion taking a problem, then by using dual simplex method, it is very difficult to say whether a problem has an unbounded solution or not.

Step 5. The rule of transformation of the tables are same as in the case of ordinary simplex method after the selection of the key element and in each table all $z_j - c_j$ remain non-negative for a maximization problem.

Note: Readers must note that in dual simplex method, one first determines the vector to leave the basis and then the vector to enter which is just opposite to the technique used in the case of ordinary simplex algorithm.

▶ **Example 15.1.1** *Solve, if possible, the linear programming problem by using dual simplex method.*

$$\textit{Minimize, } z = x_1 + 5x_2 \textit{ subject to}$$

$$\begin{aligned} 3x_1 + 4x_2 &\leq 6 \\ x_1 + 3x_2 &\geq 3\,, \quad x_1, x_2 \geq 0. \end{aligned}$$

Our problem is to maximize $(-x_1 - 5x_2)$ and Min $z = -$ Max $(-z)$ with the same solution set.

The constraints can be written as

$$\begin{aligned} 3x_1 + 4x_2 &\leq 6 \\ \text{and} \quad -x_1 - 3x_2 &\leq -3\,, \quad x_1, x_2 \geq 0. \end{aligned}$$

After adding two slack variables x_3 and x_4, one to each constraint, the constraints (converted equations) are

$$\begin{aligned} 3x_1 + 4x_2 + x_3 &\leq 6 \\ -x_1 - 3x_2 + x_4 &\leq -3\,, \quad x_1, x_2, x_3, x_4 \geq 0 \end{aligned}$$

$B = (\mathbf{a}_3, \mathbf{a}_4) = I_2$ constitute an initial unit matrix of order 2.

$$z' = -x_1 - 5x_2 = -x_1 - 5x_2 + 0x_3 + 0x_4.$$

Then

$$\mathbf{c}_B = (c_3, c_4) = (0, 0).$$

$$z_j - c_j = \mathbf{c}_B \mathbf{y}_j - c_j = -c_j.$$

Thus

$$\begin{aligned} z_1 - c_1 &= -(-1) = 1, \\ z_2 - c_2 &= -(-5) = 5, \\ z_3 - c_3 &= z_4 - c_4 = 0. \end{aligned}$$

$\mathbf{x}_B = B^{-1}\mathbf{b} = I_2\mathbf{b} = \mathbf{b} = [6 \; -3]$, at least one component of which is negative. Then it may be possible to solve the above problem by dual simplex method.

$$\mathbf{y}_i = B^{-1}\mathbf{a}_j = \mathbf{a}_j \; [j = 1, 2, 3, 4].$$

Initial table

Basis	$\mathbf{c}_B$	$\mathbf{c}$ / $\mathbf{b}$	-1 / $\mathbf{a}_1$	-5 / $\mathbf{a}_2$	0 / $\mathbf{a}_3$	0 / $\mathbf{a}_4$
$\mathbf{a}_3$	0	6	3	4	1	0
$\mathbf{a}_4^*$	0	-3^*	-1^*	-3	0	1
$z_j - c_j$		0	1	5	0	0

Step 1. Now $x_{B2} = -3 < 0$ is the only negative component of basic variables and at least one $y_{2j} < 0$. Thus the second vector $\mathbf{a}_4$ of the basis will leave the basis.

$$\max_j \left(\frac{z_j - c_j}{y_{2j}}, \, y_{2j} < 0 \right) = \max \left(\frac{z_1 - c_1}{y_{21}}, \, \frac{z_2 - c_2}{y_{22}} \right) = \max \left(\frac{1}{-1}, \frac{5}{-3} \right) = -1,$$

which occurs for $j = 1$. Then the first vector $\mathbf{a}_1$ will enter in the next basis and $y_{21} = -1$ is the key element.

2nd table

Basis	$\mathbf{c}_B$	$\mathbf{c}$ / $\mathbf{b}$	-1 / $\mathbf{a}_1$	-5 / $\mathbf{a}_2$	0 / $\mathbf{a}_3$	0 / $\mathbf{a}_4$
$\mathbf{a}_3^*$	0	-3^*	0	-5^*	1	3
$\mathbf{a}_1$	0	3	1	3	0	-1
$z_j - c_j$		-3	0	2	0	1

In the second table $x_{B1} = x_3 = -3 < 0$ with $y_{12} = -5 < 0$. Then we have to proceed further to reach at the optimal stage. As x_{B1} is the only negative component of the basic variables then the first vector $\mathbf{a}_3$ of the current basis will leave the basis.

Now

$$\max_j \left(\frac{z_j - c_j}{y_{1j}}, \, y_{1j} < 0 \right) = \left(\frac{z_2 - c_2}{y_{12}}, \, y_{12} < 0 \right) = \frac{2}{-5} = -\frac{2}{5}$$

which occurs for $j = 2$. Then the second vector $\mathbf{a}_2$ will enter in the next basis and $(\mathbf{a}_2, \mathbf{a}_1)$ will constitute the next basis.

3rd table

Basis	$\mathbf{c}_B$	$\mathbf{c}$ / $\mathbf{b}$	-1 / $\mathbf{a}_1$	-5 / $\mathbf{a}_2$	0 / $\mathbf{a}_3$	0 / $\mathbf{a}_4$
$\mathbf{a}_2$	-5	$\frac{3}{5}$	0	1	$-\frac{1}{5}$	$-\frac{3}{5}$
$\mathbf{a}_1$	-1	$\frac{6}{5}$	1	0	$\frac{3}{5}$	$\frac{4}{5}$
$z_j - c_j$		$-\frac{21}{5}$	0	0	$\frac{2}{5}$	$\frac{6}{5}$

All components of the basic solution are non-negative. Thus we have reached at the optimal stage.

$$\text{Max } (-x_1 - 5x_2) = -\tfrac{21}{5} \text{ at } x_1 = \tfrac{6}{5} \text{ and } x_2 = \tfrac{3}{5},$$

$$\text{i.e., Min } z = -(-\tfrac{21}{5}) = \tfrac{21}{5} \text{ at } x_1 = \tfrac{6}{5} \text{ and } x_2 = \tfrac{3}{5}.$$

Remark: Note that $z_j - c_j \geq 0$ for all j in all simplex tables.

▶ **Example 15.1.2** *Solve, if possible, by dual simplex method the following problem*

$$\textit{minimize,} \quad z = 6x_1 + 11x_2 \textit{ subject to}$$

$$x_1 + x_2 \geq 11$$
$$2x_1 + 5x_2 \geq 40, \quad x_1, x_2 \geq 0.$$

The problem is to maximize, $z^* = -6x_1 - 11x_2$ and min $z = -$ Max z^* $[z^* = -z]$ with the same solution set.

Considering all points discussed in Example (15.1.1) the initial table can be formed in the manner.

Initial table

		c	−6	−11	0	0
Basis	c_B	b	a_1	a_2	a_3	a_4
a_3	0	−11	−1	−1	1	0
a_4^*	0	−48	−2	-5^*	0	1
$z_j - c_j$		0	6	11	0	0

At least one component of $\mathbf{x}_B$ say x_{Bi} is negative and at least one y_{ij} is negative and all $z_j - c_j \geq 0$. Then it is possible to solve the problem by the dual simplex method.

$$\min_i (x_{Bi}, x_{Bi} < 0) = \min(-11, -48) = -48,$$

which occurs for $i = 2$, which implies $r = 2$. Then the second vector a_4 will leave the basis

$$\max_j \left(\frac{z_j - c_j}{y_{2j}}, \, y_{2j} < 0 \right) = \max \left(\frac{6}{-2}, \frac{11}{-5} \right) = \frac{11}{-5},$$

which occurs for $j = 2$. Then the second vector a_2 will enter in the next basis. The key element is $y_{22} = -5$. The rule of transformation is same as that of the ordinary simplex table.

2nd table

		c	−6	−11	0	0
Basis	c_B	b	a_1	a_2	a_3	a_4
a_3^*	0	−3	$-\frac{3}{5}^*$	0	1	$\frac{1}{5}$
a_2	−11	8	$\frac{2}{5}$	1	0	$-\frac{1}{5}$
$z_j - c_j$		−88	$\frac{46}{5}$	0	0	$\frac{11}{5}$

Here $x_{B1} = -3 < 0$ and $y_{11} = -3/5 < 0$. Then we proceed further to get an optimal solution.

Since only $x_{B1} = -3 < 0$ and $y_{11} = -3/5 < 0$ then the vector $\mathbf{a}_3$ will leave the basis and $\mathbf{a}_1$ is the vector to enter in the next basis. $y_{11} = -3/5$ is the key element.

3rd table

	c		−6	−11	0	0
Basis	$\mathbf{c}_B$	b	$\mathbf{a}_1$	$\mathbf{a}_2$	$\mathbf{a}_3$	$\mathbf{a}_4$
$\mathbf{a}_1$	−6	5	1	0	$-\frac{5}{3}$	$-\frac{1}{3}$
$\mathbf{a}_2$	−11	6	0	1	$\frac{2}{3}$	$-\frac{1}{3}$
$z_j - c_j$		−96	0	0	$\frac{8}{3}$	5

We reach at the optimal stage. Max $z^* = -96$ at $x_1 = 5$ and $x_2 = 6$, i.e., Min $z = -$ Max $z^* = -(96) = 96$ at $x_1 = 5, x_2 = 6$.

► **Example 15.1.3** *Minimize,* $z = 4x_1 + 2x_2$ *subject to*

$$\begin{aligned} 3x_1 + x_2 &\geq 27 \\ x_1 + x_2 &\geq 21 \\ x_1 + 2x_2 &\geq 30, \quad x_1, x_2 \geq 0. \end{aligned}$$

The problem is maximize $z^* = -4x_1 - 2x_2$ and Min $z = -$ Max z^*.

Introducing three surplus variables, one to each constraint, the converted equations are

$$\begin{aligned} 3x_1 + x_2 - x_3 \qquad &= 27 \\ x_1 + x_2 \qquad - x_4 \quad &= 21 \\ x_1 + 2x_2 \qquad - x_5 &= 30, \quad x_1, x_2 \geq 0. \end{aligned}$$

x_3, x_4, x_5 (all surplus variables) ≥ 0, and the objective function is

$$z^* = -4x_1 - 2x_2 + 0x_3 + 0x_4 + 0x_5.$$

If we select the basis matrix $B = (\mathbf{a}_3, \mathbf{a}_4, \mathbf{a}_5) = -I_3$ we have $B^{-1} = -I_3$ and $\mathbf{c}_B = (c_3, c_4, c_5) = (0, 0, 0)$. Now $\mathbf{c} = (-4, -2, 0, 0, 0)$. Then $z_j - c_j = \mathbf{c}_B\mathbf{y}_i - c_j = -c_j \geq 0$ for all j. Hence we can use dual simplex method to solve the problem.

$\mathbf{x}_B = B^{-1}\mathbf{b} = -I_3\mathbf{b} = [-27, -21, -30]$ at least one component of which is negative. Thus the initial basis is not an optimal basis. $\mathbf{y}_j = B^{-1}\mathbf{a}_j = -\mathbf{a}_j$ $[j = 1, 2, \cdots, 5]$.

With these data we can construct the initial Dual simplex table putting $\mathbf{x}_B$ under $\mathbf{b}$ and $\mathbf{y}_j$ under $\mathbf{a}_j$ as we have done in the case of ordinary simplex table.

Initial table

		c	−4	−2	0	0	0
Basis	$\mathbf{c}_B$	b	$\mathbf{a}_1$	$\mathbf{a}_2$	$\mathbf{a}_3$	$\mathbf{a}_4$	$\mathbf{a}_5$
$\mathbf{a}_3$	0	−27	−3	−1	1	0	0
$\mathbf{a}_4$	0	−21	−1	−1	0	1	0
$\mathbf{a}_5^*$	0	−30*	−1	−2*	0	0	1
$z_j - c_j$		0	4	2	0	0	0

$\min_i(x_{Bi}, x_{Bi} < 0) = \min(-27, -21, -30) = -30$ which occurs for $i = 3$ which implies $r = 3$. Thus the third vector $\mathbf{a}_5$ of the basis under the column "Basis" will leave the basis and third row is the key row.

$$\theta = \max_j \left(\frac{z_j - c_j}{y_{rj}}, \ y_{rj} < 0 \right) = \max \left(\frac{z_1 - c_1}{y_{31}}, \frac{z_2 - c_2}{y_{32}} \right) = \max \left(\frac{4}{-1}, \frac{2}{-2} \right) = -1,$$

which occur for $j = 2$ which implies that $\mathbf{a}_2$ is the vector to enter in the next basis and $y_{32} = -2$ is the key element. $\mathbf{a}_5$ will be replaced by $\mathbf{a}_2$ to form a new basis.

Second table: The rule transformation is exactly same as in the case of ordinary simplex method.

Basis	c_B	b	$\mathbf{a}_1$	$\mathbf{a}_2$	$\mathbf{a}_3$	$\mathbf{a}_4$	$\mathbf{a}_5$
$\mathbf{a}_3^*$		-12^*	$-\frac{5}{2}^*$	0	1	0	$-\frac{1}{2}$
$\mathbf{a}_4$		-6	$-\frac{1}{2}$	0	0	1	$-\frac{1}{2}$
$\mathbf{a}_2$		15	$\frac{1}{2}$	1	0	0	$-\frac{1}{2}$
$z_j - c_j$		-30	3	0	0	0	1

Note: From the second table and onwards, after determining the key row and the key element of the initial table it is better to calculate first, all the components of B.S. under **b** and if all of them be non-negative, calculate only the value of z which is the last element under the column **b** and we need not complete the table as we have got a clear indication that we have reached at the optimal stage with finite optimal value of the objective function. We even need not compute the values of $z_j - c_j$ $[j = 1, \cdots, n]$. Here in the second table all components of B.S. are not non-negative and at least one of them is negative. Thus we shall have to make the second table complete by using the same rule of transformation which has been done in ordinary simplex method.

$$\min_i(x_{Bi}, x_{Bi} < 0) = \min(x_{B1}, x_{B2}, \cdots) = \min(12, -6, \cdots)$$

which occur for $i = 1$ implies $r = 1$. Thus the first vector $\mathbf{a}_3$ under the column 'basis' will leave the basis and the first row will be the key row.

$$\max_j \left(\frac{z_j - c_j}{y_{1j}}, \ y_{1j} < 0 \right) = \max \left(\frac{z_1 - c_1}{y_{11}}, \frac{z_5 - c_5}{y_{15}} \right) = \max \left(\frac{3}{-\frac{5}{2}}, \frac{1}{-\frac{1}{2}} \right) = -\frac{6}{5},$$

which occurs for $j = 1$. Thus the first vector $\mathbf{a}_1$ will enter in the next basis, $y_{15} = -\frac{5}{2}$ is the key element, $\mathbf{a}_3$ will be replaced by $\mathbf{a}_1$ to form a new basis.

Third table

Basis	c_B	b	$\mathbf{a}_1$	$\mathbf{a}_2$	$\mathbf{a}_3$	$\mathbf{a}_4$	$\mathbf{a}_5$
$\mathbf{a}_1$		$\frac{24}{5}$	1	0	$-\frac{2}{5}$	0	$\frac{1}{5}$
$\mathbf{a}_4^*$		$-\frac{18}{5}^*$	0	0	$-\frac{1}{5}$	1	$-\frac{2}{5}^*$
$\mathbf{a}_2$		$\frac{63}{5}$	0	1	$\frac{1}{5}$	0	$-\frac{3}{5}$
$z_j - c_j$		$-44\frac{2}{5}$	0	0	$\frac{6}{5}$	0	$\frac{2}{5}$

$x_{B2} = -\frac{18}{5} < 0$ is the only negative component of B.S. and there exists at least one y_{2j} negative. Thus the solution is not optimal. We shall have to proceed further. The second vector $\mathbf{a}_4$ will leave the basis and the second row is the key row.

$$\max_j \left(\frac{z_j - c_j}{y_{2j}}, \, y_{2j} < 0 \right) = \max \left(\frac{z_3 - c_3}{y_{23}}, \frac{z_5 - c_5}{y_{25}} \right) = \max \left(\frac{\frac{6}{5}}{-\frac{1}{5}}, \frac{\frac{2}{5}}{-\frac{2}{5}} \right) = -1,$$

which occurs for $j = 5$. Thus the fifth vector $\mathbf{a}_5$ is the vector to enter in the next basis to replace $\mathbf{a}_4$ to form a new basis. $y_{25} = -\frac{2}{5}$ is the key element.

Fourth table

Basis	$\mathbf{c}_B$	$\mathbf{b}$	$\mathbf{a}_1$	$\mathbf{a}_2$	$\mathbf{a}_3$	$\mathbf{a}_4$	$\mathbf{a}_5$
$\mathbf{a}_1$		3					
$\mathbf{a}_5$		9					
$\mathbf{a}_2$		18					
$z_j - c_j$		-48	0	0	1	1	0

After calculating the components of B.F.S. we have seen that all components are non-negative and thus the solution is B.F.S. and hence we reach at the optimal stage. Now calculate the value of the objective function. We need not compute further. Even, we need not calculate all $z_j - c_j$ $[j = 1, 2, \cdots, 5]$ as we know it clearly that all of them will be non-negative. It will be helpful only to know whether there exists any alternative optimal solution and also to find the dual optimal solution. If we ignore those points, we can say that Max $z^* = -48$ at $x_1 = 3$, $x_5 = 9$, $x_2 = 18$, i.e., for $x_1 = 3$ and $x_2 = 18$, the original problem attains its minimum value and Min $z = -$ Max $z^* = -(-48) = 48$.

Dual optimal solution is $w_1 = 1$, $w_2 = 1$, $w_3 = 0$ which can be determined from the $z_j - c_j$ $[j = 3, 4, 5]$. The problem has no alternative optimal solution as none of $z_j - c_j$ corresponding to non-basic variables is zero.

► **Example 15.1.4** *Using dual simplex method, prove that the following problem has no feasible solution. Verify this result by graphical method.*

Minimize, $z = 7x_1 + 3x_2$ *subject to*

$$\begin{aligned} x_1 - 3x_2 &\geq 1 \\ x_1 + x_2 &\geq 2 \\ -2x_1 + 2x_2 &\geq 1 \qquad x_1, x_2 \geq 0. \end{aligned}$$

Our problem is to Maximize $-7x_1 - 3x_2$ and Min $(7x_1 + 3x_2)$

$$= -\operatorname{Max}(-7x_1 - 3x_2).$$

After the introduction of three surplus variables one to each constraint, the constraints are

$$\begin{aligned} x_1 - 3x_2 - x_3 \qquad\qquad &= 1 \\ x_1 + x_2 \qquad - x_4 \qquad &= 2 \\ -2x_1 + 2x_2 \qquad\qquad - x_5 &= 1, \quad x_j \geq 0, \; j = 1, 2, \cdots, 5. \end{aligned}$$

Objective function $= -7x_1 - 3x_2 + 0x_3 + 0x_4 + 0x_5$.

Now taking

$$B = (\mathbf{a}_3, \mathbf{a}_4, \mathbf{a}_5) = \begin{bmatrix} -1 & 0 & 0 \\ 0 & -1 & 0 \\ 0 & 0 & -1 \end{bmatrix} = -I_3$$

and

$$\mathbf{c}_B = (0, 0, 0),\ z_j - c_j = \mathbf{c}_B \mathbf{y}_j - c_j = 0,\ \mathbf{y}_j - c_j = -c_j.$$

Thus

$$z_1 - c_1 = -c_1 = -(-7) = 7,$$
$$z_2 - c_2 = -(-c_2) = 3$$
$$\text{and } z_3 - c_3 = z_4 - c_4 = 0.$$

Thus with basis $-I_3$, all $z_j - c_j \geq 0$. Hence we can start the problem by using dual simplex method.

$$\mathbf{y}_j = B^{-1}\mathbf{a}_j = -I_3\mathbf{a}_j = -\mathbf{a}_j, \quad \mathbf{x}_B = B^{-1}\mathbf{b} = -I_3\mathbf{b} = -\mathbf{b}.$$

With these data, we can start the simplex tables.

			−7	−3	0	0	0
Basis	$\mathbf{c}_B$	$\mathbf{b}$	$\mathbf{a}_1$	$\mathbf{a}_2$	$\mathbf{a}_3$	$\mathbf{a}_4$	$\mathbf{a}_5$
$\mathbf{a}_3$	−	−1	−1	3	1	0	0
$\mathbf{a}_4^*$	0	−2*	−1*	−1	0	1	0
$\mathbf{a}_5$	0	−1	2	−2	0	0	1
$z_j - c_j$		0	7	3*	0	0	0
$\mathbf{a}_3^*$	0	−7*	−4*	0	1	3	0
$\mathbf{a}_2$	−3	2	1	1	0	−1	0
$\mathbf{a}_5$	0	3	4	0	0	−2	1
$z_j - c_j$		−6	4	0	0	3	0
$\mathbf{a}_1$	−7	$\frac{7}{4}$	1	0	$-\frac{1}{4}$	$-\frac{3}{4}$	0
$\mathbf{a}_2$	−3	$\frac{1}{4}$	0	1	$\frac{1}{4}$	$-\frac{1}{4}$	0
$\mathbf{a}_5$	0	−4	0	0	0	1	1
$z_j - c_j$		−	0	0	1	6	0

One basic variable, $x_5 = -4 < 0$ in the last table, then $\mathbf{a}_5$ will leave the basis. Therefore, the third row is the key row. But in that row there is no negative element, i.e., there exists no $y_{3j} < 0$. Hence it is not possible to find the key column, i.e., the vector to enter in the next basis. Hence it is not possible to find any feasible solution and the only conclusion is that the problem has no feasible solution.

Exercise 15

1. Discuss the working principle of "Dual simplex method" in solving a linear programming problem. Why is the method called as dual simplex method?

2. Is it possible to solve each and every linear programming problem with the help of "dual simplex method"? If not, why?

3. Prove that if $\mathbf{w}' = \mathbf{c}_B B^{-1}$, a solution (feasible or not) to the dual where B is a basis matrix of the primal, Maximize, $z = \mathbf{cx}$, $A\mathbf{x} = \mathbf{b}$, $\mathbf{x} \geq \mathbf{0}$, be not an optimal solution of the dual, then at least one component x_{Bi} of the basic solution $\mathbf{x}_B = B^{-1}\mathbf{b}$ of the primal be negative.

4. If for a basis B of the primal, Maximize, $z = \mathbf{cx}$, $A\mathbf{x} = \mathbf{b}$, $\mathbf{x} \geq \mathbf{0}$, at least one x_{Bi} of $\mathbf{x}_B = B^{-1}\mathbf{b}$ be negative with all $y_{ij} \geq 0$ and $\mathbf{w}' = \mathbf{c}_B B^{-1}$ be a solution of the dual, then prove that the primal will have no feasible solution.

5. Write down the working rule of the vector to leave the current basis and the vector to enter in the next basis in dual simplex algorithm.

6. Solve, if possible, the following problems with the help of dual simplex method.

   (a) Minimize, $z = 4x_1 + 3x_2$ subject to

   $$x_1 + 2x_2 \geq 8$$
   $$3x_1 + 2x_2 \geq 12\,, \quad x_1, x_2 \geq 0.$$

   (b) Maximize, $z = -x_1 - x_2$ subject to

   $$2x_1 + x_2 \geq 4$$
   $$x_1 + 7x_2 \geq 7\,, \quad x_1, x_2 \geq 0.$$

   (c) Maximize, $z = 2x_1 - 3x_2$ subject to

   $$-x_1 + x_2 \geq -2$$
   $$-5x_1 - 4x_2 \geq -46$$
   $$7x_1 + 2x_2 \geq 32\,, \quad x_1, x_2 \geq 0.$$

   (d) Minimize, $z = 2x_1 + 3x_2$ subject to

   $$2x_1 + 7x_2 \geq 22$$
   $$x_1 + x_2 \geq 6$$
   $$5x_1 + x_2 \geq 10\,, \quad x_1, x_2 \geq 0.$$

   (e) Maximize, $z = -2x_1 - 3x_2 - x_3$ subject to

   $$2x_1 + x_2 + 2x_3 \geq 3$$
   $$3x_1 + 2x_2 + x_3 \geq 4\,, \quad x_1, x_2 \geq 0.$$

   [C.U. M.Sc. '80]

(f) Maximize, $z = -3x_1 - x_2$ subject to

$$\begin{aligned} x_1 + \ \ x_2 &\geq 1 \\ 2x_1 + 3x_2 &\geq 2, \ x_1, x_2 \geq 0. \end{aligned}$$

[Ranchi, M.Sc. (Math) '83]

(g) Maximize, $-2x_1 - x_2$ subject to

$$\begin{aligned} 3x_1 + \ \ x_2 &\geq 3 \\ 4x_1 + 3x_2 &\geq 6 \\ x_1 + 2x_2 &\geq 3, \ x_1, x_2 \geq 0. \end{aligned}$$

[Madurai, B.Sc. '81]

(h) Minimize, $4x_1 + 7x_2$ subject to

$$\begin{aligned} 2x_1 + 5x_2 &\leq 40 \\ x_1 + \ \ x_2 &\leq 11 \\ x_2 &\geq 4, \ x_1, x_2 \geq 0. \end{aligned}$$

(i) Maximize, $z = -3x_1 - 2x_2$ subject to

$$\begin{aligned} x_1 + x_2 &\geq 1 \\ x_1 + x_2 &\leq 7 \\ x_1 + x_2 &\geq 10 \\ x_2 &\leq 3, \ x_1, x_2 \geq 0. \end{aligned}$$

[Kalyani, '86]

(j) Minimize, $z = 2x_1 + 2x_2$ subject to

$$\begin{aligned} 2x_1 + 4x_2 &\geq 1 \\ x_1 + 2x_2 &\geq 1 \\ 2x_1 + \ \ x_2 &\geq 1, \text{ and } x_1, x_2 \geq 0. \end{aligned}$$

(k) Minimize, $z = 5x_1 + 6x_2$ subject to

$$\begin{aligned} 2x_1 + 3x_2 &\geq 6 \\ 5x_1 + 2x_2 &\geq 10 \\ 5x_1 + 6x_2 &\leq 30, \ x_1, x_2 \geq 0. \end{aligned}$$

(l) Maximize, $z = 6x_1 + 7x_2 + 3x_3 + 5x_4$ subject to

$$\begin{aligned} 5x_1 + 6x_2 - 3x_3 + 4x_4 &\geq 12 \\ x_2 + 5x_3 - 6x_4 &\geq 10 \\ 2x_1 + 5x_2 + \ \ x_3 + \ \ x_4 &\geq 8, \ x_1, x_2, x_3, x_4 \geq 0. \end{aligned}$$

[Madurai, B.Sc. (App. Math.) '85]

Answers

6. (a) min $z = 17$ at $x_1 = 2$, $x_2 = 3$; (b) max $z = -\frac{31}{13}$ at $x_1 = \frac{21}{13}$, $x_2 = \frac{10}{13}$; (c) not possible; (d) min $z = 14$ at $x_1 = 4$, $x_2 = 2$; (e) max $z = -\frac{11}{4}$ at $x_1 = \frac{5}{4}$, $x_2 = 0$, $x_3 = \frac{1}{4}$; (f) max $z = -1$ at $x_1 = 0$, $x_2 = 1$; (g) max $z = -\frac{12}{5}$ at $x_1 = \frac{3}{5}$, $x_2 = \frac{6}{5}$; (h) Minimum value is 32 at $x_1 = 0$, $x_2 = 8$; (i) max $z = -18$ at $x_1 = 4$, $x_2 = 3$; (j) min $z = \frac{4}{3}$ at $x_1 = \frac{1}{3}$, $x_2 = \frac{1}{3}$; (k) min $z = \frac{150}{11}$ at $x_1 = \frac{18}{11}$, $x_2 = \frac{10}{11}$; (l) min $z = \frac{258}{11}$ at $x_1 = 0$, $x_2 = \frac{30}{11}$, $x_3 = \frac{16}{11}$ and $x_4 = 0$.

Chapter 16

Sensitivity Analysis

16.1 Introduction

In Linear Programming, Sensitivity Analysis plays a vital role to discuss the nature of a problem and it gives a direct test of sensitivity of a particular problem due to the variation of a component of cost or requirement vector etc. It is in general a post-optimality test. It determines the range of a component of the cost or requirement vector when all other components remain unchanged so that the optimal solution remains unaffected, i.e., the optimality remains undisturbed. Such type of change is called a discrete change. It gives a clear idea how the nature of a particular problem changes due to the addition or deletion of a single variable and due to the removal or addition of a constraint or due to the change of an element a_{ij} of the coefficient matrix. As it is possible to discuss the effect on a problem due to the variations stated above, it is extremely helpful from the commercial point of a view and it is possible to rectify or adjust a problem suitably so that it will enhance the prospect of a problem from the economical points of view. For example, as the procedure is able to determine the extremum of a component b_i of $\mathbf{b}$ so that the optimal basis remains unchanged, this technique can be suitably used to get the benefit from the knowledge of shadow prices or accounting prices to get an overall maximum profit etc.

16.2 Variation of c_j

There are two different cases to discuss: (i) c_j is the cost component corresponding to a non-basic variable. (ii) c_j is the cost component corresponding to an optimal basic variable.

Case (i). At the optimum simplex table all $z_j - c_j \geq 0$ [maximization problem]

Let Δc_j be the amount to the added to c_j corresponding to a non-basic variable x_j. Now for that discrete change of the cost component c_j the simplex table remains optimal provided.

$z_j - (c_j + \Delta c_j) \geq 0$ [As z_j does out change due to the change of c_j]

$$\Rightarrow z_j - c_j \geq \Delta c_j. \tag{16.2.1}$$

From (16.2.1) we can say that there is no lower bound of Δc_j and any change of c_j does not change the value of the objective function since $x_j = 0$ [non-basic variable].

Case (ii). Let $x_k = x_{Br}$ be the rth component of basic variable $[r = 1, 2, \cdots, m]$ and the cost component $c_k = c_{Br}$ corresponding to x_k changes to $c_{Br} + \Delta c_{Br}$. Now the change of c_k affects all z_j, for all j in the basis. Now due to change

$$\begin{aligned} z_j^* - c_j^* &= \mathbf{c}_B \mathbf{y}_j - c_j \quad [\text{for all } j \notin B] \\ &= \sum_{\substack{i=i \\ i \neq r}}^{m} c_{Bi} y_{ij} - c_j + (c_{Br} + \Delta c_{Br}) y_{rj} \\ &= \sum_{i=i}^{m} c_{Bi} y_{ij} - c_j + \Delta c_{Br} y_{rj} \\ &= z_j - c_j + \Delta c_{Br} y_{rj}. \end{aligned}$$

Now to keep the solution optimal, we must have all $z_j^* - c_j^* \geq 0$ for all $j \notin B$.
Thus $z_j - c_j + \Delta c_{Br} y_{rj} \geq 0$.
If $y_{rj} = 0$ for some j, then $z_j - c_j + \Delta c_{Br} y_{rj} = z_j - c_j \geq 0$.

$$\left.\begin{aligned} \text{For } y_{rj} < 0,\ \Delta c_{Br} \leq \frac{(z_j - c_j)}{y_{rj}} \\ \text{For } y_{rj} > 0,\ \Delta c_{Br} \geq \frac{(z_j - c_j)}{y_{rj}} \end{aligned}\right\}. \tag{16.2.2}$$

Combining these two conditions given in (16.2.2) we can say that the optimal basis remains unaffected if $\Delta c_k = \Delta c_{Br}$ lies in the interval given below

$$\max_j \left\{ \frac{-(z_j - c_j)}{y_{rj}},\ y_{rj} > 0 \right\} \leq \Delta c_{Br} = \Delta c_k \leq \min_j \left\{ \frac{-(z_j - c_j)}{y_{rj}},\ y_{rj} < 0 \right\} \tag{16.2.3}$$

for all $j \notin B$. If no $y_{rj} > 0$, there is no upper bound of Δc_k and if no $y_{rj} < 0$, there is no lower bound of Δc_k. But here the value of the objective function changes by $\Delta c_{Br} x_{Br} = \Delta c_k x_k$.

16.3 Variation of b_i

Let us now discuss the optimality of a problem when a component of $\mathbf{b}$, say b_k changes to $b_k + \Delta b_k$ $[k = 1, 2, \cdots, m]$.

For the requirement vector $\mathbf{b}$, let the optimal basis be B and $\mathbf{x}_B = B^{-1}\mathbf{b} \geq 0$. For the change of Δb_k of the component b_k, the new basic solution corresponding to the basis B is

$$\begin{aligned} \mathbf{x}_B^* &= B^{-1}[b_1, b_2, \cdots, b_k + \Delta b_k, \cdots, b_m] \\ &= [x_{Bi} + b_{ik}\Delta b_k] \quad [i = 1, 2, \cdots, m] \end{aligned} \tag{16.3.1}$$

for all i in the basis and where b_{ik} is the element of the ith row and kth column of B^{-1}.

Now we shall have to discuss the following cases so that new B.S. $\mathbf{x}_B^*$ remains feasible.

(i) If $b_{ik} = 0$ for some i, then $x_{Bi}^* = x_{Bi} \geq 0$.

(ii) If $b_{ik} > 0$ for some i, then $x_{Bi} + b_{ik}\Delta b_k \geq 0 \Rightarrow \Delta b_k \geq \dfrac{-x_{Bi}}{b_{ik}}$ [$\Delta b_k \leq 0$ here]

(iii) If $b_{ik} < 0$, or some i then $x_{Bi} + b_{ik}\Delta b_k \geq 0 \Rightarrow \Delta b_l \leq \dfrac{-x_{Bi}}{b_{ik}}$ [$\Delta b_k \geq 0$ here].

Combining these two, we can say that the optimality remains unaffected if Δb_k lies in the interval

$$\max_i \left(\frac{-x_{Bi}}{b_{ik}},\ b_{ik} > 0 \right) \leq \Delta b_k \leq \min \left(\frac{-x_{Bi}}{b_{ik}},\ b_{ik} < 0 \right). \tag{16.3.2}$$

Here also, if no $b_{ik} > 0$, there is no upper bound and if no $b_{ik} < 0$, there is no lower bound.

Again it can be easily verified that if for $i = r$, $\max \left(\dfrac{-x_{Bi}}{b_{ik}},\ b_{ik} > 0 \right)$ occurs then x_{Br}^* will be zero for the change Δb_k of b_k. Similarly if for $i = l$, $\min\limits_i \left(\dfrac{-x_{Bi}}{b_{ik}}, b_{ik} < 0 \right)$ occurs then x_{Bl}^* will be zero which has been shown in the solved example.

Due to the new B.F.S. $\mathbf{x}_B^*$, the change in the value of the objective function is given by $\sum\limits_{i=1}^{m} c_{Bi} b_{ik} \Delta b_k$.

There are other types of 'Sensitivity Analysis' due to change of a_{ij}, addition or deletion of a single constraint, addition or deletion of a single variable, discussions of which are not so vital.

▶ **Example 16.3.1** *Given the following L.P.P.*

Maximize, $z = x_1 + 4x_2 - 2x_3 + 3x_4 + x_5$ subject to

$$\begin{aligned} x_1 - 3x_2 + x_3 + 2x_4 + 6x_5 &\leq 3 \\ 2x_1 + x_2 \qquad + 3x_4 + 2x_5 &\leq 6 \\ 4x_1 + x_2 \qquad - x_4 + x_5 &\leq 2\,, \quad x_j \geq 0,\ j = 1, 2, \cdots, 5. \end{aligned}$$

Find the ranges over which c_1, c_3, c_4 and b_3 can be changed so that the optimality of the current solution remains undisturbed [Assume that when c_1 changes, all other quantities remain unchanged etc.].

The optimal simplex table is

	c		1	4	−2	3	−1	0	0	0
Basis	c_B	b	$\mathbf{a}_1$	$\mathbf{a}_2$	$\mathbf{a}_3$	$\mathbf{a}_4$	$\mathbf{a}_5$	$\mathbf{a}_6$	$\mathbf{a}_7$	$\mathbf{a}_8$
$\mathbf{a}_6$	0	10	$\frac{25}{2}$	0	1	0	$\frac{37}{4}$	1	$\frac{1}{4}$	$\frac{11}{4}$
$\mathbf{a}_4$	3	1	$-\frac{1}{2}$	0	0	1	$\frac{1}{4}$	0	$\frac{1}{4}$	$-\frac{1}{4}$
$\mathbf{a}_2$	4	3	$\frac{7}{2}$	1	0	0	$\frac{5}{4}$	0	$\frac{1}{4}$	$\frac{3}{4}$
$z_j - c_j$		15	$\frac{23}{2}$	0	2	0	$\frac{27}{4}$	0	$\frac{7}{4}$	$\frac{9}{4}$

where $\mathbf{a}_6, \mathbf{a}_7$ and $\mathbf{a}_8$ are slack vectors.

The optimal basis

$$B = (\mathbf{a}_6, \mathbf{a}_4, \mathbf{a}_2) = \begin{bmatrix} 1 & 2 & -3 \\ 0 & 3 & 1 \\ 0 & -1 & 1 \end{bmatrix}. \tag{16.3.3}$$

Thus $\mathbf{c}_B = (c_6, c_4, c_2) = (0, 3, 4)$; $c_1, c_3 \notin \mathbf{c}_B$ and $c_4 = c_{B2} \in \mathbf{c}_B$.

Then for c_1 and c_3 the ranges are

$$\left.\begin{array}{l} \Delta c_1 \le z_1 - c_1, \quad \text{or,} \quad \Delta c_1 \le \frac{23}{2} \\ \Delta c_3 \le z_3 - c_3, \quad \text{or,} \quad \Delta c_3 \le 2 \end{array}\right\}. \tag{16.3.4}$$

$c_4 = c_{B2}$, the cost component corresponding to the second basic variable.

Hence Δc_4 lies in the interval given by the formula.

$$\max_j \left\{ \frac{-(z_j - c_j)}{y_{2j}},\ y_{2j} > 0 \right\} \le \Delta c_4 \le \min_j \left\{ \frac{-(z_j - c_j)}{y_{2j}},\ y_{2j} < 0 \right\}$$

$$\text{or,} \quad \max\left\{ -\frac{27}{4} \Big/ \frac{1}{4},\ -\frac{7}{4} \Big/ \frac{1}{4} \right\} \le \Delta c_4 \le \min\left\{ -\frac{23}{2} \Big/ -\frac{1}{2},\ -\frac{9}{4} \Big/ -\frac{1}{4} \right\},$$

$$\text{or,} \quad -7 \le \Delta c_4 \le 9 \tag{16.3.5}$$

$\mathbf{a}_6, \mathbf{a}_7$ and $\mathbf{a}_8$ constitute the initial unit basis. Hence the vectors $\mathbf{y}_6, \mathbf{y}_7, \mathbf{y}_8$ under the vectors $\mathbf{a}_6, \mathbf{a}_7, \mathbf{a}_8$ in the final simplex table constitute the optimal basis inverse and hence

$$B^{-1} = \begin{bmatrix} 1 & \frac{1}{4} & \frac{11}{4} \\ 0 & \frac{1}{4} & -\frac{1}{4} \\ 0 & \frac{1}{4} & \frac{3}{4} \end{bmatrix}. \tag{16.3.6}$$

Note: Verify this result by finding the inverse of B given in (16.3.3) by matrix inverse method.

The range of b_3 be such that the optimal basis remains unchanged is given by

$$\max_i \left(\frac{-x_{Bi}}{b_{i3}},\ b_{i3} > 0 \right) \le \Delta b_3 \le \min_i \left(\frac{-x_{Bi}}{b_{i3}},\ b_{i3} < 0 \right),$$

where b_{i3} is the element in the ith row and third column of B^{-1}.

$$\text{or,}\quad \max\left(-10\Big/\frac{11}{4}, -3\Big/\frac{3}{4}\right) \le \Delta b_3 \le \min\left(-1\Big/-\frac{1}{4}\right),$$

$$\text{or,}\quad -\frac{40}{11} \le \Delta b_3 \le 4,$$

i.e., for $-\frac{18}{11} \le b_3 \le 6$, the optimal solution remains unchanged.

► **Example 16.3.2** ***Discuss the effect of discrete changes in the requirements for the following L.P.P.***

Maximize, $z = x_1 - x_2 + 3x_3$ subject to

$$\begin{aligned} x_1 + x_2 + x_3 &\le 10 \\ 2x_1 - x_3 &\le 2 \\ 2x_1 - 2x_2 + 3x_3 &\le 0, \quad x_1, x_2, x_3 \ge 0 \end{aligned}$$

so that the optimality of this problem remains undisturbed.

The components of requirement vector $\mathbf{b} = [b_1, b_2, b_3] = [10, 2, 0]$.

The optimum simplex table is given by

Basis	c_B	b	a_1	a_2	a_3	$a_4(e_1)$	$a_5(e_2)$	$a_6(e_3)$
a_2	−1	6	$\frac{1}{5}$	1	0	$\frac{3}{5}$	0	$-\frac{1}{5}$
a_5	0	6	$\frac{14}{5}$	0	0	$\frac{2}{5}$	1	$\frac{1}{5}$
a_3	3	4	$\frac{4}{5}$	0	1	$\frac{2}{5}$	0	$\frac{1}{5}$
$z_j - c_j$		6	$\frac{6}{5}$	0	0	$\frac{3}{5}$	0	$\frac{4}{5}$

The optimal basis

$$B = (\mathbf{a}_2, \mathbf{a}_5, \mathbf{a}_3) = \begin{bmatrix} 1 & 0 & 1 \\ 0 & 1 & -1 \\ -2 & 0 & 3 \end{bmatrix}.$$

$\mathbf{a}_4(\mathbf{e}_1)$, $\mathbf{a}_5(\mathbf{e}_2)$ and $\mathbf{a}_6(\mathbf{e}_3)$ constitute the initial unit basis which are three slack vectors. Then $(\mathbf{y}_4, \mathbf{y}_5, \mathbf{y}_6)$ under $\mathbf{a}_4, \mathbf{a}_5$ and $\mathbf{a}_6$ respectively of the final table constitute final basis inverse which is given by

$$B^{-1} = \begin{array}{c} \begin{matrix} \mathbf{y}_4 & \mathbf{y}_5 & \mathbf{y}_6 \end{matrix} \\ \begin{bmatrix} \frac{3}{5} & 0 & -\frac{1}{5} \\ \frac{2}{5} & 1 & \frac{1}{5} \\ \frac{2}{5} & 0 & \frac{1}{5} \end{bmatrix} \end{array}$$

and the optimal basic solution

$$\mathbf{x}_B = \begin{bmatrix} 6 \\ 6 \\ 4 \end{bmatrix}.$$

Hence the individual range of changes of b_1, b_2 and b_3 are given by

$$\max\left(-6\Big/\frac{3}{5}, -6\Big/\frac{2}{5}, -4\Big/\frac{2}{5}\right) \le \Delta b_1 \quad [\because k = 1]$$

or, $-10 \le \Delta b_1$ [There is no upper limit]

$$\max\left(-\frac{6}{1}\right) \le \Delta b_2$$

or, $-6 \le \Delta b_2$

$$\max\left(-6\Big/\frac{1}{5}, -4\Big/\frac{1}{5}\right) \le \Delta b_3 \le \min\left(\frac{-6}{-\frac{1}{5}}\right) \quad [\because k = 3]$$

or, $-20 \le \Delta b_3 \le 30$ [maximum occurs for $i = 3$ and minimum occurs for $i = 1$]

The interpretation is that b_1, b_2 remain unchanged then the optimal solution remains optimal if b_3 lies in the interval

$$(0 - 20 \le b_3 \le 0 + 30) = (-20 \le b_3 \le 30).$$

For $b_3 = -20$, the B.F.S. corresponding to the basis B is $B^{-1}[10, 2, -20] = [10, 2, 0]$ which is a degenerate B.F.S. with third component $x_{B3} = x_3 = 0$.

For $b_3 = 30$, the B.F.S. corresponding to the basis B is $B^{-1}[10, 2, 30] = [0, 12, 10]$ which is also a degenerate B.F.S. with the first component $x_{B1} = x_2 = 0$.

16.4 Introduction of additional constraint

Sometime an additional constraint is required to introduce for some reasons or other. But this introduction is not possible always. The working principle is shown in the Chapter "Integer programming". The interested students should go through very carefully, the solved problem in that chapter.

Exercise 16

1. Write a short note on 'sesitivity analysis'. Give example of the economical use of the sensitivity analysis.

2. Describe the working rule for determination of the range of variation of the discrete change of the components of **c** such that an optimal solution remains undisturbed.

   (a) When the component c_j is the coefficient of a non-basic variables.

   (b) When the component c_j is the coefficient of a basic variable x_j belonging to the optimal basis.

3. Describe the working rule for determination of the range of a component of the requirement vector **b** such that optimality remains undisturbed. Also prove that for the extreme values of the component, the B.F.S. will be a degenerate one.

4. Solve the following L.P.P. by simplex method:

Maximize, $z = 5x_1 + 4x_2 + x_3$ subject to

$$\begin{aligned} 6x_1 + x_2 + 2x_3 &\le 12 \\ 8x_1 + 2x_2 + x_3 &\le 30 \\ 4x_1 + x_2 - 2x_3 &\le 16, \quad x_1, x_2, x_3 \ge 0. \end{aligned}$$

(a) Find the range of c_2, the coefficient of x_2 in the objective function, such that the optimality remains undisturbed. [Assume that c_1 and c_3 remain unchanged]

(b) Find the ranges of b_2 and b_3 so that optimal basis remains same. [assume that when b_2 changes, b_1 and b_3 remain unaffected etc.] Prove that there exist no definite upper bounds of Δb_2 and Δb_3. Find the value of the degenerate B.F.S. for the extreme lower value of Δb_2.

5. From the optimal simplex table of the following L.P.P.

Maximize, $z = 2x_1 - x_2 + 3x_3$ subject to

$$\begin{aligned} 3x_1 + x_2 - 2x_3 &\le 6 \\ 2x_1 + 5x_2 + x_3 &\le 14 \\ x_1 + 4x_2 + 2x_3 &\le 8, \quad x_1, x_2, x_3 \ge 0. \end{aligned}$$

Find the ranges of c_2, c_3 and b_3 such that the optimal basis remains unchanged.

6. Solve the L.P.P.

Maximize, $z = x_1 + x_2 + 3x_3$ subject to

$$\begin{aligned} x_1 + 2x_2 - x_3 &\le 10 \\ 3x_2 + 2x_3 &\le 8 \\ x_2 + 3x_3 &\le 15, \quad x_1, x_2, x_3 \ge 0. \end{aligned}$$

Taking the initial basis as $(\mathbf{a}_1, \mathbf{a}_4, \mathbf{a}_5)$ where $\mathbf{a}_4 = \mathbf{e}_2$ and $\mathbf{a}_5 = \mathbf{e}_3$ are unit slack vectors.

From the optimal simplex table, find the ranges of c_1, c_3 and b_2 such that the optimality remains unchanged. [discrete change]

7. Solve the L.P.P. by simplex method.

Maximize, $z = 2x_1 + 5x_2$ subject to

$$\begin{aligned} -2x_1 + x_2 &\le 0 \\ x_1 + 3x_2 &\le 14 \\ x_1 + x_2 &\le 8, \quad x_1, x_2 \ge 0, \end{aligned}$$

where the initial B.F.S. is a degenerate one.

Find the range of b_3 such that optimality be not violated, and find the B.F.S. for the extreme values of b_3.

8. Following is the final optimal table for an L.P.P. Assume that originally the identity matrix was under $\mathbf{a}_3$ and $\mathbf{a}_4$ (Maximization).

	c		2	2	0	0
c_B	Vectors in the basis	**b**	$\mathbf{a}_1$	$\mathbf{a}_2$	$\mathbf{a}_3$	$\mathbf{a}_4$
2	$\mathbf{a}_1$	5	1	0	$\frac{1}{2}$	$-\frac{1}{2}$
2	$\mathbf{a}_2$	4	0	1	$-\frac{1}{2}$	$\frac{3}{2}$
	$z_j - c_j$	18	0	0	0	2

Suppose c_2 was equal to 3 instead of 2. Would you still have the optimal solution? Conclude it by calculating $z_3 - c_3$ with new cost components. Verify that result by using the theory of sensitivity analysis.

Answers

4. (a) $-\frac{19}{6} \leq \Delta c_2$, (b) $-6 \leq \Delta b_3, -4 \leq \Delta b_3$. Degenerate B.F.S. is $[x_2 = 12, x_5 = 0, x_6 = 4]$ other variables non-basic, 5. $\Delta c_2 \leq \frac{61}{8}, -\frac{13}{3} \leq \Delta c_3 \leq 1$.

6. $-6 \leq \Delta b_3 \leq \frac{38}{7}, -1 \leq \Delta c_1, -4 \leq \Delta c_2, -8 \leq \Delta b_2 \leq 2$.

7. $-2 \leq \Delta b_3 \leq 6, 6 \leq b_3 \leq 14, [4, 2, 0], [0, 14, 28]$.

8. $z_3 - c_3 = -\frac{1}{2} < 0$. Hence the solution will not be optimal.

Now using 'sensitivity analysis' we get $-\frac{4}{3} \leq \Delta c_2 \leq 0$. Thus the upper limit of c_2 so that problem remains optimal is $2 + 0 = 2$, but here the value of $c_2 = 3 > 2$. Hence the problem no longer remains optimal.

Chapter 17

Parametric Programming

17.1 Introduction

Parametric programming is a more sensitive type of analysis of linear programming problem where we are able to discuss the nature of the changes that takes place due to the variation of all components of **b** or **c** simultaneously instead of a particular components of **b** or **c** etc. For simplicity we shall confine our discussion only for linear variation of **b** and **c** in terms of a scalar quantity θ or α (which is called a parameter and that is why such type of programming problem is called parametric programming.

17.2 Variation of c (parametric variation)

Let **c** be the cost vector of the original problem [maximization]. Now we assume that **c** changes to $\mathbf{c}^*$, where $\mathbf{c}^*$ can be written as

$$\mathbf{c}^* = \mathbf{c} + \theta \mathbf{c}^0, \tag{17.2.1}$$

where $\mathbf{c}^*$ is a linear function of the parameter θ keeping **c** as a constant vector (unchanged).

For example, if $z = 2x_1 + 3x_2 - x_3 + 0x_4$, then $\mathbf{c} = (2, 3, -1, 0)$.

Now if the new objective function be

$$z^* = (2 - \theta)x_1 + (2\theta + 3)x_2 + (3\theta - 1)x_3 + \theta x_4$$

then

$$\begin{aligned} \mathbf{c}^* &= (2 - \theta, 2\theta + 3, 3\theta - 1, \theta) \\ &= (2, 3, -1, 0) + \theta(-1, 2, 3, 1) \\ &= \mathbf{c} + \theta \mathbf{c}^0 \text{ where } \mathbf{c}^0 = (-1, 2, 3, 1) \text{ etc.} \end{aligned}$$

Here all the components **c** change simultaneously due to the change of the parameter θ and such a variation is called the variation of **c** in term of the parameter θ. Again putting $\theta = 0$ we get the cost vector of the original problem. We now discuss the nature of the changes that takes place due to the change of $\mathbf{c}$ to $\mathbf{c} + \theta\mathbf{c}^0$, where θ varies in the region $\phi < \theta < \delta$, where $[\phi, \delta] \in R$ in a finite domain of real number set R.

Let us assume that the original problem (maximization problem) has a finite optimal solution and let B be the optimal basis.

Hence all $z_j - c_j \geq 0$ $j = 1, \cdots, n$ in the final optimum table, or $\mathbf{c}_B\mathbf{y}_j - c_j \geq 0$.

Now due to the change of $\mathbf{c}, \mathbf{y}_j$ do not change $[j = 1, \cdots, n]$

Hence if $\mathbf{c}$ be replaced by $\mathbf{c} + \theta\mathbf{c}^0, \mathbf{c}_B\mathbf{y}_j - c_j$ changes to $(\mathbf{c}_B + \theta\mathbf{c}_B^0)\,\mathbf{y}_j - (c_j + \theta c_j^0)$ which we shall denote as $z_j^* - c_j^*$ where $\mathbf{c}_B^0$ is the associated cost vector corresponding to the vector $\mathbf{c}^0$.

Now

$$\begin{aligned} z_j^* - c_j^* &= (\mathbf{c}_B\mathbf{y}_j - \mathbf{c}_j) + \theta(\mathbf{c}_B^0\mathbf{y}_j - c_j^0) \\ &= z_j - c_j + \theta(\mathbf{c}_B^0\mathbf{y}_j - c_j^0) \\ &= \alpha_j + \theta\beta_j [\alpha_j = z_j - c_j,\ \beta_j = \mathbf{c}_B^0\mathbf{y}_j - c_j^0]. \end{aligned} \tag{17.2.2}$$

Now we shall have to discuss the following cases.

Case (i). If $\beta_j = 0$ for some j, then

$$z_j^* - c_j^* = z_j - c_j \geq 0 \quad [\theta \text{ is unrestricted here}]$$

(ii) If $\beta_j > 0$ for some j, $z_j^* - c_j^*$ will remain positive provided

$$\alpha_j + \theta\beta_j \geq 0, \quad \text{or,} \quad \theta \geq \frac{-\alpha_j}{\beta_j}, \quad \beta_j > 0[\theta \leq 0]. \tag{17.2.3}$$

(iii) If $\beta_j < 0$ for some j, $z_j^* - c_j^*$ will remain positive provided

$$\alpha_j + \theta\beta_j \geq 0, \quad \text{or,} \quad \theta \leq \frac{-\alpha_j}{\beta_j}, \quad \beta_j < 0[\theta \geq 0]. \tag{17.2.4}$$

Combining (17.2.3) and (17.3.4) we can say that the current B.F.S. remains optimum so long θ does not violate the condition

$$\max_j\left(-\frac{\alpha_j}{\beta_j}, \beta_j > 0\right) \leq \theta \leq \min_j\left(-\frac{\alpha_j}{\beta_j}, \beta_j < 0\right) \tag{17.2.5}$$

$$\max_j\left(-\frac{\alpha_j}{\beta_j}, \beta_j > 0\right) = \bar{\theta},\ \bar{\theta} \leq 0,\ \min_j\left(-\frac{\alpha_j}{\beta_j}, \beta_j < 0\right) = \underline{\theta}, \quad \underline{\theta} \geq 0$$

are called as the first lower and upper critical values respectively of the parameter θ. From the above discussion it is also clear that if there exist no positive or negative value of β_j, there exist no definite lower and upper bound of θ respectively.

Now we shall consider the case $\theta = \underline{\theta} > 0$, which is called the first critical value of $\theta > 0$.

Let us assume for $j = k$, minimum value of $\theta = \underline{\theta} = \theta^{(1)}$ occurs.

Then

$$z_k^* - c_k^* = z_k - c_k + \theta^{(1)}(\mathbf{c}_B^0\mathbf{y}_k - c_k^0) = \alpha_k + \theta^{(1)}\beta_k \quad \text{(from (17.2.2))}$$

Again

$$\theta^{(1)} = \min_j \left(-\frac{\alpha_j}{\beta_j}, \beta_j < 0\right) = -\frac{\alpha_k}{\beta_k}, \beta_k < 0.$$

From these two, we get

$$z_k^* - c_k^* = 0. \tag{17.2.6}$$

All other values of $z_j^* - c_j^*$ corresponding to non-basic variables will be all non-negative for the value of $\theta^{(1)}$ stated above.

Now, we have, as $z_k^* - c_k^* = 0$, there exists an alternative optimal solution, which will be obtained inserting $\mathbf{a}_k$ in the new basis after replacing one basis vector from the current basis.

Now taking $\theta = \theta^{(1)}$, with the new basis we get an optimum simplex table but with different optimal B.F. solution. Now from the table we can perform a second sensitivity analysis and may get a few critical value of $\theta = \theta^{(2)}$ provided there exists at least one $\mathbf{c}_B^0\mathbf{y}_j - c_j^0 < 0$ which is to be calculated from the data of the second table. Proceeding in the manner, we may get a sequence of critical values and this process will be stopped as soon as we reach at the stage, when all $\mathbf{c}_B^0\mathbf{y}_j - c_j^0 > 0$ for j corresponding to the non-basic variables. Let $\theta = \theta^{(l)}$, be the critical value at that stage. Now for $\theta \geq \theta^{(1)} + \theta^{(2)} + \cdots \theta^{(l)}$, the optimal basis will remain unchanged and for finite value of θ, we get finite value of the objective function with the same basis.

Remark: During the discussion of the problem relating to the parametric variation we are supposed to think that for $\theta = 0$, the problem will have finite value of the objective function.

17.3 Variation of B (Parametric Variation)

Let $\mathbf{b}$ be the requirement vector of the original problem [maximization]. Now we assume that $\mathbf{b}$ changes to $\mathbf{b}^*$ where $\mathbf{b}^*$ can be written as

$$\mathbf{b}^* = \mathbf{b} + \theta\mathbf{b}^0 \quad [\theta \text{ being a scalar parameter}] \tag{17.3.1}$$

Let us further assume that the original problem has a finite optimal solution and let B be the optimal basis. Then the optimal B.F.S. of the original problem is

$$\mathbf{x}_B = B^{-1}\mathbf{b}. \tag{17.3.2}$$

With the same basis

$$\begin{aligned} B^{-1}\mathbf{b}^* &= B^{-1}(\mathbf{b} + \theta\mathbf{b}^0) = B^{-1}\mathbf{b} + \theta B^{-1}\mathbf{b}^0 \\ &= \mathbf{x}_B + \theta\mathbf{d} \text{ where } \mathbf{d} = B^{-1}\mathbf{b}^0. \end{aligned} \tag{17.3.3}$$

Now the changed B.S. will also be a feasible solution if

$$\mathbf{x}_B + \theta \mathbf{d} \geq \mathbf{0}, \quad \text{or,} \quad x_{Bi} + \theta d_i \geq 0, \quad i = 1, 2, \cdots, m.$$

Now we shall have to discuss the three cases:

(i) If $d_i = 0$ for some i, then $x_{Bi} + \theta d_i = x_{Bi} \geq 0$.

(ii) If $d_i > 0$ for some i, then $x_{Bi} + \theta d_i \geq 0$.

$$\Rightarrow \theta \geq -\frac{x_{Bi}}{d_i}, \quad d_i > 0, \quad [\theta \leq 0]. \tag{17.3.4}$$

(iii) If $d_i < 0$ for some i, then $x_{Bi} + \theta d_i \geq 0$

$$\Rightarrow \theta \leq -\frac{x_{Bi}}{d_i}, \quad d_i < 0, \quad [\theta \geq 0]. \tag{17.3.5}$$

Thus if there exist both positive and negative components of $\mathbf{d}$, then the current basis will remain unchanged if θ lies in the interval

$$\max_i \left\{ -\frac{x_{Bi}}{d_i}, \ d_i > 0 \right\} \leq \theta \leq \min_i \left\{ -\frac{x_{Bi}}{d_i}, \ d_i < 0 \right\}. \tag{17.3.6}$$

The two extreme values of θ, given in (17.3.6), one of them is non-positive and the other is non-negative are called the first critical values of θ, and for those critical values of θ, the B.F.S. will be degenerate.

Again, if there is no positive component of $\mathbf{d}$, there is no finite lower bound of θ etc. Now we shall confine our discussion for $\theta > 0$.

Let $\theta = \theta^{(1)}$ be the first critical value ($\theta > 0$). For $\theta = \theta^{(1)}$ the B.F.S. will be degenerate with the original basis. Let $x_{Br} = 0$ for $\theta = \theta^{(1)}$. Now if there exists at least one $y_{rj} < 0$, taking $x_{Br} = 0$ which is a non-positive quantity, by applying dual simplex method we get a new basis by replacing the r-th vector of the basis by a vector $\mathbf{a}_k$ which will be determined by the rule

$$z_k - c_k = \max_j \left\{ \frac{(z_j - c_j)}{y_{rj}}, \ y_{rj} < 0 \right\}. \tag{17.3.7}$$

Now using the second basis, we may get a second critical value $\theta^{(2)}$ and for $\theta = \theta^{(1)} + \theta^{(2)}$ we again get a degenerate B.F.S. with the second basis. In the same way, we may proceed until we may get a problem with no feasible solution or for $\theta > \theta^*$ there will be no change of basis.

▶ **Example 17.3.1** *Solve the following L.P.P. by simplex method.*

Maximize, $z = 6x_1 + 12x_2 + 4x_3$ *subject to*

$$\begin{aligned} 3x_1 + 4x_2 + \ \ x_3 &\leq 2 \\ x_1 + 3x_2 + 2x_3 &\leq 1, \quad x_1, x_2, x_3 \geq 0 \end{aligned}$$

Determine the optimal basis and find the range of λ such that objective function $(6-\lambda)x_1 + (12-\lambda)x_2 + (4-\lambda)x_3$ also assumes the optimal value with the same optimal basis and with the same constraints of the above problem.

The optimal simplex table of the initial problem is

	c		6	12	4	0	0
Basis	c_B	b	a_1	a_2	a_3	a_4	a_5
a_1	6	$\frac{2}{5}$	1	0	−1	$\frac{3}{5}$	$-\frac{4}{5}$
a_2	12	$\frac{1}{5}$	0	1	1	$-\frac{1}{5}$	$\frac{3}{5}$
$z_j - c_j$		$4\frac{4}{5}$	0	0	2	$\frac{6}{5}$	$\frac{12}{5}$

Optimal basis $= (\mathbf{a}_1, \mathbf{a}_2) = \begin{pmatrix} 3 & 4 \\ 1 & 3 \end{pmatrix}$.

In the second problem

$$\begin{aligned} \mathbf{c}^* &= (6,12,4,0,0) + \lambda(-1,-1,-1,0,0) \\ &= \mathbf{c} + \lambda\mathbf{c}^0 \end{aligned}$$

Then

$$c_B^0 = (-1,-1).$$

Now for the non-basic variables x_3, x_4 and x_5.

$$c_B^0\mathbf{y}_3 - c_3^0 = 1, \quad c_B^0\mathbf{y}_4 - c_4^0 = -\frac{2}{5}, \quad c_B^0\mathbf{y}_5 - c_5^0 = \frac{1}{5}.$$

As there exist both positive and negative values, hence both lower and upper bound of λ exist.

Now

$$\min_j \left\{ \frac{-(z_j - c_j)}{c_B^0\mathbf{y}_j - c_j^0}, c_B^0\mathbf{y}_j - c_j^0 < 0 \right\} = \frac{-(z_4 - c_4)}{c_B^0\mathbf{y}_4 - c_4^0} = \frac{-\frac{6}{5}}{-\frac{2}{5}} = 3.$$

Then $\lambda \le 3$.

Again

$$\max_j \left\{ \frac{-(z_j - c_j)}{c_B^0\mathbf{y}_j - c_j^0}, c_B^0\mathbf{y}_j - c_j^0 > 0 \right\} = \max \frac{-2}{1}, \frac{-\frac{12}{5}}{\frac{1}{5}} = -2.$$

Hence $\lambda \ge -2$. Thus within this range $-2 \le \lambda \le 3$, the second objective function remains optimal with the basis $B = (\mathbf{a}_1, \mathbf{a}_2)$.

► **Example 17.3.2** ***Perform a complete parametric programming analysis and identify all the critical values of the parameter θ of the following L.P.P.***

Maximize, $z = 2x_1 - (3-\theta)x_2$ ***subject to***

$$\begin{aligned} 2x_1 - 5x_2 &\le 7 \\ 4x_1 + x_2 &\le 8 \\ 7x_1 + 2x_2 &\le 16, \quad x_1 \ge 0, x_2 \ge 0. \end{aligned}$$

We shall initially consider the case taking $\theta = 0$.

Then $\mathbf{c} = (2, -3, 0, 0, 0)$. The initial optimum simplex table is given below:

			2	−3	0	0	0
Basis	c_B	b	$\mathbf{a}_1$	$\mathbf{a}_2$	$\mathbf{a}_3$	$\mathbf{a}_4$	$\mathbf{a}_5$
$\mathbf{a}_3$	0	3	0	$-\frac{11}{2}$	1	$-\frac{1}{2}$	0
$\mathbf{a}_1$	2	2	1	$\frac{1}{4}$	0	$\frac{1}{4}$	0
$\mathbf{a}_5$	0	2	0	$\frac{1}{4}$	0	$-\frac{7}{4}$	1
$z_j - c_j$		4	0	$\frac{7}{2}$	0	$\frac{1}{2}$	0

Now $\mathbf{c}^* = \mathbf{c} + \theta\mathbf{c}^0 = (2, -3, 0, 0, 0) + \theta(0, 1, 0, 0, 0)$.

Obviously $\mathbf{c}_B^0 = (0, 0, 0)$. Therefore $\mathbf{c}_B^0\mathbf{y}_j - \mathbf{c}_j$ corresponding to the non-basic variables x_2 and x_4, we have

$$\mathbf{c}_B^0\mathbf{y}_2 - c_2^0 = 0 - 1 = -1, \quad \mathbf{c}_B^0\mathbf{y}_4 - c_4 = 0 - 0 = 0.$$

There exist no positive value of $\mathbf{c}_B^0\mathbf{y}_j - c_j$. Hence there exist no finite lower bound of θ and thus for any finite negative value of θ, the optimal basis $B = (\mathbf{a}_3, \mathbf{a}_1, \mathbf{a}_5)$ remains unchanged and optimal value of the objective function remains same. There exists at least one negative value of $\mathbf{c}_B^0\mathbf{y}_j - \mathbf{c}_j^0$. Then there exists a positive finite value of θ which is called the first critical value of θ, and up to that value of θ, the optimal basis remains unchanged. Now the first critical value of $\theta = \theta^{(1)}$ is given by

$$\min_j \left\{ \frac{-(z_j - c_j)}{\mathbf{c}_B^0\mathbf{y}_j - c_j^0}, \mathbf{c}_B^0\mathbf{y}_j - c_j^0 < 0 \right\} = \frac{-\frac{7}{2}}{-1} = \frac{7}{2},$$

which occurs for $j = 2$.

Now the net evaluations corresponding to the non-basic variable x_2 and x_4 for $\theta = \theta^{(1)}$ are

$$z_2^* - c_2^* = z_2 - c_2 + \theta^{(1)}(\mathbf{c}_B^0\mathbf{y}_2 - \mathbf{c}_2^0) = \frac{7}{2} - \frac{7}{2}(1) = 0 \quad \text{and}$$

$$z_4^* - c_4^* = z_4 - c_4 + \theta^{(1)}(\mathbf{c}_B^0\mathbf{y}_4 - \mathbf{c}_4^0) = \frac{1}{2} - 0 = \frac{1}{2}.$$

As $z_2^* - c_2^* = 0$, then there exists an alternative optimal basis which will be obtained by taking $\theta = \theta^{(1)} = \frac{7}{2}$ and inserting the vector $\mathbf{a}_2$ in the basis [$\because z_2^* - c_2^* = 0$] and removing either the vector $\mathbf{a}_1$ or $\mathbf{a}_5$ we have another optimum simplex table given below [taking $\theta = \theta^{(1)} = \frac{7}{2}$] [Removing $\mathbf{a}_1$ from the basis]

		$\mathbf{c}^*$	2	$\frac{1}{2}$	0	0	0
Basis	c_B	b	$\mathbf{a}_1$	$\mathbf{a}_2$	$\mathbf{a}_3$	$\mathbf{a}_4$	$\mathbf{a}_5$
$\mathbf{a}_3$	0	47	22	0	1	5	0
$\mathbf{a}_2$	$\frac{1}{2}$	8	4	1	0	1	0
$\mathbf{a}_5$	0	0	−1	0	0	−2	1
$z_j - c_j$		4	0*	0	0	$\frac{1}{2}$	0

Now we try to test the case for $\theta > \theta^{(1)} = \frac{7}{2}$.

From the second optimum table, we have

$$\mathbf{c}_B^9 = (0, 1, 0).$$

Thus

$$\mathbf{c}_B^0\mathbf{y}_1 - c_1^0 = 2, \quad \mathbf{c}_B^0\mathbf{y}_4 - c_4^0 = 1.$$

[corresponding to non-basic variables], both of them are being positive. Hence with that optimal basis there is no upper bound of θ and thus for any finite $\theta > \theta^{(1)} = \frac{7}{2}$, the optimal basis remains same and there exists finite value of the objective function for finite $\theta > \frac{7}{2}$ and for $\theta > \frac{7}{2}$, optimal basis is $B^{(1)} = (\mathbf{a}_3, \mathbf{a}_2, \mathbf{a}_5)$.

Hence from $\theta < 0$, there is no lower bound of θ and the basis is $(\mathbf{a}_3, \mathbf{a}_1, \mathbf{a}_5)$.

For $0 \le \theta \le \frac{7}{2}$, the optimal basis is $(\mathbf{a}_3, \mathbf{a}_1, \mathbf{a}_5)$ and $\mathbf{x} = \mathbf{x}^{(1)} = [x_3 = 3, x_1 = 2, x_5 = 2]$.

For $\theta \ge \frac{7}{2}$, the optimal basis is $(\mathbf{a}_3, \mathbf{a}_2, \mathbf{a}_5)$ and the optimal solution $\mathbf{x} = \mathbf{x}^{(2)} = [x_3 = 47, x_2 = 8, x_5 = 0]$, the solution being degenerate.

Remark: As the min-ratio is same for the second or third row we may also remove the vector $\mathbf{a}_5$ from the basis and can insert then $\mathbf{a}_2$ and here also the solution will be degenerate.

► **Example 17.3.3** *Given the L.P.P. of maximizing $z = 4x_1 + 12x_2$ subject to the constraints $x_1 + 2x_2 \le 20$, $x_1 + x_2 \le 13$, $3x_1 + 4x_2 \le 36$, $x_1, x_2 \ge 0$. Determine an optimal solution. If z is represented by,*

$$z = (4 + \theta)x_1 + (12 - 2\theta)x_2$$

where θ is a variable parameter such that $0 \le \theta \le 5$; then use the parametric programming to find the optimal solutions and the optimal values of z.

[Maduri, B. Sc. Math. '81]

The initial optimum table is

	c		4	12	0	0	0
Basis	$\mathbf{c}_B$	b	$\mathbf{a}_1$	$\mathbf{a}_2$	$\mathbf{a}_3$	$\mathbf{a}_4$	$\mathbf{a}_5$
$\mathbf{a}_2$	12	10	$\frac{1}{2}$	1	$\frac{1}{2}$	0	0
$\mathbf{a}_4$	0	3	$\frac{1}{2}$	0	$-\frac{1}{2}$	1	0
$\mathbf{a}_5$	0	16	2	0	-1	0	1
$z_j - c_j$		120	2	0	6	0	0

$$\mathbf{c}^* = (4, 12, 0, 0, 0) + \theta(1, -2, 0, 0, 0).$$

Hence

$$\mathbf{c}_B^0 = (-2, 0, 0), \mathbf{c}_B^0\mathbf{y}_1 - c_1^0 = -2, \mathbf{c}_B^0\mathbf{y}_B - c_3^0 = -1.$$

There exist no positive value of $\mathbf{c}_B^0\mathbf{y}_j - c_j^0$ corresponding to the non-basic variables. Thus for finite $\theta < 0$, the optimal basis remains unchanged.

The first critical value of $\theta > 0$ is

$$\theta^{(1)} = \min\left(\frac{-2}{-2}, \frac{-6}{-1}\right) = 1$$

which occurs for $j = 1$.

Now the objective function for $\theta = \theta^{(1)}$ is $5x_1 + 10x_2$. Now inserting $\mathbf{a}_1$ in the basis by replacing $\mathbf{a}_4$, we get a new optimum table which is given below.

			5	10	0	0	0
Basis	c_B	b	$\mathbf{a}_1$	$\mathbf{a}_2$	$\mathbf{a}_3$	$\mathbf{a}_4$	$\mathbf{a}_5$
$\mathbf{a}_2$	10	7	0	1	1	−1	0
$\mathbf{a}_1$	5	6	1	0	−1	2	0
$\mathbf{a}_5$	0	4	0	0	1	−4	1
$z_j - c_j$		100	0	0	5	0	0

$$\mathbf{c}_B^0 = (-2, 1, 0), \mathbf{c}_B^0\mathbf{y}_3 - c_3 = -3, \mathbf{c}_B^0\mathbf{y}_4 - c_4^0 = 4.$$

There is only one negative value. Thus the second critical value of $\theta = \theta^{(2)}$ is given by

$$\theta^{(2)} = \frac{-5}{-3} = \frac{5}{3}$$

which occurs for $j = 3$.

Then $\theta = \theta^{(1)} + \theta^{(2)} = \frac{8}{3}$ and the objective function is $\frac{20}{3}x_1 + \frac{20}{3}x_3$ etc.

Now inserting the vector $\mathbf{a}_3$ in the basis replacing $\mathbf{a}_5$ we get a new optimum simplex table which is give below.

			$\frac{20}{3}$	$\frac{20}{3}$	0	0	0
Basis	c_B	b	$\mathbf{a}_1$	$\mathbf{a}_2$	$\mathbf{a}_3$	$\mathbf{a}_4$	$\mathbf{a}_5$
$\mathbf{a}_2$	$\frac{20}{3}$	3	0	1	0	3	−1
$\mathbf{a}_1$	$\frac{20}{3}$	10	1	0	0	−2	1
$\mathbf{a}_3$	0	4	0	0	1	−4	1
$z_j - c_j$		$\frac{260}{3}$	0	0	0	$\frac{20}{3}$	0

$$\mathbf{c}_B^0 = (-2, 1, 0),\ \mathbf{c}_B^0\mathbf{y}_4 - c_4^0 = -8,\ \mathbf{c}_B^0\mathbf{y}_5 - c_5^0 = 3.$$

There exists one negative value. Thus the third critical value of $\theta = \theta^{(3)}$ is given by

$$\theta^{(3)} = \frac{-\frac{30}{3}}{-8} = \frac{5}{6},$$

which occurs for $j = 4$. Then

$$\theta = \theta^{(1)} + \theta^{(2)} + \theta^{(3)} = \frac{7}{2}.$$

Inserting the vector $\mathbf{a}_4$ replacing the vector $\mathbf{a}_2$ we get a new optimal simplex table with $z = \frac{15}{2}x_1 + 5x_2$. The new optimum table is given below.

			$\frac{15}{2}$	5	0	0	0
Basis	c_B	b	$\mathbf{a}_1$	$\mathbf{a}_2$	$\mathbf{a}_3$	$\mathbf{a}_4$	$\mathbf{a}_5$
$\mathbf{a}_4$	0	1	0	$\frac{1}{3}$	0	1	$-\frac{1}{3}$
$\mathbf{a}_1$	$\frac{15}{3}$	12	1	$\frac{2}{3}$	0	0	$\frac{1}{3}$
$\mathbf{a}_3$	0	8	0	$\frac{4}{3}$	1	0	$-\frac{4}{3}$
$z_j - c_j$		90	0	0	0	0	$\frac{5}{2}$

$$\mathbf{c}_B^0 = (0,1,0),\ \mathbf{c}_B^0\mathbf{y}_2 - \mathbf{c}_2^0 = \frac{8}{3},\ \mathbf{c}_B^0\mathbf{y}_5 - \mathbf{c}_5^0 = \frac{1}{3}.$$

There exists no negative value. Then for $\theta \geq \frac{7}{2}$ optimal basis remains unchanged. $\mathbf{x}^{(1)} = [0, 10]$ for $0 \leq \theta \leq 1$, $\mathbf{x}^{(2)} = [6, 7]$ for $1 \leq \theta \leq \frac{8}{3}$, $\mathbf{x}^{(3)} = [10, 3]$ for $\frac{8}{3} \leq \theta \leq \frac{7}{2}$, $\mathbf{x}^{(4)} = [12, 0]$ for $\theta \geq \frac{7}{2}$. For $\theta = 0$, $\max z = 120$; $\theta = 1$, $\max z = 100$; $\theta = \frac{8}{3}$, $\max z = \frac{260}{3}$; $\theta = \frac{7}{2}$, $\max z = 90$ etc.

Again for any $\theta, 0 \leq \theta \leq 1$, $\max z = (4+\theta)\times 0 + (12-2\theta)\times 10$ [$\because\ \mathbf{x}_1^{(1)} = [0,10]$] or $\max z = 120 - 20\theta$; $1 \leq \theta \leq \frac{8}{3}$, $\max z = (4+\theta)\times 6 + (12-2\theta)\times 7 = 108 - 8\theta$; $\frac{8}{3} \leq \theta \leq \frac{7}{2}$, $\max z = (4+\theta)\times 10 + (12-2\theta)\times 3 = 76 + 4\theta$; $\theta \geq \frac{7}{2}$, $\max z = 12\times(4+\theta) = 48 + 12\theta$.

▶ **Example 17.3.4** *Given the L.P.P.*

Maximize $z = 2x_1 - x_2 + 3x_3$ *subject to*

$$\begin{aligned} 3x_1 + \ x_2 - 2x_3 &\leq 6 + \theta \\ 2x_1 + 5x_2 + \ x_3 &\leq 14 - 4\theta \\ x_1 + 4x_2 + 2x_3 &\leq 8 - 5\theta \ , \quad x_1, x_2, x_3 \geq 0. \end{aligned}$$

Determine the optimal solution for $\theta = 0$ and find the range of θ for which the optimal basis remains unaltered.

The optimal simplex table is

Vectors in the basis	c_B	b	$\mathbf{a}_1$	$\mathbf{a}_2$	$\mathbf{a}_3$	$\mathbf{a}_4$	$\mathbf{a}_5$	$\mathbf{a}_6$
$\mathbf{a}_1$	2	$\frac{7}{2}$	1	$\frac{5}{4}$	0	$\frac{1}{4}$	0	$\frac{1}{4}$
$\mathbf{a}_5$	0	$\frac{19}{4}$	0	$\frac{9}{8}$	0	$-\frac{3}{8}$	1	$-\frac{7}{8}$
$\mathbf{a}_3$	3	$\frac{9}{4}$	0	$\frac{11}{8}$	1	$-\frac{1}{8}$	0	$\frac{3}{8}$
$z_j - c_j$		$13\frac{3}{4}$	0	$\frac{61}{8}$	0	$\frac{1}{8}$	0	$\frac{13}{8}$

$\mathbf{a}_4, \mathbf{a}_5, \mathbf{a}_6$ are slack vectors. Then the optimal basis inverse is

$$B^{-1} = \begin{bmatrix} \frac{1}{4} & 0 & \frac{1}{4} \\ -\frac{3}{8} & 1 & -\frac{7}{8} \\ -\frac{1}{8} & 0 & \frac{3}{8} \end{bmatrix}$$

corresponding to $\mathbf{y}_4, \mathbf{y}_5$ and $\mathbf{y}_6$ in the final table.

Now

$$\mathbf{b}^* = [6, 14, 8] + \theta[1, -4, -5] = \mathbf{b} + \theta\mathbf{b}^0 \text{ (say)}.$$

Then

$$\mathbf{b}^0 = [1, -4, -5], \quad \mathbf{x}_B = \left[\frac{7}{2}, \frac{19}{4}, \frac{9}{4}\right].$$

$$\mathbf{d} = B^{-1}\mathbf{b}^0 = \begin{bmatrix} \frac{1}{4} & 0 & \frac{1}{4} \\ -\frac{3}{8} & 1 & -\frac{7}{8} \\ -\frac{1}{8} & 0 & \frac{3}{8} \end{bmatrix} \begin{bmatrix} 1 \\ -4 \\ -5 \end{bmatrix} = \begin{bmatrix} -1 \\ 0 \\ -2 \end{bmatrix}.$$

All components of $\mathbf{d}$ are non-positive. Hence there is no lower bound of θ and for finite negative value of θ, the solution remains optimal. The maximum positive value of θ for which the present basis remains optimal is given by

$$\theta = \theta^{(1)} = \min_i \left(\frac{-x_{Bi}}{d_i}, \; d_i < 0\right) = \min\left(\frac{-\frac{7}{2}}{-1}, \frac{-\frac{9}{4}}{-2}\right) = \frac{9}{8},$$

which occurs for $i = 3$ and for that value of θ we shall get an optimal degenerate solution with $x_{B3} = x_3 = 0$ (basic). This value of $\theta = \theta^{(1)} = \frac{9}{8}$ is the first critical value of $\theta > 0$.

► **Example 17.3.5** *Solve the parametric L.P.P.*

Maximize, $z = 4x_1 + 7x_2$ *subject to*

$$\begin{aligned} 2x_1 + \;\; x_2 &\le 1000 - 5\theta \\ x_1 + \;\; x_2 &\le \;\; 600 + 2\theta \\ x_1 + 2x_2 &\le 1000 + 3\theta, \quad x_1, x_2 \ge 0. \end{aligned}$$

assuming $\theta = 0$. Now find the range of θ such that current optimal basis remains unaltered.

Perform, if possible, the critical analysis of θ, up to the second critical value (positive) and find the optimal feasible solutions and optimal values of z within the ranges. From the original problem find minimum positive value of θ such that problem will have no feasible solution.

For $\theta = 0$, the optimal simplex table is

Basis	c_B	$\mathbf{b}$	$\mathbf{a}_1$	$\mathbf{a}_2$	$\mathbf{a}_3$	$\mathbf{a}_4$	$\mathbf{a}_5$
$\mathbf{a}_3$	0	200	0	0	1	−3	1
$\mathbf{a}_1$	4	200	1	0	0	2	−1
$\mathbf{a}_2$	7	400	0	1	0	−1	1
$z_j - c_j$		3600	0	0	0	1	3

For $\theta \neq 0$, $\mathbf{b}^* = [1000, 600, 1000] + \theta[-5, 2, 3]$.

Hence $\mathbf{b}^0 = [-5, 2, 3]$.

$$\text{Optimal basis} = (\mathbf{a}_3, \mathbf{a}_1, \mathbf{a}_2) = \begin{bmatrix} 1 & 2 & 1 \\ 0 & 1 & 1 \\ 0 & 1 & 2 \end{bmatrix} \cdot \mathbf{a}_3, \mathbf{a}_4, \mathbf{a}_5 \text{ are slack vectors.}$$

$$\text{Then the optimal basis inverse} = (\mathbf{y}_3, \mathbf{y}_4, \mathbf{y}_5) = \begin{bmatrix} 1 & -3 & 1 \\ 0 & 2 & -1 \\ 0 & -1 & 1 \end{bmatrix}$$

$$\mathbf{d} = B^{-1}\mathbf{b}^0 = \begin{bmatrix} 1 & -3 & 1 \\ 0 & 2 & -1 \\ 0 & -1 & 1 \end{bmatrix} \begin{bmatrix} -5 \\ 2 \\ 3 \end{bmatrix} = \begin{bmatrix} -8 \\ 1 \\ 1 \end{bmatrix}.$$

Thee exist both positive and negative components of **d**. Thus there exist both lower ($\theta < 0$) and upper bound of $\theta(\theta > 0)$.

$$\theta = \max_i \left(\frac{-x_{Bi}}{d_i}, d_i > 0 \right) = \max \left(\frac{-200}{1}, \frac{-400}{1} \right) = -200,$$

which occurs for $i = 2$. Again,

$$\theta = \min_i \left(\frac{-x_{Bi}}{d_i}, d_i < 0 \right) = \left(\frac{-200}{-8} \right) = 25,$$

which occurs for $i = 1$. Hence for $-200 \le \theta \le 25$, the current optimal basis remains unchanged and for $\theta = -200, 25$ we will have degenerate B.F.S. with $x_{B2} = 0$ and $x_{B1} = 0$ respectively. The basis at that stage is $(\mathbf{a}_3, \mathbf{a}_1, \mathbf{a}_2)$, the optimal B.F.S. is $[200 - 8\theta, 200 + \theta, 400 + \theta]$ and the optimal value of z is $3600 + 11\theta$. Now for $\theta = \theta^{(1)} = 25$, the first positive critical value of θ, the B.F.S. is $[0, 225, 425]$, putting $\theta = 25$ in the above solution. The solution is degenerate, with $x_{B1} = 0$ which we have stated earlier. Now to get the second positive critical value of θ, we require to use dual simplex method. With the B.F.S. $\mathbf{x}_B = [0, 225, 425]$ and with the basis $B = (\mathbf{a}_3, \mathbf{a}_1, \mathbf{a}_2)$ the optimal simplex table is

Basis	$\mathbf{c}_B$	$\mathbf{b}$	$\mathbf{a}_1$	$\mathbf{a}_2$	$\mathbf{a}_3$	$\mathbf{a}_4$	$\mathbf{a}_5$
$\mathbf{a}_3^*$	0	0	0	0	1	-3^*	1
$\mathbf{a}_1$	4	225	1	0	0	2	-1
$\mathbf{a}_2$	7	425	0	1	0	-1	1
$z_j - c_j$		3875	0	0	0	1	3

Now to get an improved value of θ, we shall have to apply dual simplex algorithm. Now taking $0 = x_{B1}$ as the limiting value of negative quantity and as at least one $y_{ij} = y_{14} = -3 < 0$, we can apply dual simplex method to get a new basis by inserting $\mathbf{a}_4$ in the new basis replacing $\mathbf{a}_3$ from the current basis as

$$\max_j \left\{ \frac{(z_j - c_j)}{y_{1j}}, y_{1j} < 0 \right\} = -\frac{1}{3}$$

with basis $B = (\mathbf{a}_4, \mathbf{a}_1, \mathbf{a}_3)$ is given below.

Basis	$\mathbf{c}_B$	$\mathbf{b}$	$\mathbf{a}_1$	$\mathbf{a}_2$	$\mathbf{a}_3$	$\mathbf{a}_4$	$\mathbf{a}_5$
$\mathbf{a}_4$	0	0	0	0	$-\frac{1}{3}$	1	$-\frac{1}{3}$
$\mathbf{a}_1$	4	225	1	0	$\frac{2}{3}$	0	$-\frac{1}{3}$
$\mathbf{a}_2$	7	425	0	1	$-\frac{1}{3}$	0	$\frac{2}{3}$
$z_j - c_j$		3875	0	0	$\frac{1}{3}$	0	$\frac{10}{3}$

$$B^{-1} = \begin{bmatrix} -\frac{1}{3} & 1 & -\frac{1}{3} \\ \frac{2}{3} & 0 & -\frac{1}{3} \\ -\frac{1}{3} & 0 & \frac{2}{3} \end{bmatrix}, \ \mathbf{b}^0 = \begin{bmatrix} -5 \\ 2 \\ 3 \end{bmatrix}, \ B^{-1}\mathbf{b}^0 = \left[\frac{8}{3}, -\frac{13}{3}, \frac{11}{3}\right].$$

There exists one negative component of $\mathbf{d}$ which is $d_2 = -\frac{13}{3}$. Therefore

$$\theta^{(2)} = \max_i \left(\frac{-x_{Bi}}{d_i}, d_i < 0\right) = \frac{-225}{-\frac{13}{5}} = \frac{675}{13},$$

which is the second critical value of θ. The upper limit of θ at that stage

$$\theta = \theta^{(1)} + \theta^{(12)} = 25 + \frac{675}{13} = \frac{1000}{13}.$$

$$B^{-1}\mathbf{b} = \left[-\frac{200}{3}, \frac{1000}{3}, \frac{1000}{3}\right].$$

Hence the B.F.S. is within the interval is

$$\left[-\frac{200}{3} + \frac{8}{3}\theta, \ \frac{1000}{3} - \frac{13}{3}\theta, \ \frac{1000}{3} + \frac{11}{3}\theta\right].$$

We can summarize the results as follows.

For θ, lying in the interval $-200 \le \theta \le 25$, basis $B = (\mathbf{a}_3, \mathbf{a}_1, \mathbf{a}_2)$.

B.F.S. $[200 - 8\theta, 200 + \theta, 400 + \theta]$ and optimal value of $z = 3600 + 11\theta$.

For θ, in the interval $25 \le \theta \le \frac{1000}{3}$, basis $B = (\mathbf{a}_4, \mathbf{a}_1, \mathbf{a}_2)$.

B.F.S. $\left[-\frac{200}{3} + \frac{8}{3}\theta, \frac{1000}{3} - \frac{13}{3}\theta, \frac{1000}{3} + \frac{11}{3}\theta\right]$, optimal value of $z = \frac{11000}{3} + \frac{25}{3}\theta$.

Now for positive θ, $b_1 = 1000 - 5\theta$ which will only decrease in the increase of $\theta(\theta > 0)$. Now for $\theta = 200$, $b_1 = 0$ and for $\theta > 200$, $b_1 < 0$, which implies that there exist no feasible solution to the problem for $\theta > 200$.

Note: (1) Of course, if we continue to proceed in the above manner we shall be able to find $\theta^{(3)} = \frac{1600}{13}$, i.e., for $\frac{1000}{13} \le \theta \le 200$, the basis is $B = (\mathbf{a}_4, \mathbf{a}_5, \mathbf{a}_2)$ and for $\theta > 200$, the solution becomes infeasible.

Exercise 17

1. Write a short note on 'Parametric programming' regarding the
   (a) Linear parametric variation of the cost vector.
   (b) Linear parametric variation of the requirement vector.

2. Prove that the L.P.P. Maximize, $z = \mathbf{cx}$ subject to

$$A\mathbf{x} \leq \mathbf{b}, \ \mathbf{x} \geq \mathbf{0}$$

remains optimal with the same basis, if the objective function be changed to $z^* = \lambda\mathbf{cx}$, $\lambda > 0$ [a scalar] and the optimal value of $z^* = \lambda$ (optimal value of z).

3. Find the range of θ [θ being a scalar] such that the problem

   Maximize, $z = 14x_1 + 4x_2$ subject to

$$2x_1 + x_2 \leq 3$$
$$x_1 - x_2 \leq 1, \quad x_1, x_2 \geq 0$$

remains optimal for the new objective function $z^* = (14 - 2\theta)x_1 + (4 + 3\theta)x_2$.

4. Solve the following L.P.P.

   Maximize, $z = 4x_1 + 3x_2$ subject to

$$3x_1 + x_2 \leq 15 + \theta$$
$$3x_1 + 4x_2 \leq 24 - 5\theta, \quad x_1, x_2 \geq 0.$$

Assuming $\theta = 0$. Now find the range of θ, such that initial problem with $\theta = 0$, remains optimal with the same basis. Find the B.F.S. at the extreme values of θ.

5. Perform a parametric analysis of the following problem up to second positive critical value of θ and find the values of B.F.S. and the values of the objective function within these ranges.

   Maximize, $z = 5x_1 + 2x_2 + 3x_3$ subject to

$$x_1 + 5x_2 + 2x_3 \leq 30 + 4\theta$$
$$x_1 - 5x_2 - 6x_3 \leq 40 + 3\theta, \quad x_1, x_2, x_3 \geq 0.$$

6. Perform a parametric analysis of the following problem for $\theta \geq 0$ and find the values of B.F.S. and values of the objective function within these ranges.

   Maximize, $z = 3x_1 + 2x_2 + x_3$ subject to

$$x_1 + 2x_2 - 3x_3 \leq 8 + 4\theta$$
$$3x_2 + x_3 \leq 6 - \theta, \quad x_1, x_2, x_3 \geq 0.$$

7. Perform a parametric analysis of the following problem and find the B.F.S. and the values of the objective functions within the ranges.

Maximize, $-x_1 - x_2 + \lambda x_3 + 2\lambda x_4$ subject to

$$\begin{aligned} x_1 + \quad + x_3 + 2x_4 &= 2 \\ 2x_1 + x_2 \quad + 3x_4 &= 5\,, \quad x_j \geq 0, j = 1, 2, \cdots, 4. \end{aligned}$$

8. Prove that the following L.P.P.

Minimize, $z = \lambda x - y$ subject to

$$\begin{aligned} 3x - y &\geq 5 \\ 2x + y &\leq 3, \end{aligned}$$

x, y unrestricted in sign, has only finite optimal solution for $-2 \leq \lambda \leq 3$, a multiple solution for $\lambda = 3$ and unbounded solution for $\lambda > 3$ and $\lambda < -2$.

9. Stating the optimal basis of the following L.P.P. assuming $\lambda = 0$ find the range of λ such then the above optimal basis remains unaffected for the problem.

Maximize, $z = 3x_1 + 2x_2 + 5x_3$ subject to the constraints

$$\begin{aligned} x_1 + 2x_2 + x_3 &\leq 430 + \lambda \\ 3x_1 \quad + 2x_3 &\geq 460 - 4\lambda \\ 2x_1 + 4x_2 \quad &\leq 420 - 4\lambda; \quad x_1, x_2, x_3 \geq 0. \end{aligned}$$

Find also the value of the objective function and the B.F.S. within this range.

Answers

3. $-18 \leq \theta \leq \frac{3}{4}$. 4. $-4 \leq \theta \leq \frac{3}{2}$, $[x_1 = 0, x_2 = 11]$ for lower value of $\theta = -4$, $[x_1 = \frac{11}{2}, x_2 = 0]$ for upper value of $\theta = \frac{3}{2}$.

5. For $0 \leq \theta \leq 10$, basis $= (\mathbf{a}_1, \mathbf{a}_5)$ B.F.S. $= [30 + 4\theta = x_1, 10 - \theta = x_5]$, value of the objective function is $150 + 20\theta$. For $\theta \geq 10$, Basis $= (\mathbf{a}_1, \mathbf{a}_3)$, B.F.S. $= [\frac{65}{2} + \frac{15}{4}\theta = x_1, -\frac{5}{4} + \frac{1}{8}\theta = x_3]$ and the value of the objective function $\frac{635}{4} + \frac{153}{8}\theta$.

6. For $0 \leq \theta \leq 6$, basis is $(\mathbf{a}_1, \mathbf{a}_3)$ B.F.S $= [26 + \theta = x_1, 6 - \theta = x_3]$ and the value of the objective function is $84 + 2\theta$. For $\theta > 6$, the problem has no feasible solution.

7. For $\lambda \leq \frac{1}{2}$, the basis is $(\mathbf{a}_2, \mathbf{a}_4)$, B.F.S. $[x_2 = 2, x_4 = 1]$, value of the objective function is $-2 - 2\lambda$; For $\frac{1}{2} \leq \lambda \leq 1$, basis is $(\mathbf{a}_1, \mathbf{a}_2)$, value of the objective function is -3, For $\lambda \geq 1$, basis is $(\mathbf{a}_2, \mathbf{a}_3)$ value of the objective function $-5 + 2\lambda$.

9. Optimal basis $B = (\mathbf{a}_2, \mathbf{a}_3, \mathbf{a}_6)$, range of λ, $-\frac{200}{3} \leq \lambda \leq 10$, value of the objective function is $1350 - 7\lambda$, b.f.s. $= [100 + \frac{3}{2}\lambda = x_2, 230 - 2\lambda = x_3, 20 - 2\lambda = x_6]$.

Chapter 18

Integer Programming

18.1 Introduction

So far, we have solved the L.P.P. by algebraic method or by graphical method and the components of the optimal feasible solution/s may be all integers or some of them be integers and the remaining components fractions or all of them may be fractions. But from the practical points of view, there are problems where the components of the optimal solution must be all non-negative integers or some of its components may be fractions. When the components of the optimal feasible solution are all non-negative integers, the problem is called a pure integer programming. In this chapter we are interested to solve such problems where by proper adjustment the components of the optimal solution/s are all integers and the optimal value of the objective function may be less or at most same as that of the solutions when we ignore the constraint of being all of them integers. For examples, if we require to manufacture Radios and TV's such that the profit be maximum and let the maximum attain when the production of the Radios and the TV's be $4\frac{1}{3}$ and $7\frac{2}{3}$ unit respectively. But it is totally absurd because the manufacturer can produce either four or less or five or more [All numbers must be purely integers] Radios and not $4\frac{1}{3}$ radios. Our problem is to find an optimal integer solution for which the optimal value of the objective function may be less or at most equal to the value of the objective function, keeping in mind that the constraints of integer solution be satisfied. Below we discuss the procedure of the related developments. There are two very common methods of solving such problems: (i) Gomory's cutting plane (Hyper plane) method and (ii) Branch and bound method.

For two or three variables graphical method can also be used to find the optimal value of such problems.

18.2 Cutting plane (Hyper plane) method

Let the problem be optimize $z = \mathbf{cx}$ subject to

$$A\mathbf{x} = \mathbf{b},\ \mathbf{x} \geq \mathbf{0} \tag{18.2.1}$$

and all components of the optimal solution be non-negative integer. Here A is an $(m \times n)$ matrix and $m < n$.

Let the problem be solved initially ignoring the restriction adopted in the last constraint.

For the sake of simplicity, we assume that the optimal basis be $B = [\mathbf{a}_1, \mathbf{a}_2, \cdots, \mathbf{a}_m]$ and the remaining matrix, which is a sub-matrix of A is R where

$$R = \begin{bmatrix} a_1, m+1 & a_1, m+2 & \cdots\cdots & a_{1n} \\ a_2, m+1 & a_2, m+2 & \cdots\cdots & a_{2n} \\ \cdots\cdots & \cdots\cdots & \cdots\cdots & \cdots\cdots \\ \cdots\cdots & \cdots\cdots & \cdots\cdots & \cdots\cdots \\ a_m, m+1 & a_m, m+2 & \cdots & a_{mn} \end{bmatrix} \tag{18.2.2}$$

$$a_m, m+1 \tag{18.2.3}$$

and then $A = [B, R]$.

For the optimal solution $\mathbf{x}^*$ of the problem, we have

$$\mathbf{x}^* = [\mathbf{x}_B^*, \mathbf{x}_R^*] \quad [\mathbf{x}_R^* = 0 \text{ (non-basic)}]$$

Then $A\mathbf{x} = \mathbf{b}$ can be written as

$$[B, R] \begin{bmatrix} \mathbf{x}_B^* \\ \mathbf{x}_R^* \end{bmatrix} = \mathbf{b}$$

which implies

$$B\mathbf{x}_B^* + R\mathbf{x}_R^* = \mathbf{b}. \tag{18.2.4}$$

Pre-multiplying by B^{-1} [since B is non-singular] we get

$$B^{-1}[B\mathbf{x}_B^*] + B^{-1}[R\mathbf{x}_R^*] = B^{-1}\mathbf{b}$$

$$\text{or, } \mathbf{x}_B^* = B^{-1}\mathbf{b} - (B^{-1}R)\mathbf{x}_R^*. \tag{18.2.5}$$

Now for vectors $\mathbf{a}_j \in R$, $j = m+1, m+2, \cdots, n$.

Let $B^{-1}\Sigma\mathbf{a}_j = \Sigma\mathbf{y}_j$ and we know that

$$B^{-1}\mathbf{b} = \mathbf{x}_B^* = \mathbf{y}_0. \tag{18.2.6}$$

From (18.2.5) we get

$$\mathbf{x}_B^* = \mathbf{y}_0 - \Sigma\mathbf{y}_j x_j, \tag{18.2.7}$$

where x_j are the non-basic variables of the current solution.

Let us assume that ith component of $\mathbf{x}_B^*$ is non-integer.

We have

$$x_{Bi} = y_{i0} - \Sigma y_{ij} x_j, \tag{18.2.8}$$

where x_j's are all zero in the current solution.

x_{Bi} is not integer. Therefore y_{i0} will also be not an integer. We further assume that

$$y_{i0} = d_{i0} + f_{i0}, \tag{18.2.9}$$

where d_{i0} is a non-negative integer, the integer part of

$$y_{i0} \quad \text{and} \quad 0 < f_{i0} < 1. \tag{18.2.10}$$

Similarly let

$$y_{ij} = d_{ij} + f_{ij} \tag{18.2.11}$$

such that d_{ij} is the integer part of y_{ij}

$$\text{and} \quad (0 \leq f_{ij} < 1) \tag{18.2.12}$$

Now using (18.2.6), (18.2.7) and (18.2.8) we get

$$\begin{aligned} x_{Bi} &= d_{i0} + f_{i0} - \Sigma(d_{ij} + f_{ij})x_j \\ \Rightarrow \quad x_{Bi} &= [d_{i0} - \Sigma d_{ij}x_j] + [f_{i0} - \Sigma f_{ij}x_j]. \end{aligned} \tag{18.2.13}$$

We assume that x_{Bi} is unique with all non-basic $x_j = 0$.

If we now want to change this solution, one of the non-basic x_j's must be positive. Thus for, an integer solution the quantity $(d_{i0} - \Sigma d_{ij}x_j)$ must be an integer.

Now without going details of the theorem, we can say

$$f_{i0} - \Sigma f_{ij}x_j \leq 0, \tag{18.2.14}$$

which is the cutting plane (hyper plane) constraint of Gomory for moving towards integer programming and this (18.2.14) additional constraint is added to the final simplex method of our conventional Simplex Algorithm and the new problem (with one more constraint) is to be solved by dual simplex method.

To solve the problem as early as possible f_{i0} is taken for that value of i, where the fractional part $0 < f_{i0} < 1$ be maximum. Further if $y_{ij} = -7\frac{3}{4}$ then we divide by parts in the manner $y_{ij} = -8+\frac{1}{4}$ and $\frac{1}{4}$ is f_{ij}, which lies in the interval $(0 \leq f_{ij} < 1)$.

The following solved problem will give clear idea about the application of the theorem in a given problem.

► **Example 18.2.1** *Find all integer solution of the following L.P.P.,*

Maximize, $z = 2x_1 + 3x_2$ *subject to*

$$\begin{aligned} 3x_1 + 4x_2 &\leq 12 \\ x_1 + 3x_2 &\leq 6, \quad x_1, x_2 \geq 0. \end{aligned}$$

Solution of the problem by conventional simplex Algorithm method.

Simplex tables

		c	2	3	0	0	
Basis	c_B	b	a_1	a_2	a_3	a_4	Min ratio
a_3	0	12	3	4	1	0	$\frac{12}{4} = 3$
a_4^*	0	6	1	3^*	0	1	$\frac{6}{3} = 2^*$
$z_j - c_j$		0	-2	-3^*	0	0	
a_3^*	0	4	$\frac{5}{3}^*$	0	1	$-\frac{4}{3}$	$\frac{4}{\frac{5}{3}} = \frac{12}{5}^*$
a_2	3	2	$\frac{1}{3}$	1	0	$\frac{1}{3}$	$\frac{2}{\frac{1}{3}} = 6$
$z_j - c_j$		6	-1^*	0	0	1	
a_1	2	$\frac{12}{5}$	1	0	$\frac{3}{5}$	$-\frac{4}{5}$	
a_2	3	$\frac{6}{5}$	0	1	$-\frac{1}{5}$	$\frac{3}{5}$	
$z_j - c_j$		$\frac{42}{5}$	0	0	$\frac{3}{5}$	$\frac{1}{5}$	

Here no component of the optimal solution set is an integer.

Here $\frac{12}{5} = 2 + \frac{2}{5}$ and $\frac{6}{5} = 1 + \frac{1}{5}$.

Thew fractional part of x_{B1} > the fractional part of x_{B2}.

Then $i = 1$, $f_{10} = \frac{2}{5}, y_{13} = \frac{3}{5}, y_{14} = -\frac{4}{5} = \left(\frac{1}{5} - 1\right)$.

Thus the new additional constraint is from (18.2.14) is

$$f_{10} - \left(\frac{3}{5}x_3 + \frac{1}{5}x_4\right) \le 0, \quad \text{or,} \quad -\left(\frac{3}{5}x_3 + \frac{1}{5}x_4\right) \le -\frac{2}{5}.$$

Adding the slack variable S_1 we get

$$-\left(\frac{3}{5}x_3 + \frac{1}{5}x_4\right) + S_1 = -\frac{2}{5}.$$

We now introduce this constraint in the final table of the initial simplex table and the new table is

			2	3	0	0	0
Basis	c_B	b	a_1	a_2	a_3	a_4	a_5
a_1	2	$\frac{12}{5}$	1	0	$\frac{3}{5}$	$-\frac{4}{5}$	0
a_2	3	$\frac{6}{5}$	0	1	$-\frac{1}{5}$	$\frac{3}{5}$	0
a_5^*	0	$-\frac{2}{5}$	0	0	$-\frac{3}{5}^*$	$-\frac{1}{5}$	1
$z_j - c_j$			0	0	$\frac{3}{5}$	$\frac{1}{5}$	0

The vector a_5 corresponds to the slack vector.

The third row is the key row since only x_{B3} is negative. Now using the dual simplex method

$$\max_j \left(\frac{z_j - c_j}{y_{3j}},\ y_{3j} < 0\right) = \max\left(\frac{3/5}{-3/5}, \frac{1/5}{-1/5}\right) = -1,$$

which occurs for $j = 3$ and $j = 4$. We simply take $y_{33} = -\frac{3}{5}$ (key element) and $\mathbf{a}_5$ is the vector to leave the current basis and the new table is given below.

In this table also all components of $\mathbf{x}_B$ are not integers. We shall have to proceed further. $\frac{4}{3} = 1 + \frac{1}{3}$ and other non-integer component is $\frac{2}{3}$. Then the fractional part $x_{B3} >$ the fractional part of x_{B2}. Hence the third row is the key row. Thus $f_{30} = \frac{2}{3}$, $y_{34} = \frac{1}{3}$, $y_{35} = -\frac{5}{3}$, $(-\frac{5}{3} = \frac{1}{3} - 2)$. Then the new constraint is $-\frac{1}{3}x_4 - \frac{1}{3}x_5 + S_2 = -\frac{2}{3}$ and the new table with new constraints is given below.

		c	2	3	0	0	0
Basis	$\mathbf{c}_B$	b	$\mathbf{a}_1$	$\mathbf{a}_2$	$\mathbf{a}_3$	$\mathbf{a}_4$	$\mathbf{a}_5$
$\mathbf{a}_1$	2	2	1	0	0	-1	1
$\mathbf{a}_2$	3	$\frac{4}{3}$	0	1	0	$\frac{2}{3}$	$-\frac{1}{3}$
$\mathbf{a}_3$	0	$\frac{2}{3}$	0	0	1	$\frac{1}{3}$	$-\frac{5}{3}$
$z_j - c_j$			0	0	0	0	1

The transformed table is

		c	2	3	0	0	0	0
Basis	$\mathbf{c}_B$	b	$\mathbf{a}_1$	$\mathbf{a}_2$	$\mathbf{a}_3$	$\mathbf{a}_4$	$\mathbf{a}_5$	$\mathbf{a}_6$
$\mathbf{a}_1$	2	2	1	0	0	-1	1	0
$\mathbf{a}_2$	3	$\frac{4}{3}$	0	1	0	$\frac{2}{3}$	$-\frac{1}{3}$	0
$\mathbf{a}_3$	0	$\frac{2}{3}$	0	0	1	$\frac{1}{3}$	$-\frac{5}{3}$	0
$\mathbf{a}_6$	0	$-\frac{2}{3}$	0	0	0	$-\frac{1}{3}^*$	$-\frac{1}{3}$	1
$z_j - c_j$			0	0	0	0	1	0

$$\max_j \left(\frac{z_j - c_j}{y_{4j}}, y_{4j} < 0\right) = \max\left(\frac{0}{-1/3}, \frac{1}{-1/3}\right) = 0,$$

which occurs for $j = 4$. Therefore $\mathbf{a}_6$ will leave the basis and $\mathbf{a}_4$ will enter in the next basis. $y_{44} = -1/3$ is the key element.

The transformed table is ($\mathbf{a}_6$ corresponds to the slack variable S_2)

Basis	$\mathbf{c}_B$	b	$\mathbf{a}_1$	$\mathbf{a}_2$	$\mathbf{a}_3$	$\mathbf{a}_4$	$\mathbf{a}_5$	$\mathbf{a}_6$
$\mathbf{a}_1$	2	4	1	0	0	0	2	-3
$\mathbf{a}_2$	3	0	0	1	0	0	-1	2
$\mathbf{a}_3$	0	0	0	0	1	0	-2	1
$\mathbf{a}_4$	0	2	0	0	0	1	1	-3
		8	0	0	0	0	1	0

In this table all components the basic feasible solution are integers. Thus we reach at the optimal stage and the pure integer optimal solution is degenerate with $x_1 = 4$, $x_2 = 0$ and $\max z = 8$.

Remark: When we have ignored the integer restriction, the $\max z = \frac{42}{5} > 8$.

Exercise 18

Find the integer solution of the following problems:

1. Maximize $z = 2x_1 + 2x_2$

$$\text{subject to } 5x_1 + 3x_2 \leq 8$$
$$x_1 + 2x_2 \leq 4$$

$x_1, x_2 \geq 0$ and are integers. **[C.U. M.Sc. '89, '92]**

2. Maximize $z = 3x_1 + 4x_2$

$$\text{subject to } 3x_1 + 2x_2 \leq 8$$
$$x_1 + 4x_2 \leq 10$$

$x_1, x_2 \geq 0$ and the integers.

Answers

1. Max $z = 4$ at $x_1 = 1, x_2 = 1$ or $x_1 = 0, x_2 = 2$.
2. Max $z = 11$ at $x_1 = 1, x_2 = 2$.

Chapter 19

Theory of Games

19.1 Introduction

In this world of reality, there are always some conflicts of interest between two or more opponents engaging themselves in the same field of business. Consider a particular example. Let A and B be two business men working in the same field (ignore the presence of the third party in the field). There must be some conflicts of interest between them regarding business matters. In terms of mathematical terminology, the business men A and B are called as players and the business is called as game. Suppose A has three executives A_1, A_2 and A_3 and B has four executives B_1, B_2, B_3 and B_4 to control their respective businesses. We make a restriction that both players (business men) utilize the services of their executives only one at a time to control their whole business. For example, A may utilize the service of A_2 and B may utilize the service of B_3 only to control their business. But the selection of their executives entirely depends upon their own discretion ignoring the selection, done by the other player. The selection of a particular executive (or move) is called the strategy taken by the players and the selection of only one single move or strategy by a player, ignoring the strategy taken by his opponent is called a pure strategy. First of all, we shall consider the problems (games) relating to pure strategies taken by the players. Now we further make two assumptions: (i) The player A is in a better position. The player A is called a maximizing player (or Row player) and player B is called a minimizing player (Column player). (ii) The total gain of one player is exactly equal to the total loss of the other player, i.e., the sum total is zero.

In general, if the player A takes m pure strategies and B takes n pure strategies, then the game is called two person zero sum game or $m \times n$ rectangular game. If $m = n$ then the game is called a square game.

Below we write down a matrix which is called the pay off matrix of the maximization player A,

$$\begin{array}{cc} & \text{Player } B \\ \text{Player } A \begin{array}{c} \\ A_1 \\ A_2 \\ A_3 \end{array} & \begin{bmatrix} B_1 & B_2 & B_3 & B_4 \\ 4 & 2 & -3 & 1 \\ 0 & -3 & 4 & 6 \\ 3 & 2 & 2 & -2 \end{bmatrix} . \end{array}$$

The interpretation of the above pay-off matrix is given below.

(i) When A assumes his pure move A_1 and B assumes has pure move B_1, the gain of the player A is 4 units (which A captures from B's fields).

(ii) When A and B assume their pure moves A_1 and B_3 respectively A's gain is — 3 units, i.e., losses 3 units to B.

Similarly, for A and B's pure moves A_2 and B_1 respectively A's gain is zero unit, i.e., there is no gain or loss for both players. From the above discussions, it is clear that the pay-off matrix of the player B (minimizing payer) will be just the negative of the pay-off matrix of A. Hence, if the pay-off matrix in terms of A be C then the pay-off matrix of B is $-C$.. A pay-off matrix is always written for a maximizing player. Now considering the general $m \times n$ rectangular game, the pay-off matrix of A with A's pure moves $A_1, A_2, \cdots, A_m$ and B's pure moves $B_1, B_2, \cdots, B_n$ is given by

$$\begin{array}{cc} & \text{Player } B \\ \text{Player } A \begin{array}{c} \\ A_1 \\ A_2 \\ \\ A_m \end{array} & \begin{bmatrix} B_1 & B_2 & B_3 & B_4 \\ a_{11} & a_{12} & \cdots & a_{1n} \\ a_{21} & a_{22} & \cdots & a_{2n} \\ & \cdots & & \cdots \\ a_{m1} & a_{m2} & \cdots & a_{mn} \end{bmatrix} . \end{array}$$

There are $m \times n$ elements a_{ij} in the pay-off matrix which are the gains obtained by A from A and B's pure moves A_i and B_j respectively. $[i = 1, \cdots, m, j = 1, \cdots, n]$. The elements a_{ij} may be positive, negative or zero.

Thus we formally define the pay-off matrix in the following manner:

Pay-off matrix: A pay-off matrix is a real matrix (a_{ij}) $i, \cdots, m; j = 1, \cdots, n$. (in general a rectangular matrix) which are the element a_{ij} indicates the gain of the maximization player (Row player) for using the ith and j-th move of the row and column players respectively.

Now from the practical point of view A's aim will be to obtain a least possible gain, irrespective of the strategies taken by B. On the other hand B's aim will be to minimize his expected loss to the maximum possible extent, irrespective of the strategies taken by his opponent A. In terms of mathematical language, A's problem is to "maximize his minimum gains" and B's problem is to "minimize his maximum losses" whatever may be the strategies (pure) taken by both players. On the other hand, both players try to secure the "best of the worst results". In general, the "maximum of the minimum gains" for A is less than or equal to the 'minimum of the maximum losses" for B (we shall prove it later). When the equality condition holds, we may that the game problem is solved and the value of the game ' "maximum of the minimum gains" for A = "minimum of the maximum losses" for B. Assuming the existence of the value of the game, if the value of the game be the element a_{kl} of the pay-off matrix, the point (k, l) is called the saddle point of the pay-off matrix. From the above discussions, it is clear that there may be more than one saddle point in a pay-off matrix in a game with pure strategy. The principle of determination of the value of a game and saddle point or points is known as the *maximin minimax* principle. Again if the value of the game be a_{kl} for A and B's pure moves A_k and B_l respectively then the strategies A_k and B_l taken by both players are called the optimal strategies for the players.

Determination of the value of the game (pure strategy) by using *maximin minimax* principle.

Consider the following 3×4 pay-off matrix for A.

		Player B				
		B_1	B_2	B_3	B_4	Row minima
	A_1	4	6	−2	1	−2
Player A	A_2	3	3	4	2	2
	A_3	4	5	5	1	1
Column maxima		4	6	5	2	2|2

A's pure moves are A_1, A_2 and A_3 and B's pure moves are B_1, B_2, B_3 and B_4. Now each player has every right to select a pure move according to his own suitability.

Now for A's pure move A_1, A's minimum gain is—2 units (whatever strategy may be taken by B). Similarly for A's pure move A_2 and A_3, minimum possible gains are 2 units and 1 unit respectively. Therefore, as A has every right to select a move which will maximize his minimum gains, he will definitely select the pure move (strategy) A_2 and for that his gain will be 2 units.

His gain will not be less than 2 units at any case.

Now considering B's point of view, if B selects his pure move B_1, the maximum loss will be 4 units (independent of A's move). Similarly,. for B's pure moves B_2, B_3 and B_4 the maximum amount of losses will be 6 units, 5 units and 2 units respectively. Therefore, as the ultimate aim of B is to minimize his maximum losses, then B will definitely select the pure strategy B_4 and the minimum loss will be 2 units. His amount of loss cannot be increased any further. But it is interesting to note that the "maximum of the minimum gains" for A is equal to the "minimum of the maximum losses" for B. Hence in this pay-off matrix or game there exists the value of game and the value of game is 2 units. The optimal strategies are A_2 and B_4 for A and B respectively and (2, 4)-th position of the pay-off matrix is the saddle point of the pay-off matrix. But in general, there exists no saddle point and the value of the game cannot be determined by the above principle. Of course, there may be more than one saddle point in a particular game and in that case optimal strategies are not unique. Hence to determine the value of the game in a general $m \times n$ game by using *maximin minimax* principle we shall follow the rules step by step to obtain the final result.

Step 1. Calculate the value of the row minimum for each row and write down the m values in a column under the heading "Row minima". Find out the maximum value of the elements in that column headed by "Row minima". Let it be α and let it occur at kth row.

Step 2. Calculate the value of the column maximum for each column and write down the n values in a row under the heading "column maxima". Find the minimum value of the elements of the row under the heading "column maxima". Let it be β and let it occur at l-th column.

Step 3. In general $\beta > \alpha$ and if $\beta = \alpha$, the value of the game $\alpha = \beta$ and the optimal strategies are A_k and B_l for A and B respectively and the saddle point of the pay-off matrix is (k, l)-th position of the matrix.

Note. If the general pay-off matrix be $[a_{ij}]_{m \times n}$, then "maximum of the minimum gains" for A is mathematically expressed as $\max\limits_i \cdot \min\limits_j \cdot a_{ij}$ and "minimum of the maximum losses" for B is denoted as $\min\limits_j \cdot \max\limits_i \cdot a_{ij}$. Again taking $\max\limits_i \cdot \max\limits_j \cdot a_{ij} = \bar{v}$ and $\min\limits_j \cdot \max\limits_i \cdot a_{ij} = \underline{v}$ and the value of the game v, we have in general $\underline{v} \geq v \geq \bar{v}$.

Fair Game: If the value of the game be zero, i.e., there is no loss or gain for any player, the game is called a **fair game.**

Strictly determinate game: If the value of game is a non-zero quantity then the game is called as *strictly determinate game.* If the value of game is positive then the game is in favour of the player A and if the value of the game is negative then the game is in favour of the player B.

Theorem 19.1.1 *Let* $[a_{ij}]_{m \times n}$ *be the pay-off matrix for a two-person zero-sum game. Then prove that*

$$\min_j \cdot \max_i \cdot a_{ij} \geq \max_i \cdot \min_j \cdot a_{ij}.$$

Proof: We have $\max\limits_i (a_{ij}) \geq a_{ij}$ for any i and $\min\limits_j .(a_{ij}) \leq a_{ij}$ for any j.

Let the above maximum be attained at $i = k$ and the minimum be attained at $j = l$. Then

$$a_{kj} \geq a_{kl} \geq a_{il} \text{ for any } i \text{ and } j.$$

From this we have

$$\min_j \cdot a_{kl} \geq a_{kl} \geq \max_i \cdot a_{il} \text{ for all } i \text{ and } j.$$

Hence

$$\min_j \cdot \max_i \cdot a_{ij} \geq \max_i \cdot \min_j \cdot a_{ij}.$$

Note: It can be also proved geometrically from the scheme of the general matrix.

19.2 Analytical discussion of the theorem

Theorem 19.2.1 *$f(x,y)$ be a real valued function of x and y defined for $x \in A$ and $y \in B$, A and B being two subsets of real numbers. Now if both $\max_{x\in A}\cdot \min_{y\in B} .f(x,y)$ and $\min_{y\in B}\cdot \max_{x\in A} .(x,y)$ exist then*

$$\max_{x\in A} \min_{y\in B} f(x,y) \le \min_{y\in B}\cdot \max_{x\in A} f(x,y).$$

Proof: We have, from the definition of maximum and minimum,

$$\left.\begin{array}{l} f(x,y) \le \max\limits_{x\in A} .f(x,y) \text{ and} \\ \min\limits_{y\in B} f(x,y) \le f(x,y). \end{array}\right\} . \tag{19.2.1}$$

Hence,

$$\min_{y\in B} f(x,y) \le \max_{x\in A} f(x,y) \tag{19.2.2}$$

Now $\min_{y\in B} f(x,y)$ is independent of y [because minimum assumes for fixed value or values of y] and thus

$$\min_{y\in B} f(x,y) \le \min_{y\in B} \max_{x\in A} f(x,y). \tag{19.2.3}$$

Again the R.H.S. of (19.2.3) is independent of x and therefore we can say,

$$\max_{x\in A}\cdot \min_{y\in B} f(x,y) \le \min_{y\in B} \max_{x\in A} .f(x,y). \tag{19.2.4}$$

Note: The matrix $[a_{ij}]_{m\times n}$ ca be taken as a real valued function of the variable i and j such that,

$$f(i.j) = a_{ij} \quad [i = 1, 2, \cdots m, j = 1, 2, \cdots n]$$

Now if $A = \{1, 2, \cdots, m\}, B = \{1, 2, \cdots, n\}$ then (19.2.4) can be written

$$\max_{i}\cdot \min_{j}\cdot a_{ij} \le \min_{j}\cdot \max_{i}\cdot a_{ij}, \tag{19.2.5}$$

which is equivalent to the result of the theorem (19.2.1).

So long, we have discussed the method of solving a game by *maximin minimax* principle. Now let us try to solve a very simple 2×2 game by the above principle. Let the pay-off matrix be

$$\begin{array}{cc} & B \\ A & \begin{array}{c} \\ A_1 \\ \\ A_2 \end{array}\begin{bmatrix} B_1 & B_2 \\ 4 & -1 \\ & \\ 2 & 3 \end{bmatrix} \end{array}$$

From the pay-off matrix, it can be easily determined that

$$\max_{i}\cdot \min_{j}\cdot a_{ij} = 2 \text{ and } \min_{j}\cdot \max_{i}\cdot a_{ij} = 3.$$

Hence the problem cannot be solved and the value of the game cannot be determined by the above *maximin minimax* principle. To solve the game we shall use another technique which is called the principle of mixed strategy.

19.3 Formation of the pay-off matrix

▶ **Example 19.3.1** *Each of two players A and B shows one or two fingers simultaneously. The player B pays to A an amount equal to the total number of fingers shown; on the other hand A pays to B equal to the product of the numbers of finger shown. Form the pay-off matrix.*

Solution. The pay-off matrix is given below.

		B 1	*B* 2
A	1	$1+1=2, 1\times 1=1$ $2-1=1$	$1+2=3, 1\times 2=2$ $3-2=1$
A	2	$2+1=3, 2\times 1=2$ $3-2=1$	$2+2=4, 2\times 2=4$ $4-4=0$

The pay-off matrix is given below.

$$
\begin{array}{cc} & \quad B \\ & \begin{array}{cc} 1 & 2 \end{array} \\ A\ \begin{array}{c} 1 \\ 2 \end{array} & \begin{bmatrix} 1 & 1 \\ 1 & 0 \end{bmatrix}, \end{array}
$$

which is the pay of matrix of *A*.

▶ **Example 19.3.2** *A game is described in the following manner. There are two players A and B, A being maximizing player and B, is the minimizing player Each of the players A and B says, 'stone' or 'disk' or 'paper' which are denoted by S, D and P respectively. If one say S and the other says D, then the former wins three rupees from the later. Similarly D beats P and P beats S. On the other, if the two players mention the same item, then the game is a tie game. Form the pay-off matrix for the above-mentioned problem.*

Solution: The pay-off matrix is given below.

$$
\begin{array}{rl} & \qquad\qquad\qquad\qquad B \\ & \quad B_1 = S \qquad\qquad B_2 = D \qquad\qquad B_3 = P \\ \begin{array}{r} S = A_1 \\ A\ D = A_2 \\ P = A_3 \end{array} & \begin{bmatrix} (S,S)\to(0) & (S,D)\to(3) & (S,P)\to(-3) \\ (D,S)\to(-3) & (D,D)\to(0) & (D,P)\to(3) \\ (P,S)\to(3) & (P,D)\to(-3) & (P,P)\to(0) \end{bmatrix} \end{array}
$$

Here the table is self explanatory and the pay-off matrix is

$$\begin{array}{cc} & B \\ & \begin{array}{ccc} B_1 & B_2 & B_3 \end{array} \\ A \begin{array}{c} A_1 \\ A_2 \\ A_3 \end{array} & \begin{bmatrix} 0 & 3 & -3 \\ -3 & 0 & 3 \\ 3 & -3 & 0 \end{bmatrix} \end{array}$$

The matrix obtained is a skew symmetric matrix and such type of square game is called a symmetric game.

Exercise 19A

1. Solve the following games by using *maximin minimax* principle:

(i)

Player A \ Player B	B_1	B_2	B_3
A_1	4	−2	1
A_2	3	4	2
A_3	−3	4	0

(ii)

Player A \ Player B	B_1	B_2	B_3
A_1	6	3	−3
A_2	−2	1	2
A_3	5	4	6

(iii)

	B_1	B_2	B_3
A_1	4	−2	−1
A_2	1	5	−2
A_3	5	6	2
A_4	1	−2	−3

(iv)

	B_1	B_2	B_3
A_1	7	4	1
A_2	4	2	0
A_3	3	−1	−2
A_4	1	5	−3

2. Solve the following 4 × 4 game and prove that there are two saddle points.

A \ B	B_1	B_2	B_3	B_4
A_1	4	2	3	5
A_2	−2	−1	4	−3
A_3	5	2	3	3
A_4	4	0	0	1

[C.U. (Hons.) '90, '93]

3. Prove that following pay-off matrices do not possess any saddle point (i.e., the value of the game cannot be determined) and establish in each case $\min_j \cdot \max_i \cdot a_{ij} > \max_i \cdot \min_j \cdot a_{ij}$.

(i)

	B_1	B_2	B_3
A_1	4	5	2
A_2	1	4	6
A_3	3	2	6

(ii)

	B_1	B_2	B_3
A_1	6	0	4
A_2	−3	2	−1
A_3	4	−3	5

4. (i) Solve the following game and prove that game is a fair game. (ii) Solve the following game and prove that game is in favour of B.

(i)
$$A\begin{array}{c} \\ A_1 \\ A_2 \\ A_3 \\ A_4 \end{array}\overset{\displaystyle B}{\begin{array}{cccc} B_1 & B_2 & B_3 & B_4 \\ \hline 2 & -3 & -4 & 1 \\ 2 & 0 & 3 & 2 \\ 4 & -2 & -1 & 1 \\ 2 & -1 & 4 & 2 \end{array}}$$

(ii)
$$A\begin{array}{c} \\ A_1 \\ A_2 \\ A_3 \\ A_4 \end{array}\overset{\displaystyle B}{\begin{array}{cccc} B_1 & B_2 & B_3 & B_4 \\ \hline -2 & 1 & -3 & 4 \\ 3 & 4 & -2 & 1 \\ -2 & 3 & -3 & 4 \\ -1 & 2 & -4 & 5 \end{array}}$$

5. Solve the following games stating the optimal strategies and the saddle points etc.

(i)
$$\begin{bmatrix} 2 & 3 & 2 & 4 & 6 \\ 0 & -2 & 1 & 2 & 1 \\ -1 & 3 & 0 & -1 & 3 \\ 4 & 5 & -1 & 2 & 1 \\ 3 & 2 & -2 & 1 & -2 \end{bmatrix}$$

(ii)
$$\begin{bmatrix} 4 & -2 & -4 & -1 \\ 3 & 2 & -1 & 2 \\ 2 & 3 & -2 & -2 \\ -3 & 2 & -2 & 2 \end{bmatrix}$$

[C.U. (Hons.) '87]

Answers

1. (i) (A_2, B_3), $v = 2$; (ii) (A_3, B_2), $v = 4$; (iii) (A_3, B_3), $v = 2$; (iv) (A_1, B_3), $v = 1$. 2. $(A_1, B_2), (A_3, B_2)$, $v = 2$. 4. (i) (A_2, B_2), $v = 0$; (ii) (A_2, B_2), $v = -2$. Hence the game is in favour of B. 5. (i) $(A_1, B_3), (1, 3)$, $v = 2$. (ii) $(A_2, B_3), (2, 3)$, $v = -1$.

19.4 Mixed Strategy

We have seen that, in general, the game problem cannot be solved by using pure strategies. In that case the problem can be solved by using another technique or strategy, which is called *mixed strategy* and with this technique all game problems (two-person zero-sum games) can be solved where there exists no saddle point. The property of the existence of the value of the game is established by a theorem which is known as the *fundamental theorem of rectangular games* which we shall prove later. Now the problem before us is to explain the technique of *mixed strategy.* In the case of pure strategy, both players A and B select only a single move at a time say A_i and B_j at their discretion, irrespective of the move taken by other. In mixed strategy both players A and B select their m and n moves simultaneously. Let us define a set of m quantities $p_1, p_2, \cdots, p_i, \cdots, p_m$ such that $p_i \geq 0$, $i = 1, 2, \cdots, m$ and $\sum_{i=1}^{m} p_i = 1$. Now to obtain a least possible gain A may utilize the services of the moves $A_1, A_2, \cdots, A_m$ in such a way that A_i performs p_i times of the total work so that the total work will be done by the moves $A_1, A_2, \cdots, A_n$. Similarly if

$q_1, q_2, \cdots, q_n$ be a set of quantities such that $q_j \geq 0$, $j = 1, 2, \cdots, n$ and $\sum_{j=1}^{n} q_j = 1$ then B may utilize the services of the moves $B_1, B_2, \cdots, B_n$; $q_1, q_2, \cdots, q_n$ times of the total work respectively to serve his purpose ignoring all about the move or moves taken by his opponent A.

The quantities p_i and q_j associated with the ith or jth move of A and B respectively are called the *probabilities of the respective moves.*

We now define two variable vectors $\mathbf{p} = (p_1, p_2, \cdots, p_m)$ in E^m and $\mathbf{q} = (q_1, q_2, \cdots, q_n)$ in E^n. It is always possible to determine some particular value of $\mathbf{p}$ and $\mathbf{q}$ say $\mathbf{p}^*$ and $\mathbf{q}^*$ such that the value of the game can be determined, i.e., it is possible to determine the same value of "maximum of the minimum expected gain for A and the minimum of the maximum expected loss for B". In particular, if $p_i = 1$ for $i = k$ and $p_i = 0$ for all other values of i and $q_j = 1$ for $j = l$ and $q_j = 0$ for all other values of j, then the mixed strategy is simply reduced to a pure strategy with pure moves A_k and B_l of A and B respectively.

19.5 Definition

Pay-off function: Let $[a_{ij}]$ be any pay-off matrix of order $m \times n$. Then the pay-off function or mathematical expectation of a game which is denoted by $E(\mathbf{p}, \mathbf{q})$ is defined as

$$E(\mathbf{p}, \mathbf{q}) = \sum_{i=1}^{m} \sum_{j=1}^{n} a_{ij} p_i q_j,$$

where $\mathbf{p}$ and $\mathbf{q}$ are the mixed strategies for A and B respectively.

In matrix notation $E(\mathbf{p}, \mathbf{q})$ can be written as $E(\mathbf{p}, \mathbf{q}) = \mathbf{p}^t C \mathbf{q}$, where $C = [a_{ij}]$, the pay-off matrix.

In particular, if B takes his pure j-th move only then the expected gain of A is given by

$$E_j(\mathbf{p}) = \sum_{i=1}^{m} a_{ij} p_i, \quad j = 1, 2, \cdots, n.$$

Similarly for particular ith pure move of A only, the expected loss of B is given by

$$E_i(\mathbf{q}) = \sum_{j=1}^{n} a_{ij} q_j, \quad i = 1, 2, \cdots, m.$$

For any $\mathbf{p}$, A is sure that his expected winning will be at least $\min_{\mathbf{q}} E(\mathbf{p}, \mathbf{q})$. He then maximizes the expression over $\mathbf{p}$, so that his expected winnings will be at least $\max_{\mathbf{p}} \min_{\mathbf{q}} E(\mathbf{p}, \mathbf{q})$. That is the interpretation of the term $\max_{\mathbf{p}} \min_{\mathbf{q}} E(\mathbf{p}, \mathbf{q})$.

Theorem 19.5.1 *Let* $E(\mathbf{p}, \mathbf{q})$ *be such that both* $\min_{\mathbf{q}} \cdot \max_{\mathbf{p}} E(\mathbf{p}, \mathbf{q})$ *and* $\max_{\mathbf{p}} \min_{\mathbf{q}} E(\mathbf{p}, \mathbf{q})$ *exist, then* $\min_{\mathbf{q}} \cdot \max_{\mathbf{p}} E(\mathbf{p}, \mathbf{q}) \geq \max_{\mathbf{p}} \min_{\mathbf{q}} E(\mathbf{p}, \mathbf{q})$.

Proof. Let $\mathbf{p}^*$ and $\mathbf{q}^*$ be some arbitrarily chosen point in their given specified ranges. Then we must have,

$$\max_{\mathbf{p}} E(\mathbf{p}, \mathbf{q}^*) \geq E(\mathbf{p}^*, \mathbf{q}^*) \quad \text{and} \quad \min_{\mathbf{q}} E(\mathbf{p}^*, \mathbf{q}) \leq E(\mathbf{p}^*, \mathbf{q}^*).$$

Combining these two, we get

$$\max_{\mathbf{p}} E(\mathbf{p}, \mathbf{q}^*) \geq \min_{\mathbf{q}} E(\mathbf{p}^*, \mathbf{q}).$$

But $\mathbf{q}^*$ is any arbitrarily chosen point in the given specified range and hence the inequality holds for all values of $\mathbf{q}$. Hence if $\mathbf{q}^*$ be such a point for which $\max_{\mathbf{p}} E(\mathbf{p}, \mathbf{q})$ has the minimum value, the inequality remains true. Therefore,

$$\min_{\mathbf{q}} \cdot \max_{\mathbf{p}} E(\mathbf{p}, \mathbf{q}) \geq \min_{\mathbf{q}} E(\mathbf{p}^*, \mathbf{q}).$$

Again, since $\mathbf{p}^*$ is any point in its given range, the inequality holds even if we select that $\mathbf{p}^*$ which gives the maximum value of $\min E(\mathbf{p}, \mathbf{q})$.

Therefore,,

$$\min_{\mathbf{q}} \cdot \max_{\mathbf{p}} E(\mathbf{p}, \mathbf{q}) \geq \max_{\mathbf{p}} \min_{\mathbf{q}} E(\mathbf{p}, \mathbf{q}).$$

Note: From our previous notation of games with pure strategy if we denote that

$$\underline{v} = \min_{\mathbf{q}} \cdot \max_{\mathbf{p}} E(\mathbf{p}, \mathbf{q}) \quad \text{and} \quad \bar{v} = \max_{\mathbf{p}} \min_{\mathbf{q}} E(\mathbf{p}, \mathbf{q})$$

then $\underline{v} \geq \bar{v}$ in general.

But we shall later establish that three exists a point $(\mathbf{p}^*, \mathbf{q}^*)$ in their specified given range of $\mathbf{p}$ and $\mathbf{q}$ such that $\underline{v} = E(\mathbf{p}^*, \mathbf{q}^*) = v$ which is the value of the game in the case of mixed strategy and the point $(\mathbf{p}^*, \mathbf{q}^*)$ is called the strategic saddle point.

Solution of the 2×2 games using mixed strategy [Problem without a saddle point in the case of pure strategy in the pay-off matrix].

Let the pay-off matrix for A be

$$\text{Player } A \quad \begin{array}{c} \\ A_1 \\ A_2 \end{array} \overset{\text{Player } B}{\begin{array}{c} \begin{array}{cc} B_1 & B_2 \end{array} \\ \begin{bmatrix} a_{11} & a_{12} \\ a_{21} & a_{22} \end{bmatrix} \end{array}}$$

with the mixed strategies for A and B be $\mathbf{p} = (p_1, p_2)$ and $\mathbf{q} = (q_1, q_2)$ respectively such that $p_1 + p_2 = 1$, $q_1 + q_2 = 1$ and p_1, p_2, q_1, q_2 all lie in the open interval (0, 1). Because, in a particular case if the value of $p_1 = 0$ (say) then evidently the value of p_2 will be 1 which indicates the reduction of the problem to a problem of pure strategy. Now to satisfy the condition that there will be no saddle point in the pay-off matrix the following relations must hold good that one set of two diagonal elements must be greater than the other two diagonal elements and vice-versa or only for $(a_{11}, a_{22}) < > (a_{12}, a_{21})$ the condition will be satisfied. We assume that all pay-off matrices satisfy the given conditions stated above. Now the problem can be solved algebraically, easily. Taking the problem of mixed strategy given above if the player B moves to B_1, the net expected gain of A will be $E_1(\mathbf{p}) = a_{11}p_1 + a_{21}p_2 = a_{11}p_1 + a_{21}(1 - p_1)$ by using his mixed strategy $\mathbf{p} = (p_1, p_2)$ and $p_1 + p_2 = 1$.

Similarly for B's pure move B_2, the net expected gain of A is given by $E_2(\mathbf{p}) = a_{12}p_1 + a_{22}p_2 = a_{12}p_1 + a_{22}(1 - p_1)$.

Again if B plays his mixed strategy $\mathbf{q} = (q_1, q_2)$ with $q_1 + q_2 = 1$ then B's net expected loss will be

$$E_1(\mathbf{q}) = a_{11}q_1 + a_{12}q_2 = a_{11}q_1 + a_{12}(1 - q_1) \quad \text{if } A \text{ plays } A_1$$

and B's net expected loss will be

$$E_2(\mathbf{q}) = a_{21}q_1 + a_{22}q_2 = a_{21}q_1 + a_{22}(1 - q_1)$$

if A plays A_2. Now if v be the value of the game, then

$$E_1(\mathbf{p}) \geq v, E_2(\mathbf{p}) \geq v \quad \text{and} \quad E_1(\mathbf{q}) \leq v, E_2(\mathbf{q}) \leq v.$$

Now assuming the existence of the value of the game, and as both $0 < p_1, p_2 < 1$ and $0 < q_1, q_2 < 1$, then by the complementary slackness theorem we can say that all the inequations strictly reduce to equality [which will be shown in L.P.P. method] and we have therefore

$$E_1(\mathbf{p}) = v = E_2(\mathbf{p}) \tag{19.5.1}$$

$$\text{and} \quad E_1(\mathbf{q}) = v = E_2(\mathbf{q}). \tag{19.5.2}$$

From (19.5.1), we get

$$a_{11}p_1 + a_{21}(1 - p_1) = a_{12}p_1 + a_{22}(1 - p_1)$$

from which we get

$$p_1^* = \frac{a_{22} - a_{21}}{a_{11} + a_{22} - (a_{12} + a_{21})} \tag{19.5.3}$$

Now using the relation $p_1^* + p_2^* = 1$ we get,

$$p_2^* = \frac{a_{11} - a_{12}}{a_{11} + a_{22} - (a_{12} + a_{21})}, \tag{19.5.4}$$

which are the optimal strategies for A.

Similarly from (19.5.2) we get,

$$q_1^* = \frac{a_{22} - a_{12}}{a_{11} + a_{22} - (a_{12} + a_{21})}, \quad q_2^* = \frac{a_{11} - a_{21}}{a_{11} + a_{22} - (a_{12} + a_{21})}, \tag{19.5.5}$$

which are the optimal strategies for B and the value of the game will be

$$v = \frac{a_{11}a_{22} - a_{21}a_{12}}{a_{11} + a_{22} - (a_{12} + a_{21})}. \tag{19.5.6}$$

Now it can be easily verified that all $p_1^*, p_2^*, q_1^*, q_2^*$, lie in the open interval (0, 1). Moreover, the value of the game v will be

$$E(\mathbf{p}^*, \mathbf{q}^*) = \sum_{i=1}^{m}\sum_{j=1}^{n} a_{ij}p_i^* q_j^*.$$

Again, we can establish the fact that whatever strategies may be taken by B, the value of the pay-off function will remain same with the assumption $E_1(\mathbf{p}) = v = E_2(\mathbf{p})$. For example, if B uses the mixed strategies $q_1 = (q_1, q_2)$, with $q_1 + q_2 = 1$

$$\begin{aligned}
\text{the pay-off function} &= E(\mathbf{p}, \mathbf{q})\\
&= a_{11}p_1q_1 + a_{12}p_1q_2 + a_{21}p_2q_1 + a_{22}p_2q_2\\
&= (a_{11}p_1 + a_{21}p_2)q_1 + (a_{12}p_1 + a_{22}p_2)q_2\\
&= vq_1 + vq_2 = v(q_1 + q_2) = v \quad \text{[the value of the game]}
\end{aligned}$$

Similarly, the value of the pay-off function will remain same with the assumption $E_1(\mathbf{q}) = v = E_2(\mathbf{q})$ whatever strategies may be taken by A.

Before solving a 2×2 game problem by using the formula of mixed strategy discussed above, let us solve a problem from the stand point of theory.

Solve the game from the stand point of theory

$$\begin{array}{cc} & \overbrace{\hphantom{B_1 \quad B_2}}^{B} \\ & \begin{array}{cc} B_1 & B_2 \end{array} \\ A \begin{array}{c} A_1 \\ A_2 \end{array} & \begin{bmatrix} 4 & -1 \\ -2 & 3 \end{bmatrix} \end{array}$$

Let $\mathbf{p} = (p_1, p_2), p_1 + p_2 = 1, p_1, p_2 \geq 0$ and $\mathbf{q} = (q_1, q_2), q_1 + q_2 = 1, q_1, q_2 \geq 0$ be the mixed strategies used by player A and B respectively. Then the pay-off function

$$\begin{aligned}
E(\mathbf{p}, \mathbf{q}) &= \sum_{i=1}^{2}\sum_{j=1}^{2} a_{ij}p_iq_j = a_{11}p_1q_1 + a_{21}p_2q_1 + a_{12}p_1q_2 + a_{22}p_2q_2\\
&= 4p_1q_1 - 2p_2q_1 - p_1q_2 + 3p_2q_2\\
&\qquad [\because a_{11} = 4, a_{21} = -2, a_{12} = -1, a_{22} = 3]\\
&= 4p_1q_1 - 2(1 - p_1)q_1 - p_1(1 - q_1) + 3(1 - p_1)(1 - q_1)\\
&= 10p_1q_1 - 4p_1 - 5q_1 + 3\\
&= p_1(10q_1 - 4) - \frac{1}{2}(10q_1 - 4) + 3 - 2\\
&= \left(p_1 - \frac{1}{2}\right)(10q_1 - 4) + 1.
\end{aligned} \tag{19.5.7}$$

Under the circumstance, given in (19.5.3) the least expected gain of A and maximum expected loss of B will be 1 if A and B adhere to the strategies $p_1^* = \frac{1}{2}$ for A and $q_1^* = \frac{2}{5}$ for B respectively. Thus the value of the game $v = 1$ and the optimal strategies for A and B be $p^* = \left(\frac{1}{2}, \frac{1}{2}\right), \mathbf{q}^* = \left(\frac{2}{5}, \frac{3}{5}\right)$. [taking $p_1 - \frac{1}{2} = 0$ and $10q_1 - 4 = 0$].

Now we can verify the result by the formula discussed above detailed discussion is given below.

$$\begin{aligned} E_1(\mathbf{p}) &= 4p_1 - 2p_2 = v = -p_1 + 3p_2 = E_2(\mathbf{p}) \\ &\Rightarrow 4p_1 - 2(1 - p_1) = -p_1 + 3(1 - p_1) \\ &\Rightarrow 10p_1 = 5 \Rightarrow p_1^* = \frac{1}{2}\cdot \text{ Hence } p_2^* = \frac{1}{2}. \\ E_1(\mathbf{q}) &= 4q_1 - 2q_2 = v = -2q_1 + 3q_2 = E_2(\mathbf{q}) \\ &\Rightarrow 10q_1 = 4, \quad \text{or,} \quad q_1^* = \frac{2}{5} \text{ and } q_2^* = \frac{3}{5}. \end{aligned}$$

Hence the optimal strategies are $\mathbf{p}^* = \left(\frac{1}{2}, \frac{1}{2}\right), \mathbf{q}^* = \left(\frac{2}{5}, \frac{3}{5}\right)$ and the value of game $v = 4p_1^* - 2p_2^* = 2 - 1 = 1$.

► **Example 19.5.1** *Find the value of the 2×2 game algebraically by using mixed strategies.*

		Player A	
		B_1	B_2
Player B	A_1	2	3
	A_2	4	−1

Solution: The problem has no saddle point in the case of pure strategy in the pay-off matrix for A.

Let us try to solve the problem by using mixed strategies $\mathbf{p} = (p_1, p_2)$ and $\mathbf{q} = (q_1, q_2)$ with $p_1 + p_2 = 1$, $p_1, p_2 > 0$ and $q_1 + q_2 = 1$, $q_1, q_2 > 0$ for A and B respectively. Assuming the existence of the value of game we have

$$\begin{aligned} E_1(\mathbf{p}) &= 2p_1 + 4p_2 = 2p_1 + 4(1 - p_1) \\ E_2(\mathbf{p}) &= 3p_1 - p_2 = 3p_1 - (1 - p_1). \end{aligned}$$

To determine the optimal values of p_1 and p_2 we have,

$$2p_1 + 4(1 - p_1) = v = 3p_1 - (1 - p_1).$$

Solving we get $p_1^* = \frac{5}{6} > 0$. Therefore, $p_2^* = 1 - p_1^* = \frac{1}{6}$ and value of the game is

$$v = 2p_1^* + 4(1 - p_1^*) = 2 \times \frac{5}{6} + 4 - 4 \times \frac{5}{6} = \frac{7}{3}.$$

Again considering from the B's point of view we have,

$$\begin{aligned} E_1(\mathbf{q}) &= 2q_1 + 3q_2 = 2q_1 + 3(1 - q_1) \\ E_2(\mathbf{q}) &= 4q_1 - q_2 = 4q_1 - (1 - q_1). \end{aligned}$$

To determine the optimal values of q_1 and q_2 we have,

$$2q_1 + 3(1 - q_1) = v = 4q_1 - (1 - q_1).$$

Solving we get $q_1^* = \frac{2}{3} > 0$ and $q_2^* = 1 - q_1^* = \frac{1}{3}$ and the value of the game is

$$2q_1^* + 3(1 - q_1^*) = 2 \times \frac{2}{3} + 3 - 3 \times \frac{2}{3} = \frac{7}{3}.$$

Hence the optimal strategies are $\mathbf{p}^* = \left(\frac{5}{6}, \frac{1}{6}\right)$ and $q^* = \left(\frac{2}{3}, \frac{1}{3}\right)$ and $v = \frac{7}{3}$.

Again, it can be verified that

$$E(p^*, \mathbf{q}^*) = 2 \times \frac{5}{6} \times \frac{2}{3} + 4 \times \frac{1}{6} \times \frac{2}{3} + 3 \times \frac{5}{6} \times \frac{1}{6} - 1 \times \frac{1}{6} \times \frac{1}{3} = \frac{7}{3}$$

(value of the game).

Exercise 19B

1. Solve the following 2 × 2 games [games without saddle point] using mixed strategies:

$$\text{(i)}\quad \begin{array}{ccc} & & \text{Player } A \\ & & \begin{array}{cc} B_1 & B_2 \end{array} \\ \text{Player } B & \begin{array}{c} A_1 \\ A_2 \end{array} & \begin{bmatrix} 6 & -4 \\ -1 & 2 \end{bmatrix} \end{array} \qquad \text{(ii)}\quad \begin{array}{ccc} & & \text{Player } A \\ & & \begin{array}{cc} B_1 & B_2 \end{array} \\ \text{Player } B & \begin{array}{c} A_1 \\ A_2 \end{array} & \begin{bmatrix} -1 & 6 \\ 5 & 2 \end{bmatrix} \end{array}$$

$$\text{(iii)}\quad \begin{array}{ccc} & & \text{Player } A \\ & & \begin{array}{cc} B_1 & B_2 \end{array} \\ \text{Player } B & \begin{array}{c} A_1 \\ A_2 \end{array} & \begin{bmatrix} -2 & 4 \\ 5 & -3 \end{bmatrix} \end{array} \qquad \text{(iv)}\quad \begin{array}{ccc} & & \text{Player } A \\ & & \begin{array}{cc} B_1 & B_2 \end{array} \\ \text{Player } B & \begin{array}{c} A_1 \\ A_2 \end{array} & \begin{bmatrix} 1 & -2 \\ -1 & 3 \end{bmatrix} \end{array}$$

$$\text{(v)}\quad \begin{array}{ccc} & & \text{Player } B \\ & & \begin{array}{cc} B_1 & B_2 \end{array} \\ \text{Player } A & \begin{array}{c} A_1 \\ A_2 \end{array} & \begin{bmatrix} a & -b \\ -c & d \end{bmatrix} \end{array} \; a, b, c, d > 0 \qquad \text{(vi)}\quad \begin{array}{cc} & Q \\ P & \begin{bmatrix} 4 & 2 \\ 1 & 5 \end{bmatrix} \end{array}$$

Answers

(i) $\mathbf{p}^* = \left(\frac{3}{13}, \frac{10}{13}\right)$, $\mathbf{q}^* = \left(\frac{6}{13}, \frac{7}{13}\right)$, $v = \frac{8}{13}$. **(ii)** $\mathbf{p}^* = \left(\frac{3}{10}, \frac{7}{10}\right)$, $\mathbf{q}^* = \left(\frac{2}{5}, \frac{3}{5}\right)$, $v = \frac{16}{5}$.

(iii) $\mathbf{p}^* = \left(\frac{4}{7}, \frac{3}{7}\right)$, $\mathbf{q}^* = \left(\frac{1}{2}, \frac{1}{2}\right)$, $v = 1$. **(iv)** $\mathbf{p}^* = \left(\frac{4}{7}, \frac{3}{7}\right)$, $\mathbf{q}^* = \left(\frac{5}{7}, \frac{2}{7}\right)$, $v = \frac{1}{7}$.

(v) $\mathbf{p}^* = \left(\frac{c+d}{a+b+c+d}, \frac{a+b}{a+b+c+d}\right)$, $\mathbf{q}^* = \left(\frac{b+d}{a+b+c+d}, \frac{a+c}{a+b+c+d}\right)$, $v = \frac{ad-bc}{a+b+c+d}$.

(vi) $\mathbf{p}^* = \left(\frac{2}{3}, \frac{1}{3}\right)$, $\mathbf{q}^* = \left(\frac{1}{2}, \frac{1}{2}\right)$, $v = 3$.

2. A tosses two coins at a time. He receives Rs. 2 and Rs. 3 for two heads and two tails respectively and he losses Re. 1 for one head and one tail. Find the value of the game.

Hints: $$\begin{array}{c} \\ H \\ T \end{array}\begin{array}{c} \begin{array}{cc} H & T \end{array} \\ \begin{bmatrix} 2 & -1 \\ -1 & 3 \end{bmatrix} \end{array}$$ Ans: $\mathbf{p}^* = \left(\frac{4}{7}, \frac{3}{7}\right), \mathbf{q}^* = \left(\frac{4}{7}, \frac{3}{7}\right), v = \frac{5}{7}$.

19.6 Graphical solution of $2 \times n$ or $m \times 2$ games

In the previous section, we have discussed the method of finding out the value of the game of 2×2 pay-off matrix (without any saddle point), algebraically. But it is not possible to solve easily any rectangular game of $m \times n$ order algebraically. But by using graph, it is possible to reduce any rectangular game of order $2 \times n$ or $m \times 2$ to a 2×2 game and then it can be solved by algebraic method. Let a particular $2 \times n$ game be

$$\begin{array}{cc} & \text{Player } B \\ & \begin{array}{cccc} B_1 & B_2 & \cdots & B_n \end{array} \\ \text{Player } A \begin{array}{c} A_1 \\ \\ A_2 \end{array} & \begin{bmatrix} a_{11} & a_{12} & \cdots & a_{1n} \\ & & & \\ a_{21} & a_{22} & \cdots & a_{2n} \end{bmatrix} \end{array}$$

without any saddle point. Let the mixed strategies used by A be $\mathbf{p} = (p_1, p_2)$ and B be $\mathbf{q} = (q_1, q_2, \cdots, q_n)$. Then the net expected gain of a A when B plays his pure strategy B_j is given by

$$E_j(\mathbf{p}) = a_{1j}p_1 + a_{2j}p_2 = a_{1j}p_1 + a_{2j}(1 - p_1) \quad j = 1, 2, \cdots, n \qquad (19.6.1)$$

as $p_1 + p_2 = 1$. Now both p_1 and p_2 must lie in the open interval (0, 1) [because if either p_1 or $p_2 = 1$ the game reduces to a game of pure strategy which is against our assumption]. Hence $E_j(\mathbf{p})$ is a linear function of either p_1 or p_2. Considering $E_j(\mathbf{p})$ as a linear function of p_1 (say) we have from the limiting values (0, 1) of p_1

$$\left.\begin{aligned} E_j(\mathbf{p}) &= a_{2j} \text{ for } p_1 = 0 \\ &= a_{1j} \text{ for } p_1 = 1 \end{aligned}\right\} \text{ from (19.6.1)}$$

Hence $E_j(\mathbf{p})$ represents a line segment joining the points $(0, a_{2j})$ and $(1, a_{1j})$.

Now A expects a least possible gain u. Therefore $E_j(\mathbf{p}) \geq u$ for $\forall j$. The main object of A is to select p_1, p_2 in such a way that $p_1 + p_2 = 1$ and both lie in the open interval (0, 1). Since p_1, p_2 lie in the open interval (0, 1) using complementary slackness theorem we may assume that for optimal moves, the inequations strictly reduce to equations and with the help of graphical method, we shall be in a position to find the two particular moves (critical moves) of B which will maximize the minimum gain of A.

Let us draw two parallel vertical lines, distance between them being one unit length; first one represents the line $p_1 = 0$ and the 2nd one represents $p_1 = 1$. Now draw n line segments joining the points $(0, a_{2j})$ and $(1, a_{1j}), j = 1, 2, \cdots, n$. The lower envelope of these line segments (indicate it by thick line segments) will give the minimum expected gain of A as a function of p_1. Now the highest point[1] of the lower envelope will give the maximum of minimum gain of A. The line segments passing through the point corresponding to B's two pure moves say B_k and B_l are the critical moves for B which will maximize the minimum expected gain of A. Now the 2×2 pay-off matrix corresponding to A's moves A_1 and A_2 and B's moves B_k and B_l will produce the required result and thus solving the 2×2 pay-off matrix algebraically we shall be in a position to determine the value of the game. Using similar technique we may solve any $m \times 2$ game.

If the original $2 \times n$ pay-off matrix be

$$\begin{array}{cc} & \text{Player } B \\ & \begin{array}{cccc} B_1 & B_2 & \cdots & B_n \end{array} \\ \text{Player } A \begin{array}{c} A_1 \\ \\ A_2 \end{array} & \left[\begin{array}{cccc} a_{11} & a_{12} & \cdots & a_{1n} \\ \\ a_{21} & a_{22} & \cdots & a_{2n} \end{array}\right] \end{array}$$

then to reduce the game into 2×2 game draw to vertical lines $p_1 = 0$ and $p_1 = 1$ at unit distance apart as given in the figure. First line is $p_1 = 0$ and the second line is $p_1 = 1$. Join the points as given in the figure and name the line segments B_j for B's jth pure move.

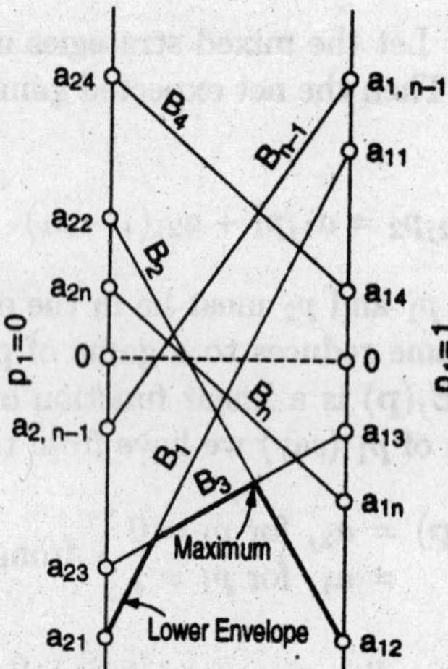

Fig. 19.1

The highest point of the lower envelope determines the two critical moves of B.

[1]If there are more than two line segments passing through the highest point (maximin) there are ties for the optimum mixed strategies for the player B. Thus any two such lines with opposite sign slopes will determine an alternative optimum for B.

Again if there are more than one maximin points alternative optimum exist corresponding to these points.

Let the critical moves be B_k and B_l. Then solving the 2 × 2 pay-off matrix

$$A\begin{matrix} \\ A_1 \\ A_2 \end{matrix}\overset{\displaystyle B}{\begin{matrix} B_k & B_l \\ \end{matrix}}\begin{bmatrix} a_{1k} & a_{1l} \\ a_{2k} & a_{2l} \end{bmatrix}$$

we get the value of the game with the corresponding probabilities taken to determine the value of the game and q_j will be all zero except for q_k and q_l.

▶ **Example 19.6.1** *Solve the following* 2 × 4 *game geometrically.*

$$\text{Player } A \begin{matrix} A_1 \\ A_2 \end{matrix} \overset{\displaystyle \text{Player } B}{\overset{\begin{matrix} B_1 & B_2 & B_3 & B_4 \end{matrix}}{\begin{bmatrix} 3 & 2 & -1 & 4 \\ 2 & 5 & 6 & -2 \end{bmatrix}}}$$

Solution: The problem does not possess a saddle point. Let the player A play with mixed strategy $\mathbf{p} = (p_1, p_2)$, $p_1 + p_2 = 1$ and both p_1 and p_3 lie in the open interval (0, 1).

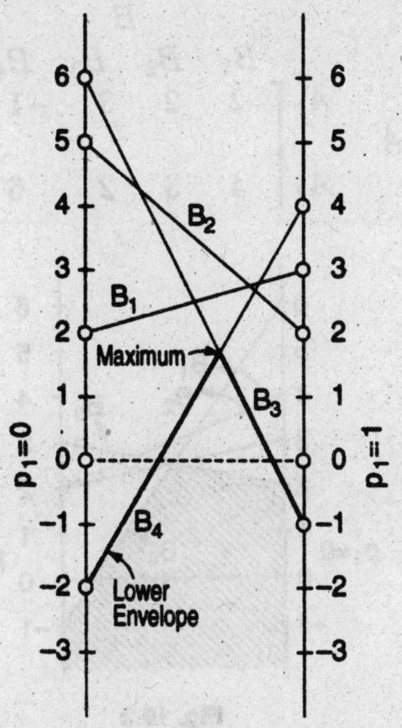

Fig. 19.2

Draw two vertical lines $p_1 = 0$ and $p_1 = 1$ at unit distance apart. Mark the lines $p_1 = 0$ and $p_1 = 1$ by using the same scale as given in the figure 19.2. Draw a line segment joining the point (0, 2) and (1, 3) which represents A's expected gain due to B's pure move B_1. Number the line segment as B_1. Similarly draw line segments joining the points (0, 5) and (1, 2) which represents A's expected gain due to B's pure move B_2. Number the line segment as B_2. Similarly draw the line segments B_3

and B_4. The maximin point of the lower envelope lies on the point of intersection of the line segments B_3 and B_4. Therefore, the game can be solved ultimately solving the 2×2 pay-off matrix given below

$$\text{Player } A \quad \begin{array}{c} \\ A_1 \\ A_2 \end{array} \overset{\text{Player } B}{\begin{array}{cc} B_3 & B_4 \end{array}} \begin{bmatrix} -1 & 4 \\ 6 & -2 \end{bmatrix}$$

with mixed strategies $\mathbf{p} = (p_1, p_2)$, $p_1 + p_2 = 1$ for A and $\mathbf{q} = (q_3, q_4)$ $q_3 + q_4 = 1$ for B. Then using the formula given in the previous article we have,

$$p_1^* = \frac{-2-6}{-1-2-(6+4)} = \frac{8}{13} \text{ therefore } p_2^* = 1 - \frac{8}{13} = \frac{5}{13}.$$

$$q_3^* = \frac{-2-4}{-1-2-(6+4)} = \frac{6}{13} \text{ therefore } q_4^* = 1 - \frac{6}{13} = \frac{7}{13}$$

and value of the game $= \dfrac{(-1)\times(-2) - 6 \times 4}{-1-2-(6+4)} = \dfrac{22}{13} = v.$

Hence ultimately the solution of the problem is given by $\mathbf{p}^* = \left(\frac{8}{13}, \frac{5}{13}\right)$ for A and $\mathbf{q}^* = \left(0, 0, \frac{6}{13}, \frac{7}{13}\right)$ for B and the value of the game is $\frac{22}{13}$.

▶ **Example 19.6.2** *Solve the following* 2×4 *game by graphical method.*

$$A \quad \begin{array}{c} \\ A_1 \\ A_2 \end{array} \overset{B}{\begin{array}{cccc} B_1 & B_2 & B_3 & B_4 \end{array}} \begin{bmatrix} 2 & 2 & 3 & -1 \\ 4 & 3 & 2 & 6 \end{bmatrix}$$

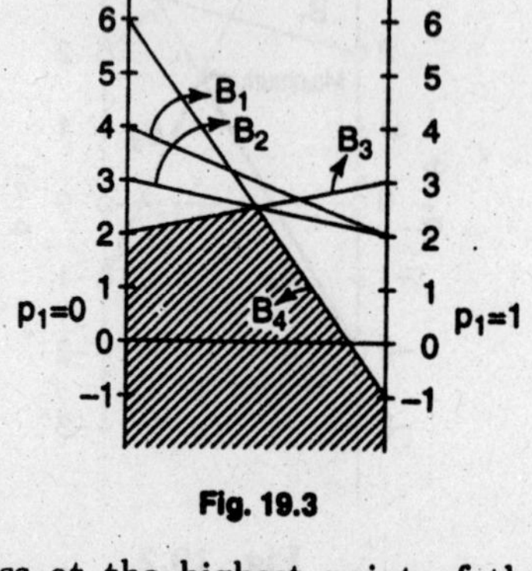

Fig. 19.3

Here the three lines pass at the highest point of the lower envelope. Thus accordingly we get $^3c_2 = 3$ square matrices of order 2 and there are three optimal solutions. But actually we shall have to select such pair of lines which have opposite sign slope. Thus one square matrix is (B_2, B_3) and other is (B_3, B_4)

and the square matrices are $\begin{array}{c} \\ A_1 \\ A_2 \end{array} \overset{\begin{array}{cc} B_2 & B_3 \end{array}}{\begin{bmatrix} 2 & 3 \\ 3 & 2 \end{bmatrix}}$ and $\begin{array}{c} \\ A_1 \\ A_2 \end{array} \overset{\begin{array}{cc} B_3 & B_4 \end{array}}{\begin{bmatrix} 3 & -1 \\ 2 & 6 \end{bmatrix}}$ and solving by usual method the value of the game is 5/2 and the optimal solutions set are $p^* = (1/2, 1/2)$, $q^*(0, 1/2, 1/2, 0)$ and $p^* = (1/2, 1/2)$, $q^*(0, 0, 7/8, 1/8)$.

Any $m \times 2$ game problem can be solved by using B's mixed strategy $\mathbf{q} = (q_1, q_2)$, $q_1 + q_2 = 1$ and both lie in the open interval (0, 1) and A's particular two critical moves can be determined graphically. In this case the problem is to determine minimum of the maximum expected loss of B. Thus we shall have to select the lowest (minimax) point of the upper envelope and A's critical moves are the moves, line segments corresponding to which pass through the minimax point. Now selecting a 2×2 pay-off matrix, the value of the game can be determined easily.

▶ **Example 19.6.3** *Solve the* 3×2 *game graphically.*

$$\text{Player } A \quad \begin{array}{c} \\ A_1 \\ A_2 \\ A_3 \end{array} \overset{\text{Player } B}{\begin{array}{cc} B_1 & B_2 \\ \end{array}} \begin{bmatrix} 2 & -3 \\ -2 & 5 \\ 0 & -1 \end{bmatrix}$$

This problem does not possess a saddle point.

To solve the problem, draw two vertical straight lines at unit distance apart; the equation of the first being $q_1 = 0$ and the second being $q_1 = 1$. Join (0, −3) and (1, 2) to get the line segment corresponding to A's pure move A_1. Similarly join (0, 5) and (1, −2) [excluding the points] to get the line segment corresponding to A's pure move A_2. From the figure we get the minimax point of the upper envelope, which is the point of intersection of A's two critical moves A_1 and A_2. Therefore the problem ultimately can be solved by solving the 2×2 pay-off matrix.

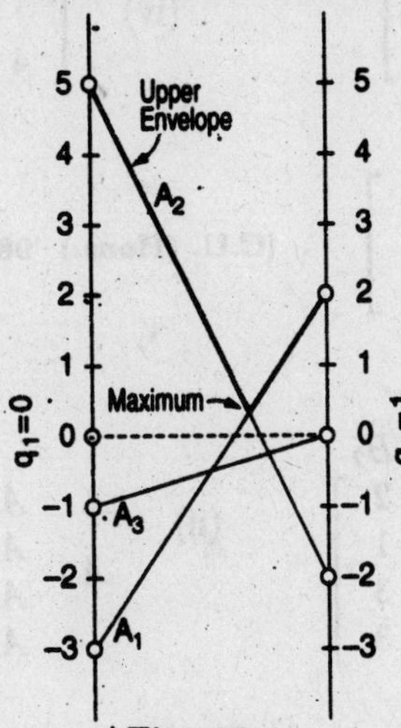

Fig. 19.4

$$\text{Player } A \quad \begin{array}{c} \\ A_1 \\ A_2 \end{array} \overset{\text{Player } B}{\begin{array}{cc} B_1 & B_2 \\ \end{array}} \begin{bmatrix} 2 & -3 \\ -2 & 5 \end{bmatrix}$$

with the mixed strategies $\mathbf{p} = (p_1, p_2)$, $p_1 + p_2 = 1$ for A and $\mathbf{q} = (q_1, q_2)$, $q_1 + q_2 = 1$ for B. Now using the formula given in (9.5.3) etc. we have,

$$p_1^* = \frac{5-(-2)}{2+5-(-2-3)} = \frac{7}{12}, \quad p_2^* = 1 - \frac{7}{12} = \frac{5}{12}, \quad p_3^* = 0.$$

$$q_1^* = \frac{5-(-3)}{2+5-(-2-3)} = \frac{2}{3}, \quad q_2^* = 1 - \frac{2}{3} = \frac{1}{3}$$

and the value of the game is

$$v = \frac{2 \times 5 - (-2) \times (-3)}{2+5-(2-3)} = \frac{1}{3}.$$

Exercise 19C

Solve the following problems graphically:

1. (i) $A \begin{matrix} \\ A_1 \\ A_2 \end{matrix} \overset{B}{\begin{matrix} B_1 & B_2 & B_3 & B_4 \\ \end{matrix}} \begin{bmatrix} 4 & 5 & 2 & -1 \\ 6 & 3 & 7 & 4 \end{bmatrix}$ (ii) $A \begin{matrix} A_1 \\ A_2 \end{matrix} \overset{B \quad B_1 \; B_2 \; B_3 \; B_4}{\begin{bmatrix} 0 & 5 & -2 & 3 \\ 2 & 3 & 4 & 1 \end{bmatrix}}$

(iii) $\begin{bmatrix} 2 & 1 & 0 & -2 \\ 1 & 0 & 3 & 2 \end{bmatrix}$ (iv) $\begin{bmatrix} 2 & 2 & 3 & -1 \\ 4 & 3 & 2 & 6 \end{bmatrix}$

[C.U. (Hons.) '81,'88,'92]

(v) $\begin{bmatrix} 1 & 2 & -3 & 7 \\ 2 & 5 & 4 & -6 \end{bmatrix}$ [C.U. (Hons.) '98]

2. (i) $A \begin{matrix} A_1 \\ A_2 \\ A_3 \\ A_4 \end{matrix} \overset{B \quad B_1 \; B_2}{\begin{bmatrix} -1 & 2 \\ 4 & 1 \\ 1 & 3 \\ 2 & 3 \end{bmatrix}}$ (ii) $A \begin{matrix} A_1 \\ A_2 \\ A_3 \\ A_4 \end{matrix} \overset{B \quad B_1 \; B_2}{\begin{bmatrix} 2 & -1 \\ 3 & 2 \\ -1 & 5 \\ -2 & 1 \end{bmatrix}}$

(iii) $A \begin{matrix} A_1 \\ A_2 \\ A_3 \end{matrix} \overset{B \quad B_1 \; B_2}{\begin{bmatrix} 2 & 7 \\ 3 & 5 \\ 11 & 2 \end{bmatrix}}$ (iv) $\begin{bmatrix} 1 & -3 \\ 3 & 5 \\ -1 & 6 \\ 4 & 1 \\ 2 & 2 \\ -5 & 0 \end{bmatrix}$

[C.U. (Hons.) '86] [C.U. (Hons.) '86]

Answers

1.(*i*) $\mathbf{p}^* = (\frac{1}{7}, \frac{6}{7})$, $\mathbf{q}^* = (0, \frac{5}{7}, 0, \frac{2}{7})$, $v = \frac{23}{7}$; (*ii*) $\mathbf{p}^* = (\frac{1}{4}, \frac{3}{4})$, $\mathbf{q}^* = (\frac{1}{2}, 0, 0, \frac{1}{2})$, $v = \frac{3}{2}$;
(*iii*) $\mathbf{p}^* = (\frac{2}{5}, \frac{3}{5})$, $\mathbf{q}^* = (0, \frac{4}{5}, 0, \frac{1}{5})$, $v = \frac{2}{5}$; (*iv*) $\mathbf{p}^* = (\frac{1}{2}, \frac{1}{2})$, $\mathbf{q}^* = (0, \frac{1}{2}, \frac{1}{2}, 0)$, $v = \frac{5}{2}$;
or $\mathbf{p}^* = (\frac{1}{2}, \frac{1}{2})$, $\mathbf{q}^* = (0, 0, \frac{7}{8}, \frac{1}{8})$.

2.(*i*) $\mathbf{p}^* = (0, \frac{1}{4}, 0, \frac{3}{4})$, $\mathbf{q}^* = (\frac{1}{2}, \frac{1}{2})$, $v = \frac{5}{2}$;
(*ii*) $\mathbf{p}^* = (0, \frac{6}{7}, \frac{1}{7}, 0)$, $\mathbf{q}^* = (\frac{3}{7}, \frac{4}{7})$, $v = \frac{17}{7}$;
(*iii*) $\mathbf{p}^* = (\frac{9}{14}, 0, \frac{5}{14})$, $\mathbf{q}^* = (\frac{5}{14}, \frac{9}{14})$, $v = \frac{73}{14}$;
(*iv*) $\mathbf{p}^* = (0, \frac{3}{5}, 0, \frac{2}{5}, 0, 0)$, $\mathbf{q}^* = (\frac{4}{5}, \frac{1}{5})$, $v = \frac{17}{5}$;

19.7 Dominance property

We have already discussed the method of solving games with pure strategies by *maximin minimax* principle. Further we have discussed the method of solving a 2×2 game (having no saddle point) algebraically by using mixed strategies and also the method of solving $2 \times n$ games and $m \times 2$ games graphically. Now we discuss a particular method. In this method, the pay-off matrix can be reduced by mere observation and in many cases it may be solved only by adopting this method. In other cases it may be reduced to 2×2 game or $2 \times n$ game or $m \times 2$ game so that it can be solved by the methods discussed above. The method of reduction of the pay-off matrix by this process is called the dominance property of the rows and columns of the pay-off matrix.

Consider a particular pay-off matrix.

$$A \quad \begin{array}{c} \\ \\ A_1 \\ A_2 \\ A_3 \end{array} \begin{array}{c} B \\ \begin{array}{cccc} B_1 & B_2 & B_3 & B_4 \end{array} \\ \begin{bmatrix} -5 & 3 & 1 & 20 \\ 5 & 5 & 4 & 6 \\ -4 & -2 & 0 & -5 \end{bmatrix} \end{array}$$

[C.U. (Hons.) '85]

We have seen that for A's pure move A_3, all elements of the third row are less than or equal to the corresponding elements of the second row. Hence A will not definitely play his pure strategy A_3. Then the pay-off matrix reduces to.

$$\begin{array}{c} \\ A_1 \\ A_2 \end{array} \begin{array}{c} \begin{array}{cccc} B_1 & B_2 & B_3 & B_4 \end{array} \\ \begin{bmatrix} -5 & 3 & 1 & 20 \\ 5 & 5 & 4 & 6 \end{bmatrix} \end{array}$$

Again for pure moves B_2 and B_4 of B, all elements of the second and fourth columns are greater than or equal to the corresponding elements of the third column.

Therefore, B definitely does not play his pure moves B_2 and B_4. Therefore, the pay-off matrix reduces to $\begin{matrix} & B_1 & B_3 \\ A_1 & -5 & 1 \\ A_2 & 5 & 4 \end{matrix}$. Again for A's pure move A_1, all elements of the first row are less than or equal to the corresponding elements of the second row. Hence A will reject his pure move A_1. Then the pay-off matrix reduces to $\begin{matrix} & B_1 & B_3 \\ A_2 & 5 & 4 \end{matrix}$. Now from the pay-off matrix, it is evident that B will select his pure move B_3 and the value of the game is 4 and the optimal strategies are A_2 and B_3 for A and B respectively.

► **Example 19.7.1** *In a rectangular game the pay-off matrix is given by*

		B_1	B_2	B_3	B_4	B_5
A	A_1	10	5	5	20	4
	A_2	11	15	10	17	25
	A_3	7	12	8	9	8
	A_4	5	13	9	10	5

State giving reasons, whether the players will use pure or mixed strategies. What is the value of the game? **[C.U. (Hons.) '89, 94]**

Solution: Initially we try to solve the problem by dominance method. A_2 dominates both the fourth and third rows. Therefore, we can ignore or drop the third and fourth rows and the resulting matrix is

	B_1	B_2	B_3	B_4	B_5
A_1	10	5	5	20	4
A_2	11	15	10	17	25

The third column dominates simultaneously first, second and fourth columns and we ignore the corresponding columns and the resulting matrix is

	B_3	B_5
A_1	5	4
A_2	10	25

A_2 dominates A_1 and thus we ignore first row and the resulting matrix is

	B_3	B_5
A_2	10	25

In this matrix B_3 dominates B_5 and thus the value of the game is 10 (A_2, B_3) and (2, 3) is the saddle point.

Now we try to solve by maximin principle.

		B_1	B_2	B_3	B_4	B_5	min row
	A_1	10	5	5	20	4	4
	A_2	11	15	10	17	25	10
A	A_3	7	12	8	9	8	7
	A_4	5	13	9	10	5	5
	Max col	11	15	10	20	25	10 (Max)
							10 Min

Thus the value of the game is $10(A_2, B_3)$ (2, 3) is saddle point.

Remark. Each and every rectangular game can be solved by using mixed strategy. Thus the first part of the question is quite inconsistent and irrelevant. Here we are able to solve the problem by pure strategy also.

19.8 General rule for dominance

(i) If all the elements of kth row are less than or equal to the corresponding elements of rth row, then we shall say that kth row is dominated by rth row and we reject the kth row and we can take p_k to be equal to zero which is obvious from logical point of view. The proof of this theorem will be given at some later stage.

(ii) If all the elements of the lth column are greater than than or equal to the corresponding elements of sth column we shall say that lth column is dominated by sth column and we shall reject the lth column and we take q_l to be zero.

The theorem (19.10.5) will establish these.

19.9 Modified dominance property

A pure move or strategy of a maximizing (minimizing) player may be dominated if it be inferior (superior) to a convex combination of two or more strategies of that player etc. This property is called the modified dominance property. The following examples will give clear idea about the use of modified dominance property.

► **Example 19.9.1** *Reduce the following game to 2×2 game by using dominance and modified dominance property and then solve the game.*

Table 1

		B_1	B_2	B_3	B_4
	A_1	1	2	−2	2
A	A_2	3	1	2	3
	A_3	−1	3	2	1
	A_4	−2	2	0	−3

Step 1. All the elements of fourth row are less than the corresponding elements of third row. Thus R_4 is dominated by R_3 and thus drop the fourth row. The resulting matrix is then given in table 2.

Table 2

$$A \begin{array}{c} \\ A_1 \\ A_2 \\ A_3 \end{array} \begin{array}{c} \begin{array}{cccc} B_1 & B_2 & B_3 & B_4 \end{array} \\ \begin{bmatrix} 1 & 2 & -1 & 2 \\ 3 & 1 & 2 & 3 \\ -1 & 3 & 2 & 1 \end{bmatrix} \end{array}$$

Step 2. All the elements of fourth column are greater or equal to all the corresponding elements of first column. Then drop the fourth column. The resulting matrix is given in table 3.

Table 3

$$\begin{array}{c} \\ A_1 \\ A_2 \\ A_3 \end{array} \begin{array}{c} \begin{array}{ccc} B_1 & B_2 & B_3 \end{array} \\ \begin{bmatrix} 1 & 2 & -1 \\ 3 & 1 & 2 \\ -1 & 3 & 2 \end{bmatrix} \end{array}$$

Step 3. By using a convex combination of 2nd row and third row we have $\frac{1}{2}R_2 + \frac{1}{2}R_3 = [1, 2, 2]$, all the elements are greater or equal to he corresponding elements of the first row. Then drop the first row and the resulting matrix is given in table 4.

Table 4

$$\begin{array}{c} \\ A_1 \\ A_2 \end{array} \begin{array}{c} \begin{array}{ccc} B_1 & B_2 & B_3 \end{array} \\ \begin{bmatrix} 3 & 1 & 2 \\ -1 & 3 & 2 \end{bmatrix} \end{array}$$

Step 4. Now one convex combination of C_1 and C_2 $[\frac{1}{2}C_1 + \frac{1}{2}C_2] = \begin{bmatrix} 2 \\ 1 \end{bmatrix}$ are less than or equal to the corresponding elements of the third column, and thus drop the third column, and the resulting matrix is given in table 5.

Now using the formula for solving the 2×2 game we have,

Table 5

$$\begin{array}{c} \\ A_1 \\ A_2 \end{array} \begin{array}{c} \begin{array}{cc} B_1 & B_2 \end{array} \\ \begin{bmatrix} 3 & 1 \\ -1 & 3 \end{bmatrix} \end{array}$$

$$p_2^* = \frac{3-(-1)}{3+3-(1-1)} = \frac{2}{3};$$
$$p_3^* = 1 - \frac{2}{3} = \frac{1}{3}.$$
$$q_1^* = \frac{3-1}{3+3-(1-1)} = \frac{1}{3};$$
$$q_2^* = 1 - \frac{1}{3} = \frac{2}{3}$$

and the value of the game is

$$\frac{3\times 3-(1\times -1)}{3+3-(1-1)} = \frac{5}{3}.$$

Thus the optimal strategies are

$$\mathbf{p}^* = \left(0, \frac{2}{3}, \frac{1}{3}, 0\right), \quad \mathbf{q}^* = \left(\frac{1}{3}, \frac{2}{3}, 0, 0\right), \quad v = \frac{5}{3}.$$

► **Example 19.9.2** ***Solve the game problem by reducing into 2 × 2 problem using the dominance property.***

$$\begin{bmatrix} 0 & 0 & 0 & 0 & 0 \\ 4 & 2 & 0 & 2 & 1 \\ 4 & 3 & 1 & 3 & 2 \\ 4 & 3 & 4 & -1 & 2 \end{bmatrix}$$

[C.U. (Hons.) '87]

Observations: We can simply drop the first and second row as they are dominated by the third row. Similarly, the first and second column can be dropped as they are dominated by the fourth or fifth column. The resultant matrix is

$$\begin{array}{cc} & \begin{array}{ccc} B_3 & B_4 & B_5 \end{array} \\ \begin{array}{c} A_3 \\ A_4 \end{array} & \begin{bmatrix} 1 & 3 & 2 \\ 4 & -1 & 2 \end{bmatrix} \end{array}$$

Now one convex combination of C_3 and C_4 give $\frac{1}{2}(C_3 + C_4) = (2, 3/2)$. Then C_5 is dominated by the convex combination of C_3 and C_4. Hence drop C_5 and the resultant matrix is

$$\begin{array}{cc} & \begin{array}{cc} B_3 & B_4 \end{array} \\ \begin{array}{c} A_3 \\ A_4 \end{array} & \begin{bmatrix} 1 & 3 \\ 4 & -1 \end{bmatrix} \end{array}$$

Now solving algebraically we have

$$p_3^* = \frac{-1-(4)}{1-1-(4+3)} = \frac{5}{7}; \quad p_4^* = 1 - \frac{5}{7} = \frac{2}{7}.$$
$$q_3^* = \frac{-1-(3)}{1-1(4+3)} = \frac{4}{7} \quad \text{and} \quad q_4^* = 1 - \frac{4}{7} = \frac{3}{7}$$

and the values of the game $= \frac{13}{7}$.

Hence the optimal strategies are

$$\mathbf{p}^* = \left(0, 0, \frac{5}{7}, \frac{2}{7}\right), \quad \mathbf{q}^* = \left(0, 0, \frac{4}{7}, \frac{3}{7}, 0\right), \quad v = \frac{13}{7}.$$

▶ **Example 19.9.3** *Can we reduce the pay-off matrix given below*

$$A \begin{array}{c} \\ A_1 \\ A_2 \\ A_3 \end{array} \begin{array}{c} B \\ \begin{array}{ccc} B_1 & B_2 & B_3 \end{array} \\ \begin{bmatrix} 3 & -2 & 4 \\ -1 & 4 & 2 \\ 2 & 2 & 6 \end{bmatrix} \end{array}$$

to a 2×2 *matrix by using dominance property? If not, reduce the problem to a* 2×2 *matrix by graphical method and solve it and discuss about the peculiar nature of the problem.*

Solution: We notice that the third column is dominated by the first column and hence we reject the third column and then we can take

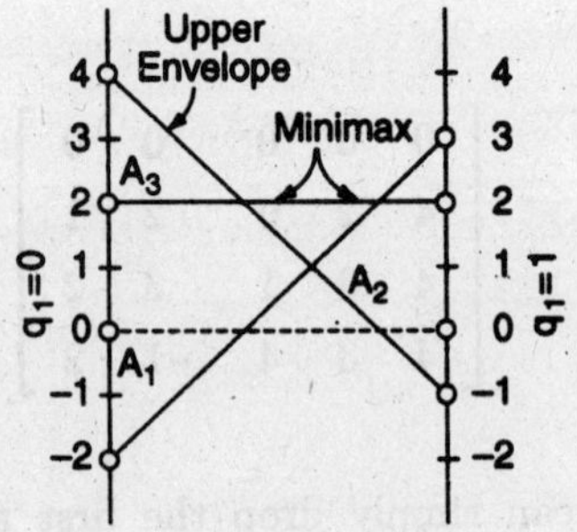

Fig. 19.5

$$\begin{array}{c} \\ A_1 \\ A_2 \\ A_3 \end{array} \begin{array}{c} \begin{array}{cc} B_1 & B_2 \end{array} \\ \begin{bmatrix} 3 & -2 \\ -1 & 4 \\ 2 & 2 \end{bmatrix} \end{array}$$

$q_3^* = 0$ and the resulting matrix is given above. Now, it is not possible to find a convex combination of any two rows which dominates the remaining row. On the other hand, it can be shown that a particular convex combination of the first and the second row is $\frac{1}{3}(3, -2) + \frac{1}{2}(-1, 4) = (1, 1)$, which is being dominated by the third row which indicates that the third move of the row player A will be an optimal move but we are not certain which one of the first two moves will be an optimal one. To determine it, we shall have to use the graphical method. From the graph, we can trace that there are two lowest points of the uppermost envelope and hence there are two sets of optimal moves for A, (A_1, A_3) and (A_2, A_3). For (A_1, A_3) moves, the multiple optimal strategies are ($p_1^* = 0$, $p_2^* = 0$, $p_3^* = 1$), ($q_1^* = 4/5$, $q_2^* = 1/5$, $q_3^* = 0$); ($p_1^* = 0$, $p_2^* = 0$, $p_3^* = 1$), ($q_1^* = 0$, $q_2^* = 1$, $q_3^* = 0$). For (A_1, A_3) moves, the multiple optimal strategies are ($p_1^* = 0$, $p_2^* = 0$, $p_3^* = 1$), ($q_1^* = 1$, $q_2^* = 0$, $q_3^* = 0$); ($p_1^* = 0$, $p_2^* = 0$, $p_3^* = 1$), ($q_1^* = 2/5$, $q_2^* = 3/5$, $q_3^* = 0$) and the value of the game is 2 which is same in all cases.

Remark: Since there are two lowest points of the upper envelope, then both (A_3, A_1) and (A_3, A_2) are optimal moves which indicates the peculiar nature of the problem. But if we consider the nature of the two pay-off matrices given below

$$\text{(i)} \begin{bmatrix} 3 & -2 & 4 \\ -1 & 4 & 2 \\ 2 & 1.5 & 6 \end{bmatrix} \text{ and (ii)} \begin{bmatrix} 3 & -2 & 4 \\ -1 & 4 & 2 \\ 1.5 & 2 & 6 \end{bmatrix}$$

we get only single set of optimal moves for each problem and for

(i) the 2×2 matrix is $\begin{bmatrix} -1 & 4 \\ 2 & 1.5 \end{bmatrix}$ and for

(ii) 2×2 matrix is $\begin{bmatrix} 3 & -2 \\ 1.5 & 2 \end{bmatrix}$

which can be determined uniquely only by the graphical method (not possible even by using modified dominance).

Exercise 19D

1. (i) Solve the problem by using dominance property.

(ii) Reduce the following pay-off matrix to 2×2 matrix and thus solve the problem.

(i)

	B_1	B_2	B_3
A_1	3	1	1
A_2	5	2	−1
A_3	4	2	−1

(ii)

	B_1	B_2	B_3	B_4
A_1	4	2	3	2
A_2	−2	4	6	4
A_3	2	1	3	5

2. Reduce the following game to 2×2 game by modified dominance property and then solve it.

	B_1	B_2	B_3	B_4
A_1	3	2	4	0
A_2	3	4	2	4
A_3	4	2	4	0
A_4	0	4	0	8

[C.U. (Hons.) '90]

3. Reduce the following games to 2×2 game and then solve the games.

(i)

	B_1	B_2	B_3
A_1	4	1	2
A_2	2	3	5
A_3	2	1	3

, (ii)

$$\begin{bmatrix} 3 & 5 & 4 & 2 \\ 5 & 6 & 2 & 4 \\ 3 & 1 & 4 & 0 \\ 3 & 3 & 5 & 2 \end{bmatrix}$$

4. Solve the following games by reducing 2×2 games using dominance and modified dominance properties.

$$\text{(i)} \begin{bmatrix} -1 & -1 & 2 & 1 \\ 2 & 2 & 0 & 1 \\ 3 & -2 & 1 & -2 \\ 2 & 1 & -3 & 2 \end{bmatrix}, \quad \text{(ii)} \begin{bmatrix} 2 & 0 & -1 & 1 \\ 3 & 3/2 & 3 & 0 \\ 2 & 3 & 3 & 1 \\ 2 & 5/2 & 3 & 3 \end{bmatrix}$$

Hints: (i) Remove C_1 and R_3, $\frac{1}{2}(C_2 + C_3)$ removes C_4. Remove R_4 and then solve.

(ii) Remove R_1 and C_3, $\frac{1}{2}(C_1 + C_4)$ removes C_2, Remove R_3 and then solve.

5. Use dominance and modified dominance property to reduce the pay-off matrix given by

$$\begin{array}{c} \\ \text{Player } A \end{array} \overset{\text{Player } B}{\begin{bmatrix} 3 & -1 & 1 & 2 \\ -2 & 3 & 2 & 3 \\ 2 & -2 & -1 & 1 \end{bmatrix}}$$

into a 2×2 matrix and find the mixed strategies for A and B and the value of the game. **[C.U. M. Com. '84]**

6. Two candidates in an election campaign selected three cities as crucial and potentially worth a last visit. A survey report indicates the estimated gain (in thousand votes) for candidate A as follows. Find the priorities of cities to be visited by both the candidates. Satisfy that the values of the game is 1/3 for either of the candidates.

Candidate B

Candidate A: Cities	1	2	3
1	12	–9	14
2	–8	7	12
3	11	–10	10

[C.U. M. Com. '88]

Answers

1. (i) (A_1, B_3), $v = 1$. (ii) $\mathbf{p}^* = (\frac{3}{4}, \frac{1}{4}, 0)$, $\mathbf{q}^* = (\frac{1}{4}, \frac{3}{4}, 0, 0)$, $v = \frac{5}{2}$. 2. $\mathbf{p}^* = (0, 0, \frac{2}{3}, \frac{1}{3})$, $\mathbf{q}^* = (0, 0, \frac{2}{3}, \frac{1}{3})$, $v = \frac{8}{3}$. 3. $\mathbf{p}^* = (\frac{1}{4}, \frac{3}{4}, 0)$, $\mathbf{q}^* = (\frac{1}{2}, \frac{1}{2}, 0)$, $v = \frac{5}{2}$, (ii) $\mathbf{p}^* = (0, \frac{3}{5}, 0, \frac{2}{5})$, $\mathbf{q}^* = (0, 0, \frac{2}{5}, \frac{3}{5})$, $v = \frac{16}{5}$. 4. (i) $\mathbf{p}^* = (\frac{2}{5}, \frac{3}{5}, 0, 0)$, $\mathbf{q}^* = (0, \frac{2}{5}, \frac{3}{5}, 0)$, $v = \frac{4}{5}$; (ii) $\mathbf{p}^* = (0, \frac{1}{4}, 0, \frac{3}{4})$, $\mathbf{q}^* = (\frac{3}{4}, 0, 0, \frac{1}{4})$. 5. $\mathbf{p}^* = (\frac{5}{9}, \frac{4}{9}, 0, 0,)$, $\mathbf{q}^* = (\frac{4}{9}, \frac{5}{9}, 0, 0)$, $v = \frac{7}{9}$. 6. $\mathbf{p}^* = (\frac{5}{12}, \frac{7}{12}, 0)$, $\mathbf{q}^* = (\frac{4}{9}, \frac{5}{9}, 0)$.

Reduction of a game problem to a L.P.P.: (A very important theorem).

Theorem 19.9.1 *Every game problem (two-person zero-sum game) can be converted to a L.P.P.*

Proof: Let the pay-off matrix of a game problem be (a_{ij}) $i = 1, \cdots, m$ and $j = 1, \cdots, n$ and with proper adjustment [by adding a suitable constant to each element of the matrix] we may assume that $a_{ij} > 0$, $\forall i$ and j, i.e., if the value of the game exists, then it will be a positive quantity. Let the mixed strategies used by A and B be

$$\mathbf{p} = (p_1, p_2, \cdots, p_m), \quad \sum_{i=1}^{m} p_i = 1, \quad p_i \geq 0, \forall i$$

$$\mathbf{q} = (q_1, q_2, \cdots, q_n), \quad \sum_{j=1}^{n} q_j = 1, \quad q_j \geq 0, \forall j, \text{ respectively.}$$

The net expected gain of A when B plays his pure move B_j is

$$E_j(\mathbf{p}) = \sum_{i=1}^{m} a_{ij} p_i, \quad j = 1, 2, \cdots, n.$$

Now A will expect a least possible gain u. Hence

$$E_j(\mathbf{p}) = \sum_{i=1}^{m} a_{ij} p_i \geq u, \quad j = 1, 2, \cdots, n.$$

As all $a_{ij} > 0$, then u will be essentially a positive quantity.

Now A's problem is to select $p_1, p_2, \cdots, p_m$ in such a way that they will satisfy the conditions stated above and u attains its maximum.

Since $u > 0$, maximization of u is equivalent to minimization of $1/u$. Now dividing $E_j(\mathbf{p}) \geq u$ by u, A's problem can be written as

Minimize $\left\{ p_0 = \dfrac{1}{u} = \dfrac{p_1 + p_2 + \cdots + p_m}{u} = p_1' + p_2' + \cdots + p_m' \right\}$

subject to $\displaystyle\sum_{i=1}^{m} a_{ij} p_i' \geq 1$, $j = 1, 2, \cdots, n$, where

$$p_i' = \frac{p_i}{u} \geq 0, \quad i = 1, 2, \cdots, m. \qquad (19.9.1)$$

Using the same logic, B's problem can be written as

Maximize $\left\{ q_0 = \dfrac{1}{w} = \dfrac{q_1 + q_2 + \cdots + q_n}{w} = q_1' + q_2' + \cdots + q_n' \right\}, w > 0$

subject to $\displaystyle\sum_{j=1}^{n} a_{ij} q_j' \leq 1$, $i = 1, 2, \cdots, m$ where

$$q_j' = \frac{q_j}{w} \geq 0, \quad j = 1, 2, \cdots, n. \qquad (19.9.2)$$

Thus considering from A and B's point of view, we find that a game problem can be reduced to a L.P.P., in the form of (19.9.1) and (19.9.2) respectively.

19.10 Fundamental theorem of a rectangular game

Theorem 19.10.1 *If mixed strategies be allowed, the value of a game exists and unique (optimal strategies may be different).*

Proof: We have already noticed in theorem (19.9.1) that each rectangular game can be reduced to a L.P.P. considering both from A and B's point of view.

A's problem is

$$\text{Minimize, } p_0 = \frac{1}{u} = p'_1 + p'_2 + \cdots + p'_m \text{ subject to}$$

$$\sum_{i=1}^{m} a_{ij} p'_i \geq 1, \quad j = 1, 2, \cdots, n,$$

$$\text{where } p'_i = \frac{p_i}{u} \geq 0, \quad i = 1, 2, \cdots, m, \quad u > 0 \qquad (19.10.1)$$

and B's problem is

$$\text{Maximize, } q_0 = \frac{1}{w} = q'_1 + q'_2 + \cdots + q'_n \text{ subject to}$$

$$\sum_{j=1}^{n} a_{ij} q'_j \geq 1, \quad i = 1, 2, \cdots, m,$$

$$\text{where } q'_j = \frac{q_i}{w} \geq 0, \quad j = 1, 2, \cdots, n, \quad w > 0. \qquad (19.10.2)$$

Now one interesting point to note is that the two problems (19.10.1) and (19.10.2) are such that (19.10.1) is the dual of (19.10.2) and vice-versa.

Now using a pure strategy for A, and taking $0 < u \leq \min(a_{ij})$ we have

$$a_{1j} \times 0 + a_{2j} \times 0 + \cdots + a_{kj} p'_k + \cdots + a_{mj} \times 0 = a_{kj} p'_k = a_{kj} \frac{p_k}{u} = \frac{a_{kj}}{u} p_k \geq 1$$

for $p_k = 1$ and $\sum_{i=1}^{m} p_i = p_k = 1$ $[\because p_i = 0, i \neq k]$

which indicates that for a pure strategy of A, $\mathbf{p} = (0, 0, \cdots, p_k = 1, \cdots, 0)$ the L.P.P. (19.10.1) has at least one feasible solution set.

Similarly, it can be shown that the problem (19.10.2) has also a feasible solution set. Then using the vital property of duality theory, we can say that both problems (19.10.1) and (19.10.2) have finite optimal solutions and Min $p_0 =$ Max $q_0 = \frac{1}{v^*}$ (say), i.e., Max $u =$ Min $w = v^*$ which is the unique value of the game and the optimal strategies are

$\mathbf{p}^* = (p_1^*, p_2^*, \cdots, p_m^*)$, $\mathbf{q}^* = (q_1^*, q_2^*, \cdots, q_n^*)$ where
$p_i^* = p_i^{*\prime} \times v^*, i = 1, 2, \cdots, m$ and $q_j^* = q_j^{*\prime} \times v^*, j = 1, \cdots, n$.

Note: (1) If a quantity l has been added to make $a_{ij} > 0, \forall i, j$, then the value of the original game will be $v - l$, but the optimal strategies remain unaltered [see theorem 19.10.2].

(2) To get the value of a game by solving a L.P.P., we shall in general prefer B's problem as it is easier to solve B's problem without introducing any artificial variable etc.

(3) As the value of the game exists and unique, then both $\max_p \cdot \min_q E(\mathbf{p}, \mathbf{q})$ and $\min_q \cdot \max_p(\mathbf{p}, \mathbf{q})$ exist and $\max_p \cdot \min_q E(\mathbf{p}, \mathbf{q}) = \min_q \cdot \max_p(\mathbf{p}, \mathbf{q}) = E(\mathbf{p}^*, \mathbf{q}^*) = v$, the value of the game. The optimal strategies $(\mathbf{p}^*, \mathbf{q}^*)$ is called as generalised saddle point or strategic saddle point. Of course, the optimal strategies may not be unique.

(4) We have $\min_q E(\mathbf{p}^*, \mathbf{q}) = E(\mathbf{p}^*, \mathbf{q}^*)$ and hence $E(\mathbf{p}^*, \mathbf{q}) \geq E(\mathbf{p}^*, \mathbf{q}^*)$ and similarly we can prove that $E(\mathbf{p}^*, \mathbf{q}^*) \geq E(\mathbf{p}, \mathbf{q}^*)$ and hence $E(\mathbf{p}^*, \mathbf{q}) \geq E(\mathbf{p}^*, \mathbf{q}^*) = v \geq E(\mathbf{p}, \mathbf{q}^*)$.

1. Convert the following game problem to a L.P.P. we consider B's problem and add the number 3 to each element of the pay-off matrix to make the value of game positive and the pay-off matrix then is given in table (2). [$\because$ in table (2), $1 \leq v \leq 3$]

Table 1

	B_1	B_2	B_3
A_1	4	2	0
A_2	−1	1	0
A_3	4	2	−2

Table 2

	B_1	B_2	B_3
A_1	7	5	3
A_2	2	4	3
A_3	7	5	1

Now considering B's problem, the problem is

Maximize, $q_0 = q_1' + q_2' + q_3'$ subject to

$$\begin{aligned} 7q_1' + 5q_2' + 3q_3' &\leq 1 \\ 2q_1' + 4q_2' + 3q_3' &\leq 1 \\ 7q_1' + 5q_2' + q_3' &\leq 1, \quad q_1' \geq 0, q_2' \geq 0, q_3' \geq 0. \end{aligned}$$

Now if Max $q_0 = \frac{1}{v^*}$ (say) then the value of the game of original problem is $v^* - 3$ and the optimal strategy for B, is

$$q_j^* = q_j^{*'} \times v^* \quad [j = 1, 2, 3].$$

Convert the following game problems to L.P.P.

(i)
$$\begin{array}{c} \begin{matrix} B_1 & B_2 & B_3 \end{matrix} \\ \begin{bmatrix} 2 & 1 & 4 \\ 3 & 0 & 2 \\ 5 & 1 & -1 \end{bmatrix} \end{array}$$

(ii)
	B: 1	2	3
A_1	3	−1	−3
A_2	−3	3	−1
A_3	−4	−3	3

[C.U. (Hons.) '86, '93]

► **Example 19.10.1** *The effectivity of advertisements in different media by competitive farms A and B is given by a pay-off matrix* (Rs. 000) *as follows:*

$$\begin{array}{c c} & B \\ & \begin{array}{ccc} \text{Radio} & \text{T.V.} & \text{Press} \end{array} \\ A \begin{array}{c} R \\ T \\ P \end{array} & \begin{bmatrix} 5 & 3 & 2 \\ 2 & -2 & 5 \\ -1 & -5 & 1 \end{bmatrix} \end{array}$$

[C.U. (Hons.) '87]

(i) Satisfy that there is no saddle point and no pure strategy for optimum returns, with reasons.

(ii) Use dominance property to evaluate, the mixed strategies of A and B and the value of the game. Test the results starting from A's and B's point of view.

(iii) Write down the problem in L.P.P. for maximization and minimization of the value of the game from A's and B's angles respectively.

Solution: (i) Using minimax principle we find that there is no saddle point by using pure strategy and value of the game v,

		B: R	T	P	Row minima
A	R	5	3	2	2
	T	2	−2	5	−2
	P	−1	−5	1	−5
	Column maxima	5	3	5	3/2 (max) min

lies in the interval $2 \le v \le 3$ and then the value of the game cannot be determined by using pure strategy as $\underline{v} = 3 > 2 = \bar{v}$.

(ii) By using dominance property we see that third row is dominated by the first row and thus we drop the third row and the resulting matrix is

$$\begin{array}{c c} & B \\ & \begin{array}{ccc} R & T & P \end{array} \\ A \begin{array}{c} R \\ T \end{array} & \begin{bmatrix} 5 & 3 & 2 \\ 2 & -2 & 5 \end{bmatrix} \end{array}$$

Now we observe that first column is dominated by the second column and thus we drop the first column and the resulting matrix is given by

$$\begin{array}{c c} & B \\ & \begin{array}{cc} T & P \end{array} \\ A \begin{array}{c} R \\ T \end{array} & \begin{bmatrix} 3 & 2 \\ -2 & 5 \end{bmatrix} \end{array}$$

Now using mixed strategy for 2 × 2 game we have if $\mathbf{p}^* = (p_1^*, p_2^*, p_3^*)$ and $\mathbf{q}^* = (q_1^*, q_2^*, q_3^*)$ then $p_3^* = 0$ and $q_1^* = 0$ and p_1, p_2 etc. are obtained by using the formula (19.5.3) and (19.5.5) etc.

$$p_1^* = \frac{5-(-2)}{(3+5)-(-2+2)} = \frac{7}{8} \quad \text{and then} \quad p_2^* = 1 - \frac{7}{8} = \frac{1}{8}.$$

$$q_2^* = \frac{5-2}{3+5-(-2+2)} = \frac{3}{8} \quad \text{and} \quad q_3^* = 1 - \frac{3}{8} = \frac{5}{8}$$

and the value of the game is

$$v = \frac{3 \times 5 - \{2 \times -2\}}{3+5-(-2+2)} = \frac{19}{8} \text{ (in thousand Rs.)}$$

Thus optimal strategies are

$$\mathbf{p}^* = \left(\frac{7}{8}, \frac{1}{8}, 0\right) \quad \text{and} \quad \mathbf{q}^* = \left(0, \frac{3}{8}, \frac{5}{8}\right).$$

Thus A will not at all use the press medium and B will not use the radio medium. Since q_2^* and $q_3^* > 0$, then from complementary slackness theorem we can say that the second and third constraints of A's problem will be strictly equations and hence to test the result from A's point of view, we have

$$v = a_{12}p_1^* + a_{22}p_2^* + a_{32}p_3^* = 3 \times \frac{7}{8} - 2 \times \frac{1}{8} = \frac{19}{8}.$$

Similarly,

$$v = a_{13}p_1^* + a_{23}p_2^* + a_{33}p_3^* = 2 \times \frac{7}{8} + 5 \times \frac{1}{8} = \frac{19}{8}.$$

In the same way, the result can be verified from B's point of view.

From the point (i) we know that the value of game is positive.

(iii) Hence the L.P.P., from B's point of view is

$$\text{Maximize, } q_0 = q_1' + q_2' + q_3' \text{ subject to}$$

$$\begin{aligned} 5q_1' + 3q_2' + 2q_3' &\le 1 \\ 2q_1' - 2q_2' + 5q_3' &\le 1 \\ -q_1' + 5q_2' + q_3' &\le 1, \quad q_1', q_2', q_3' \ge 0. \end{aligned}$$

Now if max $q_0 = 1/v^*$, the v^* is the value of the game and $q_j^* = q_j'^* v^*$ etc. from A's point of view is

$$\text{Minimize, } p_0 = p_1' + p_2' + p_3', \text{ subject to}$$

$$\begin{aligned} 5p_1' + 2p_2' + p_3' &\ge 1 \\ 3p_1' - 2p_2' - 5p_3' &\ge 1 \\ 2p_1' + 5p_2' + p_3' &\ge 1, \quad p_1', p_2', p_3' \ge 0 \end{aligned}$$

and max q_0 = min $p_0 = 1/v^*$, where v^* is the value of the game, and $p_i^* = p_i'^* \times v^*$ etc.

▶ **Example 19.10.2** *Solve the game problem by converting it into a L.P.P.*

$$\begin{array}{c} \\ A_1 \\ \\ A_2 \end{array}\begin{array}{c} \begin{array}{cc} B_1 & B_2 \end{array} \\ \left[\begin{array}{cc} 3 & 1 \\ & \\ -1 & 2 \end{array}\right] \end{array}$$

The value of the game may not be positive. Adding 2 to each element we get a pay-off matrix whose value will be essentially positive and solving the second problem we can find the value of the original problem. The pay-off matrix after such addition is $\begin{bmatrix} 5 & 3 \\ 1 & 4 \end{bmatrix}$.

Let the optimal strategies for A and B is $\mathbf{p}^* = (p_1^*, p_2^*)$ and $\mathbf{q}^* = (q_1^*, q_2^*)$.

Now considering B's problem, it can be reduced to

$$\text{Maximize, } q_0 = q_1' + q_2' \text{ subject to}$$

$$5q_1' + 3q_2' \leq 1$$
$$q_1' + 4q_2' \leq 1\,, \quad q_1' \geq 0 \text{ and } q_2' \geq 0.$$

Now if Max $q_0 = 1/v^*$, the value of the game of the original problem will be $v^* - 2$ and the optimal solution $q_j = q_j^{*\prime} \times v^*[j = 1, 2]$, $p_i^* = p_i^{'*} \times v^*[i = 1, 2]$.

Simplex tables

		c	1	1	0	0	
Basic	$\mathbf{c}_B$	$\mathbf{b}$	$\mathbf{a}_1$	$\mathbf{a}_2$	$\mathbf{a}_3(\mathbf{e}_1)$	$\mathbf{a}_4(\mathbf{e}_2)$	
$\mathbf{a}_3^*$	0	1	5^*	3	1	0	$\frac{1}{5}^*$
$\mathbf{a}_4$	0	1	1	4	0	1	$\frac{1}{1} = 1$
$z_j - c_j$		0	-1^*	-1	0	0	
$\mathbf{a}_1$	1	$\frac{1}{5}$	1	$\frac{3}{5}$	$\frac{1}{5}$	0	$\frac{1}{3}$
$\mathbf{a}_4^*$	0	$\frac{1}{5}$	0	$\frac{17}{5}^*$	$-\frac{1}{5}$	1	$\frac{4}{17}^*$
$z_j - c_j$		$\frac{1}{5}$	0	$-\frac{2}{5}^*$	$\frac{1}{5}$	0	
$\mathbf{a}_1$	1	$\frac{1}{17}$	1	0	$\frac{4}{17}$	$-\frac{3}{17}$	
$\mathbf{a}_2$	1	$\frac{4}{17}$	0	1	$-\frac{1}{17}$	$\frac{5}{17}$	
$z_j - c_j$		$\frac{5}{17}$	0	0	$\frac{3}{17}$	$\frac{2}{17}$	

Then

$$\max . q_0 = \frac{5}{17} = \frac{1}{v^*} \text{ (say) at } q_1^{'*} = \frac{1}{17} \text{ and } q_2^{'*} = \frac{4}{17}.$$

Then the value of the original game is

$$v^* - 2 = \frac{17}{5} - 2 = \frac{7}{5}$$

at

$$q_1^* = q_1^{*'} \times v^* = \frac{1}{17} \times \frac{17}{5} = \frac{1}{5}$$

$$\text{and } q_2^* = q_2^{'*} \times v^* = \frac{4}{17} \times \frac{17}{5} = \frac{4}{5}.$$

Now using duality theory we have

$$p_1^{'*} = \frac{3}{17} \quad \text{and} \quad p_2^{'*} = \frac{2}{17}$$

[which will be obtained from $z_j - c_j$ corresponding to slack vectors $\mathbf{a}_3$ and $\mathbf{a}_4$ respectively and hence

$$p_1^* = \frac{3}{17} \times \frac{17}{5} = \frac{3}{5} \quad \text{and} \quad p_2^* = \frac{2}{17} \times \frac{17}{5} = \frac{2}{5}.$$

Thus the optimal strategies are

$$\mathbf{p}^* = \left(\frac{3}{5}, \frac{2}{5}\right), \quad \mathbf{q}^* = \left(\frac{1}{5}, \frac{4}{5}\right), \quad v = \frac{7}{5}.$$

▶ **Example 19.10.3** *Solve the game problem with the help of L.P. method*

Table 1 — **Table 2**

$$A \overset{B}{\begin{bmatrix} -1 & -2 & 8 \\ 7 & 5 & -1 \\ 6 & 0 & 12 \end{bmatrix}} \qquad \overset{B}{\begin{bmatrix} 1 & 0 & 10 \\ 9 & 7 & 1 \\ 8 & 2 & 14 \end{bmatrix}}$$

The original pay-off matrix is given in Table 1.

We add 2 to each element, to make the value of the game essentially positive. [$\because$ in table 2, $2 \le v \le 7$].

The converted matrix is given in table 2.

Let the optimal strategies for A and B be $\mathbf{p} = (p_1, p_2, p_3)$ and $q = (q_1, q_2, q_3)$.

Now consider the problem of B, it can be reduced to

Maximize, $q_0 = q_1' + q_2' + q_3'$ subject to

$$\begin{aligned} q_1' + 0q_2' + 10q_3' &\le 1 \\ 9q_1' + 7q_2' + \quad q_3' &\le 1 \\ 8q_1' + 2q_2' + 14q_3' &\le 1, \quad q_1' \ge 0, q_2' \ge 0, q_3' \ge 0. \end{aligned}$$

Now if max $q_0 = 1/v^*$, the value of the original game is $v^* - 2$ and optimal solution is $q_j^* = q_j^{'*} \times v^*$ [$j = 1, 2, 3$].

Adding slack variables q_4', q_5', q_6' simplex tables are

		c	1	1	1	0	0	0
Basis	c_B	b	$\mathbf{a}_1$	$\mathbf{a}_2$	$\mathbf{a}_3$	$\mathbf{a}_4(\mathbf{e}_1)$	$\mathbf{a}_5(\mathbf{e}_2)$	$\mathbf{a}_6(\mathbf{e}_3)$
$\mathbf{a}_4$	0	1	1	0	10	1	0	0
$\mathbf{a}_5^*$	0	1	9	7*	1	0	1	0
$\mathbf{a}_6$	0	1	8	2	14	0	0	1
$z_j - c_j$		0	−1	−1*	−1	0	0	0
$\mathbf{a}_4$	0	1	1	0	10	1	0	0
$\mathbf{a}_2$	1	$\frac{1}{7}$	$\frac{9}{7}$	1	$\frac{1}{7}$	0	$\frac{1}{7}$	0
$\mathbf{a}_6^*$	0	$\frac{5}{7}$	$\frac{38}{7}$	0	$\frac{96}{7}$*	0	$-\frac{2}{7}$	1
$z_j - c_j$		$\frac{1}{7}$	$\frac{2}{7}$	0	$\frac{-6}{7}$*	0	$\frac{1}{7}$	0
$\mathbf{a}_5$		$\frac{23}{48}$						
$\mathbf{a}_2$		$\frac{13}{96}$						
$\mathbf{a}_3$		$\frac{5}{96}$						
$z_j - c_j$		$\frac{3}{16}$	$\frac{5}{8}$	0	0	0	$\frac{1}{8}$	$\frac{1}{16}$

Solution is optimal.

$$\max q_0 = \frac{3}{16} \text{ at } q_1'^* = 0, \quad q_2'^* = \frac{13}{96} \quad \text{and} \quad q_3'^* = \frac{5}{96}.$$

Hence the value of the game is $1/\frac{3}{16} - 2 = \frac{16}{3} - 2 = \frac{10}{3}$ and $q_1^* = 0$,

$$q_2^* = q_2'^* \times \frac{16}{3} = \frac{13}{96} \times \frac{16}{3} = \frac{13}{18} \quad \text{and} \quad q_3^* = \frac{5}{96} \times \frac{16}{3} = \frac{5}{18}.$$

Now for the last simplex table we get

$$p_1'^* = 0, \quad p_2'^* = \frac{1}{8} \quad \text{and} \quad p_3'^* = \frac{1}{16}.$$

Therefore,

$$p_1^* = 0, \quad p_2^* = \frac{1}{8} \times \frac{16}{3} = \frac{2}{3} \quad \text{and} \quad p_3^* = \frac{1}{16} \times \frac{16}{3} = \frac{1}{3}.$$

Therefore the solution of the game; the value of game $= \frac{10}{3}$ at

$$\mathbf{p}^* = \left(0, \frac{2}{3}, \frac{1}{3}\right) \quad \text{and} \quad \mathbf{q}^* = \left(0, \frac{13}{18}, \frac{5}{18}\right).$$

Now it can be easily verified that by applying dominance property the pay-off matrix can be reduced to

$$\begin{array}{cc} & \quad B \\ & \begin{array}{cc} B_1 & B_2 \end{array} \\ A \begin{array}{c} A_1 \\ \\ A_2 \end{array} & \begin{bmatrix} 5 & -1 \\ & \\ 0 & 12 \end{bmatrix} \end{array}$$

which can be solved easily by algebraically method.

► **Example 19.10.4** *Solve the game problem by using L.P. method*

$$\begin{array}{cc} & \text{Player } B \\ & \begin{array}{ccc} B_1 & B_2 & B_3 \end{array} \\ \text{Player } A \begin{array}{c} A_1 \\ A_2 \end{array} & \begin{bmatrix} 1 & 0 & -2 \\ 0 & 3 & 2 \end{bmatrix} \end{array}$$

The problem has no saddle point. Now adding a sufficiently large positive quantity 2 we get a new pay-off matrix with some elements 0 and the value of the game is now essentially a positive quantity.

$$\begin{array}{cc} & \begin{array}{ccc} B_1 & B_2 & B_3 \end{array} \\ \begin{array}{c} A_1 \\ A_2 \end{array} & \begin{bmatrix} 3 & 2 & 0 \\ 2 & 5 & 4 \end{bmatrix} \end{array}$$

Now considering the problem of B, we have to determine q_1, q_2, q_3 so as to maximize, $q_0 = q_1' + q_2' + q_3'$ subject to

$$\begin{aligned} 3q_1' + 2q_2' + \quad\quad &\leq 1 \\ 2q_1' + 5q_2' + 4q_3' &\leq 1\,, \quad q_1', q_2', q_3' \geq 0. \end{aligned}$$

Now if $\max q_0 = \frac{1}{v^*}$, the value of the original game is $v^* - 2$ and optimal solution is $q_j^{'*} \times v^* = q_j^*(j = 1, 2, 3)$.

Let us try to solve B's maximization problem by simplex method.

Introducing slack variables q_4' and q_5' one in each constraint, the final simplex table is

Basis	c_B	b	a_1	a_2	a_3	$a_4(e_1)$	$a_5(e_2)$
a_1	1	$\frac{1}{3}$	1	$\frac{2}{3}$	0	$\frac{1}{3}$	0
a_3	1	$\frac{1}{12}$	0	$\frac{11}{12}$	1	$-\frac{1}{6}$	$\frac{3}{4}$
$z_j - c_j$		$\frac{5}{12}$	0	$\frac{7}{12}$	0	$\frac{1}{6}$	$\frac{1}{4}$

Hence

$$\max q_0 = \frac{5}{12} \text{ at } q_1' = \frac{1}{3}, \quad q_2^{'*} = 0 \text{ and } q_3^{'*} = \frac{1}{12}.$$

The value of the game in the original problem is

$$v = \frac{12}{5} - 2 = \frac{2}{5}.$$

The optimal strategy for B is given by

$$q_1^* = q_1^{'*} \times \frac{12}{5} = \frac{4}{5}, \quad q_2^* = 0, \quad q_3^* = \frac{1}{12} \times \frac{12}{5} = \frac{1}{5}.$$

To get the optimal strategy for A, we have

$$\max q_0 = \min p_0 = \frac{5}{12} \text{ at } p_1'^* = \frac{1}{6} \text{ and } p_2'^* = \frac{1}{4}$$

[which are obtained from the components corresponding to slack vectors $\mathbf{a}_4$ and $\mathbf{a}_5$ in the $z_j - c_j$ row]

$$p_1^* = \frac{1}{6} \times \frac{12}{5} = \frac{2}{5}, \quad p_2^* = \frac{1}{4} \times \frac{12}{5} = \frac{3}{5}.$$

Hence the solution of the original problem is, value of the game $= \frac{2}{5}$ with the optimal strategies taken by A and B are

$$\mathbf{p}^* = \left(\frac{2}{5}, \frac{3}{5}\right) \text{ and } \mathbf{q}^* = \left(\frac{4}{5}, 0, \frac{1}{5}\right).$$

▶ **Example 19.10.5** *Determine the optimum strategies of A and B from the following pay-off matrix using Linear Programming technique.* **[C.U. M. Com. '90]**

$$\begin{array}{c} \quad B \\ A \begin{pmatrix} 1 & -4 \\ -3 & 4 \\ 2 & 2 \end{pmatrix} \end{array}$$

Solution: To make the value of the game, essentially a positive quantity we add 4 to each element of the pay-off matrix and the resulting matrix is

$$\begin{array}{c} \quad B \\ A \begin{pmatrix} 5 & 0 \\ 1 & 8 \\ 6 & 6 \end{pmatrix} \end{array}$$

Let us try to solve the problem considering from B's point of view. The linear programming problem in term of B is

$$\text{Maximize, } q_0 = q_1' + q_2' \text{ subject to}$$

$$\begin{aligned} 5q_1' + 0q_2' &\le 1 \\ q_1' + 8q_2' &\le 1 \\ 6q_1' + 6q_2' &\le 1, \quad q_1', q_2' \ge 0, \end{aligned}$$

and if v^* be the value of the game then $q_1^* = q_1'^* \times v^*, q_2^* = q_2{}'^* \times v^*$, etc.

Solving the above problem in usual simplex method, the optimal simplex table is

Basis	c_B	b	$\mathbf{a}_1$	$\mathbf{a}_2$	$\mathbf{a}_3(\mathbf{e}_1)$	$\mathbf{a}_4(\mathbf{e}_2)$	$\mathbf{a}_5(\mathbf{e}_3)$
$\mathbf{a}_3$	0	$\frac{16}{21}$	0	0	1	$\frac{5}{7}$	$-\frac{20}{21}$
$\mathbf{a}_2$	1	$\frac{5}{42}$	0	1	0	$\frac{1}{7}$	$-\frac{1}{42}$
$\mathbf{a}_1$	1	$\frac{1}{21}$	1	0	0	$-\frac{1}{7}$	$\frac{4}{21}$
$z_j - c_j$		$\frac{1}{6}$	0	0	0	0	$\frac{1}{6}$

Max $q_0 = \frac{1}{6}$, then $v^* = 6$ and the value of the original game is $6 - 4 = 2$ and the optimal strategies for B, is

$$q_1^* = q_1^{*'} \times 6 = \frac{1}{21} \times 6 = \frac{2}{7}, \quad q_2^* = q_2^{*'} \times 6 = \frac{5}{42} \times 6 = \frac{5}{7}.$$

The dual optimal variables are $p_1^{*'} = 0$, $p_2^{*'} = 0$ and $p_3^{*'} = \frac{1}{6}$ which are nothing but the $z_j - c_j$ corresponding to the slack vectors, i.e., for $j = 3, 4, 5$ we get the dual optimal variables. Hence

$$p_1^* = 0 \times 6 = 0, \quad p_2^* = 0 \times 6 = 0 \quad \text{and} \quad p_3^* = \frac{1}{6} \times 6 = 1.$$

Thus the optimal strategies $\mathbf{p}^* = (0, 0, 1), \mathbf{q}^* = (\frac{2}{7}, \frac{5}{7})$ and the value of the game of the original problem is 2.

Note: (1) As $z_4 - c_4 = 0$ corresponding to a non-basic variable q_4', then alternative optimal solutions exist.

(2) You may solve A's problem, where there are artificial variables but the basis matrix will be a 2×2 matrix and the problem may be solved easily.

(3) Here the values of the primal optimal solution components are both positive and only the third component of dual optimal solution is positive. Therefore using complementary slackness theorem we can say that both the constraints of A's problem which are

$$\begin{aligned} 5p_1' + p_2' + 6p_3' &\geq 1 \\ 0p_1' + 8p_2' + 6p_3' &\geq 1 \end{aligned}$$

will be strictly equations for the optimal solution $p_1'^* = 0$, $p_2'^* = 0$, $p_3'^* = 1/6$ which can be easily verified. Similarly, as the third dual optimal variable is positive, then the third constraint of the primal will be strictly an equation for the optimal solution. Other constraint may be equation or inequation. Here the second constraint is an equation but the first one is an inequation. This knowledge will be extremely helpful in solving $n \times n$ game problems by using algebraically method.

Exercise 19E

Solve the game problem by using L.P.P. method:

1. (i) $\begin{array}{c} \\ A \end{array} \overset{B}{\begin{bmatrix} 4 & 2 \\ 1 & 3 \end{bmatrix}}$; (ii) $\begin{array}{c} \\ A \end{array} \overset{B}{\begin{bmatrix} -1 & 5 \\ 4 & 3 \end{bmatrix}}$; (iii) $\begin{bmatrix} 2 & -2 \\ -1 & 0 \end{bmatrix}$.

2. $\begin{array}{c} \\ A \end{array} \overset{B}{\begin{bmatrix} 1 & -1 & -2 \\ -1 & 1 & 1 \\ 2 & -1 & 0 \end{bmatrix}}$; 3. (i) $\begin{array}{c} \\ A \end{array} \overset{B}{\begin{bmatrix} -1 & 2 & 1 \\ 1 & -2 & 2 \\ 3 & 4 & -3 \end{bmatrix}}$;

[C.U. (Hons.) '97]

(ii) $A \overset{B}{\begin{bmatrix} 1 & -1 & -1 \\ -1 & -1 & 3 \\ -1 & 2 & -1 \end{bmatrix}}$; [C.U. (Hons.) '91]

4. $A \overset{B}{\begin{bmatrix} 3 & 2 & 4 \\ 4 & 3 & 1 \\ -1 & 2 & 5 \end{bmatrix}}$; 5. $A \overset{B}{\begin{bmatrix} 2 & 3 & 4 \\ 5 & 4 & 4 \\ -1 & 6 & 7 \end{bmatrix}}$; 6. $A \overset{B}{\begin{bmatrix} -1 & -2 & 1 \\ 1 & -3 & 0 \\ 2 & 3 & -2 \end{bmatrix}}$;

7. In a two-person game, each player simultaneously shows either one or two fingers. If the number of fingers match then the player A wins a rupee from the player B. Otherwise A pays a rupee to B. Show that the pay-off matrix for the game is

$$A \begin{matrix} 1 \\ 2 \end{matrix} \overset{\begin{matrix} & B & \\ 1 & & 2 \end{matrix}}{\begin{bmatrix} 1 & -1 \\ -1 & 1 \end{bmatrix}}.$$

[C.U. (Hons.) '82]

Solve the game by reducing it to a L.P.P.

Answers

1. (i) $\mathbf{p}^* = (\frac{1}{2}, \frac{1}{2})$, $\mathbf{q}^* = (\frac{1}{4}, \frac{3}{4})$, $v = \frac{5}{3}$.

1. (ii) $\mathbf{p}^* = (\frac{1}{7}, \frac{6}{7})$, $\mathbf{q}^* = (\frac{2}{7}, \frac{5}{7})$, $v = \frac{23}{7}$.

1. (iii) $\mathbf{p}^* = (\frac{1}{5}, \frac{4}{5})$, $\mathbf{q}^* = (\frac{2}{5}, \frac{3}{5})$, $v = \frac{2}{5}$.

2. $\mathbf{p}^* = (0, \frac{3}{5}, \frac{2}{5})$, $\mathbf{q}^* = (\frac{2}{5}, \frac{3}{5}, 0)$, $v = \frac{1}{5}$.

3. (i) $\mathbf{p}^* = (\frac{17}{46}, \frac{20}{46}, \frac{9}{46})$, $\mathbf{q}^* = (\frac{7}{23}, \frac{6}{23}, \frac{10}{23})$, $v = \frac{15}{23}$.

3. (ii) $\mathbf{p}^* = (\frac{6}{13}, \frac{3}{13}, \frac{4}{13})$, $\mathbf{q}^* = (\frac{6}{13}, \frac{4}{13}, \frac{3}{13})$, $v = -\frac{1}{13}$.

4. $\mathbf{p}^* = (\frac{3}{16}, \frac{9}{16}, \frac{1}{4})$, $\mathbf{q}^* = (\frac{1}{16}, \frac{11}{16}, \frac{1}{4})$, $v = \frac{41}{46}$.

5. (i) $\mathbf{p}^* = (0, \frac{7}{8}, \frac{1}{8})$, $\mathbf{q}^* = (\frac{1}{4}, \frac{3}{4}, 0)$, $v = \frac{17}{4}$.

6. $\mathbf{p}^* = (\frac{5}{8}, 0, \frac{3}{8})$, $\mathbf{q}^* = (0, \frac{3}{8}, \frac{5}{8})$, $v = -\frac{1}{8}$.

Theorem 19.10.2 *If we add a fixed number to each element of a pay-off matrix, the optimal strategies remain same, while the value of the game will be increased by that number.*

Proof: Let the elements of $m \times n$ pay-off matrix be $[a_{ij}]_{m\times n}$ and let the mixed strategies taken by A and B be

$$\mathbf{p} = (p_1, p_2, \cdots, p_m) \text{ with } \sum_{i=1}^{m} p_i = 1, p_i \geq 0.\ \forall_i \text{ and}$$

$$\mathbf{q} = (q_1, q_2, \cdots, q_n) \text{ with } \sum_{j=1}^{n} q_i = 1, q_i \geq 0.\ \forall_j \text{ respectively.}$$

Then the pay-off function of the game is given by

$$E(\mathbf{p},\mathbf{q}) = \sum_{i=1}^{m}\sum_{j=1}^{n} a_{ij}p_i q_j. \tag{19.10.3}$$

Now let a number k be added [k positive or negative] to each element a_{ij}. Hence the pay-off function of the new game is

$$\begin{aligned} E^k(\mathbf{p}\cdot\mathbf{q}) &= \sum_{i=1}^{m}\sum_{j=1}^{n}(a_{ij}+k)p_i q_j \\ &= \sum_{i=1}^{m}\sum_{j=1}^{n} a_{ij}p_i q_j + k\sum_{i=1}^{m}\sum_{j=1}^{n} p_i q_j \\ &= E(\mathbf{p},\mathbf{q}) + k \quad [\Sigma p_i = 1, \Sigma q_j = 1]. \end{aligned} \tag{19.10.4}$$

Without any loss of generality we can assume that $(a_{ij}+k) > 0, \forall i, j$.

Then the value of the second game exists and unique and if the optimal strategies be $\mathbf{p}^*, \mathbf{q}^*$, for the second game, then its value of the game is

$$E^k(\mathbf{p}^*,\mathbf{q}^*) = \max_{\mathbf{p}}\cdot\min_{\mathbf{q}}.E^k(\mathbf{p},\mathbf{q}) = E(\mathbf{p}^*,\mathbf{q}^*) + k \cdots \tag{19.10.5}$$

Again

$$\max_{\mathbf{p}}\cdot\min_{\mathbf{q}}.E^k(\mathbf{p},\mathbf{q}) = \max_{\mathbf{p}}\cdot\min_{\mathbf{q}}.E(\mathbf{p},\mathbf{q}) + k. \tag{19.10.6}$$

From (19.10.5) and (19.10.6), we get

$$\max_{\mathbf{p}}\cdot\min_{\mathbf{q}}.E(\mathbf{p},\mathbf{q}) = E(\mathbf{p}^*,\mathbf{q}^*). \tag{19.10.7}$$

Similarly we get

$$\min_{\mathbf{q}}\cdot\max_{\mathbf{p}}.E(\mathbf{p},\mathbf{q}) = E(\mathbf{p}^*,\mathbf{q}^*). \tag{19.10.8}$$

Hence, from (19.10.7) and (19.10.8) we can say that the value of original game also exists and unique, the optimal strategies for both games remain same and from (19.10.4) we can say that value of the second game is only increased by the number k.

Note: The relation is also valid if $a_{ij} + k \leq 0$ for $\forall_{i,j}$.

Theorem 19.10.3 *The mixed strategies* $\mathbf{p}^*, \mathbf{q}^*$ *satisfying the necessary conditions will be an optimal strategy of a game problem iff the following relation* $E(\mathbf{p}^*,\mathbf{q}) \geq E(\mathbf{p}^*,\mathbf{q}^*) = v \geq E(\mathbf{p},\mathbf{q}^*)$ *holds good.*

Proof: Part : I. We assume that $\mathbf{p}^*, \mathbf{q}^*$ is the optimal strategy of a game problem.

We are to prove that $E(\mathbf{p}^*,\mathbf{q}) \geq E(\mathbf{p}^*,\mathbf{q}^*) \geq E(\mathbf{p},\mathbf{q}^*)$ $\mathbf{p}^*, \mathbf{q}^*$ is an optimal strategy of a game. We have,

$$\max_{\mathbf{p}}\cdot\min_{\mathbf{q}}.E(\mathbf{p},\mathbf{q}) = E(\mathbf{p}^*,\mathbf{q}^*) = \min_{\mathbf{q}}\cdot\max_{\mathbf{p}} E(\mathbf{p},\mathbf{q}). \tag{19.10.9}$$

From (19.10.9) $\min_{\mathbf{q}} E(\mathbf{p}^*, \mathbf{q}) = E(\mathbf{p}^*, \mathbf{q}^*)$ and hence

$$\left.\begin{array}{r} E(\mathbf{p}^*, \mathbf{q}) \geq E(\mathbf{p}^*, \mathbf{q}^*) \\ \text{Similarly,} \quad E(\mathbf{p}, \mathbf{q}^*) \leq E(\mathbf{p}^*, \mathbf{q}^*) \end{array}\right\}. \tag{19.10.10}$$

From (19.10.10), we get

$$E(\mathbf{p}^*, \mathbf{q}) \geq E(\mathbf{p}^*, \mathbf{q}^*) \geq E(\mathbf{p}, \mathbf{q}^*). \tag{19.10.11}$$

Part 2. We assume that the relation (19.10.11) holds good, we are to prove that

$$\max_{\mathbf{p}} \cdot \min_{\mathbf{q}} . E(\mathbf{p}, \mathbf{q}) = E(\mathbf{p}^*, \mathbf{q}^*) = \min_{\mathbf{p}} \cdot \max_{\mathbf{q}} . E(\mathbf{p}, \mathbf{q}).$$

From (19.10.11), since $E(\mathbf{p}^*, \mathbf{q}) \geq E(\mathbf{p}^*, \mathbf{q}^*)$ for all $\mathbf{q}$, then

$$\min_{\mathbf{q}} E(\mathbf{p}^*, \mathbf{q}) \geq E(\mathbf{p}^*, \mathbf{q}^*).$$

Again,

$$\max_{\mathbf{p}} \cdot \min_{\mathbf{q}} E(\mathbf{p}, \mathbf{q}) \geq \min_{\mathbf{q}} E(\mathbf{p}^*, \mathbf{q}).$$

Therefore

$$\max_{\mathbf{p}} \cdot \min_{\mathbf{q}} E(\mathbf{p}, \mathbf{q}) \geq E(\mathbf{p}^*, \mathbf{q}^*). \tag{19.10.12}$$

Using the same argument we can establish

$$E(\mathbf{p}^*, \mathbf{q}^*) \geq \min_{\mathbf{q}} \cdot \max_{\mathbf{p}} . E(\mathbf{p}, \mathbf{q}). \tag{19.10.13}$$

From (19.10.12) and (19.10.13) we can write

$$\max_{\mathbf{p}} \cdot \min_{\mathbf{q}} \cdot E(\mathbf{p}, \mathbf{q}) \geq E(\mathbf{p}^*, \mathbf{q}^*) \geq \min_{\mathbf{q}} \cdot \max_{\mathbf{p}} . E(\mathbf{p}, \mathbf{q}). \tag{19.10.14}$$

Again we from the theorem 19.5.1

$$\min_{\mathbf{q}} \cdot \max_{\mathbf{p}} . E(\mathbf{p}, \mathbf{q}) \leq \max_{\mathbf{p}} \cdot \min_{\mathbf{q}} . E(\mathbf{p}, \mathbf{q}). \tag{19.10.15}$$

Combining (19.10.14) and (19.10.15) we get

$$\max_{\mathbf{p}} \cdot \min_{\mathbf{q}} . E(\mathbf{p}, \mathbf{q}) = E(\mathbf{p}^*, \mathbf{q}^*) = v = \min_{\mathbf{q}} \cdot \max_{\mathbf{p}} . E(\mathbf{p}, \mathbf{q}). \tag{19.10.16}$$

Theorem 19.10.4 *If all the elements of k-th row of the pay-off matrix be less than or equal to the corresponding elements of r-th row then we shall say that the k-th row is dominated by the r-th row and we can discard the k-th row, i.e., p_k will be zero.*

Proof:

$$a_{kj} \le a_{rj} \text{ for all } j \text{ and}$$
$$a_{kj} < a_{rj} \text{ at least for one } j,\ j = 1, 2, \cdots, n. \tag{19.10.17}$$

Let $\mathbf{p}^* = (p_1^*, p_2^*, p_k^*, \cdots, p_m^*)$ and $\mathbf{q}^* = (q_1^*, q_2^*, q_k^*, \cdots, q_n^*)$ be the optimal strategies for the row and column player respectively.

From (19.10.17) we get

$$\sum_{j=1}^{n} a_{kj} q_j^* < \sum_{j=1}^{n} a_{rj} q_j^*, \quad \text{or,} \quad E(\mathbf{e}_k, \mathbf{q}^*) < E(\mathbf{e}_r, \mathbf{q}^*).$$

Thus we have v = value of the game

$$\ge E(\mathbf{e}_r, \mathbf{q}^*) > E(\mathbf{e}_k, \mathbf{q}^*). \tag{19.10.18}$$

Let us assume that $p_k^* \ne 0$, $p_k > 0$.

Then from (19.10.18) we get

$$p_k^* v > p_k^* E(\mathbf{e}_k, \mathbf{q}^*). \tag{19.10.19}$$

Again we know that

$$v = E(\mathbf{p}^*, \mathbf{q}^*) = \sum_{l=1}^{n} p_l^* E(\mathbf{e}_l, \mathbf{q}^*)$$
$$= p_k^* E(\mathbf{e}_k, \mathbf{q}^*) + \sum_{l \ne k} p_l^* E(\mathbf{e}_l, \mathbf{q}^*)$$
$$< p_k^* v + v \sum_{l \ne k} p_l^* = v \sum_{l=1}^{m} p_l^* = v, \tag{19.10.20}$$

which is a contradiction and then p_k^* will be zero.

Definition:

Symmetric game: A two-person, zero-sum game is said to be a **symmetric game** if its pay-off matrix be skew-symmetric (which obviously implies that the rectangular matrix is a square matrix). It is so termed as the roles of player A and B can be interchanged without changing the pay-off matrix. For a skew-symmetric, pay-off matrix C, $C^t = -C$ [From the definition of skew-symmetric matrix]. We now establish a very important property of a symmetric game.

Theorem 19.10.5 *The value of a symmetric game is zero and the optimal strategies be same for both players or, the value of a game with skew-symmetric pay-off matrix vanishes and the optimal strategies for both players be same.*

Proof: For a square matrix $E(\mathbf{p}, \mathbf{p})$ can be written as

$$E(\mathbf{p}.\mathbf{p}) = \sum_{i=1}^{n} \sum_{j=1}^{n} a_{ij} p_i p_j = \mathbf{p}^t C \mathbf{p}, \tag{19.10.21}$$

which is a scalar quantity [$\mathbf{p}$ and a_{ij} have their usual meanings].

$C = [a_{ij}]$. Therefore,

$$\mathbf{p}^t C\mathbf{p} = (\mathbf{p}^t C\mathbf{p})^t = \mathbf{p}^t C^t \mathbf{p} = -\mathbf{p}^t C\mathbf{p} \quad [\because C^t = -C]$$
$$\Rightarrow 2\mathbf{p}^t C\mathbf{p} = 0 \Rightarrow \mathbf{p}^t C\mathbf{p} = 0. \tag{19.10.22}$$

Similarly, we have

$$E(\mathbf{q}, \mathbf{q}) = 0. \tag{19.10.23}$$

Hence, if $\mathbf{p}^*, \mathbf{q}^*$ be the optimal strategies of the problem [which always exists] then

$$E(\mathbf{p}^*, \mathbf{p}^*) = E(\mathbf{q}^*, \mathbf{q}^*) = 0. \tag{19.10.24}$$

Now the equality given in (19.10.24) will not be disturbed if we assume $\mathbf{p}^* = \mathbf{q}^*$ or $\mathbf{q}^* = \mathbf{p}^*$ and by that choice we have

$$E(\mathbf{p}^*, \mathbf{q}^*) = E(\mathbf{p}^*, \mathbf{q}^*) = 0. \tag{19.10.25}$$

From (19.10.25), now we can say that the optimal strategies be same for both players and the value of the game is

$$v = E(\mathbf{p}^*, \mathbf{q}^*) = 0.$$

Note: (i) Verify the above theorem for the pay-off matrices

$$\text{(i)} \begin{bmatrix} 0 & 1 & -2 \\ -1 & 0 & 3 \\ 2 & -3 & 0 \end{bmatrix} \text{ and (ii) } \begin{bmatrix} 0 & -3 & 4 \\ 3 & 0 & -2 \\ -4 & 2 & 0 \end{bmatrix}$$

(iii) Two players independently select one of stone, paper and scissors and simultaneously disclose their choices. The rules of the game are that the stone beats and scissors, the scissors beat the paper and the paper beats and stone. Suppose that there is a score of 1 point for winning, −1 point for losing and 0 if both the players select the same object. Find the pay-off matrix and prove that the game is symmetric and find also the optimal strategies for each player. **[C.U. (Hons.) '91]**

Ans. Pay-off matrix

$$\begin{array}{c} \\ S \\ S \\ P \end{array} \begin{array}{c} \begin{array}{ccc} S & S & P \end{array} \\ \begin{bmatrix} 0 & 1 & -1 \\ -1 & 0 & 1 \\ 1 & -1 & 0 \end{bmatrix} \end{array}$$

which is a symmetric game and $p_1^* = q_1^* = \frac{1}{3}$, $p_2^* = q_2^* = \frac{1}{3}$, $p_3^* = q_3^* = \frac{1}{3}$, $v = 0$.

(2) Symmetric game is always a fair game.

Theorem 19.10.6 *The set of all optimal strategies for a particular player is a convex set.*

Proof: We shall prove the theorem for the player A (row player).

If there is only one set of optimal strategy, we have nothing to prove.

Let there be at least two optimal strategies $\mathbf{p}_1^*$ and $\mathbf{p}_2^*$ which are distinct.

We have

$$\left.\begin{array}{r} E(\mathbf{p}_1^*, \mathbf{q}) \geq v,\ \mathbf{p}_1^* \geq 0,\ \sum_{i=1}^{m} p_{1i} = 1 \\ \text{and } E(\mathbf{p}_2^*, \mathbf{q}) \geq v,\ \mathbf{p}_2^* \geq 0,\ \sum_{i=1}^{m} p_{2i} = 1 \end{array}\right\}. \tag{19.10.26}$$

The convex combination of $\mathbf{p}_1^*$ and $\mathbf{p}_2^*$ is

$\mathbf{p}^* = \lambda\mathbf{p}_1^* + (1-\lambda)\mathbf{p}_2^*, \quad 0 \leq \lambda \leq 1.$

$\mathbf{p}^* \geq 0$ since $\mathbf{p}_1^* \geq 0$, $\mathbf{p}_2^* \geq 0$ and $\lambda, 1-\lambda \geq 0.$

$$\sum_{i=1}^{m} p_i^* = \lambda\Sigma p_{1i}^* + (1-\lambda)\Sigma p_{2i}^* = \lambda + 1 - \lambda = 1$$

and finally

$$\begin{aligned} E(\mathbf{p}^*, \mathbf{q}) &= E\{\lambda\mathbf{p}_1^* + (1-\lambda)\mathbf{p}_2^*, \mathbf{q}\} \\ &= \lambda E(\mathbf{p}_1^*, \mathbf{q}) + (1-\lambda)E(\mathbf{p}_2^*, \mathbf{q}) \\ &\geq \lambda v + (1-\lambda)v = v \quad \text{[from (19.10.25)]} \end{aligned}$$

which indicate that the convex combination of any two points $\mathbf{p}_1^*, \mathbf{p}_2^*$ belonging to the set is also a point of the set. Hence the set is a convex set.

Solution of any n × n game by algebraically method

We have already solved 2 × 2 games, having no saddle point in the case of pure strategy by using algebraically method, and which can only be solved using mixed strategy. In that case the elements of the pay-off matrix $\begin{bmatrix} a_{11} & a_{12} \\ a_{21} & a_{22} \end{bmatrix}$ satisfy the relation $(a_{11}, a_{22}) >< (a_{12}, a_{21})$. There both p_1, p_2 and $q_1, q_2 > 0$ and $p_1 + p_2 = q_1 + q_2 = 1$. As all p_1, p_2 and q_1, q_2 are positive, then by using complementary slackness theorem we can easily and solving the equations together with the relations $p_1 + p_2 = 1$ or $q_1 + q_2 = 1$ we can at once find the optimal solutions and the value of the game. But there are problems where some of p_1, p_2, q_1, q_2 be zero and in that case some constraints of the primal and dual may be strictly inequality and if we try to solve the problems assuming all constraints to be equations, we may get some of $p_1^*, p_2^*, q_1^*, q_2^*$ negative which are quite unnatural. In that case we shall have to find the optimal solutions by 'trial and error' method which is not so an easy task. We shall here discuss such problem where all p_i and $q_j > 0$ and the matrix is a square matrix. Here we consider a very simple case.

► **Example 19.10.6** *Solve the following L.P. problem assuming that mixed strategy exists.*

$$\begin{array}{cc} & B \\ A & \begin{bmatrix} 3 & 3 \\ 2 & -1 \end{bmatrix} \end{array}$$

Let p_1, p_2 be the mixed strategy for the row player A, and if v be the value of the game,

$$4p_1 + 2p_2 = v = 3p_1 - p_2 \text{ and } p_1 + p_2 = 1, p_1, p_2 \geq 0.$$

Solving we get $p_1 = \frac{3}{2}$ and $p_2 = -\frac{1}{2}$ which leads to a erroneous conclusion. Applying dominance property, we get the optimal strategies which are $\mathbf{p}^* = (1, 0)$ and $\mathbf{q}^* = (0, 1)$ and the value of the game is 3. As $q_2^* = 1 \neq 0$, then the second constraint will be strictly an equation for the optimal solution $\mathbf{p}^*$ which can easily be verified as $3p_1^* - p_2^* = 3 = v$; but the first constraint may not be an equation which can also be verified as $4p_1^* + 2p_2^* = 4 > v$.

As all constraints are not strict equations we may get this type of unnatural result which can only be solved only by trial and error basis. We shall here solve only such type of problems, where all p_i and $q_j > 0$.

► **Example 19.10.7** *Solve the problem given in* (19.10.3) *by algebraically method after applying the dominance property (if it is possible at all).*

After applying the dominance property, the resulting matrix is,

$$A \begin{array}{c} \\ A_2 \\ A_3 \end{array} \begin{array}{c} \begin{array}{cc} B_2 & B_3 \end{array} \\ \begin{bmatrix} 5 & -1 \\ 0 & 12 \end{bmatrix} \end{array},$$

which can be solved easily by algebraically method.

Assuming the constraints as strictly equations

$$5p_2 - 0p_3 = v = p_2 + 12p_3 \text{ and } p_2 + p_3 = 1.$$

Solving we get $p_2^* = \frac{2}{3}$ and $p_3^* = \frac{1}{3}$ and the value of the game $\frac{10}{3}$.

Here the optimal strategies are $\mathbf{p}^* = (0, \frac{2}{3}, \frac{1}{3})$ and $\mathbf{q}^* = (0, \frac{13}{18}, \frac{5}{18})$ and the value of the game is $\frac{10}{3}$.

► **Example 19.10.8** *Solve the following* 3 × 3 *game by algebraical method:*

$$\begin{array}{c} \\ A_1 \\ A_2 \\ A_3 \end{array} \begin{array}{c} \begin{array}{ccc} B_1 & B_2 & B_3 \end{array} \\ \begin{bmatrix} 1 & -1 & -1 \\ -1 & -1 & 3 \\ -1 & 2 & -1 \end{bmatrix} \end{array},$$

The problem has no saddle point and the matrix has no dominance effect. Hence it may be solved easily by algebraically method. Let the strategies taken by A and B be $\mathbf{p} = (p_1, p_2, p_3), \mathbf{q} = (q_1, q_2, q_3)$ and $p_i \geq 0, i = 1, 2, 3, q_j \geq 0$ for $j = 1, 2, 3$. Then assuming that optimal strategies $\mathbf{p}^* > 0$ and $\mathbf{q}^* > 0$ all the constraints will be strictly equations and with that assumption we have [by using the same method as in 2 × 2 game] A's expected gain for B's pure move B_1 is

$$p_1 - p_2 - p_3 = v \text{ (value of the game)} \tag{19.10.27}$$

Similarly,

$$-p_1 - p_2 + 2p_3 = v \tag{19.10.28}$$
$$\text{and} \quad -p_1 + 3p_2 - p_3 = v \tag{19.10.29}$$
$$\text{and} \quad p_1 + p_2 + p_3 = 1 \tag{19.10.30}$$

From (19.10.26) (19.10.26)–(19.10.27) we get

$$2p_1 - 3p_3 = 0 \tag{19.10.31}$$

(19.10.26)–(19.10.28) we get

$$2p_1 - 4p_2 = 0. \tag{19.10.32}$$

Solving (19.10.29), (19.10.30) and (19.10.31) we get

$$p_1^* = \frac{6}{13}, \quad p_2^* = \frac{3}{13}, \quad p_3^* = \frac{4}{13}$$

and the value of the game

$$v = 1 \times \frac{6}{13} - \frac{4}{13} - 1 \times \frac{3}{13} = -\frac{1}{13}.$$

Hence our assumption is correct.
Similarly considering from B's point of view we get

$$q_1 - q_2 - q_3 = v \tag{19.10.33}$$
$$-q_1 - q_2 + 3q_3 = v \tag{19.10.34}$$
$$-q_1 - 2q_2 - q_3 = v \tag{19.10.35}$$
$$\text{and} \quad q_1 + q_2 + q_3 = 1. \tag{19.10.36}$$

Solving we get

$$q_1^* = \frac{6}{13}, \quad q_2^* = \frac{4}{13} \quad \text{and} \quad q_3^* = \frac{3}{13}.$$

Hence the solution of the problem is; value of the game $= -\frac{1}{13}$ at

$$\mathbf{p}^* = \left(\frac{6}{13}, \frac{3}{13}, \frac{4}{13}\right) \quad \text{and} \quad \mathbf{q}^* = \left(\frac{6}{10}, \frac{4}{13}, \frac{3}{13}\right).$$

Note: If some of p_i^* and q_j^* be zero all the constraints may not be strictly equations, then we have to solve the equations assuming the constraints as strict inequations and by using trial and error method.

Exercise 19F

Solve the following games algebraically:

1. $\begin{array}{c} \\ A_1 \\ A_2 \\ A_3 \end{array} \begin{array}{c} \begin{array}{ccc} B_1 & B_2 & B_3 \end{array} \\ \begin{bmatrix} 3 & 2 & 4 \\ 4 & 3 & 1 \\ -1 & 2 & 5 \end{bmatrix} \end{array}.$ 2. $\begin{array}{c} \\ A_1 \\ A_2 \\ A_3 \end{array} \begin{array}{c} \begin{array}{ccc} B_1 & B_2 & B_3 \end{array} \\ \begin{bmatrix} 4 & 1 & 2 \\ 3 & 2 & 1 \\ 1 & 3 & 4 \end{bmatrix} \end{array}.$

3. $\begin{array}{c} \\ A_1 \\ A_2 \\ A_3 \end{array} \begin{array}{c} \begin{array}{ccc} B_1 & B_2 & B_3 \end{array} \\ \begin{bmatrix} -1 & 2 & 1 \\ 1 & -2 & 2 \\ 3 & 4 & -3 \end{bmatrix} \end{array}.$

Answers

1. $\mathbf{p}^* = \left(\frac{3}{16}, \frac{9}{16}, \frac{1}{4}\right)$, $\quad \mathbf{q}^* = \left(\frac{1}{16}, \frac{11}{16}, \frac{1}{4}\right)$, $\quad v = \frac{41}{16}$.
2. $\mathbf{p}^* = \left(\frac{1}{10}, \frac{1}{2}, \frac{2}{5}\right)$, $\quad \mathbf{q}^* = \left(\frac{2}{5}, \frac{1}{2}, \frac{1}{10}\right)$, $\quad v = \frac{23}{10}$.
3. $\mathbf{p}^* = \left(\frac{17}{46}, \frac{10}{23}, \frac{9}{46}\right)$, $\quad \mathbf{q}^* = \left(\frac{7}{23}, \frac{5}{23}, \frac{10}{23}\right)$, $\quad v = \frac{15}{23}$.

19.11 Solution of a game problem by matrix method

We can able to solve a game problem by taking a square sub-matrix from the original pay-off matrix by 'trial and error' basis and the process is extremely labourious. We know that at the optimal stage, number of positive components of $\mathbf{p}^*$ and $\mathbf{q}^*$ will same [of course, by assuming that we may ultimately get some of them be zero]. With that assumption we take a square sub-matrix B of order ≥ 2 and try to find out the value of the game and the optimal strategies by using the formula [proof is beyond the scope of the book].

$$\left.\begin{aligned} v &= \frac{|B| = \det B}{U_r'(Adj B)U_r} \\ \mathbf{p}^* &= \frac{U_r'(Adj B)}{U_r'(Adj B)U_r} \\ \mathbf{q}^* &= \frac{U_r'(Adj B)'}{(Adj B)U_r} \end{aligned}\right\},$$

where B is the square matrix of order r, $(Adj B)'$ etc. is the transpose of the matrix $(Adj B)$ and $U_r = [1; 1, \cdots, 1]$, r component column vector of the type.

To solve the problem, we shall have to choice the square sub-matrices at random and try to proceed to find $\mathbf{p}^*$ and $\mathbf{q}^*$ where $\mathbf{p}^*$ and $\mathbf{q}^*$ contains the components of $\mathbf{p}$ and $\mathbf{q}$ corresponding to the rows and columns of the original matrix which forms the square sub-matrix B and if at least one component of $\mathbf{p}^*$ and $\mathbf{q}^*$ be negative

we discard that matrix and select another matrix and ultimately we get a matrix which will satisfy the condition $\mathbf{p}^* \geq 0$ and $\mathbf{q}^* \geq 0$ and we get an unique value of the game [as the value of the game of each rectangular problem exists and unique] but there may exist alternative optimal strategies.

► **Example 19.11.1** ***Solve the following problem using matrix method:***

$$\text{Player } A \overset{\text{Player } B}{\begin{bmatrix} 2 & 7 \\ 3 & 5 \\ 11 & 2 \end{bmatrix}}$$

The three square sub-matrices of order 2 are

$$B_1 = \begin{bmatrix} 2 & 7 \\ 3 & 5 \end{bmatrix}, \quad B_2 = \begin{bmatrix} 2 & 7 \\ 11 & 2 \end{bmatrix} \quad \text{and} \quad B_3 = \begin{bmatrix} 3 & 5 \\ 11 & 2 \end{bmatrix}$$

$$|B_1| = \det B_1 = -11, \quad U_2 = \begin{bmatrix} 1 \\ 1 \end{bmatrix}, \quad U_2' = (1\ 1).$$

$$Adj B_1 = \begin{bmatrix} 5 & -7 \\ -3 & 2 \end{bmatrix},$$

$$U_2'(Adj B_1) = (1,\ 1) = \begin{bmatrix} 5 & -7 \\ -3 & 2 \end{bmatrix} = (2, -5)$$

$$U_2'(Adj B_1)U_2 = [2 \quad -5] \begin{bmatrix} 1 \\ 1 \end{bmatrix} = -3.$$

$\mathbf{p}^* = -\frac{1}{3}(2, -5) = \left(-\frac{2}{3}, \frac{5}{3}\right)$ at least one component of which negative.
Thus we simply reject the matrix B_1.
Now,

$$|B_2| = -73, \quad Adj B_2 = \begin{bmatrix} 2 & -7 \\ -11 & 2 \end{bmatrix}.$$

$$U_2'(Adj B_2) = (1, 1) \begin{bmatrix} 2 & -7 \\ -11 & 2 \end{bmatrix} = (-9, -5)$$

$$U_2'(Adj B_2)U_2 = (-9, -5) \begin{bmatrix} 1 \\ 1 \end{bmatrix} = -14.$$

$$\mathbf{p}^* = -\frac{1}{14}(-9, -5) = \frac{9}{15}, \frac{5}{14}.$$

Now both $\frac{9}{14}, \frac{5}{14} > 0$ and $\frac{9}{14} + \frac{5}{14} = 1$. Thus we may try to find the components of $\mathbf{q}^*$.

$$[Adj B_2]' = \begin{bmatrix} 2 & -11 \\ -7 & 2 \end{bmatrix}.$$

$$U_2'[Adj B_2]' = (1, 1)\begin{bmatrix} 2 & -11 \\ -7 & 2 \end{bmatrix} = [-5, -9].$$

$$U_2'[Adj B_2]'U_2 = [-5 \ -9]\begin{bmatrix} 1 \\ 1 \end{bmatrix} = -14.$$

$$\mathbf{q}^* = \frac{1}{-14}[-5, \ -9] = \left[\frac{5}{14}, \ \frac{9}{14}\right] \text{ both of which are positive and } \frac{5}{14} + \frac{9}{14} = 1.$$

The value of the game

$$v = \frac{|B_2|}{U_2'[Adj B]U_2} = \frac{-73}{-14} = \frac{73}{14}.$$

One of the optimal strategies is

$$\mathbf{p}^* = \left(\frac{9}{14}, 0, \frac{5}{14}\right); \quad \mathbf{q}^* = \left(\frac{5}{14}, 0, \frac{9}{14}\right)$$

and value of the game is $\frac{73}{14}$ which can be verified in the manner

$$\mathbf{E}(\mathbf{p}^*, \mathbf{e}_1) = 2 \times \frac{9}{14} + 0 \times \frac{3}{14} + 11 \times \frac{5}{14} = \frac{73}{14} \text{ etc.}$$

Now try to find the value of the game with the help of matrix B_3 and prove that it is not possible to find the value of the game $[\because \ v^* = \frac{49}{11} < \frac{73}{14}]$ with the matrix and thus the optimal strategies are unique.

Exercise 19G

Solve the following game problems using matrix method:

1. (i) $\begin{bmatrix} 5 & 2 & -1 \\ 3 & 7 & 4 \end{bmatrix}$; (ii) $\begin{bmatrix} 0 & 5 & 3 \\ 2 & 3 & 1 \end{bmatrix}$; (iii) $\begin{bmatrix} 2 & 3 & -1 \\ 3 & 2 & 6 \end{bmatrix}$; (iv) $\begin{bmatrix} 1 & 0 & -2 \\ 0 & 3 & 2 \end{bmatrix}$.

2. (i) $\begin{bmatrix} 4 & 1 \\ 1 & 3 \\ 2 & 3 \end{bmatrix}$; (ii) $\begin{bmatrix} 3 & 2 \\ -1 & 5 \\ -2 & 1 \end{bmatrix}$. 3. (i) $\begin{bmatrix} 1 & -1 & -2 \\ -1 & 1 & 1 \\ 2 & -1 & 0 \end{bmatrix}$; (iv) $\begin{bmatrix} -1 & 2 & 1 \\ 1 & -2 & 2 \\ 3 & 4 & -3 \end{bmatrix}$.

Answers

1.(*i*) $\mathbf{p}^* = (\frac{1}{7}, \frac{6}{7})$, $\mathbf{q}^* = (\frac{5}{7}, 0, \frac{2}{7})$, $v = \frac{23}{7}$; (*ii*) $\mathbf{p}^* = (\frac{1}{4}, \frac{3}{4})$, $\mathbf{q}^* = (\frac{1}{2}, 0, \frac{1}{2})$, $v = \frac{3}{2}$;
(*iii*) $\mathbf{p}^* = (\frac{1}{2}, \frac{1}{2})$, $\mathbf{q}^* = (\frac{1}{2}, \frac{1}{2}, 0)$, or, $\mathbf{p}^* = (\frac{1}{2}, \frac{1}{2})$, $\mathbf{q}^* = (0, \frac{7}{8}, \frac{1}{8})$, $v = \frac{5}{2}$;
(*iv*) $\mathbf{p}^* = (\frac{2}{5}, \frac{3}{5})$, $\mathbf{q}^* = (\frac{4}{5}, 0, \frac{1}{5})$; $v = \frac{2}{5}$.

2.(*i*) $\mathbf{p}^* = (\frac{1}{4}, 0, \frac{3}{4})$, $\mathbf{q}^* = (\frac{1}{2}, \frac{1}{2})$, $v = \frac{5}{2}$; (*ii*) $\mathbf{p}^* = (\frac{6}{7}, \frac{1}{7}, 0)$, $\mathbf{q}^* = (\frac{3}{7}, \frac{4}{7})$, $v = \frac{17}{7}$;

3.(*i*) $\mathbf{p}^* = (0, \frac{3}{5}, \frac{2}{5})$, $\mathbf{q}^* = (\frac{2}{5}, \frac{3}{5}, 0)$, $v = \frac{1}{5}$; (*ii*) $\mathbf{p}^* = (\frac{17}{46}, \frac{20}{46}, \frac{9}{46})$, $\mathbf{q}^* = (\frac{7}{23}, \frac{6}{23}, \frac{10}{23})$, $v = \frac{15}{23}$.

19.12 Solution of the game problem by iterative method (An approximate method)

We have already established that the solution of game problems by L.P.P. method or matrix method [which are most scientific and accurate] are very laborious and time consuming process. Here we discuss an iterative method which gives almost an accurate result of a game problem and the degree of accuracy depends generally on the number of iterations. But there is no definite rule to establish when shall we stop the iterative procedure and how much accurate is the result there? The following worked out example will give the clear idea about the procedure and it expresses the value of the game in a closed interval.

▶ **Example 19.12.1** ***Find the value of the game of the following pay-off matrix by the iterative method [value will be determined after tenth iteration]. Verify the result with the actual value of the game and the optimal strategies which will be deduced by any known accurate method.***

Player B

Player A				
	0	2	−3	1
	−1	2	1	2
	2	1	0	1

Player B

Player A														
	0	2	−3	1	−3	−6	−6	−6	−9	−12	−15	−18	−21	−24
	−1	2	1	2	①	②	1	0	1	2	3	④	⑤	⑥
	2	1	0	1	0	0	②	④	④	④	④	4	4	4
	2	1	⓪	1										
	1	3	①	3										
	⓪	5	2	5										
	②	6	2	6										
	4	7	②	7										
	6	8	②	8										
	8	9	②	9										
	10	10	②	10										
	9	12	③	12										
	8	14	④	14										

Explanation: Initially the third strategy will be considered as the best choice for A, because for that choice minimum gain will be zero unit and minimum gain will be less than zero unit for the choice of other strategies. The value 0 is being encircled. For A's such choice B will be compelled to choice his third strategy and maximum loss will be one unit which is also being encircled. Due to such choice of B, A will have to select the second strategy, total gain for which is shown in the second row, the minimum of which is 1 unit which occurs at the third columns and for that, in the same way we can find the total loss of B which is shown in the second column, the maximum of which is 2 unit which occurs at the second row. In the same way the iterative process will continue. But one thing to note is that, at the fourth row, minimum occurs at the first and third column and in that case choice will be done arbitrarily and here we have selected 2 in the first column which is being encircled. Similarly at the end of eight iteration, maximum value 4 occurs at the second and third position of the eight column. We have selected arbitrarily four (4) of the second position and the iteration is being stopped after the end of the tenth iteration. From that table we can easily find that v lies in the interval $\frac{4}{10} \le v \le \frac{6}{10}$ and and the optimal strategies are

$$\mathbf{p}^* = \left(0, \frac{5}{10}, \frac{5}{10}\right) \quad \text{and} \quad \mathbf{q}^* = \left(\frac{2}{10}, 0, \frac{8}{10}, 0\right)$$

as there are no elements encircled in the first row and second and fourth column. Probabilities are determined in the manner

$$p_2^* = \frac{\text{number of encircled elements in the second row}}{\text{number of iterations}}$$

and the value of the game is determined in the manner

$$\frac{\text{minimum element of the last row}}{\text{number of iterations}} \le v \le \frac{\text{maximum element of the last column}}{\text{number of iterations}}.$$

Now using dominance and modified dominance, we can determine the actual value of the game = 1/2 and the optimal strategies are

$$\mathbf{p}^* = \left(0, \frac{1}{2}, \frac{1}{2}\right), \quad \mathbf{q}^* = \left(0, \frac{1}{4}, 0, \frac{3}{4}, 0\right).$$

Note: (1) If we try to find the value of the game after the end of the fifth iteration, we get the approximate value of the game in the manner $\frac{2}{5} \le v \le \frac{4}{5}$ and the optimal strategies are

$$\mathbf{p}^* = \left(0, \frac{2}{5}, \frac{3}{5}\right) \quad \text{and} \quad \mathbf{q}^* = \left(\frac{2}{5}, 0, \frac{3}{5}, 0\right),$$

which is not so accurate as in the result obtained after tenth iteration. Thus it is clear that accuracy depends on the number of iterations and the approximate value generally approaches to the accurate result as we gradually increase the number of iterations.

(2) After eight iteration, the approximate value of the game is

$$\frac{2}{4} \le v \le \frac{4}{8}, \quad \text{i.e.,} \quad v = \frac{1}{2}$$

but the optimal strategies are

$$\mathbf{p}^* = \left(0, \frac{3}{8}, \frac{5}{8}\right) \quad \text{and} \quad \mathbf{q}^* = \left(\frac{1}{4}, 0, \frac{3}{4}, 0\right),$$

where the exact value of the game has been determined but $\mathbf{q}^*, \mathbf{p}^*$ differ to some extent.

Solve the following games by using iterative method:

Exercise 19H

1. Find the range of the value of the games and the optimal strategies after completing sixth iteration:

$$\text{(i)} \begin{bmatrix} 2 & 4 & 1 & 2 \\ 3 & 2 & 0 & 2 \\ 1 & 0 & -1 & 2 \end{bmatrix}; \text{ (ii)} \begin{bmatrix} 4 & 1 & 0 & 2 \\ 0 & 2 & 1 & 3 \\ 0 & 1 & 2 & 1 \end{bmatrix}.$$

2. Find the range of the value of the game and the optimal strategies after completing the eight iteration.

$$\text{(i)} \begin{bmatrix} -1 & -2 & 0 & -1 \\ 2 & -1 & -1 & -2 \\ 0 & 1 & 2 & -1 \end{bmatrix}; \text{ (ii)} \begin{bmatrix} 0 & 2 & 0 & 1 \\ 3 & 0 & 2 & 1 \\ 0 & -1 & 2 & 1 \end{bmatrix}.$$

Answers

1.(*i*) $\mathbf{p}^* = (1,0,0)$, $\mathbf{q}^* = (0,0,1,0)$, $v = 1$;

(*ii*) $\mathbf{p}^* = (\frac{1}{5}, \frac{2}{5}, \frac{2}{5})$, $\mathbf{q}^* = (\frac{1}{5}, 0, \frac{3}{5}, \frac{1}{5})$, $\frac{7}{5} \le v \le \frac{11}{5}$;

2.(*i*) $\mathbf{p}^* = (\frac{1}{4}, 0, \frac{3}{4})$, $\mathbf{q}^* = (0, \frac{1}{8}, 0, \frac{7}{8})$, $\frac{15}{8} \le v \le \frac{3}{2}$;

(*ii*) $\mathbf{p}^* = (\frac{3}{8}, \frac{5}{8}, 0)$, $\mathbf{q}^* = (\frac{1}{8}, \frac{5}{8}, \frac{2}{8}, \frac{1}{8})$, $\frac{6}{8} \le v \le \frac{9}{8}$.

Exercise 19 (Miscellaneous)

1. Define a two-person zero-sum or a rectangular game. What is the pay-off matrix of a game? If the pay-off matrix of a game be $[a_{ij}]_{\substack{i=1\cdots m \\ j=1\cdots n}}$ expressed in terms of row or maximizing player what will be pay-off matrix in terms of the column player?
2. Define (i) pure strategy, (ii) player (iii) pay-off matrix (iv) saddle point. Is the saddle point unique in the case of solving a problem by using pure strategy?
3. Define the optimal strategies and the value of the game. Are the optimal strategies unique? Is the value of the game unique?
4. Define (i) Fair game and (ii) Strictly determinate game.
5. Define the working principle of solving a game by using minimax or max-min principle. Is it possible to solve all rectangular games by using that principle?
6. Define the mixed strategies in the case of a rectangle game. Under what conditions mixed strategies be reduced to a pure strategy?
7. Prove that $\min_j \max_i a_{ij} \ge \max_i \min_j a_{ij}$, where a_{ij} is the ijth element of the pay-off matrix.

8. $f(x, y)$ be a real valued function of x and y defined for $x \in A$, $y \in B$, A and B being two sub-sets of real numbers. If both $\max_{x \in A} \cdot \min_{y \in B} .f(x, y)$ and $\min_{y \in B} \cdot \max_{x \in A} .f(x, y)$ exist prove $\max_{x \in B} \cdot \max_{y \in B} .f(x, y) \leq \min_{y \in B} \cdot \max_{x \in A} f(x, y)$.

9. Define the pay-off function of a rectangular game.

10. What will be the unique relation among the elements of the matrix $\begin{bmatrix} a_{11} & a_{12} \\ a_{21} & a_{22} \end{bmatrix}$ such that the problem will have no saddle point in the case of pure strategy.

11. Discuss the working principle of solving a $2 \times n$ game problem by using graphical method.

12. State the principle of solving a game problem by using dominance and modified dominance effect. Is it possible to solve all game problems by using dominance property?

13. Prove that every rectangular game problem can be reduced to L.P.P. considering both from A and B's point of view where A and B are row and column player respectively. What problem will you select to solve the game? Give reasons in support of your opinion.

14. State and prove the fundamental theorem of a rectangular game. Is the value of game unique? Are the optimal strategies unique?

15. If a number be added to the each element of the pay-off matrix, the value of the game of the new pay-off matrix be increased by that quantity but the optimal strategies remain same. Prove that.

16. Prove that if $\mathbf{p}^*$ and $\mathbf{q}^*$ be the optimal strategies then $E(\mathbf{p}^*, \mathbf{q}) \geq v \geq E(\mathbf{p}, \mathbf{q}^*)$ where v is the value of the game.

17. Define a symmetric game. Why is it called so? Prove that the value of the symmetric game is zero and the optimal strategies be same for both players.

18. Prove that the set of all optimal strategies of a player form a convex set.

19. Discuss the working principle of solving a game by iterative method. Can we determine the value of the game and the optimal strategies exactly by using this method? Can we expect a good result and if so under what condition can we achieve that one?

20. Discuss the method of solving a game problem by using algebraically method. What are the difficulties we may face during the procedure? Can this method be considered as a straight forward one? Give an example of 2×2 matrix which leads to erroneous result if we assume all the constraints as strict equations.

21. Explain the working principle of solving a game problem by using matrix method. Established that though the procedure is labourious but it can give exact values of the optimal strategies and the value of the game.

Chapter 20

Queuing (Deterministic and probabilistic) Models

The queuing or waiting line system discussed in this chapter may be described in the manner. A facility or a group of facilities is/are kept ready to meet the demands for service created by a population of individuals (population may be finite or infinite, i.e., the number of individuals demanding service may be finite or infinite, at least theoretically). Each and every individual is considered as a unit in the system. These units form a waiting line system or queue and receive service in accordance with a predetermined waiting line discipline, mainly in the rule 'first come, first served' basis. There are, of course, other disciplines to follow. We ignore all those cases. In our daily life, we are often facing such problems, such as, in the railway ticket counters, before a traffic signal, in doctor's chamber and in many other situations. In each case, the main object is to determine the capacity of the service facility in the light of the relevant costs and the nature of the arrival patterns so that the sum of all costs associated with the queuing system be minimum.

20.1 Queuing system: Single channel system and multiple channel system

If the service facility or facilities is/are available in a single counter, the system is known as single channel system and if the facilities be available in more than one counter the system is known as multiple channel system. We are interested only in a single channel queuing system.

20.2 The arrival population and the service facility

We have already stated that the number of units demanding service may form a queue and units present in the queue are always finite in number. But if the number

of units present in the queue be considerably large, then it may be theoretically treated as an infinite population and the discussion will be easier if the arrival population be considered as an infinite population.

Service may be thought of as the process of providing the activities required by the units in the waiting line. In each case, the act of providing the service to the units present in the queue causes a unit decrease in the waiting line. Both the arrival population and the service completion are discrete in nature since items are processed on a unit basis.

20.3 Waiting line

Depending upon the arrival patterns and the service capacity available in the channel or counter there may be more than one units in the system and in that case some units will have to wait in a line [one unit is being provided service during that time] and a queue or waiting line is formed. Formation of a queue is a discrete process. Individuals or units joining the waiting line and the units present in the waiting line are all integer values.

20.4 Waiting cost and service facility cost

When one or more units will have to wait in the queue or is being serviced waiting cost is incurred. Generally waiting cost depends on the number of units present in the system; vehicles waiting in queue at a toll gate incur a waiting cost due to interruption of trip progress. Customers waiting in check-post suffer badly due to the loss of time there.

To prove service to the units present in the system some arrangements [manual or mechanical] are to be made and for that some cost will incur which is known as service facility cost. This cost depends on the nature of the arrangements required to provide service to the units smoothly and as quickly as possible.

20.5 The arrival and the service mechanism

The analysis required to develop the theory of queuing models in known as Monte Carlo Analysis. To develop the theory we require to make certain assumptions regarding the arrival pattern and the rate at which units requiring service are being serviced. There are two specified models:

(i) Deterministic models (ii) Probabilistic models.

The deterministic model is considered to be simple in nature and in that case the future demand for the service can be predicted with certainty and also the duration of each service is known with certainty. Due to its simple nature we ignore all about the problems of the deterministic models and we try to confine our attention to the problem relating to the probabilistic models.

In the case of probabilistic models both the number of arrivals per unit period to the queue and the time required to service each unit are random variables with some known probability distributions [the parameter or parameters of the distributions are known quantities]. From our previous discussions, it is clear enough that the distribution should be discrete in nature. But considering other points of view, we may take such distributions which may not be a discrete distribution.

In general, we assume that the number of arrivals per period to the queue is a random variable which follows a Poisson distribution [discrete distribution] with known parameter $1/\lambda, \lambda > 0, 1/\lambda$ being the mean or expected value of the distribution and the time required to service each unit is a random variable which follows Exponential distribution [continuous distribution] with known parameter $\mu > 0, 1/\mu$ being the mean or expected value of the distribution and we may theoretically assume that the population to be infinite. We may also assume that both the number of arrivals per period and the time required to service each unit follow Poisson distribution with known parameters. For the sake of convenience we define the probability mass function of a Poisson distribution and the probability density function of an Exponential distribution.

(1) Poisson distribution:

$$P(X = x) = \frac{e^{-\frac{1}{\lambda}} \left(\frac{1}{\lambda}\right)^x}{\underline{|x}}, \quad \lambda > 0 \quad [x = 0, 1, 2, \cdots, n, \cdots].$$

Mean or expectation or expected value

$$E(X) = \frac{1}{\lambda}, \quad \text{or, } \quad \text{for } E(X) = \lambda, \quad P(X = x) = e^{-\lambda} \frac{(\lambda)^x}{\underline{|x}}.$$

(2) Exponential distribution:

The probability density function

$$f(x) = \mu e^{-\mu x}, \quad \mu > 0, x \geq 0$$
$$= 0, \text{ elsewhere.}$$

Mean or expected value $E(X) = \dfrac{1}{\mu}$.

The assumptions made above are mathematically convenient and have sound practical basis in many practical situations. In both the cases, the probability of occurrence tends to zero as $x \to \infty$, i.e., as x increases gradually after certain range, the probability of occurrence gradually decreases.

Under the assumption of Poisson arrival and Exponential service the rate of arrival and the service completion be λ and μ respectively. We now proceed to develop the mathematical formalities of the probabilistic queuing models under the assumption of the Poisson arrival and Exponential service scheme.

20.6 The probability of n units in the system

To find the probability of n units in the system, which is denoted by $P_n(t)$, t represents 'time,' we shall have to make the following assumptions : n = number of units in the system at time t including the unit being served, if any].

(i) The probability that an arrival occurs between the time t and $t + \Delta t$ is λdt, $\lambda > 0$.

(ii) The probability that a service completion occurs between t and $t+\Delta t$, given that one unit being serviced at time t, is $\mu\Delta t$, $\mu > 0$.

(iii) As the time interval Δt is small, it can be assumed that the probability of more than one arrival or service completion during the interval Δt is almost nil and with the above pre-conditions we shall now try to find $P_n(t)$.

There are two cases to discuss separately (i) $n \geq 1$ and (ii) $n = 0$.

For $n \geq 1$, the event n units in the system at time $t + \Delta t$ can be decomposed into three mutually exclusive events and

Event $\{n$ units in the system at time $t + \Delta t\}$

= Event $\{n$ units in the system at time t, no arrival during the interval Δt and no service completion during that interval $\Delta t\}$ or

Event $\{(n+1)$ units in the system at time t, no arrival during the interval Δt, one service completion during the interval $\Delta t\}$ or

Event $\{(n-1)$ unit in the system at time t, one arrival during the interval Δt, no service completion during the interval $\Delta t\}$.

Hence, mathematically we can write

$$\begin{aligned} P_n(t+\Delta t) &= P_n(t)(1-\lambda\Delta t)(1-\mu\Delta t) \\ &\quad + P_{n+1}(t)(1-\lambda\Delta t)\mu\Delta t \\ &\quad + P_{n-1}(t)\lambda\Delta t(1-\mu\Delta t). \end{aligned} \tag{20.6.1}$$

On simplification we get

$$\begin{aligned} P_n(t+\Delta t) &= P_n(t) - (\lambda+\mu)P_n(t)\Delta t + \lambda\mu P_n(t)(\Delta t)^2 + \mu P_{n+1}(t)\Delta t \\ &\quad - \lambda\mu P_{n+1}(t)(\Delta t)^2 + \lambda P_{n-1}(t)\Delta t - \lambda\mu P_{n-1}(t)(\Delta t)^2. \end{aligned}$$

Neglecting the terms containing $(\Delta t)^2$, we have

$$P_n(t+\Delta t) = P_n(t) - (\lambda+\mu)P_n(t)\Delta t + \mu P_{n+1}(t)\Delta t + \lambda P_{n-1}(t)\Delta t$$

$$\text{or,}\quad \frac{P_n(t+\Delta t) - P_n(t)}{\Delta t} = -(\lambda+\mu)P_n(t) + \mu P_{n+1}(t) + \lambda P_{n-1}(t). \tag{20.6.2}$$

Now with the assumption that $\Delta t \to 0$, we have

$$\lim_{\Delta t\to 0} \frac{P_n(t+\Delta t) - P_n(t)}{\Delta t} = \frac{dP_n(t)}{dt} = -(\lambda+\mu)P_n(t) + \mu P_{n+1}(t) + \lambda P_{n-1}(t). \tag{20.6.3}$$

Now considering the special case for $n = 0$,

the Event $\{0$ unit in the system at time $t + \Delta t\}$

= Event $\{0$ unit in the system at time t, no arrival during the interval $\Delta t\}$ or

Event {one unit in the system at time t, no arrival during the interval Δt, one service completion during the interval $\Delta t\}$ and the events are mutually exclusive.

Hence

$$\begin{aligned} P_0(t+\Delta t) &= P_0(t)(1-\lambda\Delta t) + P_1(t)(1-\lambda\Delta t)\mu\Delta t \\ &= P_0(t) - \lambda P_0(t)\Delta t + \mu P_1(t)\Delta t - \lambda\mu P_1(t)(\Delta t)^2. \end{aligned}$$

Neglecting the term containing $(\Delta t)^2$, we have

$$P_0(t+\Delta t) = P_0(t) - \lambda P_0(t)\Delta t + \mu P_1(t)\Delta t$$

Applying limit $\Delta t \to 0$, we have

$$\lim_{\Delta t\to 0} \frac{P_0(t+\Delta t) - P_0(t)}{\Delta t} = \frac{dP_0(t}{dt'} = -\lambda P_0(t) + \mu P_1(t). \tag{20.6.4}$$

The equations (20.6.3) and (20.6.4) are known as the governing equations of a Poisson arrival and Exponential service, single channel queue.

20.7 Solution of the governing equations

The governing equations are differential equations and the finding out the general solutions [which are infinite in number] is a very difficult task. By making some assumptions which will be available from practical experience, we can able to solve the differential equations.

It has been found in practice, that with the increase of time t, the rate of change of $P_n(t)$ gradually decreases and ultimately the rate of change of $P_n(t)$ is almost zero and obviously $P_n(t)$ is independent of time t and we can think it simply as P_n when $\frac{d}{dt}P_n(t) = 0$. With the assumption, the governing equations (20.6.3) and (20.6.4) assume the form

$$\left.\begin{aligned} (\lambda+\mu)P_n &= \mu P_{n+1} + \lambda P_{n-1},\ n \geq 1 \\ \text{and } \lambda P_0 &= \mu P_1,\ n = 0 \end{aligned}\right\}. \tag{20.7.1}$$

Let us assume $P_n = P_{n-1}\rho$ [where ρ is a constant] as a trial solution of the equations given in (20.7.1).

Putting the values of P_n, P_{n+1} etc. in the first equation we get

$$\begin{aligned} &(\lambda+\mu)P_{n-1}\rho = \mu P_{n-1}\rho^2 + \lambda P_{n-1} \\ \text{or, } &(\lambda+\mu)\rho = \mu\rho^2 + \lambda \quad [\text{assuming } P_{n-1} \neq 0] \\ \text{or, } &\mu\rho(1-\rho) = \lambda(1-\rho) \\ \text{or, } &\rho = \frac{\lambda}{\mu} > 0 \quad [\text{assuming } 1-\rho \neq 0] \end{aligned} \tag{20.7.2}$$

ρ is sometime called as 'traffic intensity'.

Now with this value of ρ, the second equation for $n = 0$ is being satisfied and this trial solution is a particular solution of the equations and we have then

$$P_1 = P_0\rho, \quad P_2 = P_1\rho = P_0\rho^2, \cdots, \quad P_n = P_0\rho^n. \tag{20.7.3}$$

Again from the axioms of the probability theory that the probability of a 'Certain event' is 1, we have

$$1 = P_0 + P_1 + P_2 + \cdots + P_n + \cdots \infty$$
$$\Rightarrow 1 = P_0(1 + \rho + \rho^2 + \cdots + \rho^n + \cdots + \infty).$$

As the above series is a convergent series, then ρ must satisfies the condition $0 < \rho < 1$ and

$$1 = P_0 \frac{1}{1-\rho} \quad \text{[sum of G.P. series]} \tag{20.7.4}$$

which gives $P_0 = 1 - \rho$.

Hence our previous assumption that $1 - \rho \neq 0$ is true and

$$P_n = P_0 \rho^n = (1-\rho)\rho^n \quad \text{where } \rho = \frac{\lambda}{\mu}. \tag{20.7.5}$$

Now, as $0 < \rho = \frac{\lambda}{\mu} < 1$, then $\lambda < \mu$ which implies that to reach at the steady state of the queue, the arrival rate must be less than the service rate.

Note: All the assumptions are almost valid physically in the case of Poisson arrival and Poisson service completion and the above probability given in (20.7.5) is almost remains valid for this arrival and service mechanism.

20.8 The average or mean number of units in the system

Let N denote the 'random variable' number of units in the system in time t. The mean number of units or the expected number of units in the system is nothing but the expected value of the random variable N and

$$E(N) = n_m = \sum_{n=0}^{\alpha} nP_n = \sum_{n=0}^{\alpha} n(1-\rho)\rho^n$$

$$= \rho(1-\rho)\sum_{n=0}^{\alpha} n\rho^{n-1} = \rho(1-\rho)(1-\rho)^{-2} \quad \text{[from binomial expansion]}$$

$$= \frac{\rho}{1-\rho} \quad [\because 0 < \rho < 1] \tag{20.8.1}$$

$$= \frac{\lambda/\mu}{1 - \dfrac{\lambda}{\mu}} \quad \text{[Putting the value of } \rho]$$

$$= \frac{\lambda}{\mu - \lambda} > 0. \tag{20.8.2}$$

20.9 The average length of the waiting line

The average length of the waiting (queue) which is denoted by m_m can be expressed as the mean number of units in the system less the mean number of units being serviced.

Mean number of units being serviced = ρ = traffic intensity.

Therefore,

$$m_m = \frac{\lambda}{\mu - \lambda} - \frac{\lambda}{\mu} = \frac{\lambda^2}{\mu(\mu - \lambda)}. \quad (20.9.1)$$

20.10 The probability of the queue being non-empty

If the queue be non-empty, then there must have at least one unit in the line and if one unit is being serviced at that time, the probability [at least two units in the system] is

$$\begin{aligned} P(N > 1) &= P(N \geq 2) = 1 - P(N = 0) - P(N = 1) \\ &= 1 - P_0 - P_1 = 1 - (1 - \rho) - (1 - \rho)\rho \doteq \rho^2. \end{aligned} \quad (20.10.1)$$

If M be the random variable 'the length of the waiting line',

$$P(N > 1) = P(M > 0) = \rho^2.$$

20.11 The average length of the non-empty queue

The average length of the non-empty queue is nothing but the conditional mean of M on the hypothesis that the queue is non-empty which is denoted by

$$E(M/M > 0) = (m/m > 0)_m$$

and defined by,

$$(m/m > 0)_m = \frac{\text{The average length of the waiting line}}{\text{Probability that the queue is not empty}}$$

$$= \frac{m_m}{P(M > 0)} = \frac{\dfrac{\lambda^2}{\mu(\mu - \lambda)}}{\dfrac{\lambda^2}{\mu^2}} = \frac{\mu}{\mu - \lambda}. \quad (20.11.1)$$

► **Example 20.11.1** *Suppose that arrivals are distributed accoraing to the Poisson distribution with a mean rate of* 7.5 *per hour and the service duration is distributed exponentially with mean of four minutes. Find the probability of n units in the system. What is the probability of there being not more than three units in the system?*

Solution: Arrival rate $= \dfrac{7.5}{60}$ per minute $= \dfrac{1}{8} = 0.125 = \lambda$.

Service rate $= \dfrac{1}{\text{mean}} = \dfrac{1}{4} = 0.25 = \mu$.

Therefore ρ = traffic intensity $= \dfrac{\lambda}{\mu} = \dfrac{0.125}{0.25} = \dfrac{1}{2}$.

Hence the probability of n units in the system is

$$P_n = (1-\rho)\rho^n = \left(1-\frac{1}{2}\right)\cdot\left(\frac{1}{2}\right)^n = \left(\frac{1}{2}\right)^{n+1} \quad \text{[using (20.7.5)]}$$

P (Not more than three units in the system).

$$= P(N=0)+P(N=1)+P(N=2)+P(N=3)$$

$$= \left(\frac{1}{2}\right)^{0+1} + \left(\frac{1}{2}\right)^{1+1} + \left(\frac{1}{2}\right)^{2+1} + \left(\frac{1}{2}\right)^{3+1}$$

$$= \frac{1}{2} + \left(\frac{1}{2}\right)^2 + \left(\frac{1}{2}\right)^3 + \left(\frac{1}{2}\right)^4 = \frac{15}{16}.$$

► **Example 20.11.2** *Customers arrive at a post office manned by a single person, according to a Poisson input process with a mean rate of 10 per hour. The time required to serve a customer has an exponential distribution with a mean of 4 minutes. Find* (i) *The average number in the system.* (ii) *The probability that there would be 2 customers in the queue.* **[C.U. M.Com. '87]**

Solution: The Poisson distribution is distributed with the rate of $\frac{10}{60} = \frac{1}{6}$ per minute.

Hence ρ = traffic intensity $= \frac{1/6}{1/5} = \frac{4}{6} = \frac{2}{3}$.

The average number in the system $= \frac{\rho}{1-\rho} = \frac{2/3}{1-\frac{2}{3}} = 2$ [from (20.8.1)].

The probability that there would be two customers in the queue

$$= P_2 = (1-\rho)\rho^2 = \left(1-\frac{2}{3}\right)\left(\frac{2}{3}\right)^2 = \frac{4}{27}.$$

20.12 Probability distribution of the waiting time

In a probabilistic queuing model, the waiting time can be considered a random variable W which depends on the nature of the system at the time of the arrival and also on the time required to service the units already waiting in the system for service. In a single channel system an arriving unit can be put into service immediately only if there are no other units in the system; otherwise the arriving unit will have to wait. We shall find the probability distribution of the random variable W, with the assumption of Poisson arrival and Poisson service system.

Here the total event space can be decomposed in the manner:

(i) Event $\{W = 0\}$ which is same as the event {no unit the system}.

(ii) Event $(w < W \leq w + dw)$ which is a compound event of the independent events which are: n units in the system at the time of arrival, $(n-1)$ services being completed during the time w and the last service being completed within the interval w and $w + dw$. $[n \geq 1]$

Hence the probability distribution is given by

$$P(W=0)=P_0=1-\rho=1-\frac{\lambda}{\mu} \text{ for } W=0$$

[$\because$ the rule is valid for Poisson arrival and Poisson service.]

$$P(w<W\leq w+dw)=f(w)dw$$

$$=\sum_{n=1}^{\infty}[P_n.P\{(n-1) \text{ service completion within the time } w\}$$

$$P \text{ (one service completion in the time } dw \text{ after time } w)$$

$$=\sum_{n=1}^{\infty}\left(1-\frac{\lambda}{\mu}\right)\left(\frac{\lambda}{\mu}\right)^n .e^{-\mu w}\frac{(\mu w)^{n-1}}{\underline{|n-1}}\cdot \mu dw$$

[$\because P_n=\left(1-\frac{\lambda}{\mu}\right)\cdot\left(\frac{\lambda}{\mu}\right)^n$; $P\{(n-1)$ service completion in time $w\}=e^{-\mu w}\frac{(\mu w)^{n-1}}{\underline{|n-1}}$, where μw is the parameter of the Poisson distribution, and P (one service completion in time dw after the time w) is μdw.]

$$=\sum_{n=1}^{\infty}\left(\frac{\lambda}{\mu}\right)^n\cdot\frac{(\mu w)^{n-1}}{\underline{|n-1}}e^{-\mu w}\left(1-\frac{\lambda}{\mu}\right)\mu dw$$

$$=\sum_{k=0}^{\infty}\frac{(\mu w)^k}{\underline{|k}}\left(\frac{\lambda}{\mu}\right)\cdot\mu\left(\frac{\lambda}{\mu}\right)^k\cdot\mu\left(\frac{\lambda}{\mu}\right)^k e^{-\mu w}\left(1-\frac{\lambda}{\mu}\right)dw \quad [\text{Putting } n-1=k]$$

$$=\lambda\left(1-\frac{\lambda}{\mu}\right)\sum_{k=0}^{\infty}\frac{(\lambda w)^k}{\underline{|k}}e^{-\mu w}dw$$

$$=\lambda\left(1-\frac{\lambda}{\mu}\right)e^{-\mu w}\cdot e^{w\lambda}dw \quad \left[\because \sum_{k=0}^{\infty}\frac{(\lambda w)^k}{k!}=e^{\lambda w}\right]$$

$$=\lambda\left(1-\frac{\lambda}{\mu}\right)e^{-(\mu-\lambda)w}dw.$$

Therefore

$$f(w)=\lambda\left(1-\frac{\lambda}{\mu}\right)e^{-(\mu-\lambda)w}. \tag{20.12.1}$$

Thus the probability distribution is

$$\left.\begin{aligned}&P(W=0)=1-\frac{\lambda}{\mu}\\ &\text{and } f(w)=\lambda\left(1-\frac{\lambda}{\mu}\right)e^{-(\mu-\lambda)w} \text{ for } w>0\end{aligned}\right\}. \tag{20.12.2}$$

The distribution is partly discrete and partly continuous.

Again

$$
\begin{aligned}
P(W=0)+\int_{w>0}^{\infty} f(w)dw &= 1-\frac{\lambda}{\mu}+\int_0^{\infty}\lambda\left(1-\frac{\lambda}{\mu}\right)e^{-(\mu-\lambda)w}dw \\
&= 1-\frac{\lambda}{\mu}+\lambda\left(1-\frac{\lambda}{\mu}\right)\left[\frac{e^{-(\mu-\lambda)w}}{-(\mu-\lambda)}\right]_0^{\infty} \\
&= 1-\frac{\lambda}{\mu}+\frac{\lambda}{\mu}(\mu-\lambda)\left\{\frac{1}{(\mu-\lambda)}\right\} \\
&= 1-\frac{\lambda}{\mu}+\frac{\lambda}{\mu}=1,
\end{aligned}
$$

which satisfies the condition of its being a probability density function of a random variable.

20.13 The average waiting time in the system

The average or mean waiting time is nothing but the expectation of the random variable W which is denoted by $E(W) = w_m$ defined by

$$
\begin{aligned}
E(W) = w_m &= 0\left(1-\frac{\lambda}{\mu}\right)+\int_0^{\infty} w\lambda\left(1-\frac{\lambda}{\mu}\right)e^{-(\mu-\lambda)w}dw \\
&= \lambda\left(1-\frac{\lambda}{\mu}\right)\left[w\frac{e^{-(\mu-\lambda)w}}{-(\mu-\lambda)}\right]_0^{\infty}+\lambda\left(1-\frac{\lambda}{\mu}\right)\int_0^{\infty}\frac{e^{-(\mu-\lambda)^w}}{-(\mu-\lambda)}dw \\
&= 0-\frac{\lambda}{\mu}\frac{(\mu-\lambda)}{(\mu-\lambda)^2}\left[e^{-(\mu-\lambda)w}\right]_0^{\infty}=\frac{\lambda}{\mu(\mu-\lambda)}. \qquad (20.13.1)
\end{aligned}
$$

20.14 The average time in the system

The average time that a unit spends in the waiting line system is equal to the sum of the average time in the system and the average time required for the completion of service of a unit in the system. Then the average time is the system which is denoted by t_m, given by

$$
t_m = \frac{\lambda}{\mu(\mu-\lambda)}+\frac{1}{\mu}=\frac{1}{\mu-\lambda}. \qquad (20.14.1)
$$

► **Example 20.14.1** *If in a railway reservation counter (assume only one counter is available to handle reservation and enquiry) customers arrive at the rate of 32 persons per hour and the clerk can handle and provide assistance using a computer terminal to 40 persons per hour. If the arrival and the services follow Poisson process.*

(i) *Calculate the average number of customers waiting for service.*

(ii) *Average waiting time for a customer before getting a service.*

(iii) *The probability that there would be other customers in the queue.*

[C.U. M.Com. '88]

Solution: $\rho =$ traffic intensity $= \dfrac{\lambda}{\mu} = \dfrac{32}{40} = \dfrac{4}{5}.$

(i) Average number of customers (units) waiting for service

$$= \frac{\rho}{1-\rho} = \frac{4/5}{1-\frac{4}{5}} = 4.$$

(ii) Average waiting time $= \dfrac{\lambda}{\mu(\mu-\lambda)} = \dfrac{32}{40(40-32)} = \dfrac{1}{5}$ hour.

(iii) The probability that there would be other customer in the queue, i.e., the waiting line will be non-empty $= \rho^2 = \left(\dfrac{4}{5}\right)^2 = \dfrac{16}{25}.$

► **Example 20.14.2** *The arrival rate for a certain waiting-line system obeys a Poisson distribution with the average value of* 1/2 *unit per period. The probability of one or more units in the system does not exceed* 0.25. *What is the minimum service rate that must be provided if the service rate if the service duration follows an exponential distribution?*

Solution: The Poisson distribution has the mean 1/2.

Therefore the arrival rate $\lambda = \dfrac{1}{1/2} = 2.$

$\rho =$ traffic intensity $= \lambda/\mu$ where μ is the service rate.

P (one or more units in the system) $= P(N \geq 1) = 1 - P_0 = 1 - (1-\rho) = \rho.$

From the condition of the problem

$$\rho \leq 0.25, \quad \text{or,} \quad \frac{\lambda}{\mu} \leq 0.25, \quad \text{or,} \quad \mu \geq 4\lambda = 8.$$

Hence the minimum service rate will be 8 units per period.

► **Example 20.14.3** *The expected number of individuals in a queue having Poisson arrivals with a rate of* 0.4 *units per period must not exceed* 5. *What is the minimum service rate if the service duration be distributed exponentially.*

Solution: $\rho =$ arrival rate $= 0.4$, $\mu =$ service rate (say) and then the traffic intensity $\rho = \frac{.4}{\mu}$.

The mean or expected number of individuals in the system

$$= \frac{\rho}{1-\rho} = \frac{\lambda}{\mu-\lambda} = \frac{0.4}{\mu-0.4}.$$

From the condition of the problem

$$\frac{0.4}{\mu-0.4} \leq 5 \Rightarrow 0.4 \leq 5\mu - 2 \Rightarrow 2.4 \leq 5\mu, \quad \text{or,} \quad \mu \geq \frac{2.4}{5} = 0.48.$$

Therefore the minimum value of the service rate = 0.48.

▶ **Example 20.14.4** *The arrival patterns of a waiting line system follows a Poisson distribution and the service completion follow an exponential distribution with service rate 1 unit per period. If the average length of the waiting line be* 0.5 *unit find the arrival rate per period.*

Solution: Let λ be the arrival rate per period and the given service rate $\mu=1$. The average length of the waiting line is

$$= \frac{\lambda^2}{\mu(\mu-\lambda)}.$$

From the condition of the problem

$$\frac{\lambda^2}{1(1-\lambda)} = 0.5, \text{ or, } \lambda^2 = 0.5 - 0.5\lambda,$$
$$\text{or, } 2\lambda^2 + \lambda - 1 = 0,$$
$$\text{or, } (\lambda+1)(2\lambda-1) = 0 \Rightarrow \lambda = -1, \text{ or, } \lambda = 0.5.$$

But λ cannot be negative. Hence $\lambda = 0.5 =$ arrival rate.

20.15 The minimum cost service rate

The expected total cost of the system per period is the sum of the expected waiting cost per period and the expected facility cost per period.

Mathematically, we can write

$$TC_m = WC_m + FC_m, \tag{20.15.1}$$

where TC_m is the expected total cost of the system per period

WC_m is the expected waiting cost of the system per period

and FC_m is the expected service facility of the system per period.

Now, WC_m is the product of the waiting cost per period and the mean number of units present in the system during that period. Hence

$$WC_m = C_w(n_m) = \frac{C_m\lambda}{\mu-\lambda},$$

where C_w is the waiting cost per period.

Again, the expected service cost per period is the product of the cost of service one unit and the service rate in unit per period or $FC_m = C_{f\mu}$ where C_f is the cost of service of one unit.

Then from (15.18),

$$TC_m = \frac{C_w\lambda}{\mu-\lambda} + C_{f\mu} = \phi(\mu) \text{ say.} \tag{20.15.2}$$

To obtain minimum cost service rate, we shall have to differentiate $\phi(u)$, w.r.t. μ. Now differentiating $\phi(\mu)$ we get

$$\phi'(\mu) = C_f - \frac{C_w \lambda}{(\mu - \lambda)^2}$$

$\phi'(\mu) = 0$ gives

$$0 = C_f - \frac{C_w \lambda}{(\mu - \lambda)^2}, \text{ or, } (\mu - \lambda)^2 = \frac{C_w \lambda}{C_f},$$

$$\text{or, } \mu_{\min} = \lambda + \sqrt{\frac{C_w \lambda}{C_f}} \quad \text{[Taking + ve square root]} \quad (20.15.3)$$

► **Example 20.15.1** *In a waiting line system if the arrival and service mechanism follows Poisson and exponential distribution respectively with the mean 4 units for the Poisson distribution and the cost of waiting per unit per period be* Rs. 10 *and the service facility cost for servicing one unit be* Rs. 5 *find the minimum cost service rate.*

Solution: We have

$$\mu_{\min} = \lambda + \sqrt{\frac{C_w \lambda}{C_f}} \quad \text{[From (20.15.3)]}$$

Again λ = arrival rate $= \frac{1}{4} = 0.25$, C_w = Rs. 10, C_f = Rs. 5.

Hence,

$$\mu_{\min} = 0.25 + \sqrt{\frac{10 \times 0.25}{5}} = 0.25 + 0.707 = 0.957.$$

Exercise

1. Name and briefly describe the components of queuing or waiting line system.

2. Describe and differentiate the arrival and service mechanism in the case of deterministic queuing models and probabilistic queuing modes.

3. Define the arrival rate and service rate in the case of Poisson arrival and exponential service mechanism in a single channel queuing system.

4. For a set of infinite population, prove that the probability of n units in the system is $P_n = (1 - \rho)\rho^n$ which is independent of time where ρ = 'traffic intensity' in the case of Poisson arrival and exponential service mechanism. What is the assumption made to solve the governing equations?

5. If λ and μ be the arrival rate and service rate in the case of Poisson arrival and service mechanism which one of the relation is true? (i) $\lambda > \mu$, (ii) $\lambda = \mu$ and (iii) $\lambda < \mu$ and explain that the condition is required to reach at a steady state.

6. Find out (i) the mean number of units in the system (ii) the mean length of the waiting line (iii) the probability that the queue is non-empty and (iv) the average length of the non-empty queue in the case of Poisson arrival and exponential service mechanism in in single channel queuing model [Population being assumed to be infinite].

7. Prove that the probability $P_n = (1-\rho)\rho^n$, here P_n and ρ have their usual meanings is also valid for Poisson arrival and Poisson service mechanism.

8. If the arrival and service [in a single channel queuing model] follows Poisson distribution then find the probability distribution of the random variable 'waiting time'. Prove that the distribution is partly discrete and partly continuous. Find the value of the mean waiting time.

9. Show that model $(M/M/1 : \alpha/F1F0)$ (i.e.,) single server Poisson arrival, exponential serviced, the probability that n calling units are in the queuing system is $P_n = (1-\rho)\rho^n$, $n \geq 0$ [where ρ is the traffic intensity].

10. The arrivals are distributed which follow Poisson distribution with a mean of 0.125 units per period and the service duration is distributed exponentially with a mean of four periods. Find the probability that

    (i) there are exactly three units in the system.

    (ii) the probability of there being more than 6 units in the system.

11. In a waiting line system with Poisson arrival and exponential service mechanism, the mean number of units in the system does not exceed 3. If the arrival rate be 0.5 unit per period find the maximum value of the mean of the exponential distribution.

12. The arrival pattern and the service mechanism in a waiting line system follows Poisson distribution and exponential distribution with means 4 units and 3 units respectively. Find the probability that the queue be non-empty.

13. If both the arrival pattern and the service mechanism follows Poisson distribution with the rate of 0.4 and 0.6 unit per period respectively, find the average value of the waiting time and the average time in the system.

14. The average length of the waiting line system be 1/6 where both arrival pattern and service mechanism follow Poisson distribution. If the service rate be 0.75 unit per period find the arrival rate and the mean of the distribution.

Answers

10. $\frac{1}{16}, \frac{1}{128}$. 11. $\frac{3}{2}$. 12. $\frac{9}{16}$. 13. $3\frac{1}{3}$, 5 Units. 14. 0.25,4.

Appendix A

Development of the Simplex Theory

A.1 Introduction

In the simplex theory, the initial basic problem is to find a basis (which always exists there) for which the solution is feasible. Let the constraints be

$$\mathbf{a}_1x_1 + \mathbf{a}_2x_2 + \cdots + \mathbf{a}_nx_n = \mathbf{b} \tag{A.1.1}$$

and the objective function $z = \mathbf{cx}$, where $\mathbf{c} = (c_1, c_2 \cdots, c_n)$

Let $\mathbf{x}_B = B^{-1}\mathbf{b} \geq \mathbf{0}$ and

$$\mathbf{x}_B = [x_{\beta 1}, x_{\beta 2}, \cdots, x_{\beta m}],$$
$$\mathbf{c}_\beta = (c_{B1}, c_{B2}, \cdots, c_{Bm})$$

as $B = (\beta_1, \beta_2 \cdots, \beta_m)$, where $\beta_i (i = 1, 2, \cdots, m)$ are some of the vectors $\in A$ $[j = 1, 2, \cdots, n]$. Now after finding such basis, we shall calculate z_B corresponding to such basis and next problem is to test whether this is an optimal solution and if not, we shall have to develop the theories such that the optimal condition be satisfied or the problem has said to have an unbounded solution etc. The developments of the theories are given below.

A.2 Determination of the improved value of the objective function

A.2.1 Problem of Maximization

Theorem A.2.1 *Let $\mathbf{x}_B = B^{-1}\mathbf{b}$ be a B.F.S to a L.P.P. and the corresponding value of the objective function $z_B = \mathbf{c}_B\mathbf{x}_B$. If for some jth columns corresponding to non-basic variables, the condition $z_j - c_j < 0$ holds with at least one $y_{ij} > 0$, $[i = 1, 2, \cdots, m]$ then it is possible to obtain a new B.F.S. which improves the value of the objective function by replacing one vector of B by $\mathbf{a}_j$.*

Proof. Let

$$\begin{aligned} A &= (\mathbf{a}_1, \mathbf{a}_2, \cdots, \mathbf{a}_n) \\ \text{and} \quad B &= (\beta_1, \beta_2, \cdots, \beta_r, \cdots, \beta_m) \text{ be a basis matrix} \\ \mathbf{x}_B &= B^{-1}\mathbf{b} \text{ is the B.F.S.} \\ \text{and} \quad z_B &= \mathbf{c}_B \mathbf{x}_B = \sum_{i=1}^{m} \mathbf{c}_{Bi} x_{Bi}, \end{aligned}$$

the value of the objective function corresponding to the B.F.S. $\mathbf{x}_B = B^{-1}\mathbf{b}$. We have

$$A\mathbf{x} = \mathbf{b} = B\mathbf{x}_B. \tag{A.2.1}$$

As $\mathbf{a}_j$ is a column vector corresponding to a non-basic variable x_j therefore

$$\mathbf{a}_j = B\mathbf{y}_j = \beta_1 y_{1j} + \beta_2 y_{2j} + \cdots + \beta_r y_{rj} + \cdots + \beta_m y_{mj}.$$

Assuming $y_{rj} \neq 0$

$$\beta_r = \frac{1}{y_{rj}}[\mathbf{a}_j - \beta_1 y_{1j} - \cdots - \beta_{r-1,j} y_{r-1,j} - \beta_{r+1,j} y_{r+1,j} - \cdots - \beta_m y_{mj}] \tag{A.2.2}$$

From (A.2.1)

$$\beta_1 x_{B1} + \beta_2 x_{B2} + \cdots + \beta_r x_{Br} + \cdots + \beta_m x_{Bm} = \mathbf{b}$$

$$\text{or,} \quad \sum_{\substack{i=1 \\ i \neq r}}^{m} \beta_i x_{Bi} + \beta_r x_{Br} = \mathbf{b}. \tag{A.2.3}$$

Putting the value of β_r from (A.2.2) in (A.2.3) we get

$$\sum_{\substack{i=1 \\ i \neq r}}^{m} \beta_i x_{Bi} + x_{Br} \left(\frac{\mathbf{a}_j}{y_{rj}} - \frac{\beta_1 y_{ij}}{y_{rj}} - \cdots \frac{y_{ij}}{y_{rj}} \beta_i - \cdots - \frac{\beta_m y_{mj}}{y_{rj}} \right) = \mathbf{b}$$

$$\text{or} \quad \sum_{\substack{i=1 \\ i \neq r}}^{m} \left(x_{Bi} - \frac{y_{ij}}{y_{rj}} x_{Br} \right) \beta_i + \frac{x_{Br}}{y_{rj}} \mathbf{a}_j = \mathbf{b}. \tag{A.2.4}$$

As $\{\beta_1, \beta_2, \beta_r, \cdots, \beta_m\}$ is a basis set of vectors $\in A$ and

$$\mathbf{a}_j = \beta_i y_{ij} + \beta_2 y_{2j} + \cdots + \beta_r y_{rj} + \cdots + \beta_m y_{mj}$$

with $y_{rj} \neq 0$ then $\mathbf{a}_j$ can replace β_r to form a new basis and hence from (A.2.3) and (A.2.4) a new B.S.

$\mathbf{x}'_B = [x'_{B1}, x'_{B2}, \cdots, x'_{Bm}]$ is obtained which is given by

$$\left.\begin{aligned} x'_{Bi} &= x_{Bi} - \frac{y_{ij}}{y_{rj}} x_{Br}, \quad i = 1, 2, \cdots, m, \quad i \neq r \\ x'_{Br} &= \frac{x_{Br}}{y_{rj}}, \quad i = r \end{aligned}\right\} \tag{A.2.5}$$

Now we put some restrictions to make (A.2.5) a feasible solution. As $x_{Br} \geq 0$, therefore $x'_{Br} \geq 0$ if and only if $y_{rj} > 0$. Again for $y_{rj} > 0$ and $y_{ij} \leq 0$, $x_{Bi} \geq 0$ for $[i = 1, 2, \cdots, m$ and $i \neq r]$.

But if $y_{rj} > 0$ and $y_{ij} > 0$ then $x'_{Bi} \geq 0$ provided

$$x_{Bi} - \frac{y_{ij}}{y_{rj}} x_{Br} \geq 0, \quad \text{or,} \quad \frac{x_{Br}}{y_{rj}} \leq \frac{x_{Bi}}{y_{ij}}. \tag{A.2.6}$$

If we select

$$\frac{x_{Br}}{y_{rj}} = \min_i \left(\frac{x_{Bi}}{y_{ij}}, y_{ij} > 0 \right) \tag{A.2.7}$$

then the condition (A.2.6) will be satisfied and solution given in (A.2.5) will be a new B.F.S. Now we are to prove that for this B.F.S. the value of the objective function will be improved.

Let the value of the objective function be z' for the B.F.S. given in (A.2.5). Then

$$z' = \mathbf{c}'_B \mathbf{x}'_B = \sum_{i=1}^{m} c'_{Bi} x'_{Bi},$$

where c'_{Bi} is the co-efficient of the corresponding basic variable x'_{Bi}.

Obviously $c'_{Bi} = c_{Bi}$ for $i = 1, 2, \cdots, m, i \neq r$ and $c'_{Br} = c_j$ for $i = r$. [As β_r is replaced by $\mathbf{a}_j$]

$$\begin{aligned}
z' &= \sum_{i=1}^{m} c'_{Bi} x'_{Bi} \\
&= \sum_{\substack{i=1 \\ i \neq r}}^{m} c_{Bi} \left(x_{Bi} - \frac{y_{ij}}{y_{rj}} x_{Br} \right) + c_j \frac{x_{Br}}{y_{rj}} \quad \text{[putting the values from (A.2.5)]} \\
&= \sum_{i=1}^{m} c_{Bi} \left(x_{Bi} - \frac{y_{ij}}{y_{rj}} x_{Br} \right) + c_j \frac{x_{Br}}{y_{rj}} \quad \left[\text{as } x_{Bi} - \frac{y_{ij}}{y_{rj}} x_{Br} = 0 \text{ for } i = r \right] \\
&= \sum_{i=1}^{m} c_{Bi} x_{Bi} + \frac{x_{Br}}{y_{rj}} \left(c_j - \sum_{i=1}^{m} c_{Bi} y_{ij} \right) \\
&= z_B - \frac{x_{\beta r}}{y_{rj}} (z_j - c_j) \quad \left[\text{from (8.2.5) since } \sum_{i=1}^{m} c_{Bi} y_{ij} = z_j \right].
\end{aligned} \tag{A.2.8}$$

Now as $(x_{\beta r}/y_{rj}) \geq 0$, and $z_j - c_j < 0$ therefore $z' \geq z_B$ and if $(x_{\beta r}/y_{rj}) = 0$ then $z' = z_B$.

Thus under the conditions stated above the value of the objective function is being improved or at least remain same.

[In a problem of minimization, the value of the objective function will be diminished, provided $z_j - c_j > 0$ for columns corresponding to the non-basic variables with at least one $y_{ij} > 0$.]

The quantity y_{rj} is known as the *key element* or *pivot element.* The value of y_{rj} is determined from the condition that in the jth column,

$$\min_i \left(\frac{x_{Bi}}{y_{ij}}, y_{ij} > 0 \right) = \frac{x_{Br}}{y_{rj}}. \tag{A.2.9}$$

The jth column vector $\mathbf{a}_j$ is the *vector to enter in the next basis.* The rule for determination of the key element is known as *Minimum ratio rule.* The rule for determination of $\mathbf{a}_j$ will be discussed later.

Now if in a current B.F.S. $x_{Br} = 0$ for which $y_{rj} > 0$ then from mini-ratio rule

$$\min_i \left(\frac{x_{Bi}}{y_{ij}}, y_{ij} > 0 \right) = \frac{x_{Br}}{y_{rj}} = 0$$

and then $z' = z_B$, i.e., the value of the objective function will not be improved due to transformation. As $x_{Br} = 0$, the current B.F.S. is degenerate and the next B.F.S. will be also degenerate because $x'_{Br} = \frac{x_{Br}}{y_{rj}} = 0$.

But if $x_{Br} = 0$, and $y_{rj} \leq 0$ then from the mini-ratio rule it is evident that this variable will not enter in the calculation of $\frac{x_{Bi}}{y_{ij}}$ and the new B.F.S. may not be degenerate, though the current B.F.S. is degenerate.

If y_{rj} be the key element then β_r, the vector in the rth position of the basis matrix B is being replaced by $\mathbf{a}_j$ and the vector β_r is the *vector* to leave the current basis.

The new basis matrix $\hat{B} = (\beta_1, \beta_2, \cdots, \beta_{r-1}, \mathbf{a}_j, \beta_{r+1}, \cdots, \beta_m)$.

A.2.2 Conditions for a solution to be optimal

Theorem A.2.2 *Given a basic feasible solution* $\mathbf{x}_B = B^{-1}\mathbf{b}$ *with* $z_B = \mathbf{c}_B\mathbf{x}_B$ *to the L.P.P. Maximize* $\mathbf{z} = \mathbf{cx}$ *subject to* $A\mathbf{x} = \mathbf{b}$, $\mathbf{x} \geq \mathbf{0}$ *such that* $z_j - c_j \geq 0$ *for every column of* $\mathbf{a}_j$ *in* A. *Then* z_B *is the maximum value and* $\mathbf{x}_B$ *is the optimal B.F.S.*

Proof.

$$A = (\mathbf{a}_1, \mathbf{a}_2, \cdots, \mathbf{a}_n)$$
$$B = (\beta_1, \beta_2, \cdots, \beta_m)$$

where $\beta_1, \beta_2, \cdots, \beta_m$ are some of

$$\mathbf{a}_1, \mathbf{a}_2, \cdots, \mathbf{a}_n \text{ and } A\mathbf{x} = \mathbf{b}. \tag{A.2.10}$$

$\mathbf{x}_B = B^{-1}\mathbf{b}$ is the B.F.S. and the value of objective function is $z_B = \mathbf{c}_B\mathbf{x}_B$, $z_j - c_j \geq 0$ for all column vectors $\mathbf{a}_j$. It is required to prove that z_B is the optimal value of the objective function corresponding to the B.F.S. $\mathbf{x}_B$.

Let $\mathbf{x}' = [x'_1, x'_2, \cdots, x'_n]$ be any F.S. (may be a B.F.S.) of the problem which makes the objective function z, to

$$z' = \sum_{j=1}^{n} c_j x'_j.$$

We are to prove that $z' \leq z_B$.

We have

$$
\begin{aligned}
Ax' &= \mathbf{b} = B\mathbf{x}_B \quad [\because \mathbf{x}' \text{ is a solution set of } A\mathbf{x} = \mathbf{b}] \\
\text{or,}\quad \mathbf{x}_B &= B^{-1}(A\mathbf{x}') = (B^{-1}A)\mathbf{x}' \\
&= \mathbf{Y}\mathbf{x}' \quad [\mathbf{Y} = \{\mathbf{y}_1, \mathbf{y}_2, \cdots, \mathbf{y}_n\}] \quad [\text{as } B^{-1}\mathbf{a}_j = \mathbf{y}_j] \\
&= (\mathbf{y}_1, \mathbf{y}_2, \cdots, \mathbf{y}_n)[x'_1, x'_2 \cdots, x'_n] \qquad \text{(A.2.11)}
\end{aligned}
$$

Therefore

$$
[x_{B1}, x_{B2}, \cdots, x_{Bm}] = \begin{bmatrix} y_{11} & y_{12} & \cdots & y_{1j} & y_{1n} \\ y_{21} & y_{22} & \cdots & y_{2j} & y_{2n} \\ \cdots & \cdots & \cdots & \cdots & \cdots \\ y_{i1} & y_{i2} & \cdots & y_{ij} & y_{in} \\ y_{m1} & y_{m2} & \cdots & y_{mj} & y_{mn} \end{bmatrix} \begin{bmatrix} x'_1 \\ x'_2 \\ \vdots \\ x'_j \\ x'_n \end{bmatrix}
$$

Equating we get

$$
x_{Bi} = \sum_{j=1}^{n} y_{ij} x'_j. \qquad \text{(A.2.12)}
$$

Now $z_j - c_j \geq 0$ for all columns $\mathbf{a}_j$ of A. Therefore $c_j \leq z_j$ for all j

$$
\begin{aligned}
\text{or,}\quad & c_j x'_j \leq z_j x'_j \quad \text{as } x'_j \geq 0 \\
\text{or,}\quad & \sum_{j=1}^{n} c_j x'_j \leq \sum_{j=1}^{n} x'_j \left(\sum_{i=1}^{m} c_{Bi} y_{ij} \right) \quad \left[\because z_j = \sum_{i=1}^{m} c_{Bi} y_{ij} \right] \\
\text{or,}\quad & z' \leq \sum_{i=1}^{m} c_{Bi} \sum_{j=1}^{n} x'_j y_{ij} \\
\text{or,}\quad & z' \leq \sum_{i=1}^{m} c_{Bi} x_{Bi} \quad [\text{from (A.2.12)}] \\
\text{or,}\quad & z' \leq \mathbf{c}_B \mathbf{x}_B \\
\text{or,}\quad & z' \leq z_B \qquad \text{(A.2.13)}
\end{aligned}
$$

Hence z_B is the optimal value of the objective function corresponding to B.F.S. $\mathbf{x}_B$. Hence the theorem is proved.

[For a problem of minimization, the necessary and sufficient condition is $z_j - c_j \leq 0$ for all columns $\mathbf{a}_j$ in A.]

A.2.3 Another Proof

We know

$$
z' = z_B - \frac{x_{Br}}{y_{rj}}(z_j - c_j) \text{ from (A.2.8).}
$$

Since $\frac{x_{Br}}{y_{rj}} \geq 0$ and all $z_j - c_j \geq 0$ then $z' \leq z_B$ and will be equal only when $\frac{x_{Br}}{y_{rj}} = 0$ or $z_j - c_j = 0$ for all values of j. Hence for all $z_j - c_j \geq 0$, z_B is the optimal value of the L.P.P.

Note: Consider the value of $z_j - c_j$ for a column vector $\beta_r (r = 1, 2, \cdots, m)$ of the basic variables. We know

$$\beta_i = B\mathbf{y}_j \qquad [\because \mathbf{a}_j = B\mathbf{y}_j]$$
$$\text{or,} \quad \beta_i = \beta_1 y_{1j} + \beta_2 y_{2j} + \cdots + \beta_i y_{ij} + \cdots + \beta_m y_{mj}$$

from which we get $\mathbf{y}_j = (0, 0, \cdots, 1, 0, \cdots, 0) = \mathbf{e}_i$ (an unit vector with i-th component unity)

$$c_{Bi} = c_j.$$

Then

$$\begin{aligned} z_j - c_j &= \mathbf{c}_B \mathbf{y}_j - \mathbf{c}_{Bi} \\ &= \mathbf{c}_B \mathbf{e}_i - \mathbf{c}_{Bi} \\ &= \mathbf{c}_{Bi} - \mathbf{c}_{Bi} = 0. \end{aligned} \qquad \text{(A.2.14)}$$

Thus $z_j - c_j = 0$ for all columns corresponding to basic variables. [a very important property to remember].

A.2.4 Condition for the non-existence of finite optimal value of the objective function

Theorem A.2.3 *In a linear programming problem,*

$$\textit{Maximize,} \quad z = \mathbf{cx}, \qquad \textit{subject to } A\mathbf{x} = \mathbf{b}, \mathbf{x} \geq \mathbf{0}$$

there exists no finite optimal value of objective function, if for at least one column corresponding to a non-basic variable $z_j - c_j < 0$ and $y_{ij} \leq 0$ for all i.

Proof.

$$A = (\mathbf{a}_1, \mathbf{a}_2, \cdots, \mathbf{a}_n), \text{ co-efficient matrix}$$
$$B = (\beta_1, \beta_2, \cdots, \beta_m), \text{ a basis matrix}$$
$$\text{and let} \quad \mathbf{x}_B = B^{-1}\mathbf{b} = [x_{B1}, x_{B2}, \cdots, x_{Bm}] \text{ is a B.F.S.}$$

Therefore

$$B\mathbf{x}_B = \mathbf{b}, \quad \text{or,} \quad \sum_{i=1}^{m} \beta_i x_{Bi} = \mathbf{b}. \qquad \text{(A.2.15)}$$

The value of the objective function corresponding to the B.F.S. is $z_B = \mathbf{c}_B \mathbf{x}_B$.

Let $\mathbf{a}_j$ be a column vector of a non-basic variable with $z_j - c_j < 0$ and $y_{ij} \leq 0$ for all i.

Then for any $\lambda > 0$

$$\sum_{i=1}^{m} \beta_i x_{Bi} - \lambda \mathbf{a}_j + \lambda \mathbf{a}_j = \mathbf{b}. \qquad \text{(A.2.16)}$$

As $\mathbf{a}_j$ is a column vector of a non-basic variable therefore

$$\mathbf{a}_j = \sum_{i=1}^{m} \beta_i y_{ij}. \tag{A.2.17}$$

Putting the value of $\mathbf{a}_j$ from (A.2.17) in (A.2.16), it may be expressed as

$$\sum_{i=1}^{m} \beta_i x_{Bi} - \lambda \sum_{i=1}^{m} \beta_i y_{ij} + \lambda \mathbf{a}_j = \mathbf{b}$$

$$\text{or} \quad \sum_{i=1}^{m} \beta_i (x_{Bi} - \lambda y_{ij}) + \lambda \mathbf{a}_j = \mathbf{b} \tag{A.2.18}$$

which indicates that

$$\mathbf{x}' = [x_{B1} - \lambda y_{1j}, x_{B2} - \lambda y_{2j}, \cdots, x_{Bm} - \lambda y_{mj}, \lambda]$$

is a solution set of the equation $A\mathbf{x} = \mathbf{b}$ [with all other variables zero].

As $\lambda > 0$ and $y_{ij} \leq 0$ for all i, then $x_{Bi} - \lambda y_{ij} \geq 0$ for all i and hence $\mathbf{x}'$ is a feasible solution of $A\mathbf{x} = \mathbf{b}$. But the solution set $\mathbf{x}'$ is a non-basic solution with $(m + 1)$ variables non-negative.

Let z' be the value of objective function corresponding to the solution set $\mathbf{x}'$. Then

$$z' = \mathbf{c}'_B \mathbf{x}' = \sum_{i=1}^{m} c_{Bi} x'_{Bi} + c_j \lambda \quad [\text{where } x'_{Bi} = x_{Bi} - \lambda y_{ij}]$$

$$= \sum_{i=1}^{m} c_{Bi}(x_{Bi} - \lambda y_{ij}) + c_j \lambda = \sum_{i=1}^{m} c_{Bi} x_{Bi} + \lambda \left(c_j - \sum_{i=1}^{m} c_{Bi} y_{ij} \right)$$

$$= z_B - \lambda (z_j - c_j) \tag{A.2.19}$$

As $\lambda > 0, z_j - c_j < 0$ therefore $z' \to \infty$ if $\lambda \to \infty$.

Hence no finite value of the objective function exists under the conditions stated above. The solution set of this type is known as an *unbounded solution* to the problem.

[**Note:** Some of the components of $\mathbf{x}'$ may be zero. For example if $x_{Br} = 0$ and $y_{rj} = 0$, then $x'_{Br} = 0$. But still the solution is non-basic. Because the column vectors, $\beta_1, \beta_2, \cdots, \beta_{r-1}, \mathbf{a}_j, \beta_{r+1}, \cdots, \beta_m$ do not form a new basis as y_{rj}, the co-efficient of β_r in the expression of $\mathbf{a}_j = \sum_{i=1}^{m} \beta_i y_{ij}$, is zero.]

A.2.5 Condition for the existence of multiple optimal solutions

1. In a simplex table, at the optimal stage, if there exist some columns corresponding to non-basic variables, where $z_j - c_j = 0$, $y_{ij} > 0$ for at least one i then there are multiple optimal B.F.S. of the problem.

Let

$$\mathbf{x}_B = [x_{B1}, x_{B2}, \cdots, x_{Bm}]$$

be an optimal B.F.S. and $\hat{z}$ be the optimal value of the objective function. The value of B.F.S. in the next iteration is

$$\mathbf{x}'_B = [x'_{B1}, x'_{B2}, \cdots, x'_{Bm}]$$

where

$$x'_{Br} = \frac{x_{Br}}{y_{rj}} \quad \text{and} \quad x'_{Bi} = x_{Bi} - \frac{y_{ij}}{y_{rj}} x_{Bi}, \quad i = 1, 2, \cdots, m, \; i \neq r.$$

$$\text{and} \quad z' = z_B - \frac{x_{Br}}{y_{rj}}(z_j - c_j),$$

where z' is the value of the objective function in the next iteration. As $\mathbf{x}_B$ is an optimal B.F.S. then there is no column vector $\mathbf{a}_j$ for which $z_j - c_j < 0$. Now $z_j - c_j = 0$ for some columns of the non-basic variables.

Then

$$z' = z_B - \frac{x_{Br}}{y_{rj}} \times 0 = z_B.$$

Then z' is also the optimal value of objective function corresponding to the B.F.S. $\mathbf{x}'_B$.

Then an alternative optimal B.F.S. exists. Hence there exist multiple optimal solutions of the problem under the conditions stated above.

Note: The above discussion is related to a problem of maximization.

2. At the optimal stage, if at least for one column corresponding to non-basic variables, the condition $z_j - c_j = 0$ with $y_{ij} \leq 0$ for all i, holds, then there exists an alternative non-basic optimal solution of the problem.

Let

$$\mathbf{x}_B = [x_{B1}, x_{B2}, \cdots, x_{Bm}]$$

be an optimal B.F.S. of the problem with the optimal value of the objective function z_B. As $y_{ij} \leq 0$, for all i, in a j-th column vector of the non-basic variable, then there exists a non-basic feasible solution,

$$\mathbf{x}' = [x_{B1} - \lambda y_{1j}, x_{B2} - \lambda y_{2j}, \cdots, x_{Bm} - \lambda y_{mj}, \lambda], \quad \lambda > 0$$

and the value of the objective function is

$$z' = z_B - \lambda(z_j - c_j)$$

corresponding to the solution set $\mathbf{x}'$.

As $z_j - c_j = 0$, then $z' = z_B$. Therefore z' is also the optimal value of the objective function for a non-basic F.S. $\mathbf{x}'$ with $(m+1)$ variables. Hence there exists an alternative non-basic optimal solutuion to the problem.

A.2.6 Selection of the vector to enter in the next basis, the key element and the vector to leave the current basis

We have

$$\begin{aligned} z' &= z_B - \frac{x_{Br}}{y_{rj}}(z_j - c_j), \quad y_{rj} > 0 \\ &= z_B - \frac{x_{Br}}{y_{rj}}(z_j - c_j). \end{aligned}$$

From the above deduction, it is evident that for $z_j - c_j < 0$, the value of the objective function will be improved provided $\frac{x_{Br}}{y_{rj}} > 0$. If $z_j - c_j < 0$, only in one column, say kth column, the kth vector $\mathbf{a}_k$ is taken as the incoming vector. Now if in that column,

$$\min_i \left(\frac{x_{Bi}}{y_{ik}}, y_{ik} > 0 \right) = \frac{x_{Br}}{y_{rk}} \tag{A.2.20}$$

which occurs at $i = r$ then y_{rk} is the key element and β_r, the rth vector in the basis, is the departing vector. But if the minimum ratio is not unique, then the selection of the key element and the departing vectors are not unique. An arbitrary choice of key row corresponding to minimum ratio may be done and almost all problems can be solved using this technique. The rth row of the simplex table is known as *key* or *pivot row* and the kth column is known as *key or pivot column* and the intersection element of the key row and key column which is y_{rk} is the key on pivot element.

If $z_j - c_j < 0$ for more than one column and

$$z_k - c_k = \min_j (z_j - c_j, z_j - c_j < 0) \tag{A.2.21}$$

then $\mathbf{a}_k$ will be the entering vector. *If the minimum is not unique, select arbitrarily any vector as incoming vector corresponding to the minimum.* Then the selection of key element and departing vector are to be done by the previous method. This method is widely applied. No difficulty arises here in selecting the entering vector and as it is easier to select the entering vector, this method is effectively used by the electronic computers to solve the problems. *But in this method once a vector enters in the basis may leave the basis during subsequent iterations.*

In the case of degenerate problems, the selection of entering vector can be done by other methods which is not discussed here.

A.2.7 Important Observations

To calculate a new B.F.S., i.e., to construct a new basis matrix, a vector β_r of B is to be replaced by a vector $\mathbf{a}_k$ of A, which is one of the column vectors associated with the non-basic variables. The vector β_r which is being removed, is the *vector to leave the current basis* and the vector $\mathbf{a}_k$ which is being inserted in the new basis matrix, is the *vector to enter in the next basis.* The method of selection of $\mathbf{a}_k$ and β_r has been discussed previously. The improved B.F.S. is given by (A.2.5) of theorem

(A.2.1) [Putting $j = k$]

$$\left.\begin{aligned} x'_{Bi} &= x_{Bi} - \frac{y_{ik}}{y_{rk}} x_{Br}, \quad i = 1, 2, \cdots, m, \; i \neq r \\ x'_{Br} &= \frac{x_{Br}}{y_{rk}}, \quad y_{rk} > 0, i = r. \end{aligned}\right\} \tag{A.2.22}$$

The element y_{rk} is the key element.

Due to the transformation all the elements y_{ij} change to y'_{ij} $[i = 1, 2, \cdots, m,$ $j = 1, 2, \cdots, n]$ and y'_{ij} can be expressed by the relations

$$\left.\begin{aligned} \text{(i)} \quad y'_{ij} &= y_{ij} - \frac{y_{ik}}{y_{rk}} y_{rj} \quad [i = 1, 2, \cdots, m; i \neq r] \\ \text{(ii)} \quad y'_{rj} &= \frac{y_{rj}}{y_{rk}}, \quad y_{rk} > 0, i = r. \end{aligned}\right\} \tag{A.2.23}$$

In particular when $i = k$

$$\text{(iii)} \quad y'_{ik} = y_{ik} - \frac{y_{ik}}{y_{rk}} y_{rk} = 0 \quad [i = 1, 2, \cdots, m; i \neq r]$$

$$\text{and (iv)} \quad y'_{rk} = \frac{y_{rk}}{y_{rk}} = 1, \; i = r.$$

Hence due to the transformation, the kth column vector y'_k in the next table will be a unit vector $\mathbf{e}_r$.

The proofs for the transformed formulae are given below.

Proof. Let the basis $B = (\beta_1, \beta_2, \cdots, \beta_r, \cdots, \beta_m)$ at any stage. Then for any non-basis vector $\mathbf{a}_j$,

$$\mathbf{a}_j = y_{1j}\beta_1 + y_{2j}\beta_2 + \cdots + y_{rj}\beta_r + \cdots + y_{mj}\beta_m = \sum_{i=1}^{m} y_{ij}\beta_i \tag{A.2.24}$$

and for $\mathbf{a}_k$ (the vector to enter in the next basis by replacing β_r), we have

$$\mathbf{a}_k = \sum_{i=1}^{m} y_{ik}\beta_i. \tag{A.2.25}$$

As $y_{rk} > 0$, therefore expressing β_r in terms of $\mathbf{a}_k$ etc. we have

$$\begin{aligned} \beta_r &= \frac{1}{y_{rk}}\mathbf{a}_k - \left(\frac{y_{1k}}{y_{rk}}\beta_1 + \frac{y_{2k}}{y_{rk}}\beta_2 + \cdots + \frac{y_{r-1\cdot k}}{y_{rk}}\beta_{r-1} + \frac{y_{r+1\cdot k}}{y_{rk}}\beta_{r+1} + \cdots + \frac{y_{mk}}{y_{rk}}\beta_m\right) \\ &= -\sum_{\substack{i=1 \\ i\neq r}}^{m} \frac{y_{ik}}{y_{rk}}\beta_i + \frac{1}{y_{rk}}\mathbf{a}_k \end{aligned} \tag{A.2.26}$$

Putting the value of β_r from (A.2.23) in $\mathbf{a}_j$ in (A.2.24) we have

$$\mathbf{a}_j = \sum_{\substack{i=1 \\ i\neq r}}^{m} \left(y_{ij} - y_{rj}\frac{y_{ik}}{y_{rk}}\right)\beta_i + \frac{y_{rj}}{y_{rk}}\mathbf{a}_k = \sum_{i=1}^{m} y'_{ij}\hat{\beta}_i, \tag{A.2.27}$$

where $\hat{\beta}_i = \beta_i[i \neq r]$ and $\hat{\beta}_r = \mathbf{a}_k$.

The next basis is

$$\hat{B} = (\beta_1, \beta_2, \cdots, \beta_{r-1}, \mathbf{a}_k, \beta_{r+1}, \cdots, \beta_m).$$

Comparing (A.2.24) and (A.2.27) we get

$$\left.\begin{aligned} y'_{ij} &= y_{ij} - \frac{y_{ik}}{y_{rk}} y_{rj}, \quad i \neq r \\ \text{and} \quad y'_{rj} &= \frac{y_{rj}}{y_{rk}} \end{aligned}\right\} \cdot \tag{A.2.28}$$

Due to the transformation the values of $z_j[j = 1, 2, \cdots, n]$ also change and the changed value

$$z'_j = z_j - \frac{y_{rj}}{y_{rk}}(z_k - c_k).$$

Proof. When $\mathbf{a}_k$ is the vector to enter in the basis by replacing β_k, we have $c'_{Bi} = c_{Bi}$ for $i \neq r$ and $c'_{Br} = c_k$.

Hence,

$$\begin{aligned} z'_j &= \sum_{i=1}^{m} c_{Bi} y'_{ij} = \sum_{\substack{i=1 \\ i \neq r}}^{m} c_{Bi} y'_{ij} + c_k y'_{rj} \\ &= \sum_{\substack{i=1 \\ i \neq r}}^{m} c_{Bi} \left(y_{ij} - \frac{y_{ik}}{y_{rk}} y_{rj} \right) + c_k \frac{y_{rj}}{y_{rk}} \quad \text{[from (A.2.27)]} \\ &= \sum_{i=1}^{m} c_{Bi} \left(y_{ij} - \frac{y_{ik}}{y_{rk}} y_{rj} \right) + c_k \frac{y_{rj}}{y_{rk}} \left[\because\ y_{rj} - \frac{y_{rk}}{y_{rk}} y_{rj} = 0 \right] \\ &= \sum_{i=1}^{m} c_{Bi} y_{ij} - \frac{y_{rj}}{y_{rk}} \left(\sum_{i=1}^{m} c_{Bi} y_{ik} - c_k \right) = z_j - \frac{y_{rj}}{y_{rk}} (z_k - c_k). \end{aligned} \tag{A.2.29}$$

Hence,

$$z'_j - c_j = z_j - c_j - \frac{y_{rj}}{y_{rk}}(z_k - c_k) \tag{A.2.30}$$

and again we have from (A.2.8)

$$z' = z_B - \frac{x_{Br}}{y_{rk}}(z_k - c_k) \quad [\text{putting } j = k] \tag{A.2.31}$$

Now a very interesting observation can be made. Suppose that we write

$$x_{Bi} = y_{i0} (i = 1, \cdot, m), \quad z_B = y_{m+1,0}, z_j - c_j = y_{m+1,j}$$

then all the formulae's of (A.2.31), (A.2.22), (A.2.30), (A.2.28) can be written in the compact form

$$\left.\begin{aligned} y'_{ij} &= y_{ij} - y_{ik} \frac{y_{rj}}{y_{rk}} \quad j = 0, 1, \cdots, n, \ i = 1, \cdots, m+1, i \neq r \\ y'_{rj} &= \frac{y_{rj}}{y_{rk}} \end{aligned}\right\} \tag{A.2.32}$$

[A very important property to remember.]

Now if the initial basis matrix be a unit matrix $B = (\mathbf{e}_1, \mathbf{e}_2, \cdots, \mathbf{e}_m)$ where $\mathbf{e}_i$ is a unit vector with ith component unity and $\mathbf{b} \geq 0$, then

$$\mathbf{x}_B = B^{-1}\mathbf{b} = \mathbf{b} \geq 0$$
$$\text{or,} \quad \mathbf{x}_B = [b_1, b_2, \cdots, b_m]$$

which is the initial B.F.S. to the problem and $z_B = \mathbf{c}_B\mathbf{b}$ and $\mathbf{y}_j = B^{-1}\mathbf{a}_j = \mathbf{a}_j$. Hence $y_{ij} = a_{ij}$ for all i and j as

$$\mathbf{a}_j = [a_{ij}, a_{2j}, a_{ij}, \cdots, a_{mj}]$$
$$\text{and} \quad z_j - c_j = \mathbf{c}_B\mathbf{y}_j - c_j = \mathbf{c}_B\mathbf{a}_j - c_j.$$

With this knowledge, the initial simplex table can be constructed. The next table can also be obtained from the initial table by transforming it, using the formula (A.2.32) in such a way that the kth column vector $\mathbf{y}'_k$ of next table be a unit vector $\mathbf{e}_r$.